No. 1499
$24.95

750 PRACTICAL ELECTRONIC CIRCUITS

EDITED BY ROLAND S. PHELPS

TAB BOOKS Inc.
BLUE RIDGE SUMMIT, PA. 17214

FIRST EDITION

FOURTH PRINTING

Printed in the United States of America

Library of Congress Cataloging in Publication Data

750 practical electronic circuits.

Includes index.
1. Electronics—Amateurs' manual. 2. Electronic circuits—Amateurs' manuals. I. Phelps, Roland S. II. Title: Seven hundred fifty practical electronic circuits.

TK9965.A13 1983 621.3815'3 82-5988
ISBN 0-8306-2499-6 AACR2
ISBN 0-8306-1499-0 (pbk.)

Contents

Acknowledgments

I would like to express my gratitude to the many fine individuals and firms that helped by providing information for this book. These firms are identified in the caption that accompanies each circuit.

AEG-Telefunken Corp.
Route 22, Orr Drive
Somerville, NJ 08876

Analog Devices Inc.
Route 1, Industrial Park
P.O. Box 280
Norwood, MA 02062

Exar Integrated Systems Inc.
750 Palomar Avenue
P.O. Box 62229
Sunnyvale, CA 94088

LSI Computer Systems Inc.
1235 Walt Whitman Road
Melville, NY 11747

Motorola Inc.
P.O. Box 20924
Phoenix, AZ 85036

National Semiconductor Corp.
2900 Semiconductor Drive
Santa Clara, CA 95051

Optical Electronics Inc.
P.O. Box 11140
Tucson, AZ 85734

Precision Monolithics Inc.
1500 Space Park Drive
Santa Clara, CA 95050

Raytheon Company
350 Ellis Street
Mountain View, CA 94042

SGS Semiconductor Corp.
7070 East Third Avenue
Scottsdale, AZ 85251

Signetics Corporation
811 East Arques Avenue
Sunnyvale, CA 94086

Silicon Systems Inc.
14351 Myford Road
Tustin, CA 92680

Sprague Electric Company
Marshall Street
No. Adams, MA 01247

Supertex Inc.
1225 Bordeaux Drive
Sunnyvale, CA 94086

Workman Electronic Products
P.O. Box 3828
Sarasota, FL 33578

Credit within the captions uses the following key:

AEG—AEG-Telefunken Corp.
AD—Analog Devices Inc.
E—Exar Integrated Systems Inc.
LSI—LSI Computer Systems Inc.
M—Motorola Inc.
NS—National Semiconductor Corp.
OE—Optical Electronics Inc.
PM—Precision Monolithics Inc.
R—Raytheon Company
SGS—SGS Semiconductor Corp.
S—Signetics Corporation
SIL—Silicon Systems Inc.
SE—Sprague Electric Company
SU—Supertex Inc.
W—Workman Electronic Products

Introduction

Late one evening in 1835, Michael Faraday was in his laboratory watching an experiment. He had connected a battery to two electrodes that he had immersed in a solution of silver nitrate. He watched as the silver was mysteriously deposited on the negative electrode. Since the battery, wires, electrodes, and silver nitrate solution formed a continuous path for the electricity from the battery, Michael Faraday was working with a simple electrical circuit. From this experiment Faraday reasoned that the electricity must also be moving from the positive electrode to the negative electrode. This turned out to be incorrect, but it did not affect the results of the experiment. From these experiments Faraday originated many of the terms used in modern electronics. Words like electrolyte, electrode, and electrolysis were all invented by Faraday to describe his observations.

Since the time of Faraday, much has been learned about electrical circuits besides the fact that electrons tend to flow from negative to positive! This book reflects that great increase in the knowledge of electronics that has been gained in the past 150 years. Here, in this book, you will find circuits that would have amazed the most brilliant minds of past generations. Many of these circuits use the most modern of integrated circuits; while others are simpler, they are often used in ways that were undreamed of until very recently. These circuits should provide you with many hours of pleasant study and the basis of some original experimentation of your own.

Chapter 1

Power Supplies and Voltage Regulators

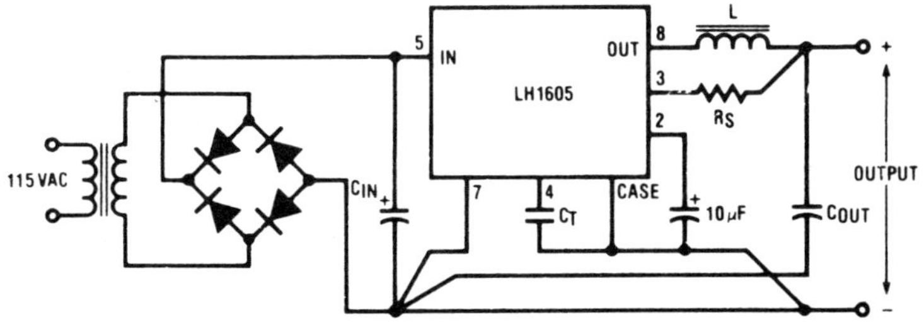

Typical power supply system (NS).

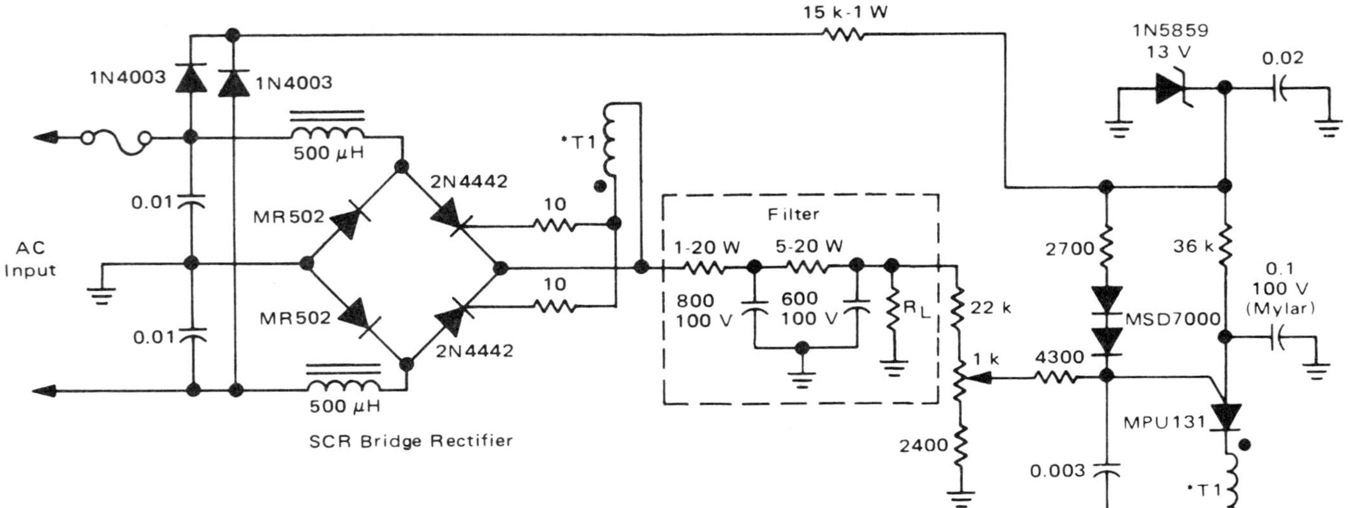

*T1 = 1 ● 1 pulse transformer
Resistors — in ohms, 1/2 W, unless specified
Capacitors — in μF, 400 V, unless specified.

80V—1.5 A power supply (M).

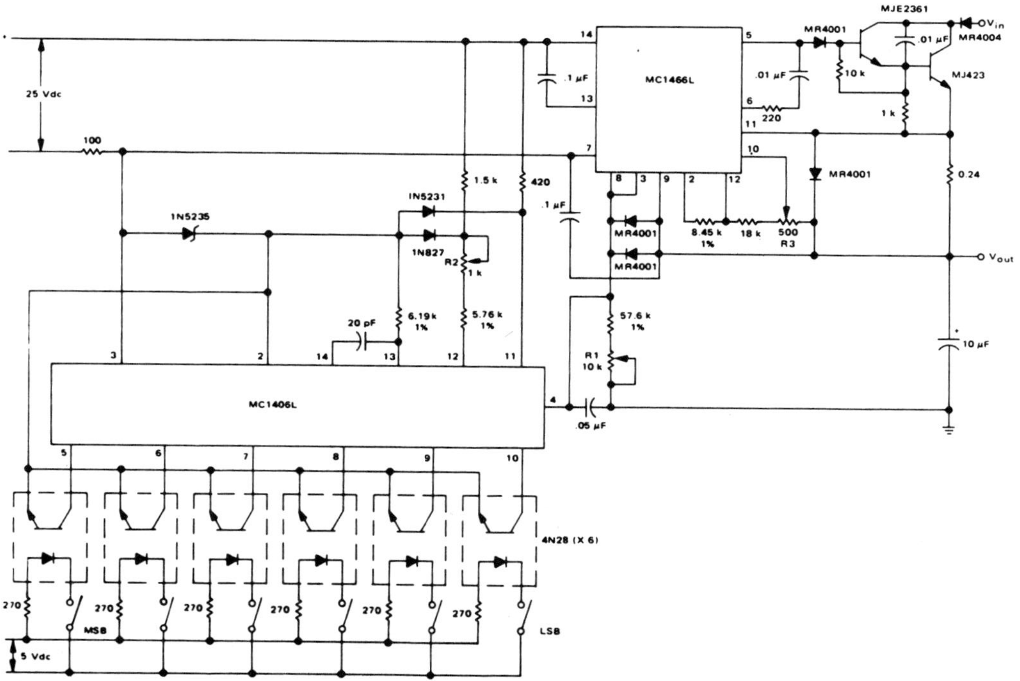

3 High voltage power supply schematic (M).

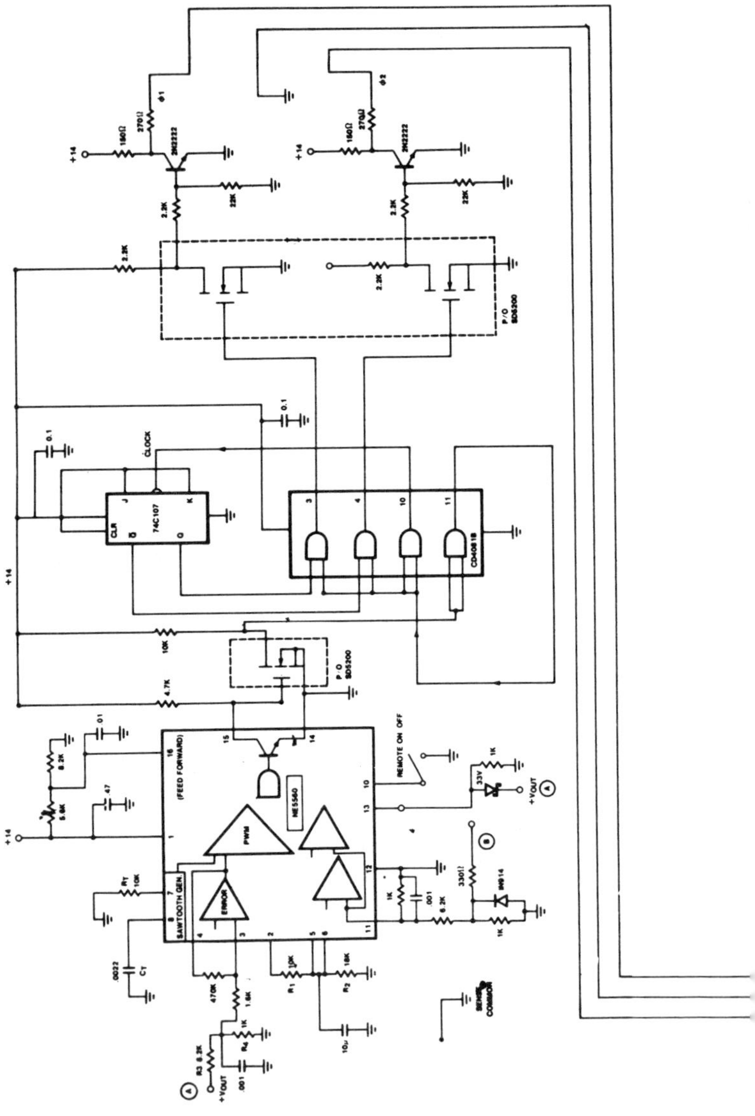
2N2222
SD5200
74C107
CLOCK
CD4081B
NE5560
SAWTOOTH GEN.
ERROR
PWM
(FEED FORWARD)
REMOTE ON OFF
SENSE COMMON

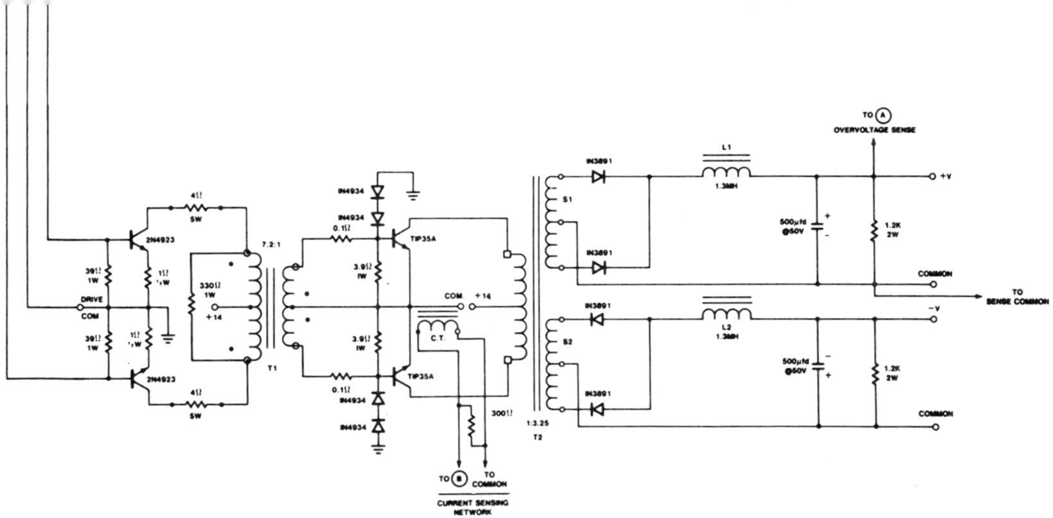

NOTE
Power ground and signal ground must be kept separated.

T_1, Primary = 130T (C.T.) #26
Secondary = 18T (C.T.) #22
Core = Ferroxcube 3622
3C8 material
C.T. = 50T #26 on
Ferroxcube 2616 core (3C8)
F2D bobbin

T_2, Primary = 16T (C.T.) #18 Secondaries (each) 52T (C.T.) #22,
Core = Ferroxcube 4229 3C8 material
L_1, L_2 120T #20 on single gapped EC35 Ferroxcube core. 3C8 material

5 NE/SE5560 push-pull switch mode regulated supply with CMOS drive conversion logic (S).

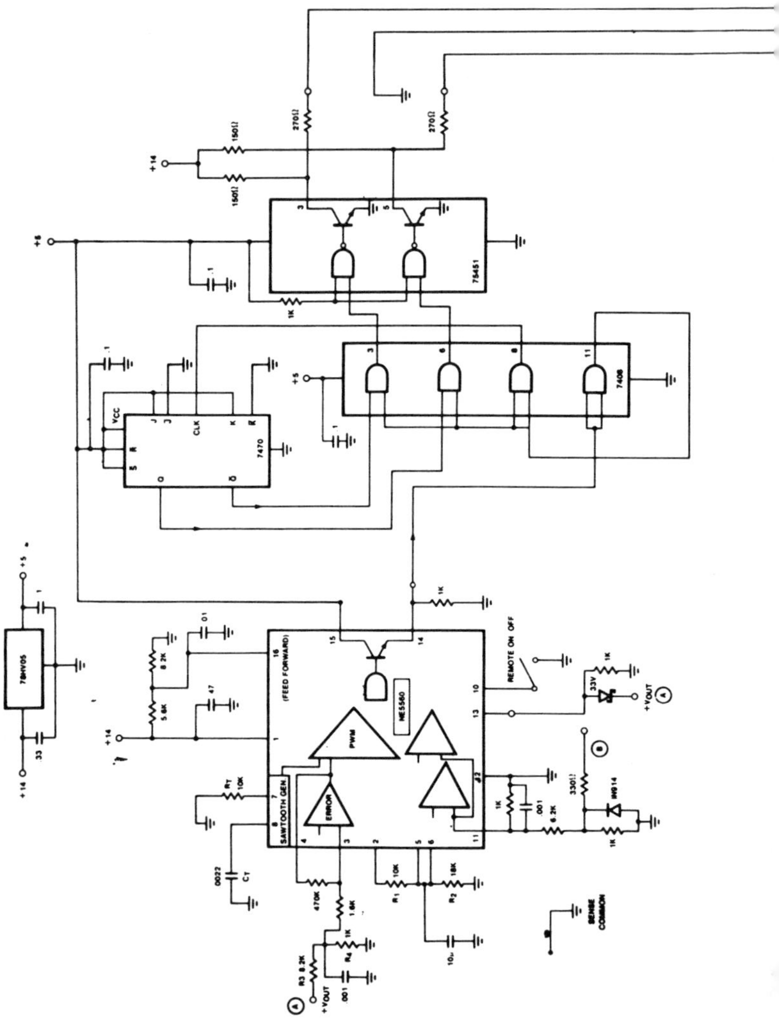
78HV05
7470
7408
75451
NE5560
PWM
ERROR
SAWTOOTH GEN
(FEED FORWARD)
REMOTE ON OFF
SENSE COMMON

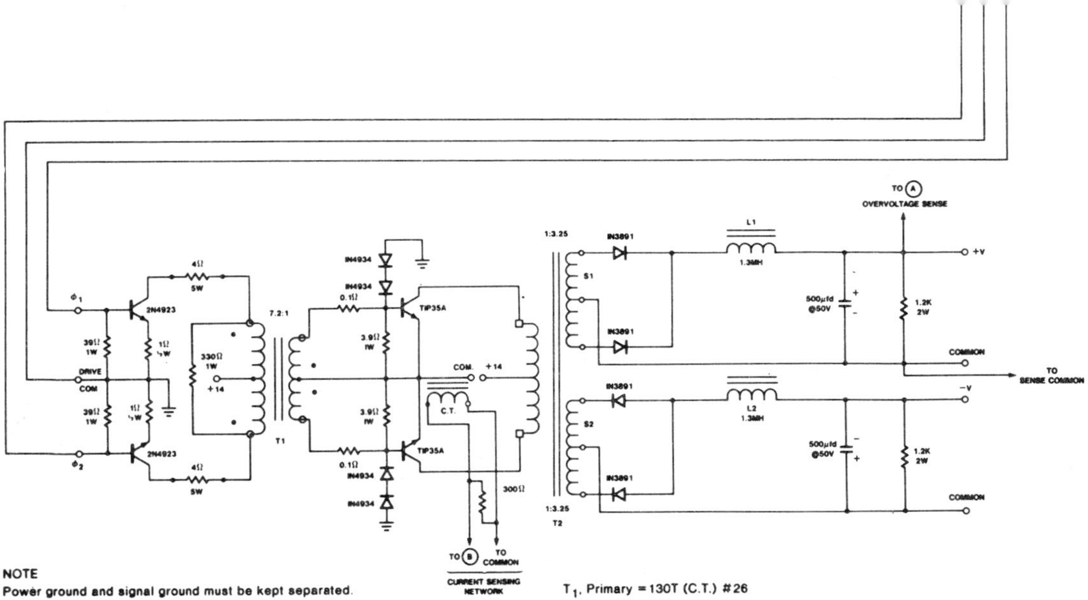

NOTE
Power ground and signal ground must be kept separated.

T_1, Primary = 130T (C.T.) #26
Secondary = 18T (C.T.) #22
Core = Ferroxcube 3622
3C8 material
C.T. = 50T #26 on
Ferroxcube 2616 core (3C8)
F2D bobbin

T_2, Primary = 16T (C.T.) #18 Secondaries (each) 52T (C.T.) #22,
Core = Ferroxcube 4229 3C8 material
L_1, L_2 120T #20 on single gapped EC35 Ferroxcube core. 3C8 material

7 NE/SE5560 push-pull switch mode regulated supply with TTL drive conversion logic (S).

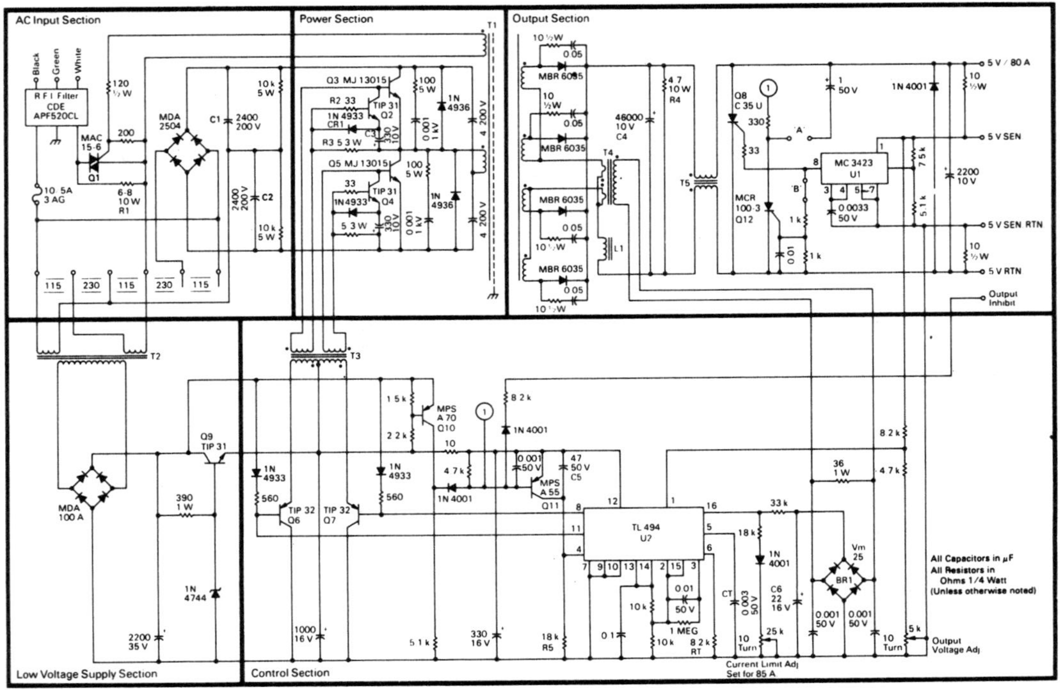

400 watt SWITCHMODE power supply (M).

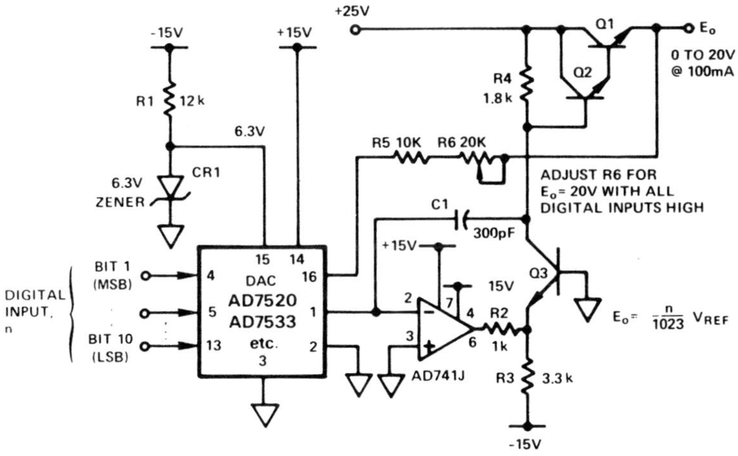

A digitally programmable power-supply (AD).

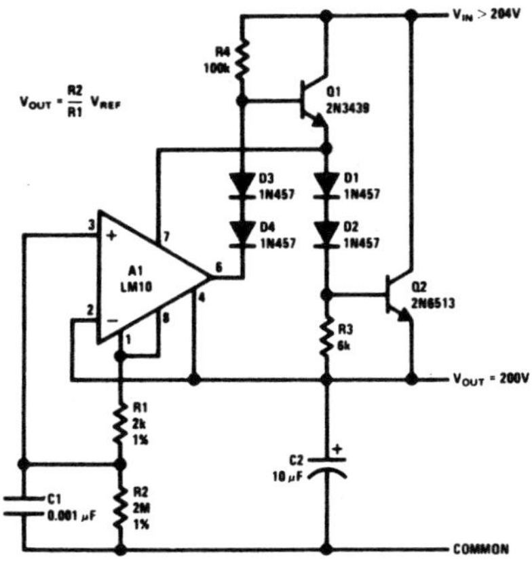

Hv regulator (NS).

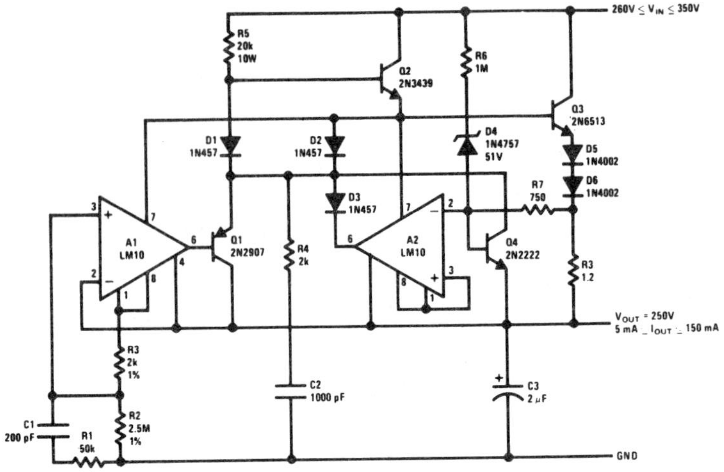

Protected hv regulator (NS).

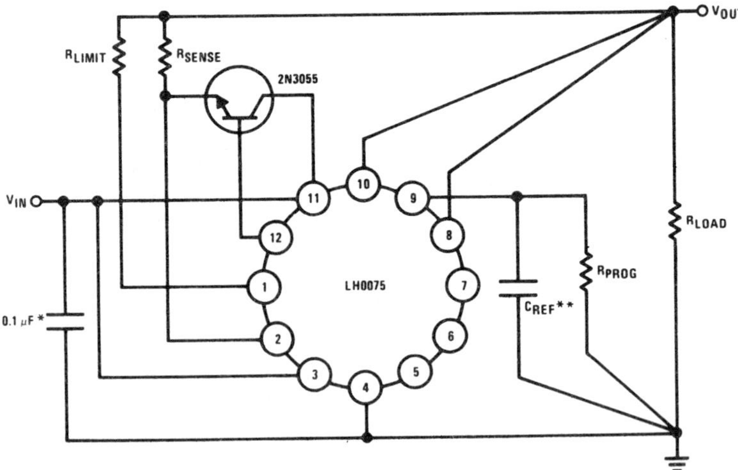

2A regulator with current limit (NS).

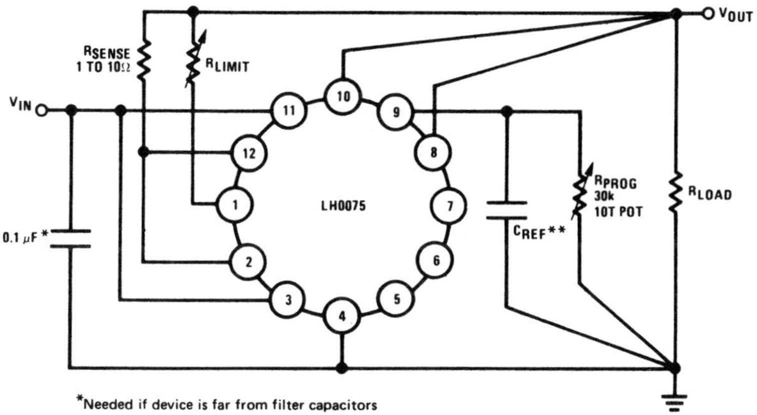

*Needed if device is far from filter capacitors
**Optional—improves transient response

$$R_{PROG} = \frac{V_{OUT}\text{ Desired}}{1\text{ mA}} \qquad I_{OUT(MAX)} = \left[\frac{R_{LIMIT}}{R_{SENSE}} + 1\right] \times 100\,\mu A$$

$I_{OUT} \leq 200$ mA

Variable voltage reference with current limit (NS).

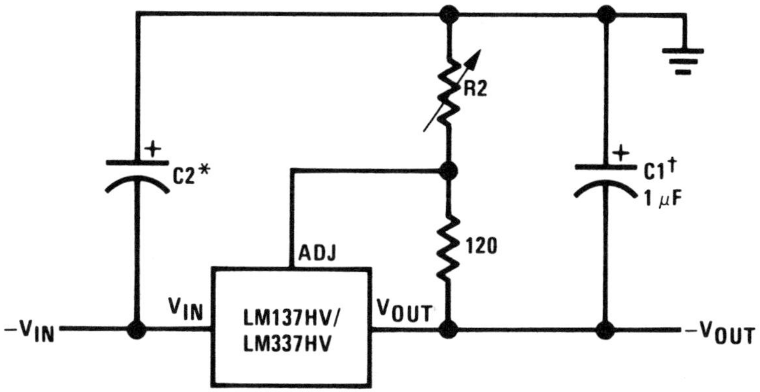

$$-V_{OUT} = -1.25V\left(1 + \frac{R2}{120\Omega}\right)$$

†C1 = 1 μF solid tantalum or 10 μF aluminum electrolytic required for stability. Output capacitors in the range of 1 μF to 1000 μF of aluminum or tantalum electrolytic are commonly used to provide improved output impedance and rejection of transients.

*C2 = 1 μF solid tantalum is required only if regulator is more than 4" from power-supply filter capacitor

Adjustable negative voltage regulator (NS).

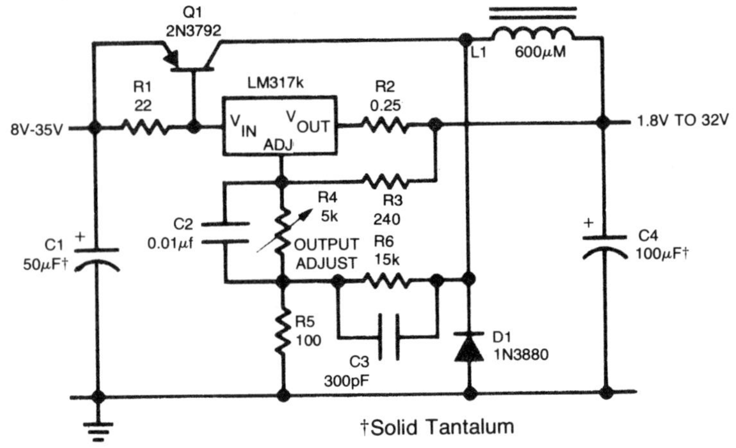

†Solid Tantalum

*Core—Arnold A-254168-2 60 turns

Low cost 3A switching regulator (NS).

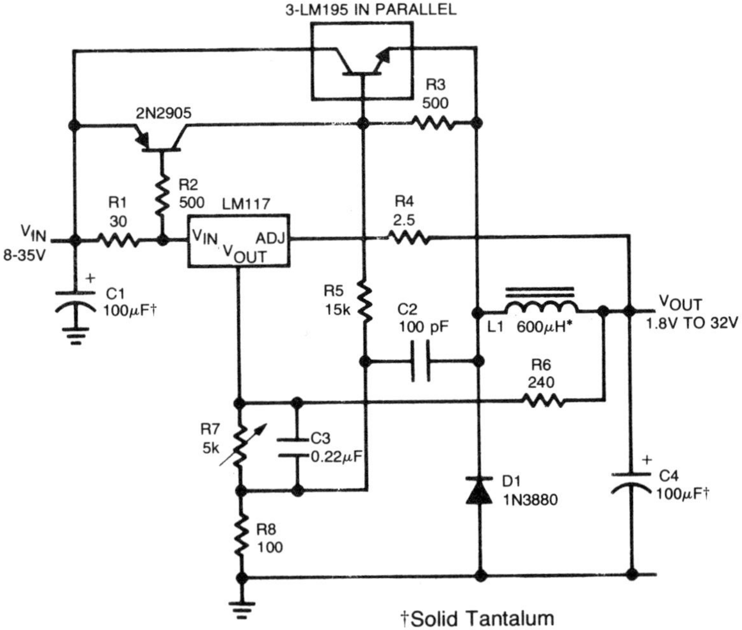

†Solid Tantalum

*Core—Arnold A-254168-2 60 turns

4A switching regulator with overload protection (NS).

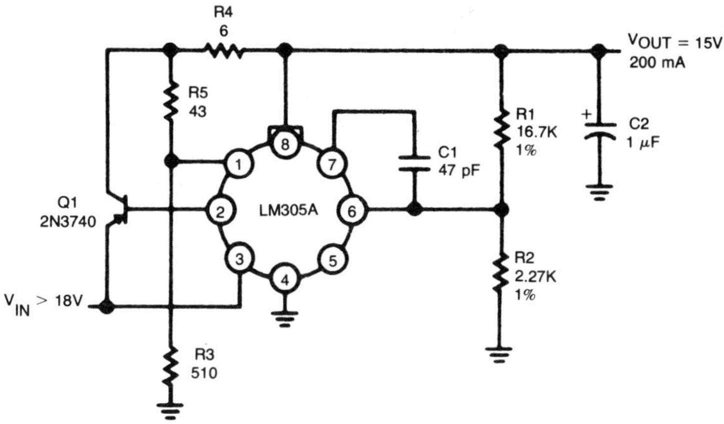

Linear regulator with foldback current limiting (NS).

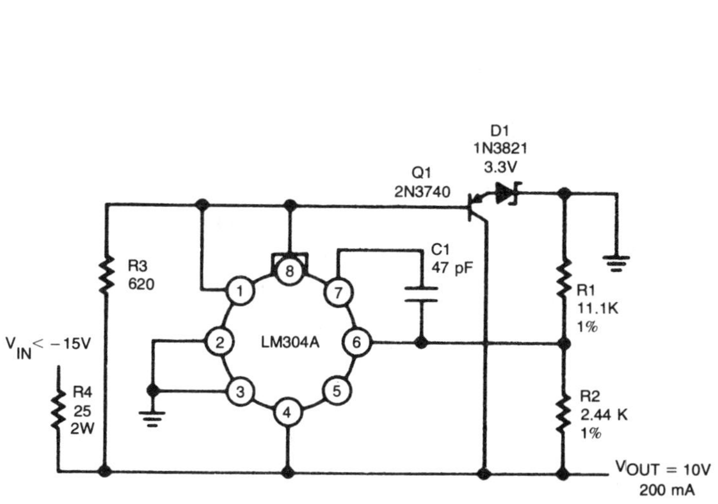

Shunt regulator (NS).

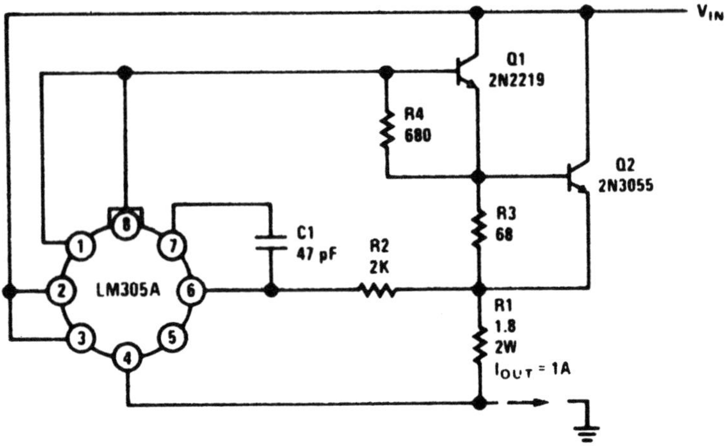

Current regulator (NS).

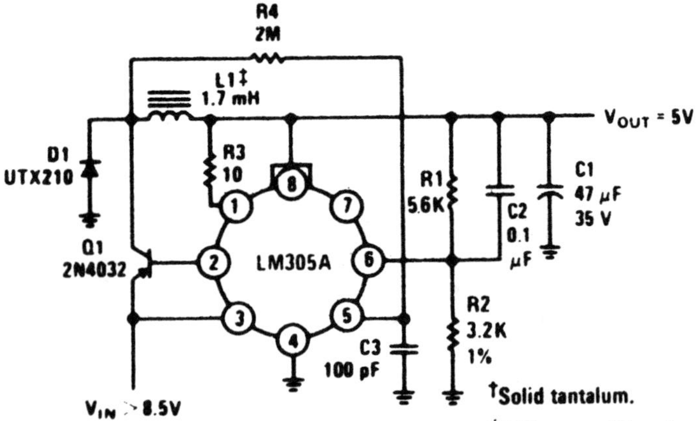

†Solid tantalum.

‡125 turns #22 on Arnold Engineering A262123-2 molybdenum permally core.

Switching regulator (NS).

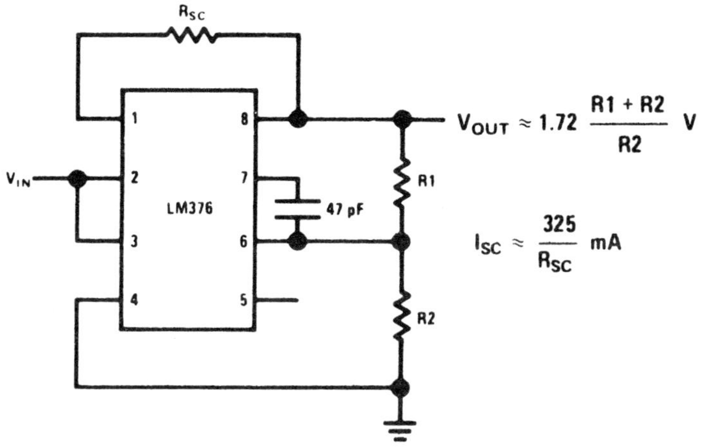

Basic positive regulator with current limiting (NS).

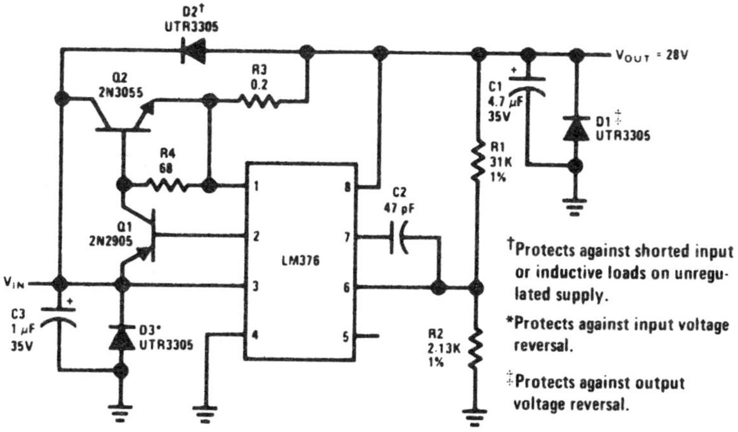

1.0A regulator with protective diodes (NS).

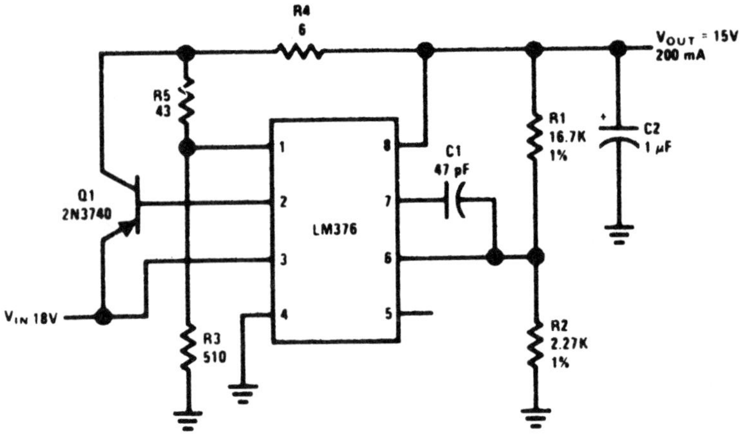

Linear regulator with foldback current limiting (NS).

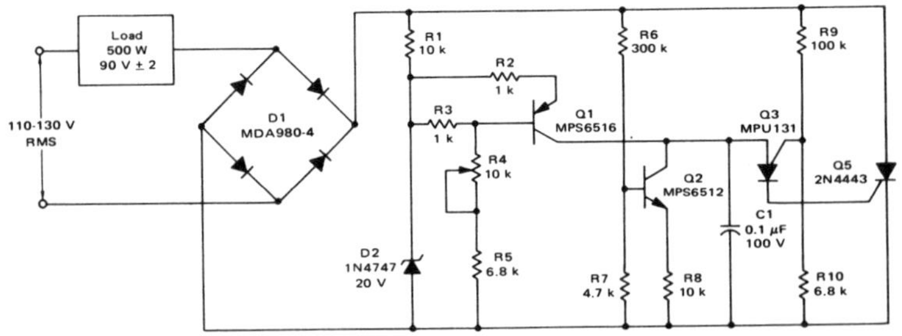

RMS voltage regulator (M).

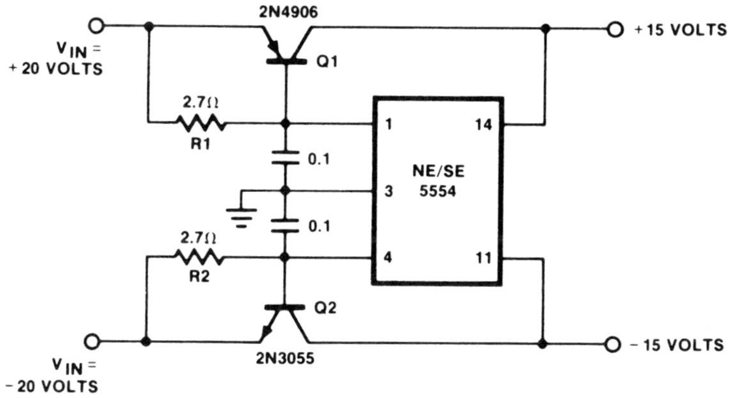

$$\beta(Q) \geq \frac{I_{OUT}(MAX)}{I_{REQ}(MAX)}$$

$$R1 = R2 = \frac{\beta V_{BE}}{I_{REG(MAX)}\,(\beta + 1) - I_{OUT(MAX)}}$$

High current dual tracking regulator (S).

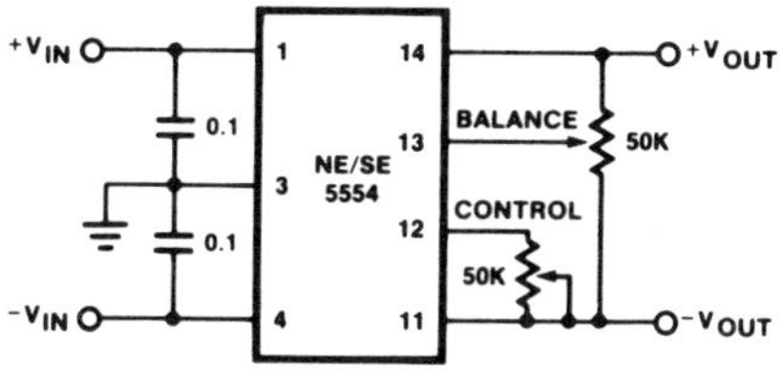

Variable output tracking regulator (S).

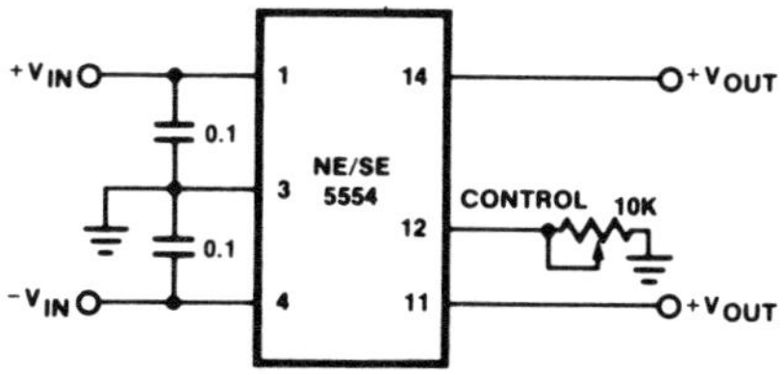

Dual tracking regulator with $\pm V_{out} > 15$ volts (S).

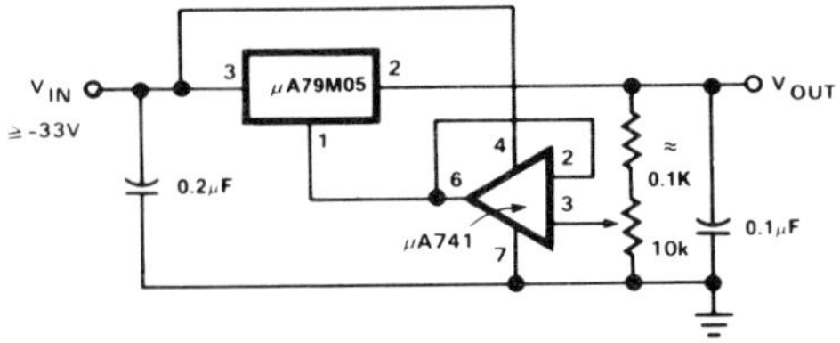

Variable output voltage, −30V to −7V (S).

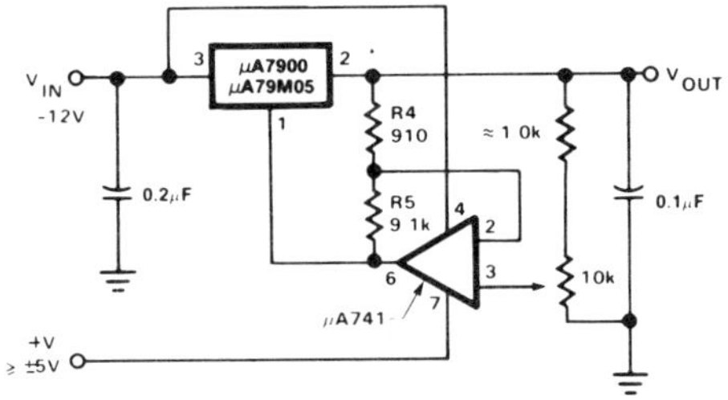

Variable output voltage, −0.5V to −10V (S).

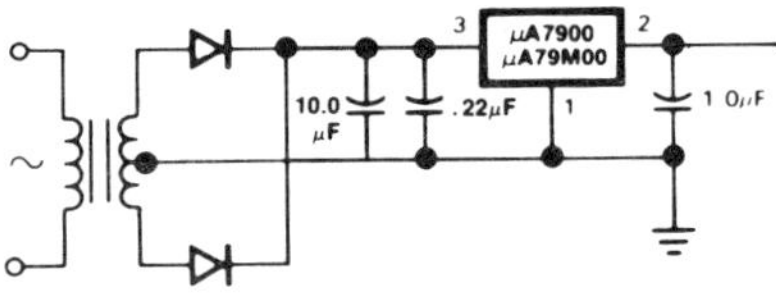

Negative output voltage circuit (S).

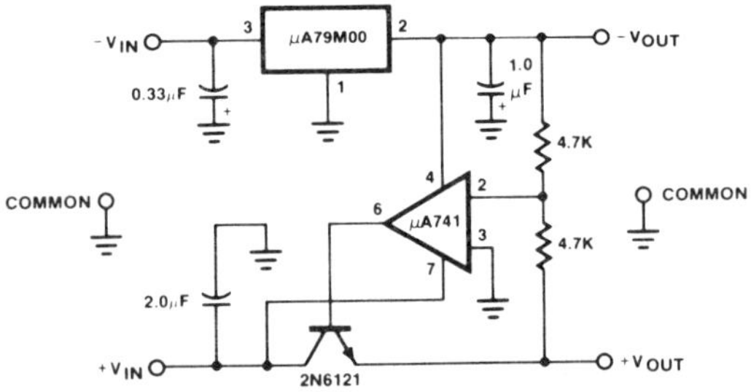

Positive and negative tracking voltage regulator (S).

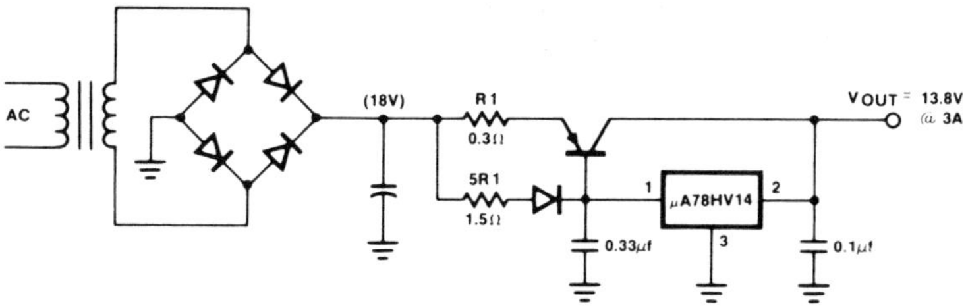

Battery substitution power supply (S).

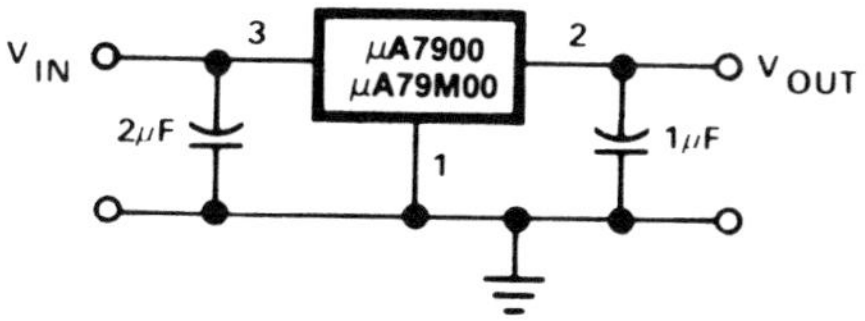

Fixed output regulator (S).

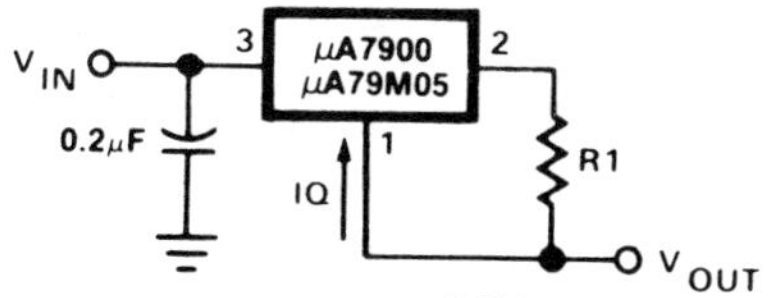

$$\text{Output current} = \frac{5.0\text{V}}{\text{R1}} + I_Q$$

Basic current regulator (S).

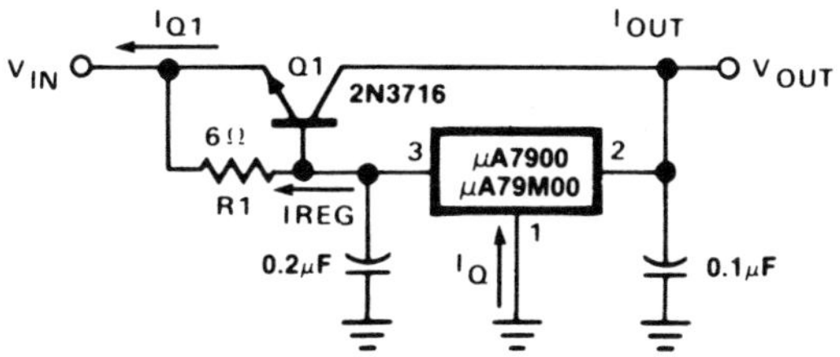

$$\text{R1} = \frac{V_{BE}(Q1)}{I_{REG}} = \frac{\beta V_{BE}\,(Q1)}{I_{REQ}(MAX)\,(\beta + 1) - I_{OUT}(MAX)}$$

$$I_{Q1} = \beta(Q1)\, I_{REG}$$

High current voltage regulator (S).

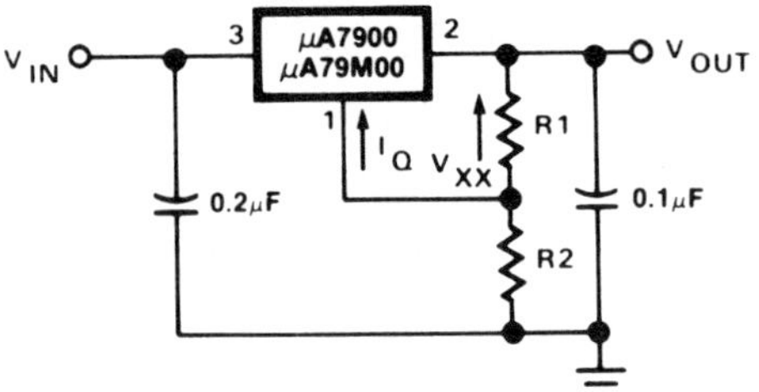

$$|V_{OUT}| = V_{XX}\left(1 + \frac{R2}{R1}\right) + I_Q R2$$

Variable output voltage regulator (S).

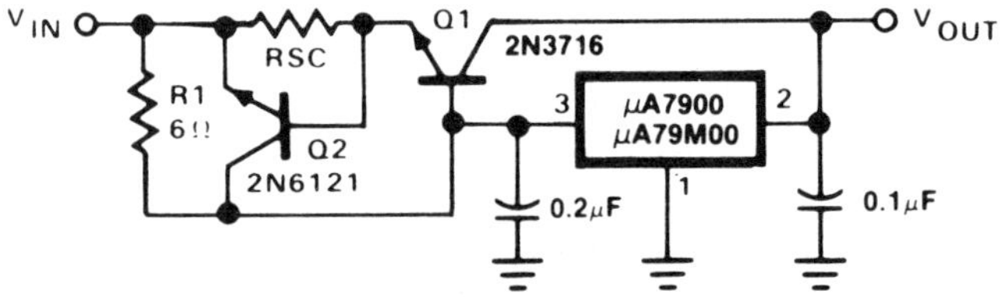

$$R1 = \frac{\beta V_{BE}\ (Q1)}{I_{REQ}(MAX)\ (\beta + 1) - I_{OUT}(MAX)}$$

$$R_{SC} = \frac{V_{BE(Q2)}}{I_{SC}}$$

High output current, short circuit protected (S).

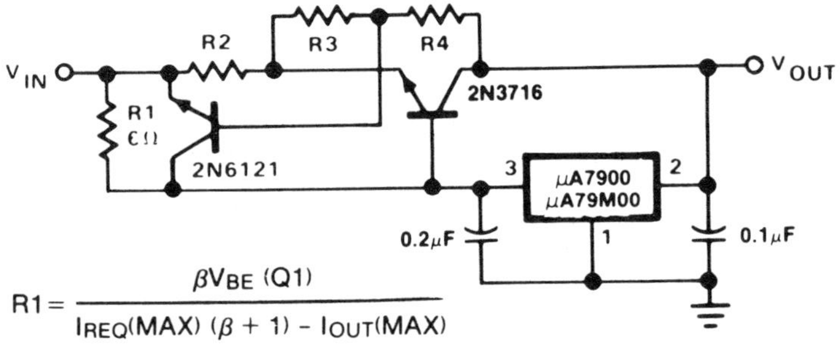

$$R1 = \frac{\beta V_{BE}\ (Q1)}{I_{REQ}(MAX)\ (\beta + 1) - I_{OUT}(MAX)}$$

High output current, foldback current limited (S).

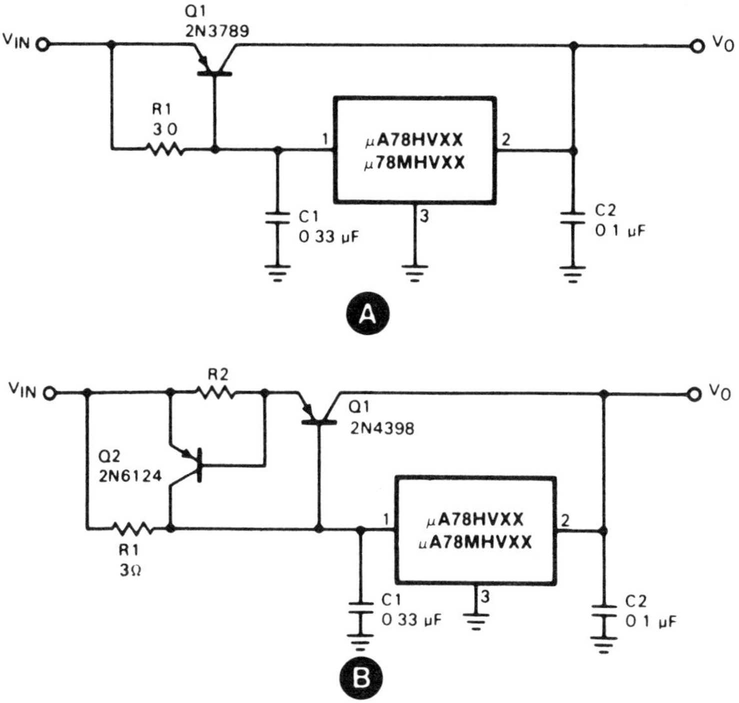

High current regulators (S).

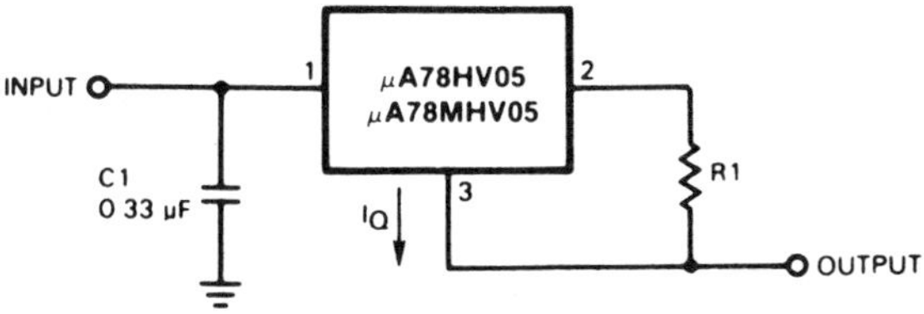

$$\text{Output Current} = \left(\frac{5.0\ V}{R1}\right) + I_Q$$

Basic current regulator (S).

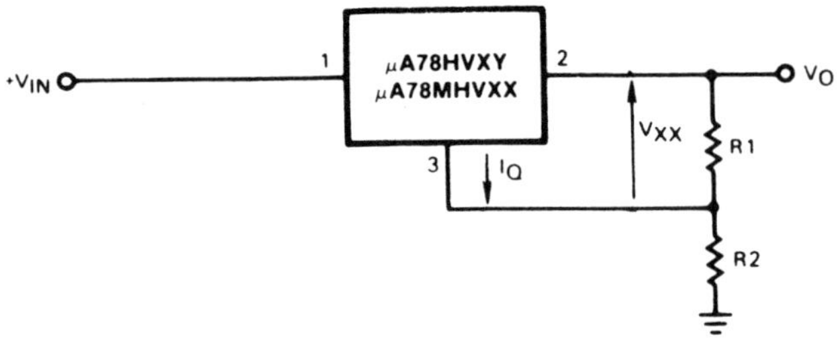

Variable output regulator (S).

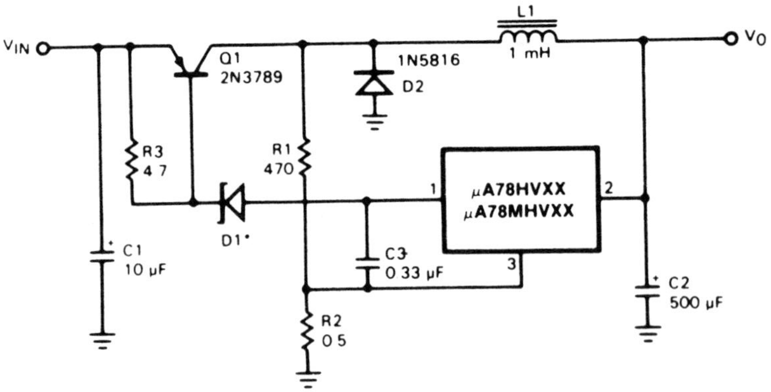

Switching regulator (S).

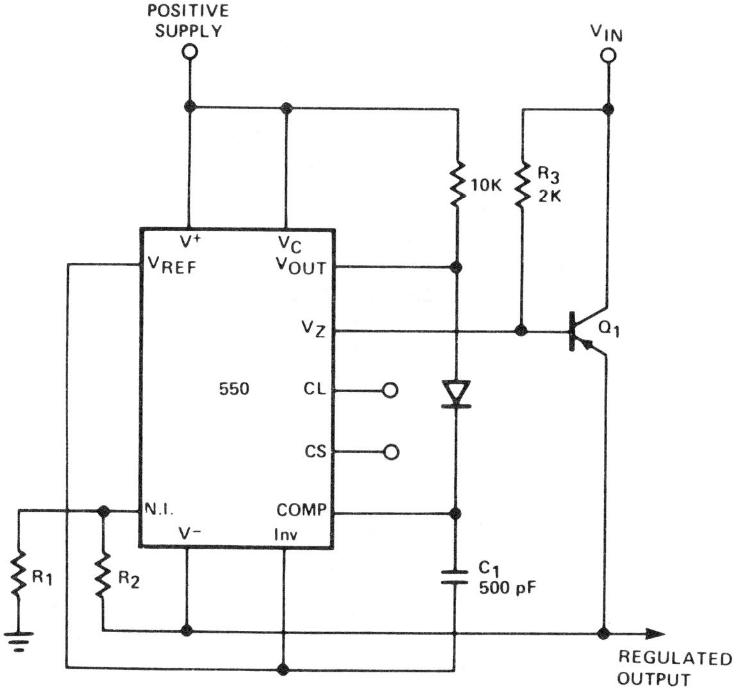

Negative regulator (less than −8.5 volts) (S).

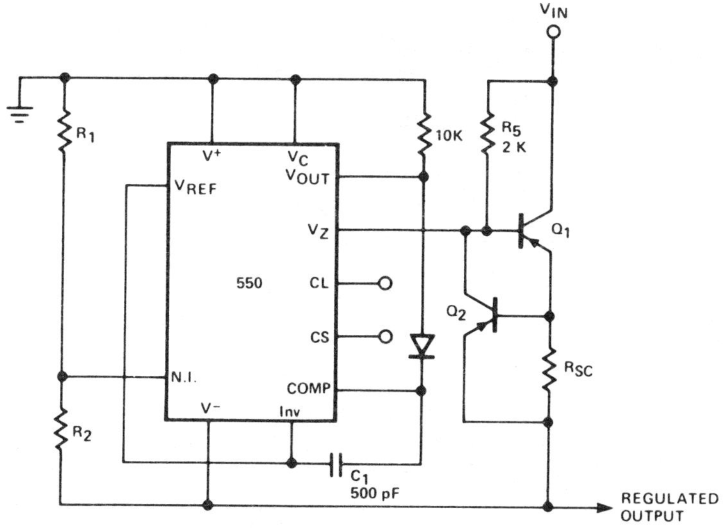

Providing negative regulator overload protection (S).

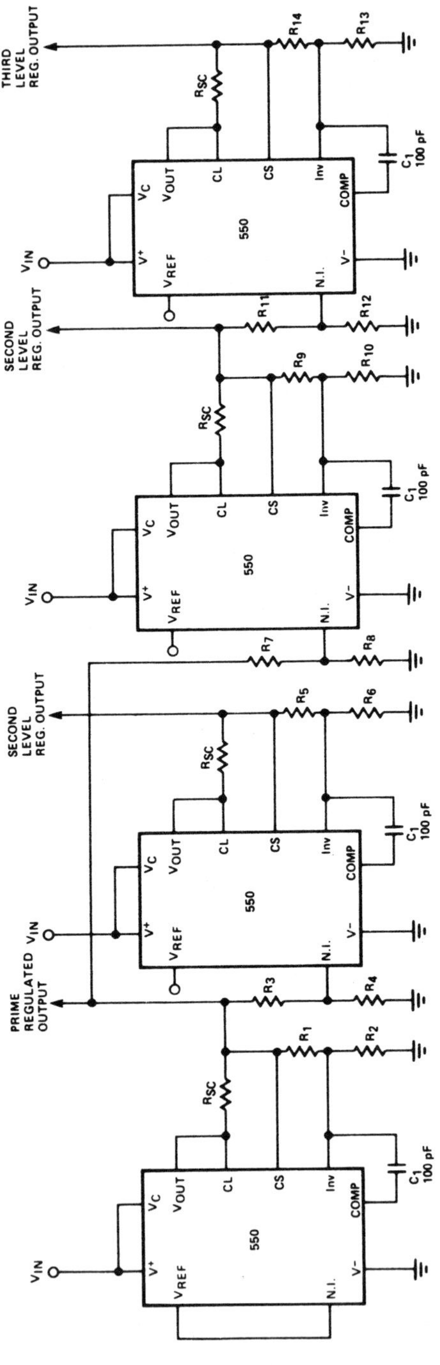

Cascading regulators (S).

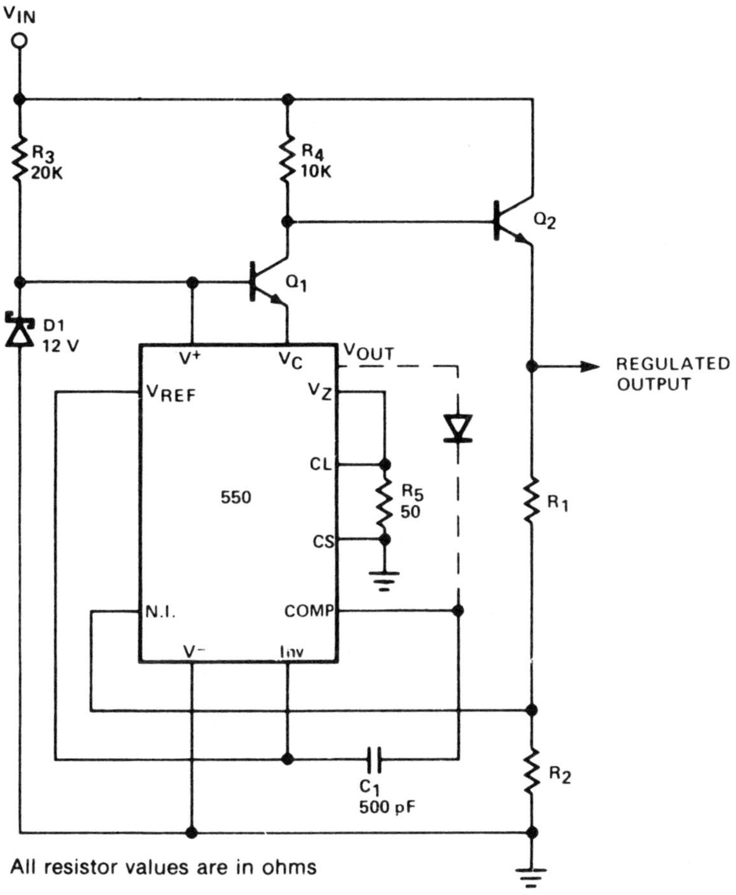

All resistor values are in ohms

Positive floating regulator (S).

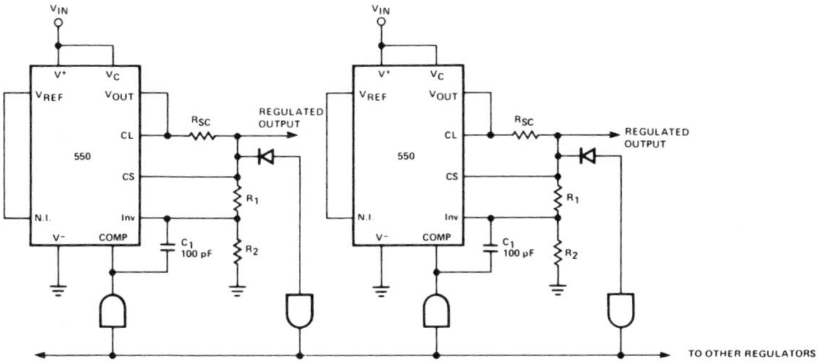

Remote slaving of multiple supplies (S).

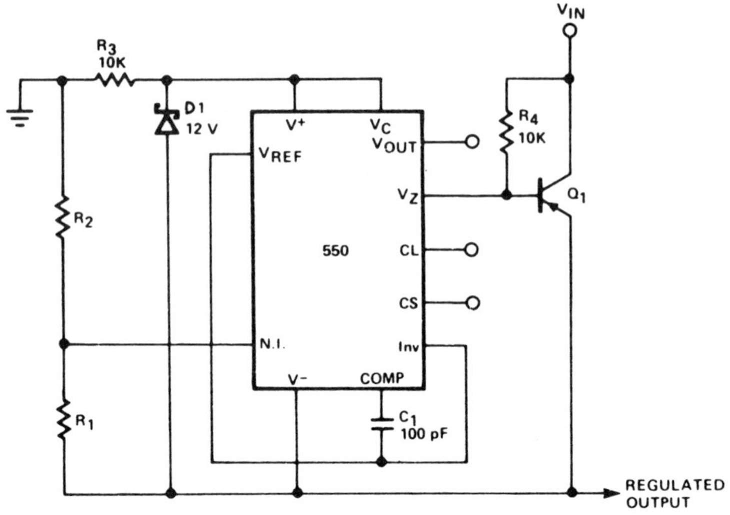

Negative floating regulator (S).

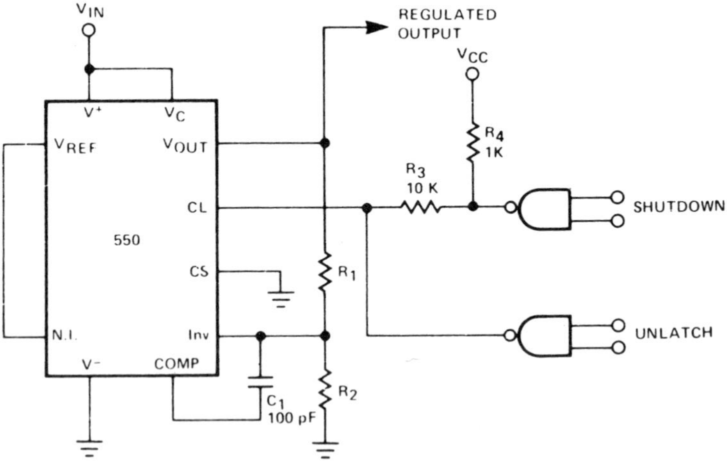

Latching remote shutdown regulator (S).

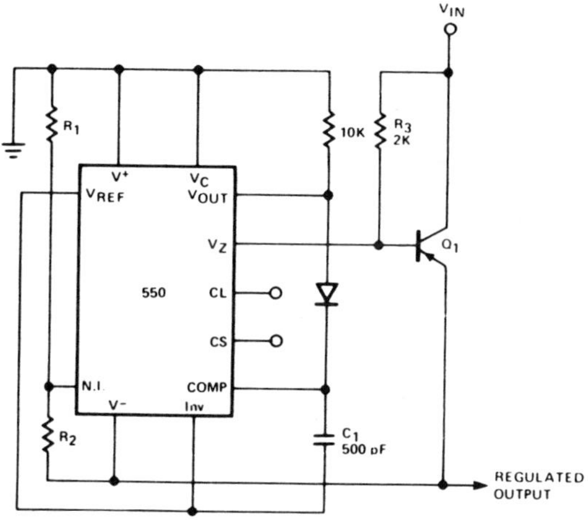

$$V_{OUT} = -V_{REF} \times \frac{R1 + R2}{R2}$$

$$\frac{R1 \times R2}{R1 + R2} = 2k\Omega \text{ for minimum temperature drift}$$

All resistor values are in ohms

Basic negative regulator (S).

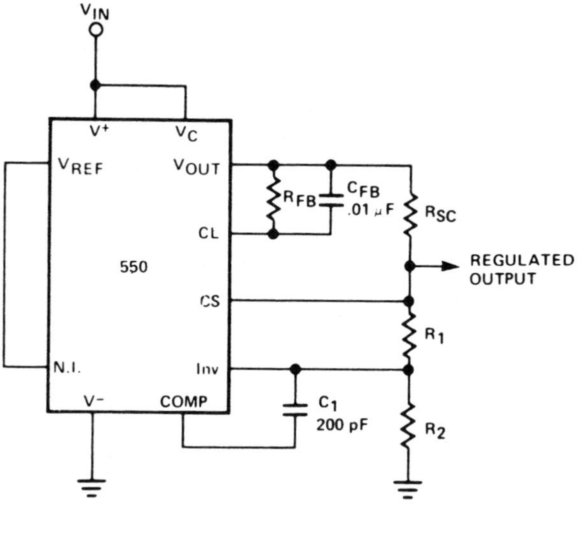

Fold back current limiting (S).

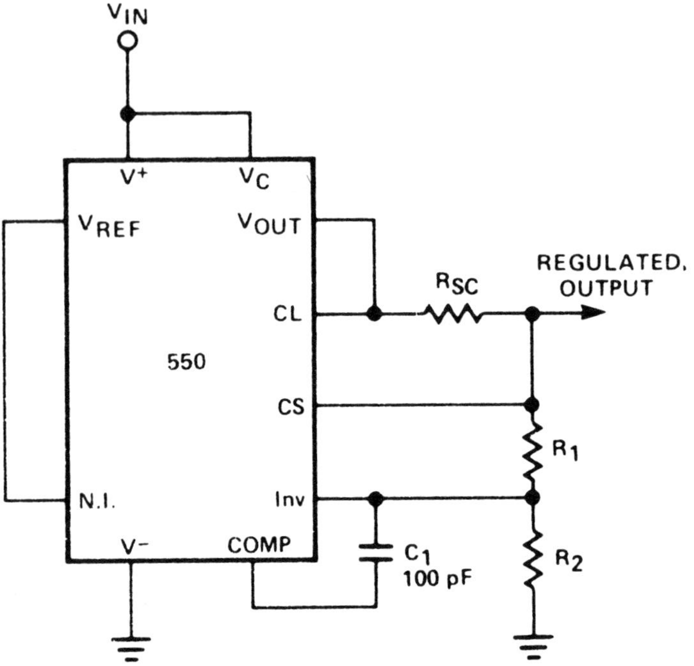

Basic regulator configuration (S).

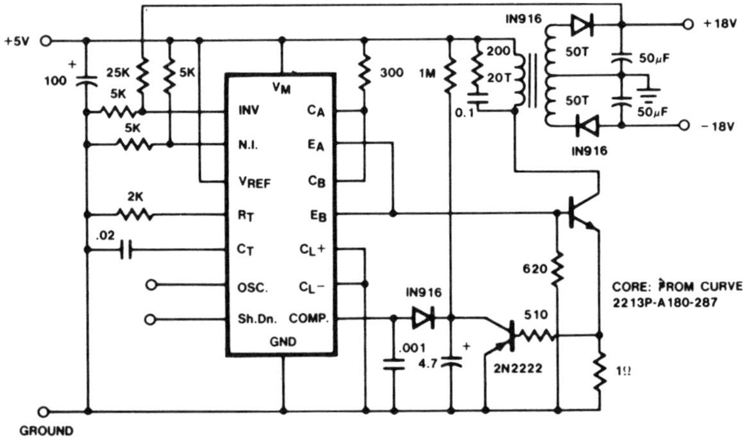

Another low-current supply is the flyback converter used here to generate ± 15 volts at 20mA from a +5 volt regulated line. The reference generator in the SG1524 is unused with the input voltage providing the reference. Current limiting in a flyback converter is difficult and is accomplished here by sensing current in the primary line and resetting a soft-start circuit (S).

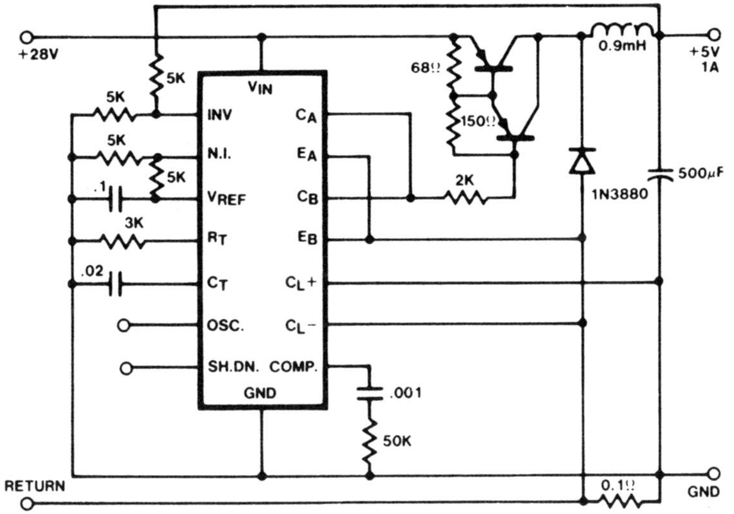

In this conventional single-ended regulator circuit, the two outputs of the SG1524 are connected in parallel for effective .0-90% duty-cycle modulation. The use of an output inductor requires an R-C phase compensation network for loop stability (S).

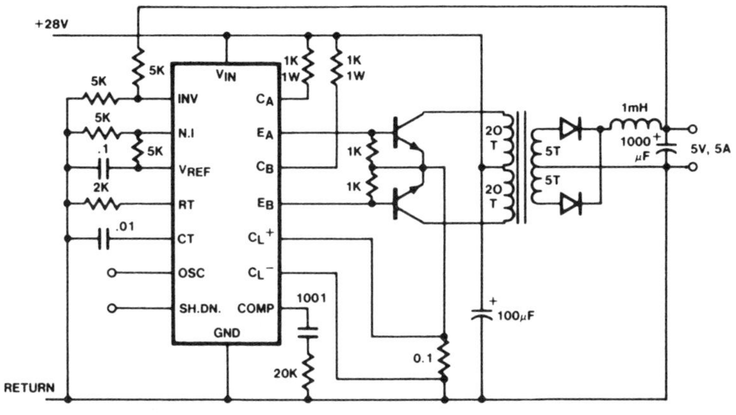

Push-pull outputs are used in this transformer-coupled dc-dc regulating converter. Note that the oscillator must be set at twice the desired output frequency as the SG1524's internal flip-flop divides the frequency by 2 as it switches the P.W.M. signal from one output to the other. Current limiting is done here in the primary so that the pulse width will be reduced should transformer saturation occur (S).

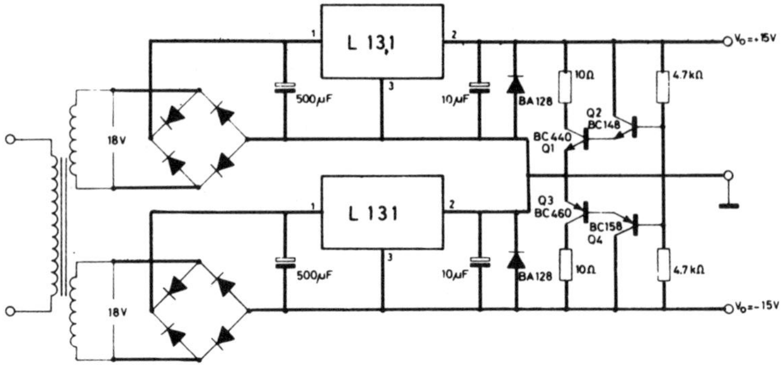

Symmetrical ±15V voltage regulator circuit (SGS).

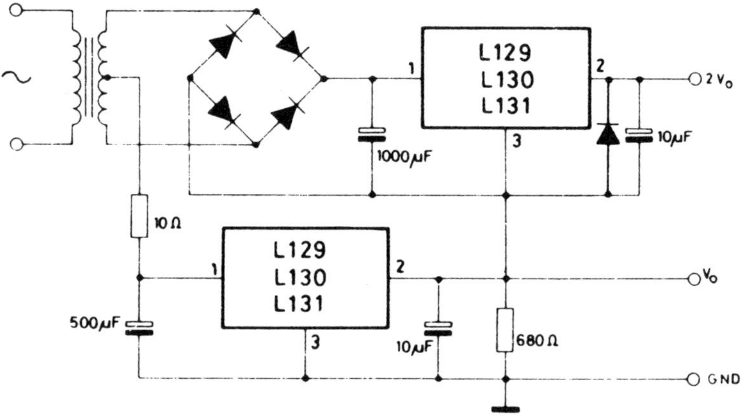

Series regulators circuit connection (SGS).

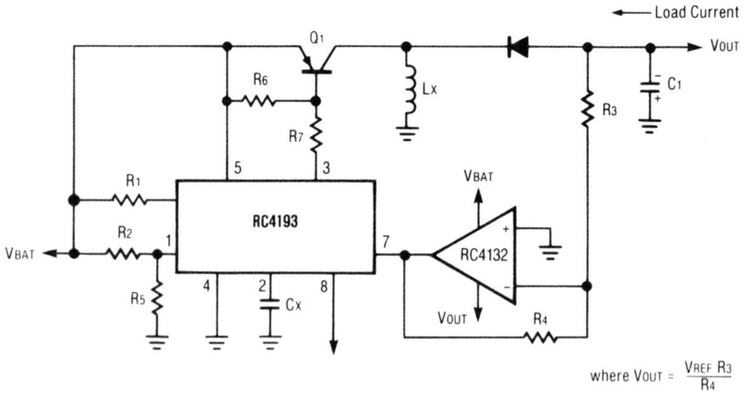

Inverting switching regulator (R).

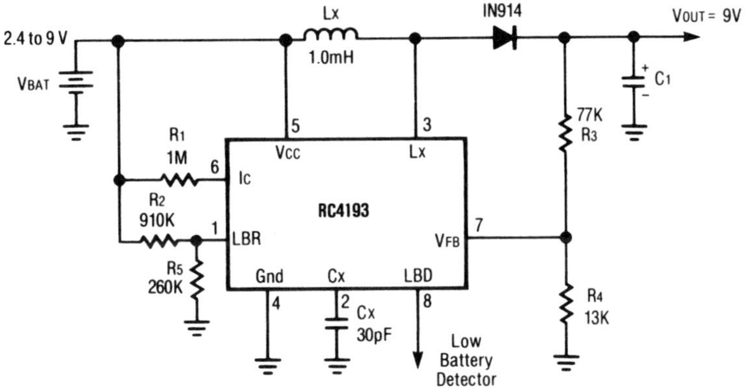

Step-up switching regulator (low current) (R).

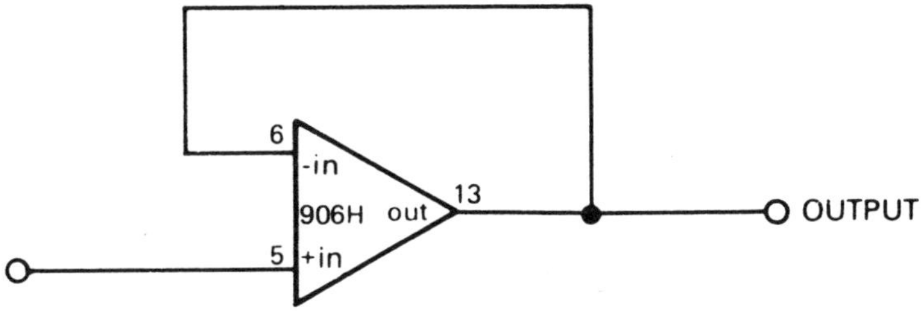

Voltage follower (OE).

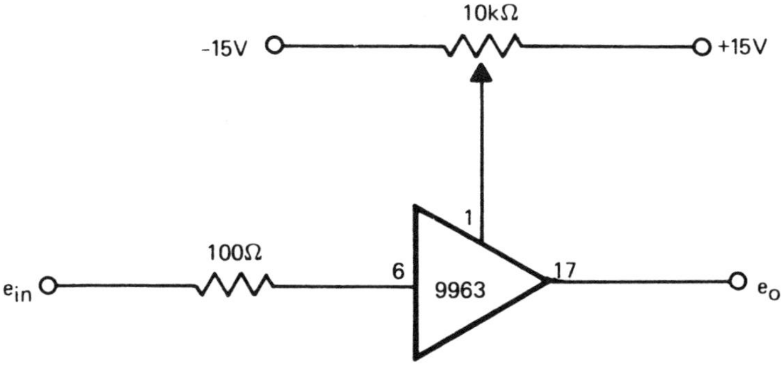

Follower with offset adjustment (OE).

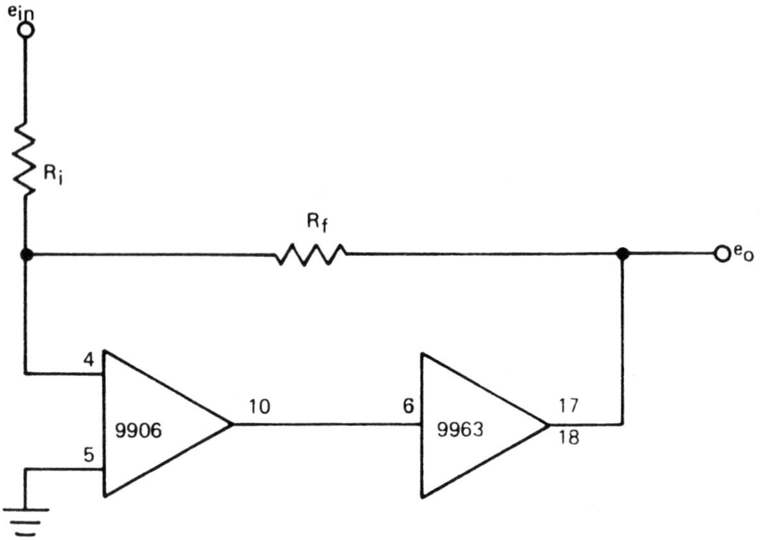

Current booster (OE).

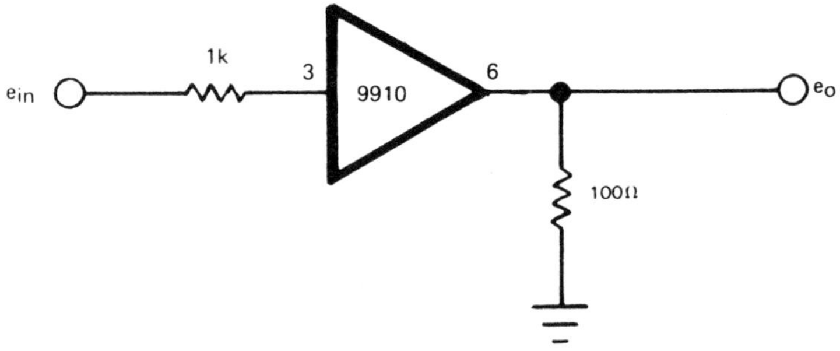

Voltage follower (OE).

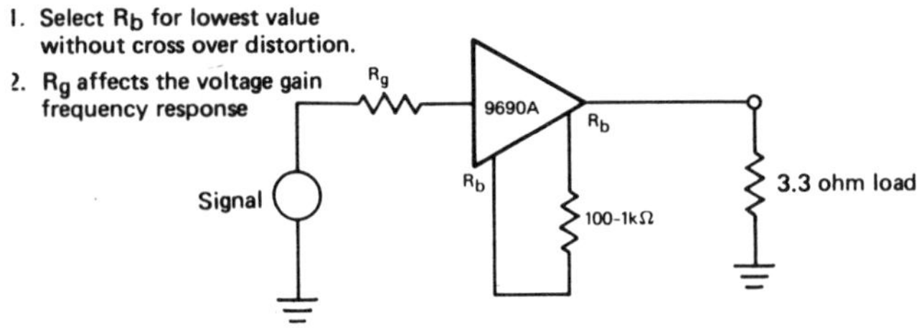

Linear buffer—follower amplifier (OE).

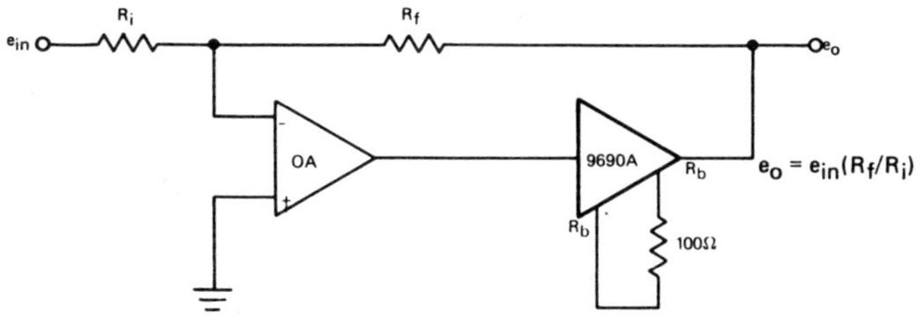

Current booster (OE).

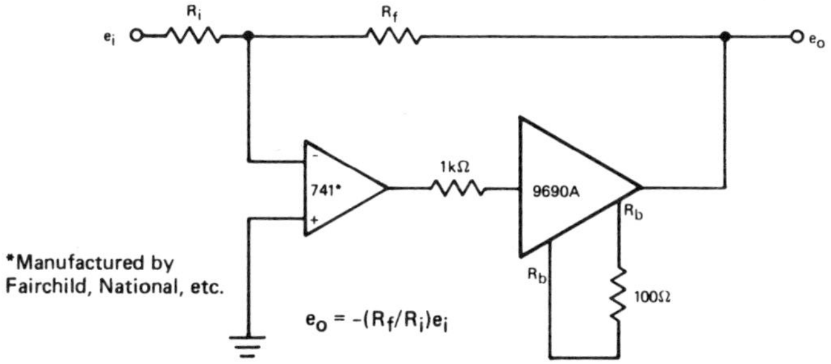

Current booster (alternate) (OE).

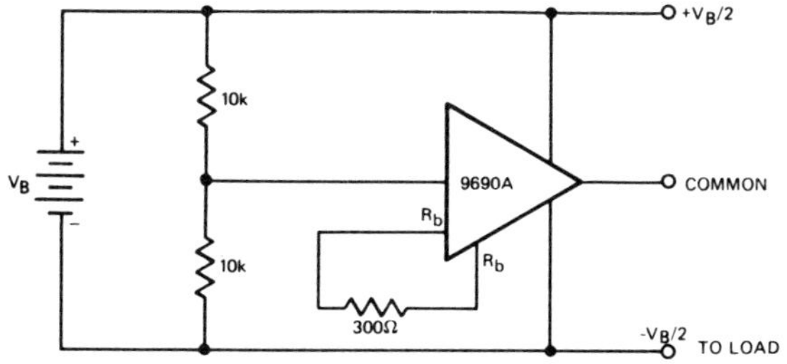

Power supply splitter (OE).

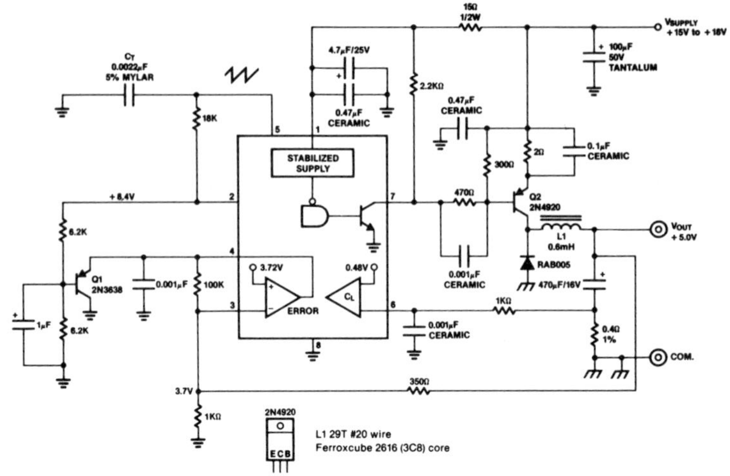

NE5561 5V/0.5A buck converter (S).

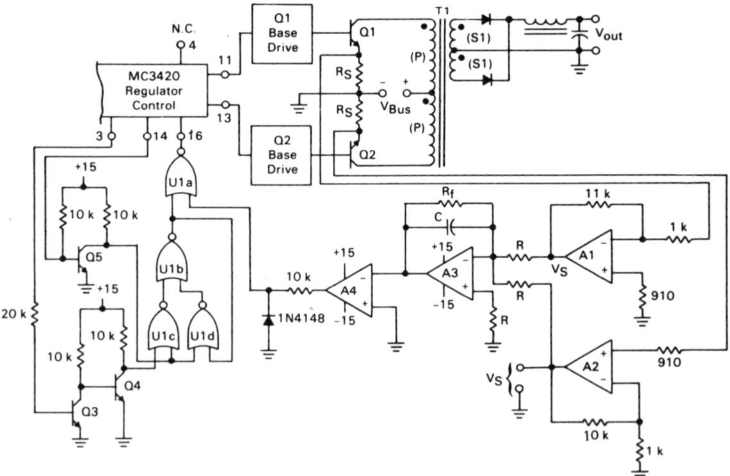

Q1, Q2 — Inverter Switching Transistors
Q3-Q5 — 2N4401
T1 — Inverter Power Transformer
U1 — MC14001 CMOS Quad 2-Input NOR Gate

Symmetry correction circuit (M).

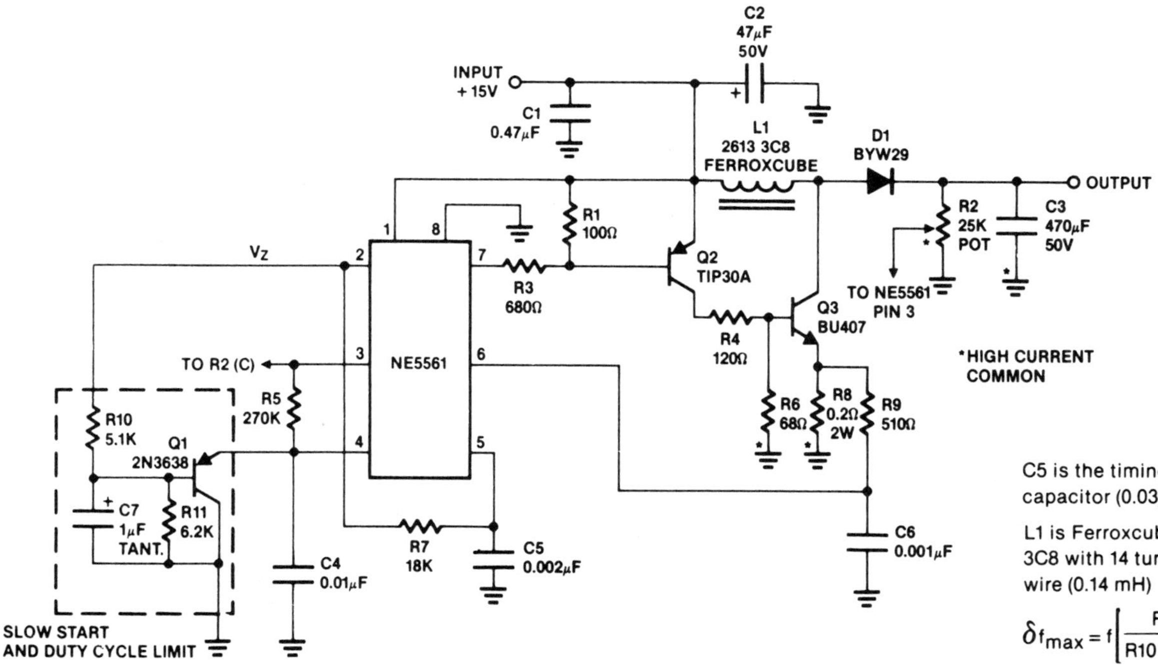

C5 is the timing capacitor (0.03μF for 22 kHz)

L1 is Ferroxcube 2613-3C8 with 14 turns of #18 enameled wire (0.14 mH)

$$\delta f_{max} = f\left[\frac{R11}{R10 + R11}\ (V_Z) + V_{BE}\right]$$

Dc to dc SMPS using NE5561 /variable 18V to 30V out at 0.2A (S).

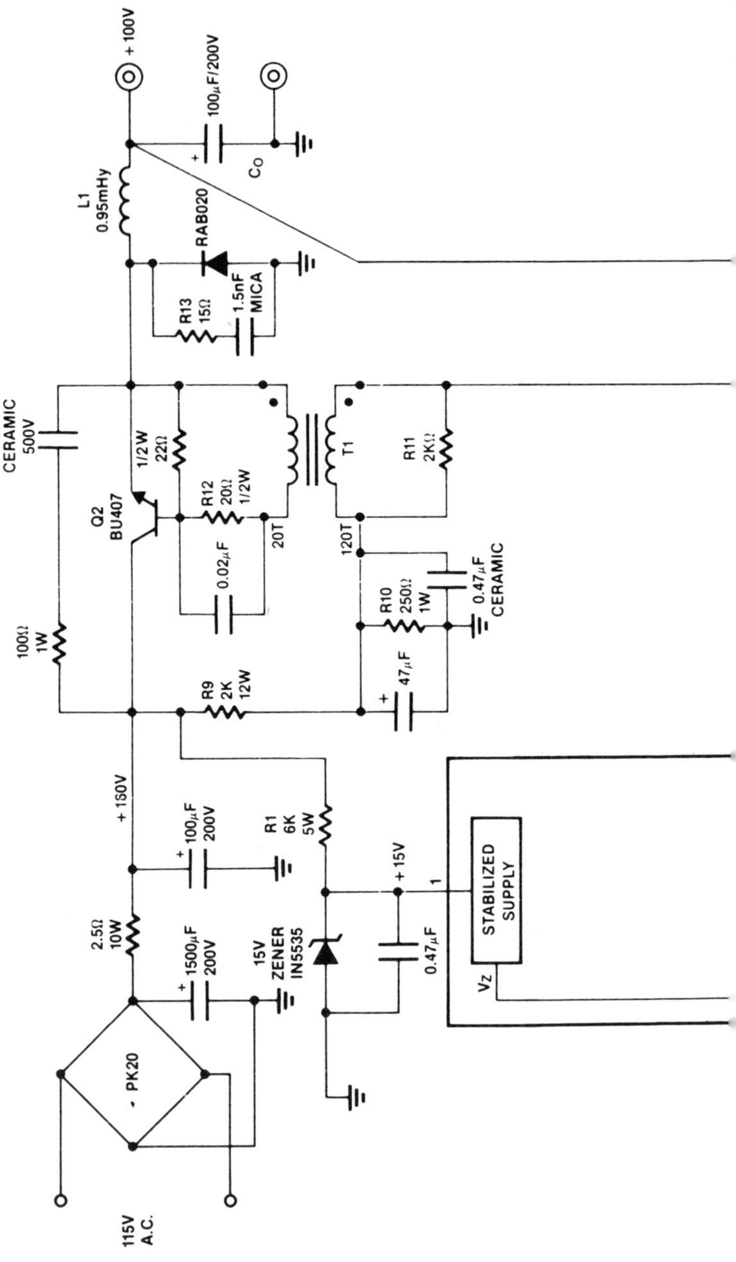

+100V
100μF/200V
C_O
L1
0.95mHy
RAB020
R13
15Ω
1.5nF
MICA
1nF
CERAMIC
500V
1/2W
22Ω
R12
20Ω
1/2W
Q2
BU407
0.02μF
20T
120T
T1
R11
2KΩ
R10
250Ω
1W
0.47μF
CERAMIC
100Ω
1W
R9
2K
12W
47μF
+180V
100μF
200V
R1
6K
5W
+15V
1
STABILIZED
SUPPLY
V_Z
0.47μF
2.5Ω
10W
1500μF
200V
15V
ZENER
IN5535
PK20
115V
A.C.

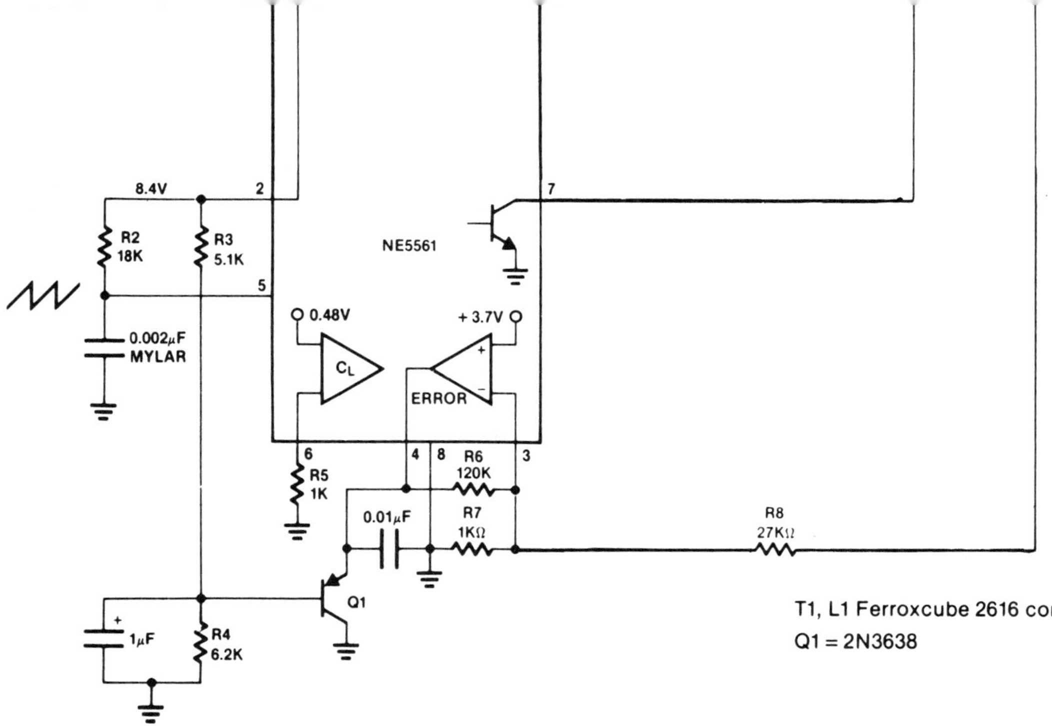

T1, L1 Ferroxcube 2616 core (3C8)
Q1 = 2N3638

37 NE5561 100V/0.25A switch mode converter (S).

Chapter 2

Test Equipment

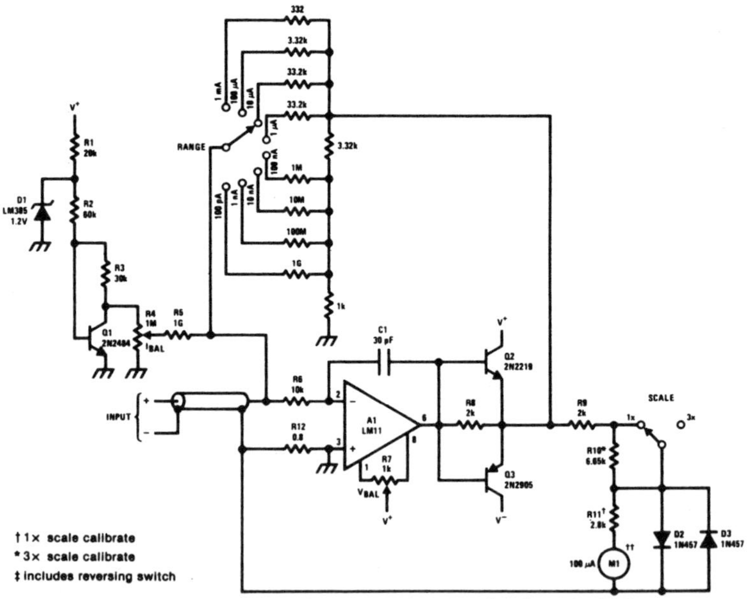

Ammeter current meter ranges from 100 pA to 3 mA full scale. Voltage across input is 100 μV at lower ranges rising to 3 mV at 3 mA. Buffers on op amp are to remove ambiguity with high-current overload. Output can also drive DVM or DPM (NS).

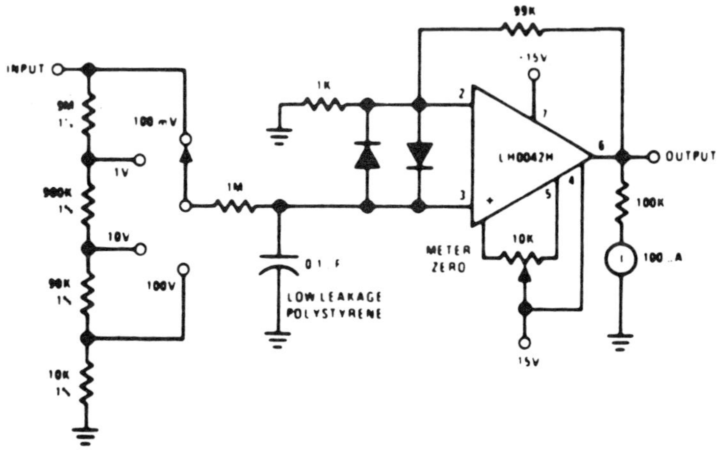

Sensitive Low Cost "VTVM" (NS).

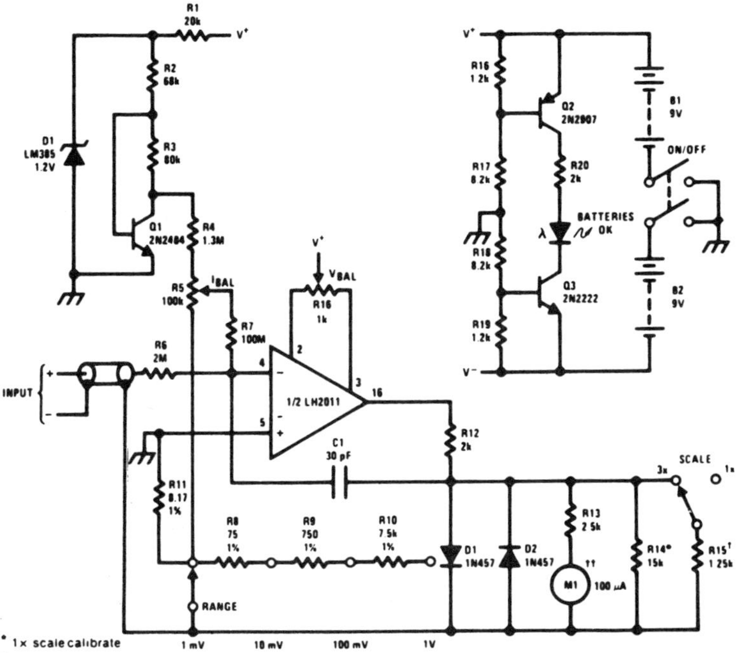

* 1x scale calibrate

† 3x scale calibrate

‡ includes reversing switch

Voltmeter: high-input-impedance millivoltmeter. Input current is proportional to input voltage, about 10 pA at full-scale. Reference could be used to make direct reading linear ohmmeter (NS).

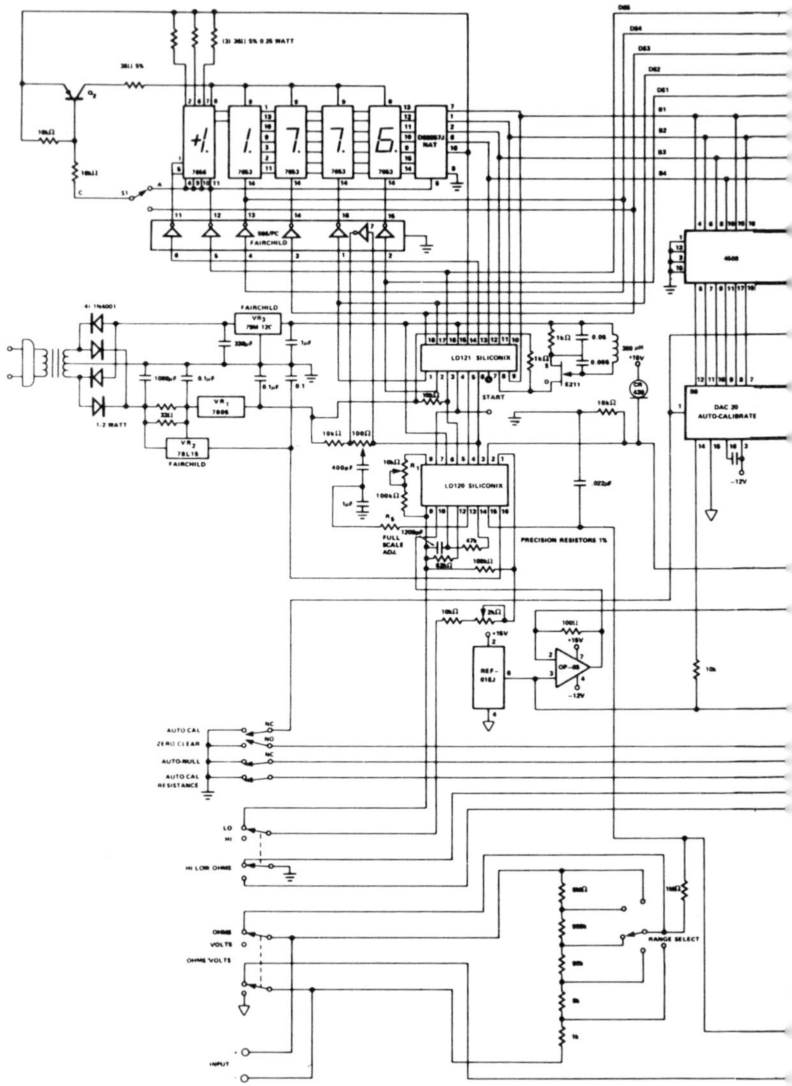

Complete Smart DMM (PM).

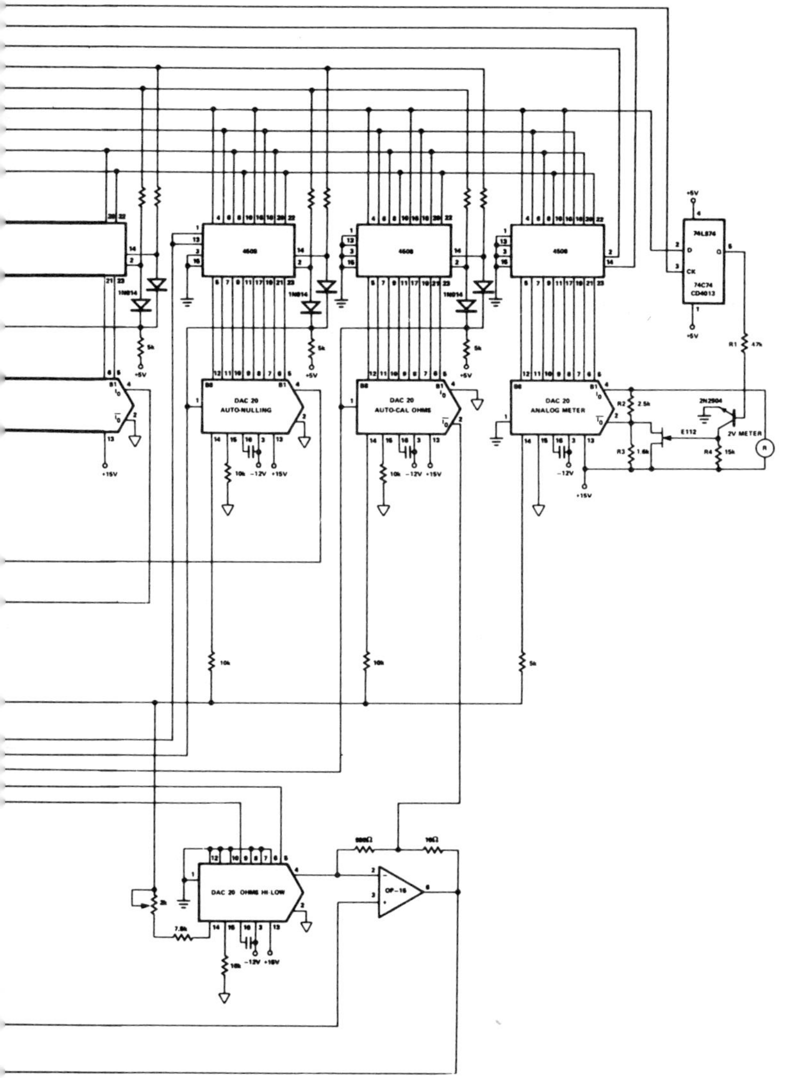
DAC 20
AUTO-NULLING
DAC 20
AUTO-CAL OHMS
DAC 20
ANALOG METER
DAC 20 OHMS HI-LOW
OP-16
74LS74
74C74
CD4013
2N2904
E112
2V METER
1N914
R1
R2
R3
R4
+5V
+15V
-12V

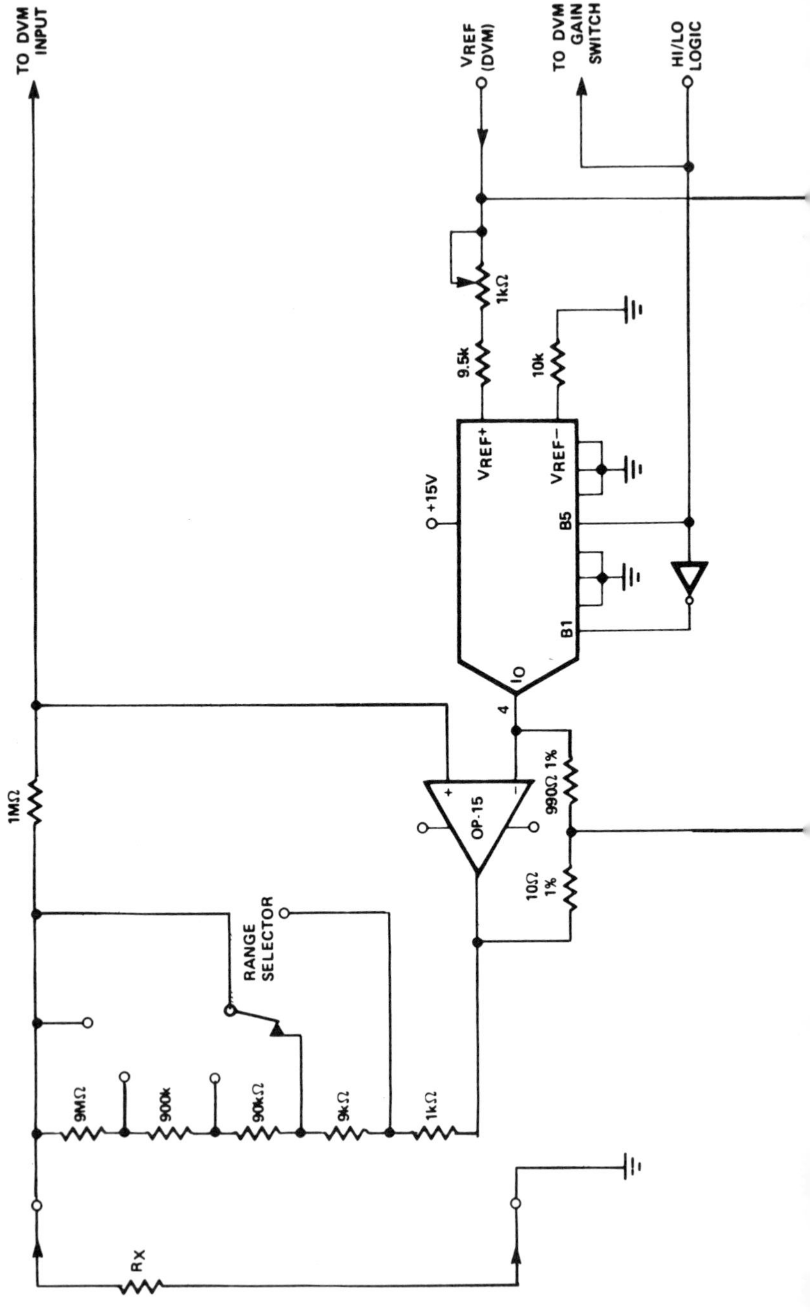

TO DVM INPUT
VREF (DVM)
TO DVM GAIN SWITCH
HI/LO LOGIC
1kΩ
9.5k
10k
+15V
VREF+
VREF−
B5
B1
IO
4
990Ω 1%
10Ω 1%
OP-15
1MΩ
RANGE SELECTOR
9MΩ
900k
90kΩ
9kΩ
1kΩ
RX

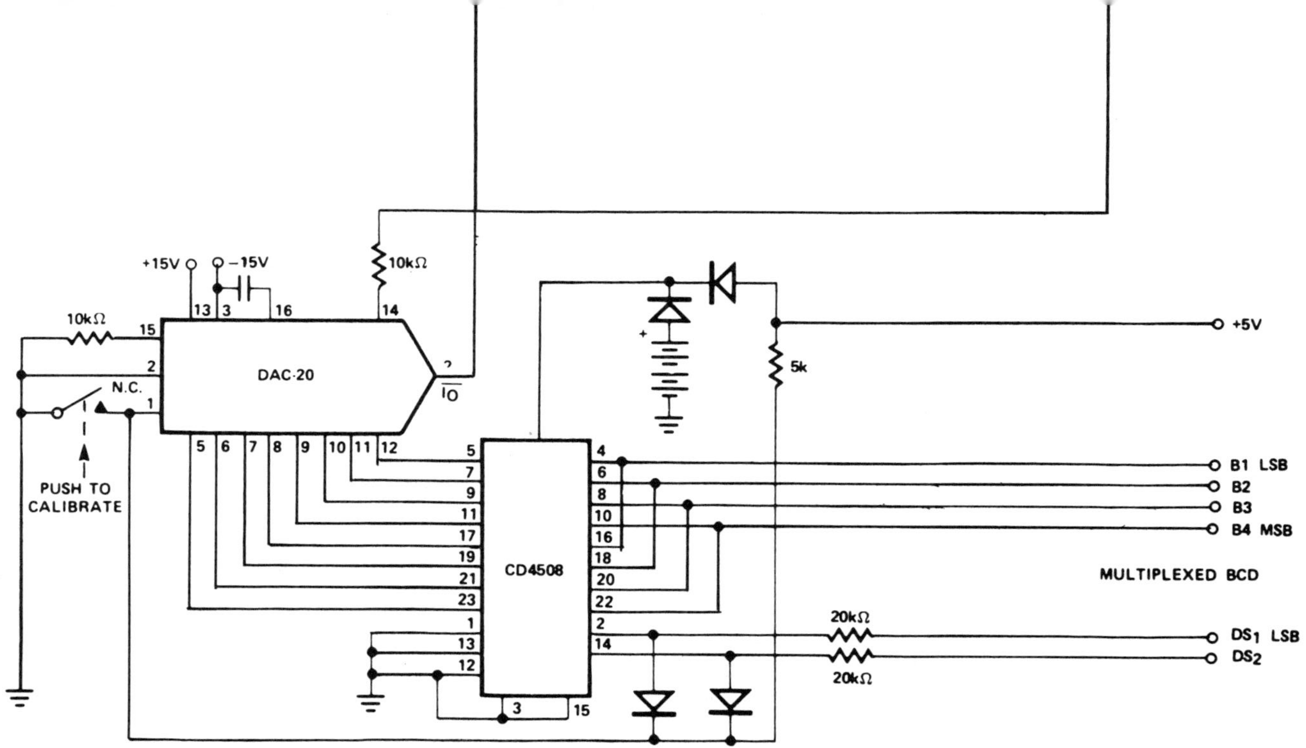

43 Auto-calibrate resistance measurement (PM).

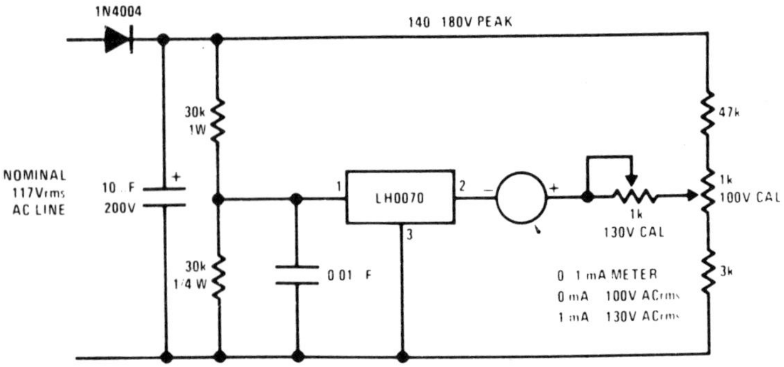

Expanded scale ac voltmeter (NS).

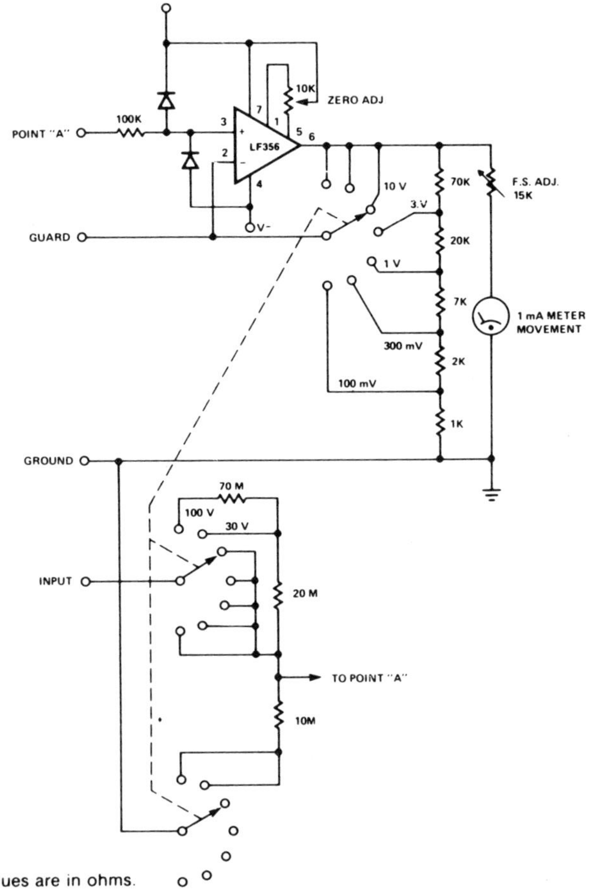

All resistor values are in ohms.

Voltmeter (S).

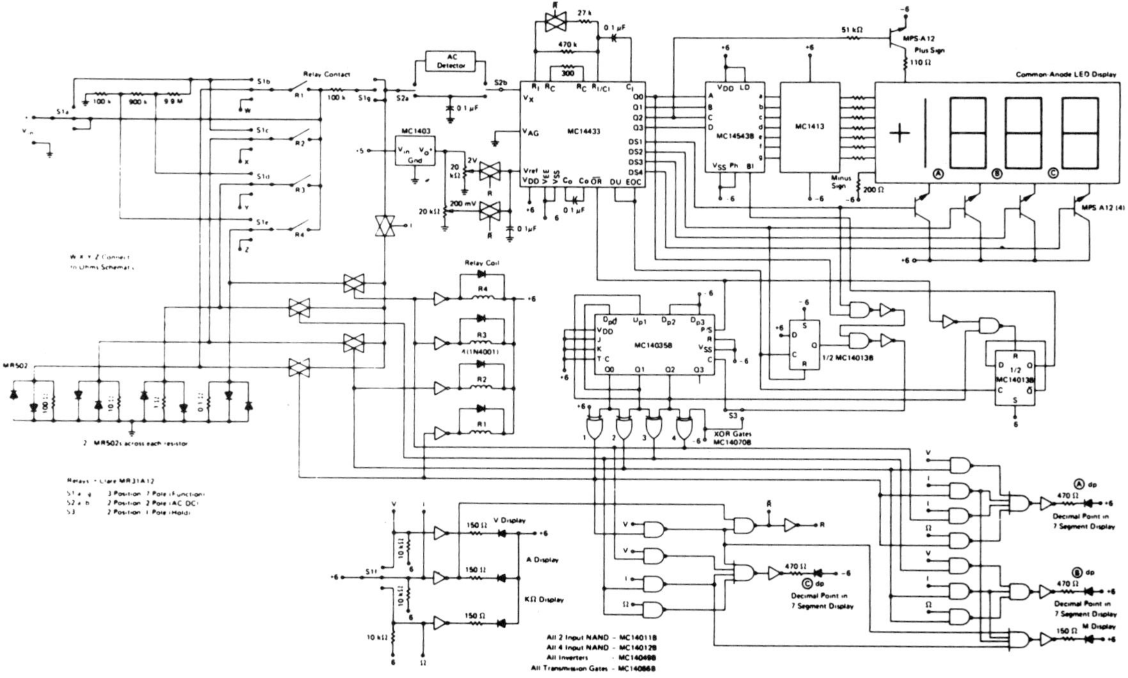

3-½-digit autoranging multimeter (M).

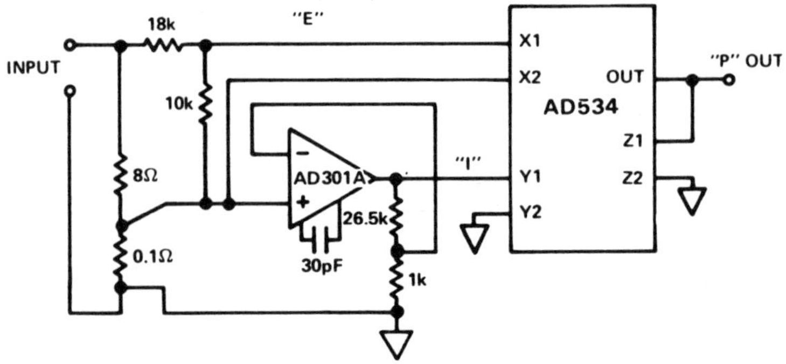

Wattmeter measures audio-amplifier output power into dummy load (AD).

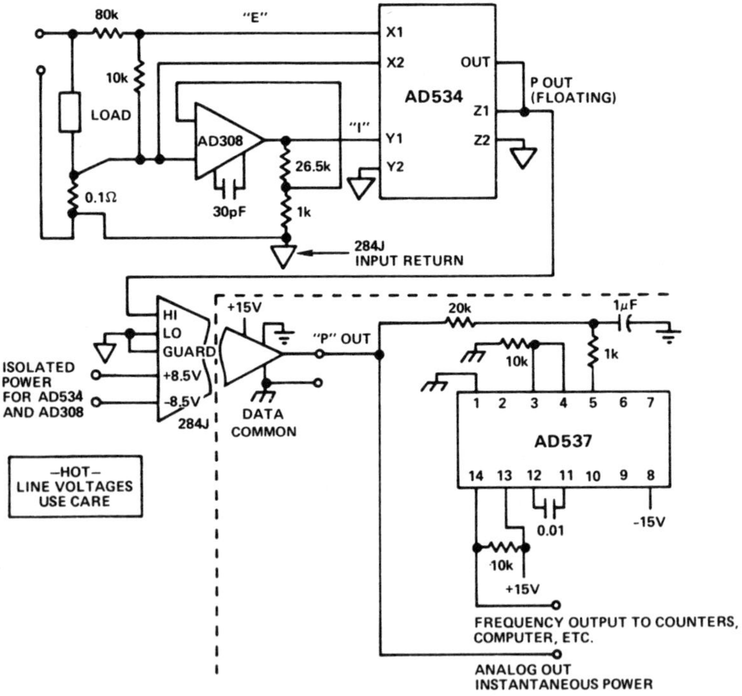

Line wattmeter—watt-hour meter (AD).

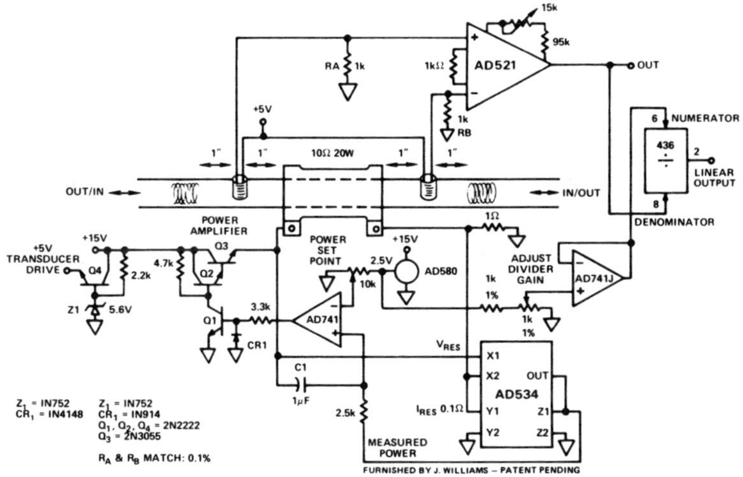

Flowmeter circuit (AD).

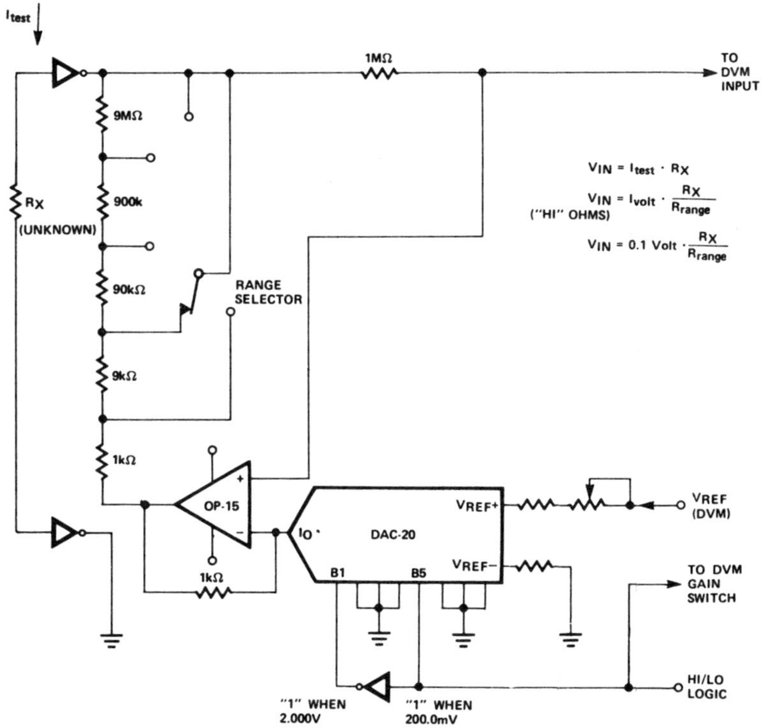

Hi-lo resistance-to-voltage converter (PM).

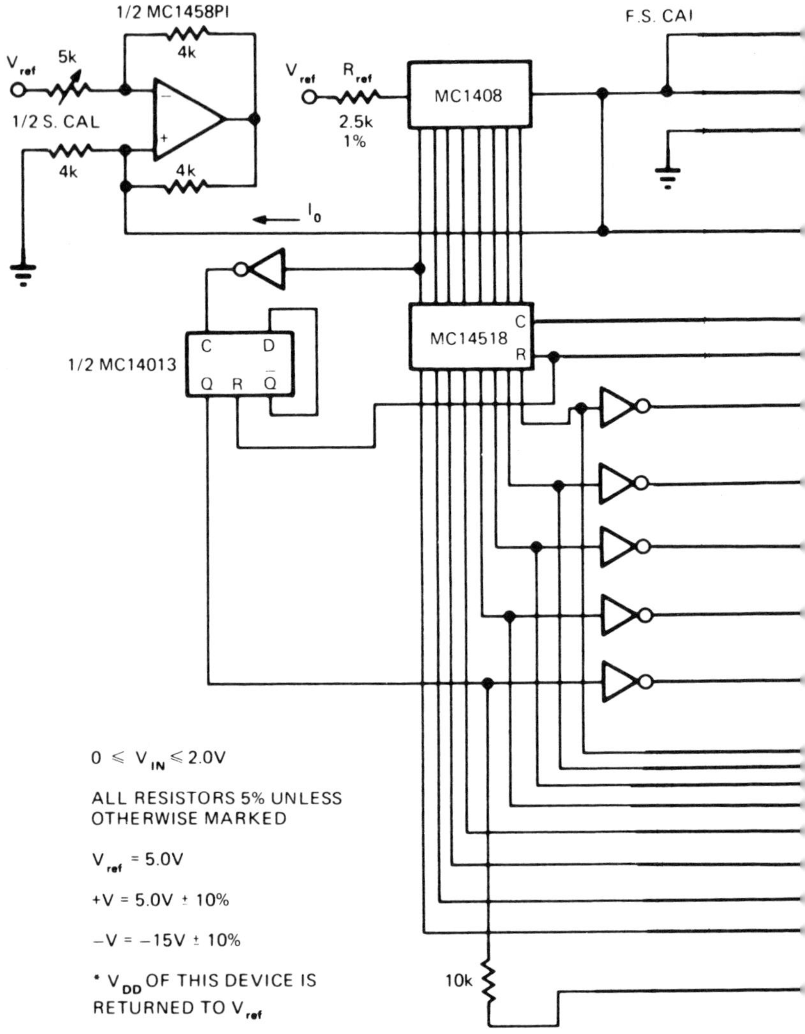

2-½-digit DVM uses a binary D/A converter to accomplish conversion of a BCD digital input signal (M).

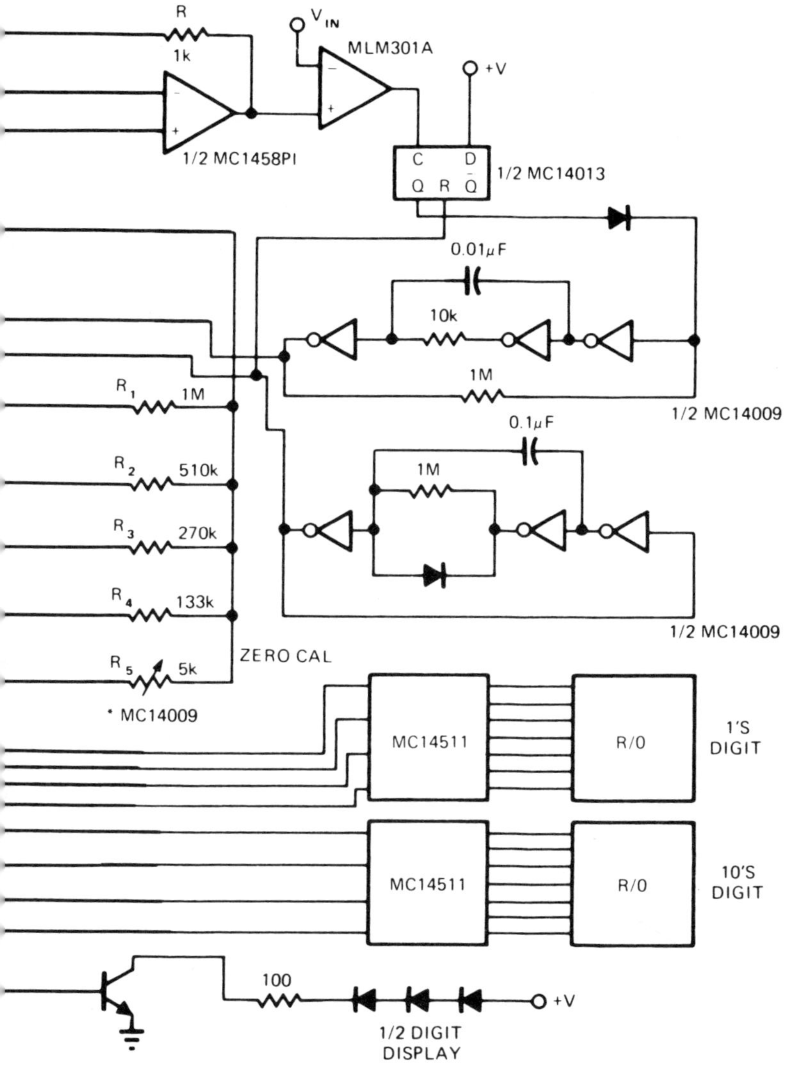

R
1k
V IN
MLM301A
+V
1/2 MC1458PI
C D
Q R Q
1/2 MC14013
0.01μF
10k
1M
R 1 1M
1/2 MC14009
0.1μF
R 2 510k
1M
R 3 270k
R 4 133k
1/2 MC14009
R 5 5k
ZERO CAL
* MC14009
MC14511
R/0
1'S DIGIT
MC14511
R/0
10'S DIGIT
100
+V
1/2 DIGIT DISPLAY

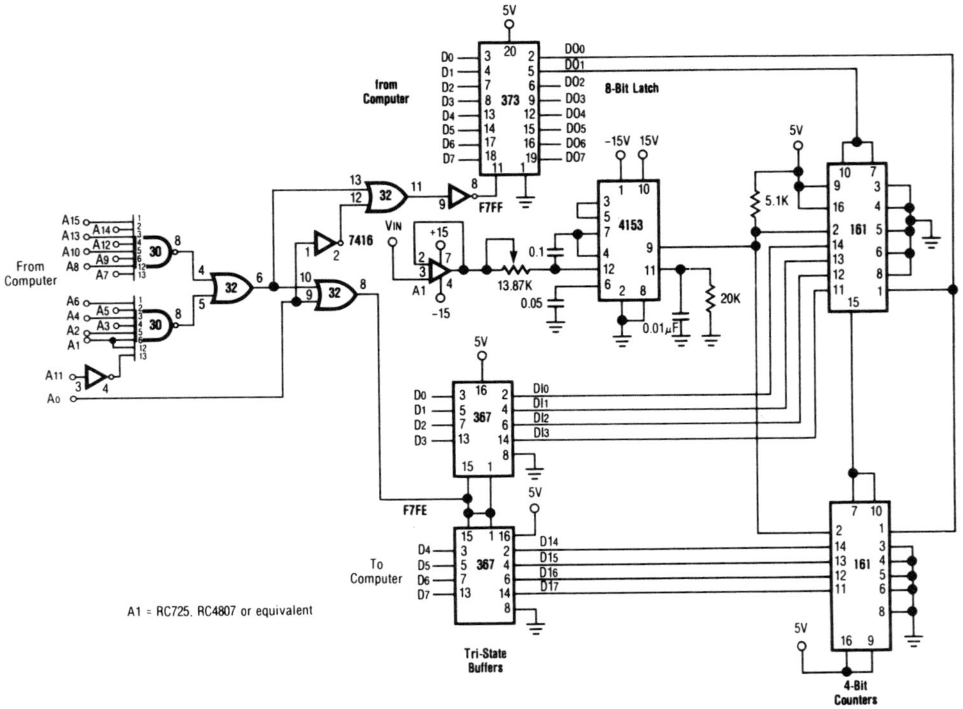

Complete OSI Microprocessor DVM (R).

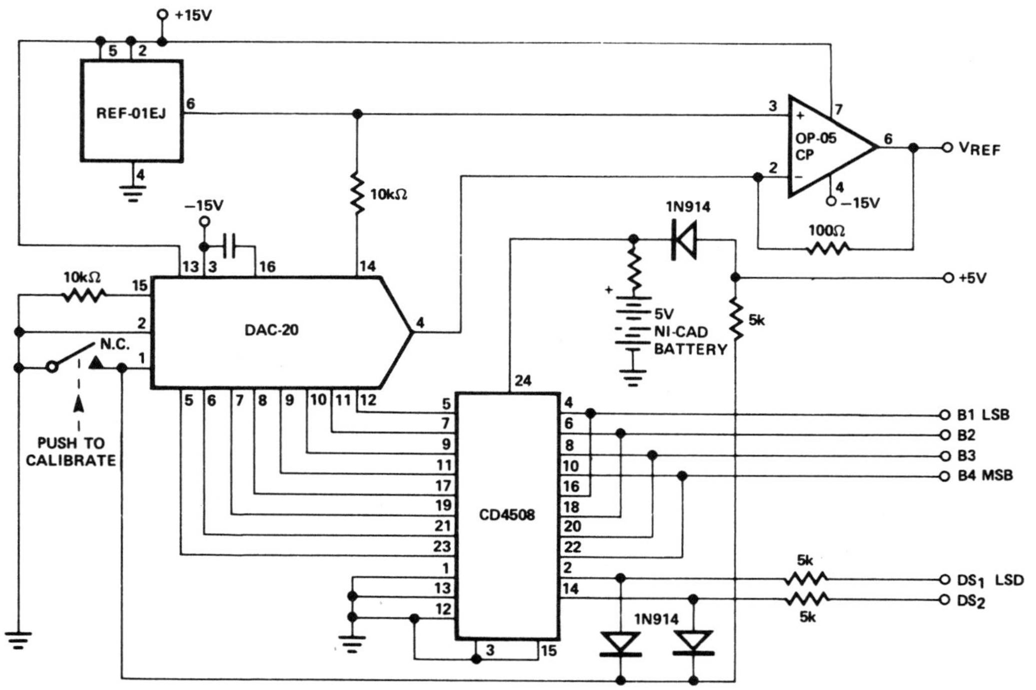

 DVM auto-calibrate circuit (PM).

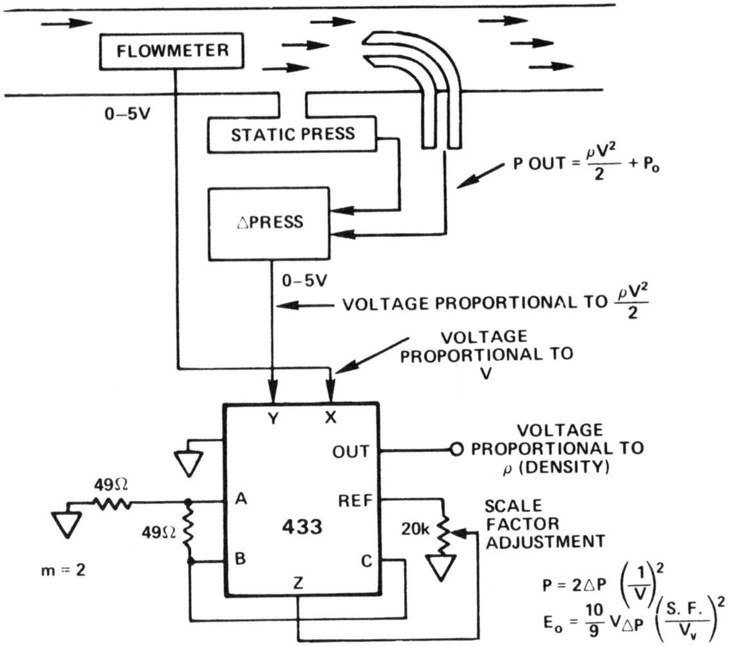

Block diagram of densitometer (AD).

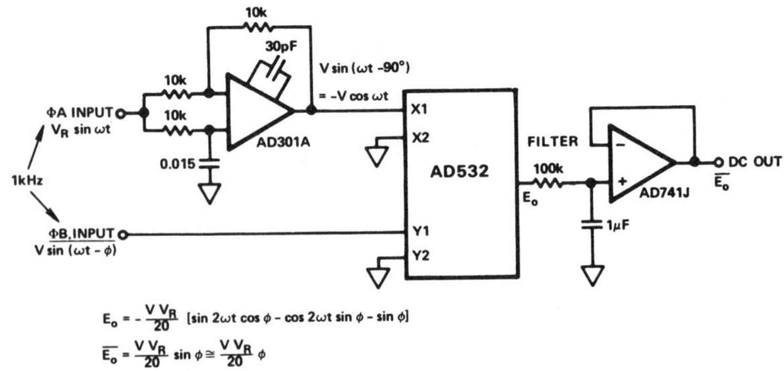

Phasemeter for sinusoidal signals (AD).

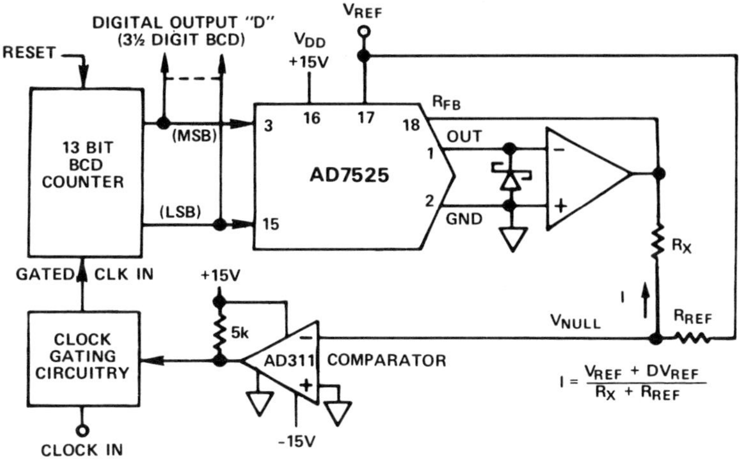

EXAMPLES OF RATIO MEASUREMENTS

"D" DIGITAL OUTPUT	"D" DECIMAL EQUIVALENT	$\frac{R_X}{R_{REF}}$ (%)
1 0000 0000 0000	1.000	100.0%
1 1001 1001 1001	1.999	199.9%
0 1000 1000 0001	0.881	88.1%
0 0010 0101 0000	0.250	25.0%
0 0000 0001 0000	0.010	1.0%

Precision R measurement (AD).

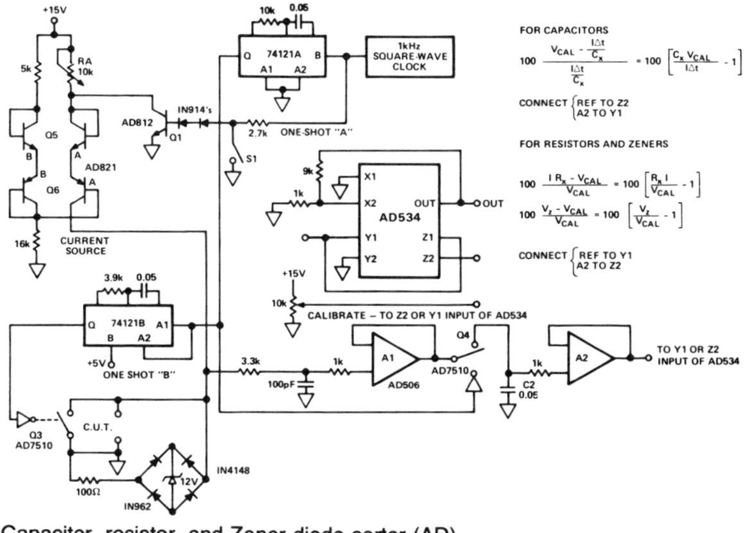

Capacitor, resistor, and Zener-diode sorter (AD).

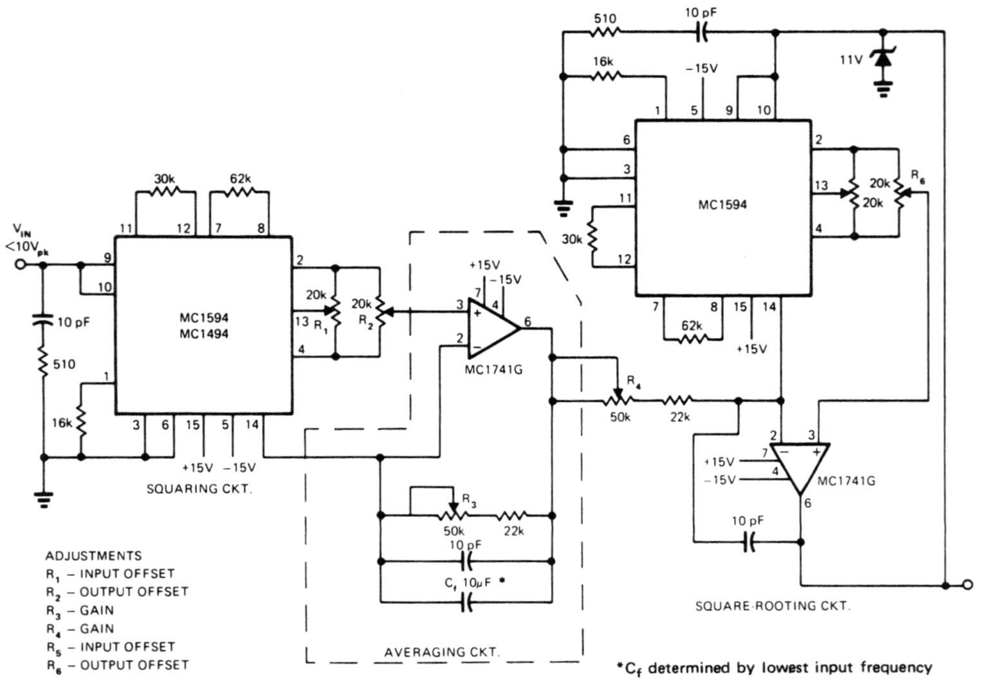

Schematic diagram of true RMS detector (M).

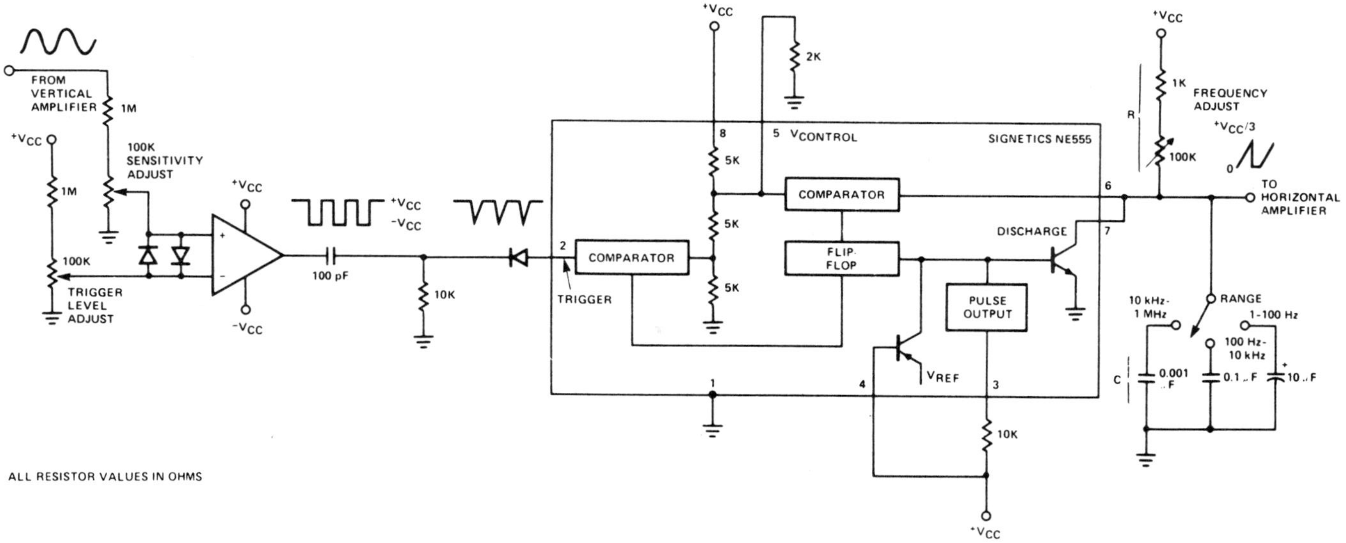

55 Schematic of triggered sweep (S).

Chapter 3

Op Amp Applications

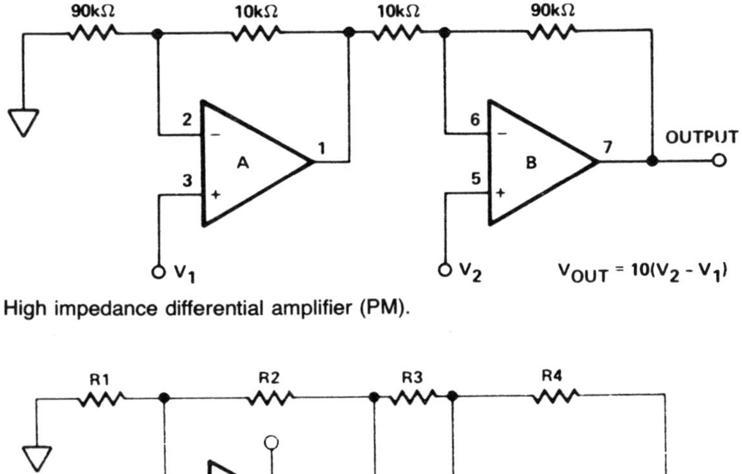

High impedance differential amplifier (PM).

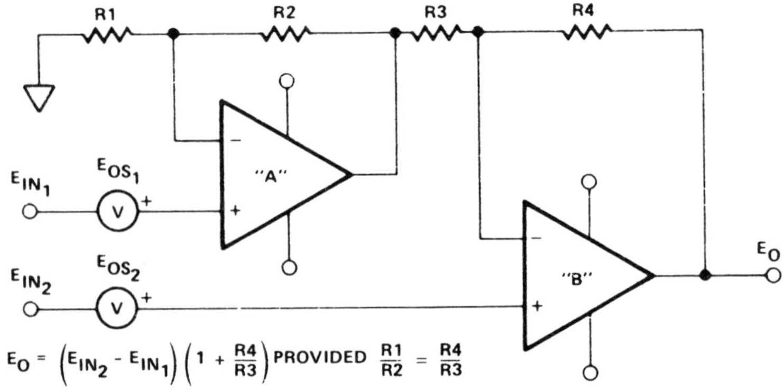

$E_O = \left(E_{IN_2} - E_{IN_1}\right)\left(1 + \frac{R4}{R3}\right)$ PROVIDED $\frac{R1}{R2} = \frac{R4}{R3}$

Instrumentation amplifier 2 op-amp design (PM).

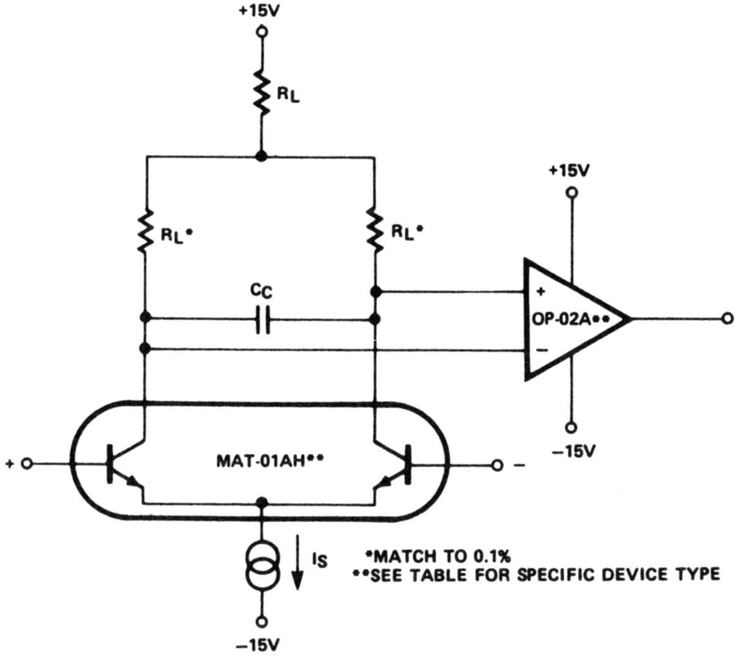

THIS CONFIGURATION CAN ALSO BE USED WITH THE LOW POWER OP-21 OR MICROPOWER OP-20 TO ACHIEVE A LOW NOISE AND LOW POWER PRECISION OP-AMP.

	MAT-01AH OP-02A	MAT-01AH OP-02A	MAT-01GH OP-02	MAT-01GH OP-02
V_{OS} Maximum	0.15mV	0.27mV	0.65mV	1.2mV
TCV_{OS} Maximum	0.6μV/°C	1μV/°C	2μV/°C	4μV/°C
I_{OS} Maximum	0.8nA	0.1nA	3.2nA	0.32nA
I_B Maximum	20nA	2nA	40nA	4nA
Gain Minimum	2,000,000	2,000,000	800,000	800,000
I_S	20μA	2μA	20μA	2μA
R_L	100kΩ	1MΩ	100kΩ	1MΩ

Precision operational amplifier (PM).

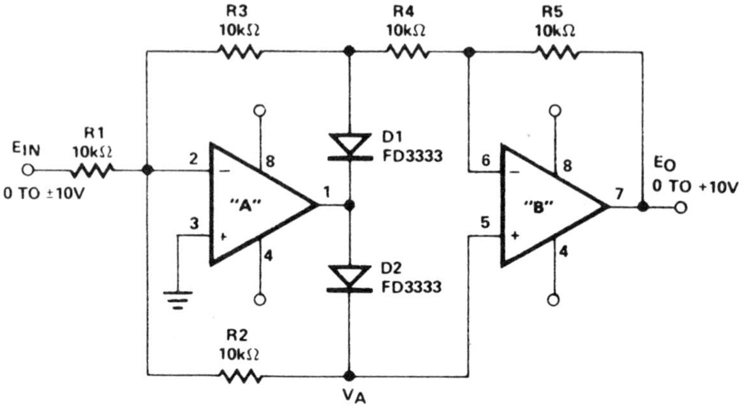

Positive Input

1) $V_A = 0$, D2 off, D1 on

2) $E_O = E_{IN}\left(-\frac{R3}{R1}\right)\left(-\frac{R5}{R4}\right)$

$= E_{in}\frac{R3\,R5}{R1\,R4}$

3) **With R1 = R3 = R4 = R5:**

$E_o = E_{in}$

4) V_{os} error included:

$E_O = E_{IN} + 2V_{OS_B}$

Negative Input

1) D1 off, D2 on

2) $\frac{-E_{in}}{R1} = \frac{V_A}{R2} + \frac{V_A}{R3 + R4}$

3) $E_o = V_A\left(1 + \frac{R5}{R3 + R4}\right)$

4) With R3 = R4 = R5:

$E_o = 1.5V_A$

5) $E_o = -\frac{(R2)\,(R3 + R4)\,(1.5)\,E_{in}}{R1\,(R2 + R3 + R4)}$

6) With R1 = R2 = R3 = R4:

$E_o = -E_{in}$

7) **V_{OS} error included:**

$E_O = -E_{IN} + 1.5\,V_{OS_B} - 0.5\,V_{OS_A}$

8) For Both Inputs:

$E_o = +\,|E_{in}|$

Absolute value circuit (PM).

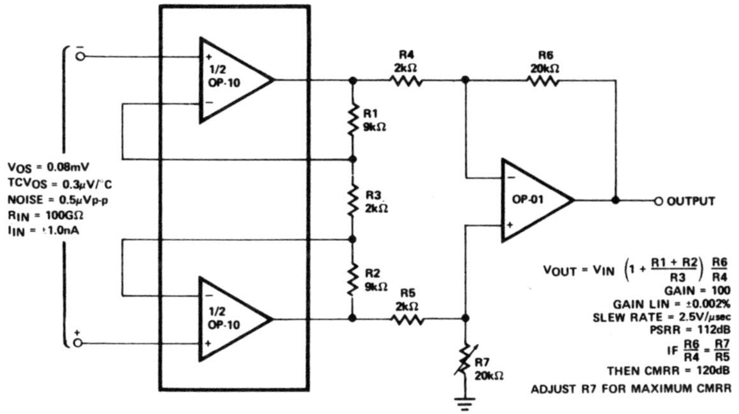

Triple op-amp instrumentation amplifier (PM).

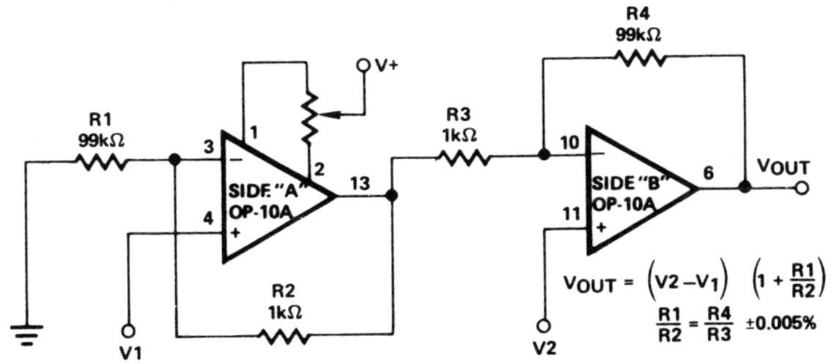

Instrumentation amplifier (2 op-amp design) (PM).

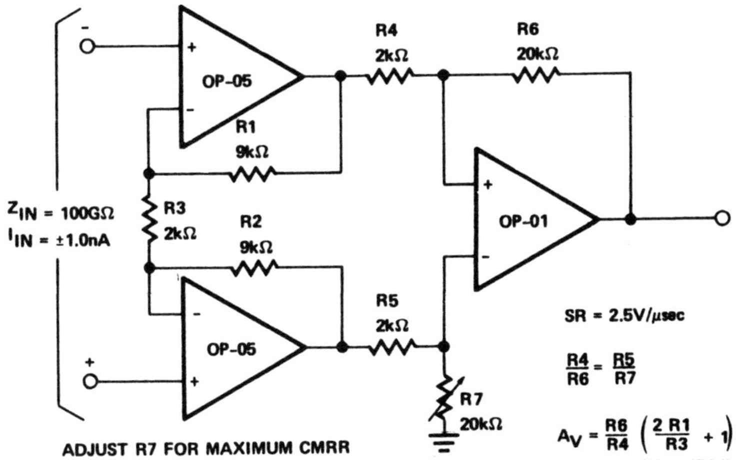

High impedance, high common mode rejection instrumentation amplifier (PM).

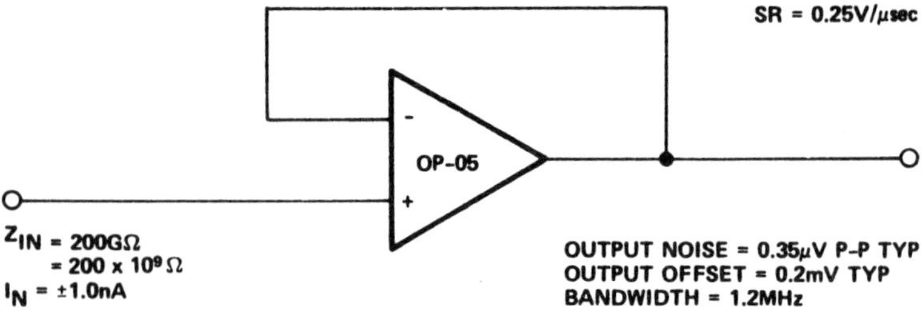

Stable, high impedance buffer (PM).

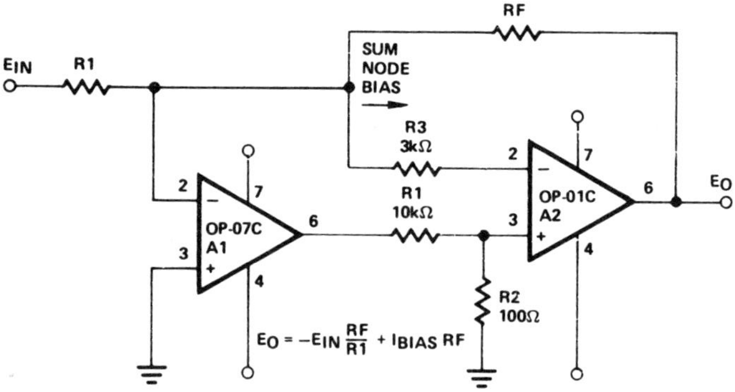

*
High speed, low V_{os} composite amplifier (PM).

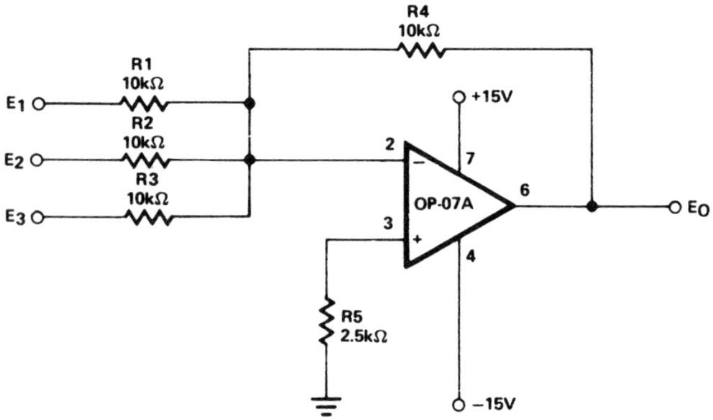

*
Adjustment-free precision summing amplifier (PM).

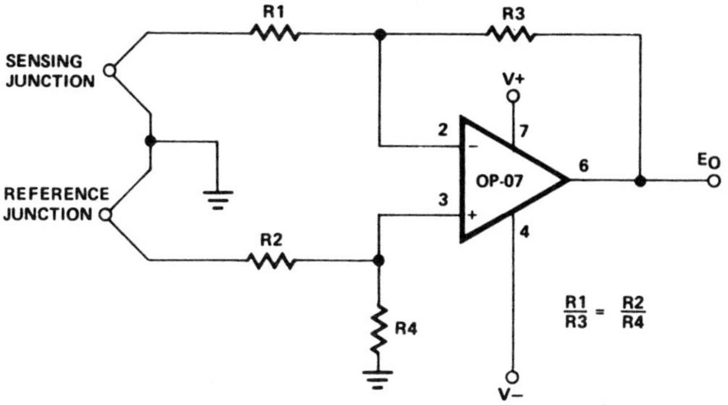

*

High stability thermocouple amplifier (PM).

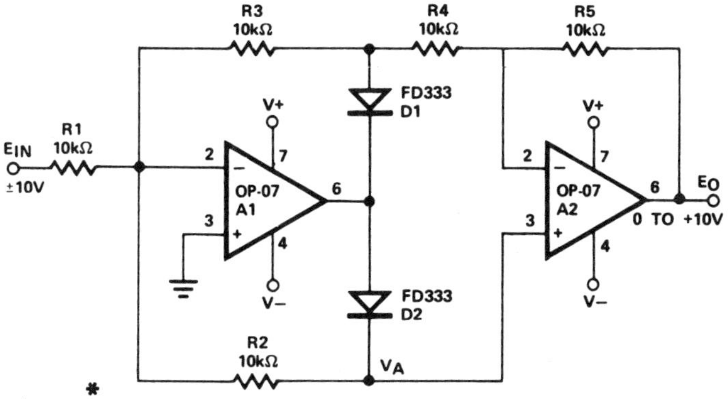

*

Precision absolute value circuit (PM).

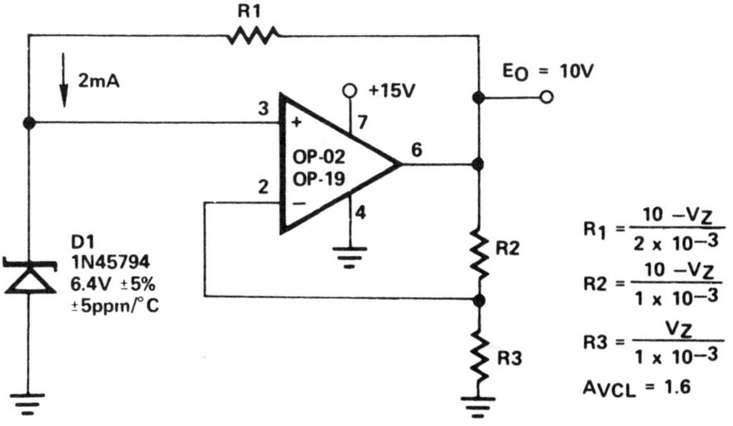

High stability voltage reference (PM).

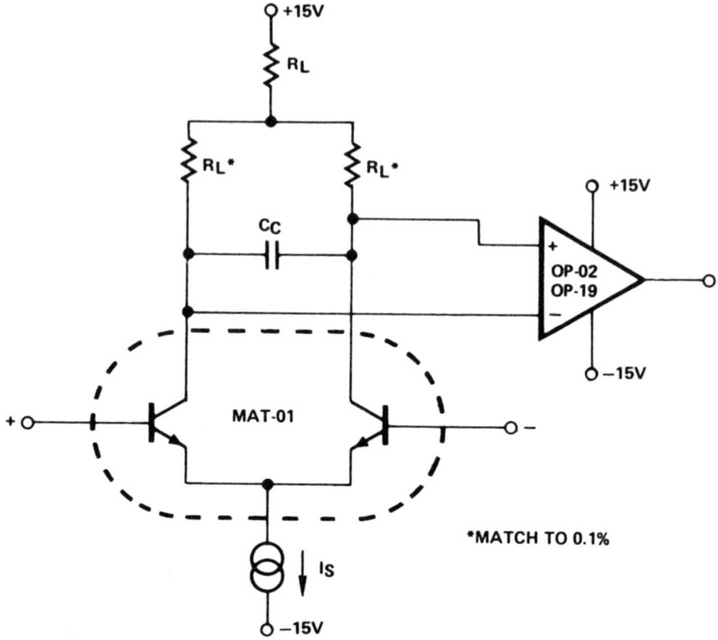

Precision operational amplifier (PM).

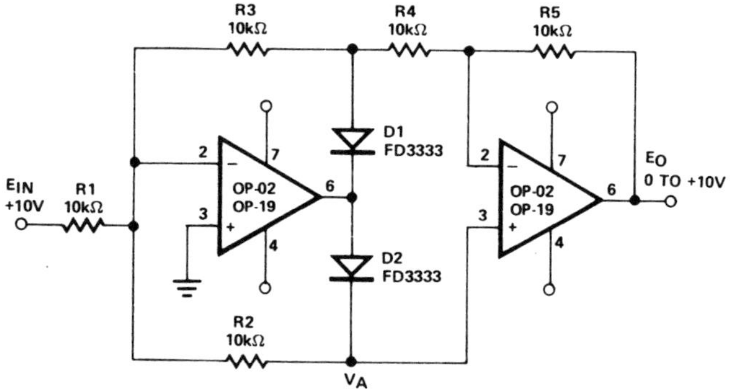

SCHEMATIC EQUATIONS

POSITIVE INPUT

1. $V_A = 0$, D2 OFF, D1 ON
2. $E_O = \left(\frac{-E_{IN}R3}{R1}\right) \cdot \left(\frac{-R5}{R4}\right) = E_{IN}\frac{R3\ R5}{R1\ R4}$
3. With R1 = R3 = R4 = R5: $E_O = E_{IN}$
4. V_{OS} error included: $E_O = E_{IN} + 2V_{OS2}$

NEGATIVE INPUT

1. D1 OFF, D2 ON
2. $\frac{-E_{IN}}{R1} = \frac{V_A}{R2} + \frac{V_A}{R3 + R4}$
3. $E_O = V_A\left(1 + \frac{R5}{R3 + R4}\right)$
4. With R3 = R4 = R5: $E_O = 1.5V_A$
5. $E_O = \frac{-(R2)\ (R3 + R4)\ (1.5)\ E_{IN}}{R1\ (R2 + R3 + R4)}$
6. With R1 = R2 = R3 = R4: $E_O = -E_{IN}$
7. V_{OS} error included:
   $E_O = -E_{IN} + 1.5V_{OS2} - 0.5V_{OS1}$
8. For both inputs: $E_O = +|E_{IN}|$

Absolute value circuit (PM).

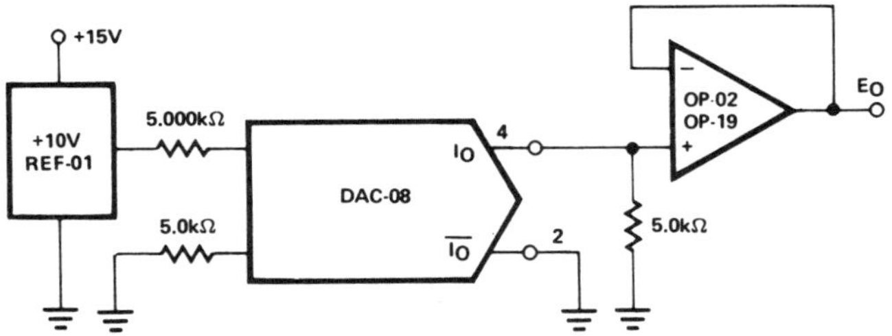

FOR COMPLEMENTARY OUTPUT (OPERATION AS A NEGATIVE LOGIC DAC) CONNECT NON-INVERTING INPUT OF OP-AMP TO $\overline{I_O}$ (PIN 2), CONNECT I_O (PIN 4) TO GROUND.

INPUT/OUTPUT TABLE

	B1	B2	B3	B4	B5	B6	B7	B8	I_OmA	E_O
FULL SCALE −1LSB	1	1	1	1	1	1	1	1	1.992	−9.960
FULL SCALE −2LSB	1	1	1	1	1	1	1	0	1.984	−9.920
HALF SCALE +LSB	1	0	0	0	0	0	0	1	1.008	−5.040
HALF SCALE	1	0	0	0	0	0	0	0	1.000	−5.000
HALF SCALE −LSB	0	1	1	1	1	1	1	1	0.992	−4.960
ZERO SCALE +LSB	0	0	0	0	0	0	0	1	0.008	−0.040
ZERO SCALE	0	0	0	0	0	0	0	0	0.000	0.000

63 DAC-08 output amplifier negative low impedance output operation (PM).

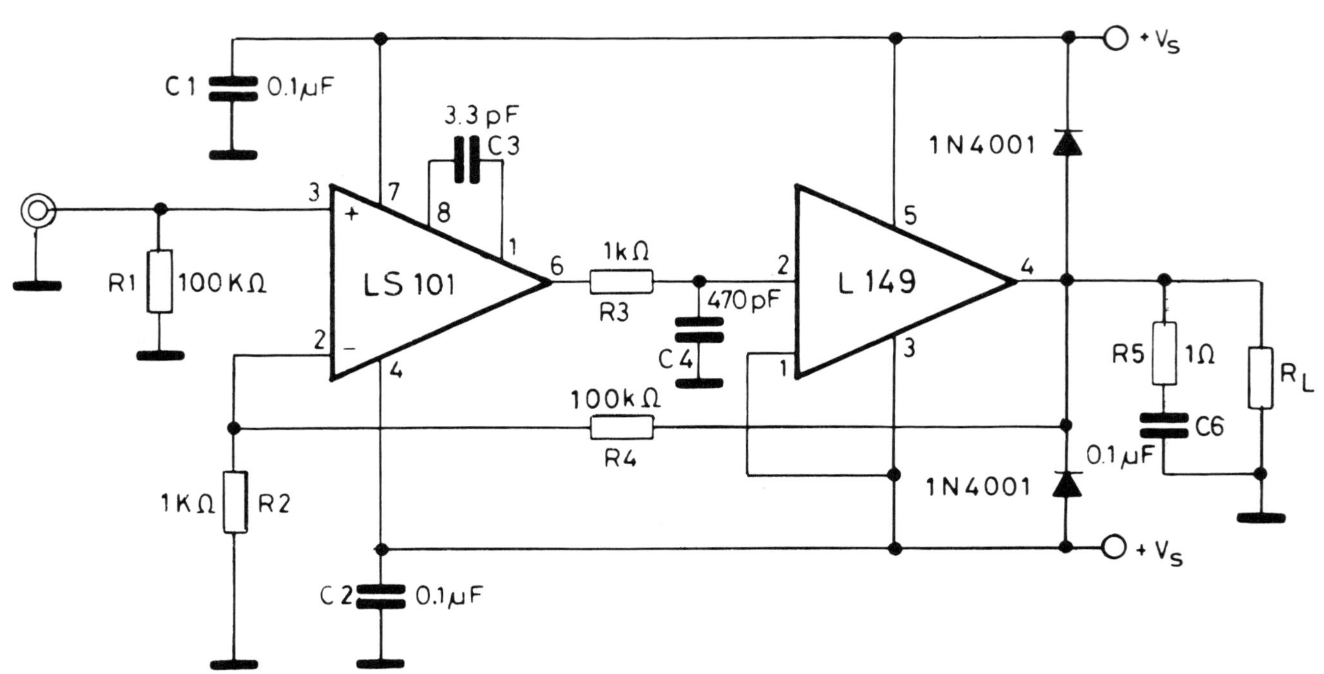

Power operational amplifier (SGS).

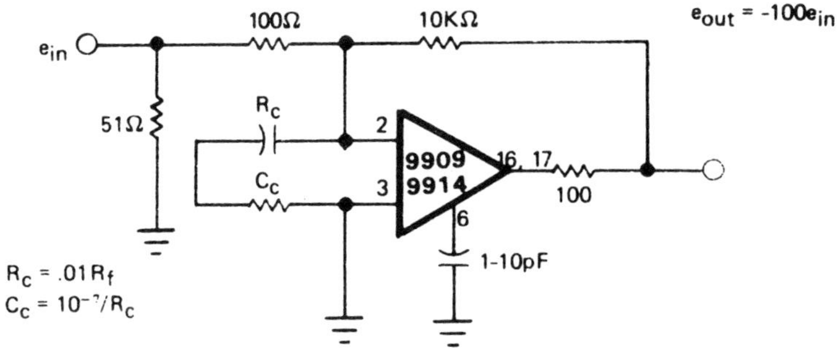

NOTE: Generally an Rc network will not be required since the 100 ohms in the loop and the compensation capacitor stabilize the unit.

X100 inverting buffer amplifier with compensation (OE).

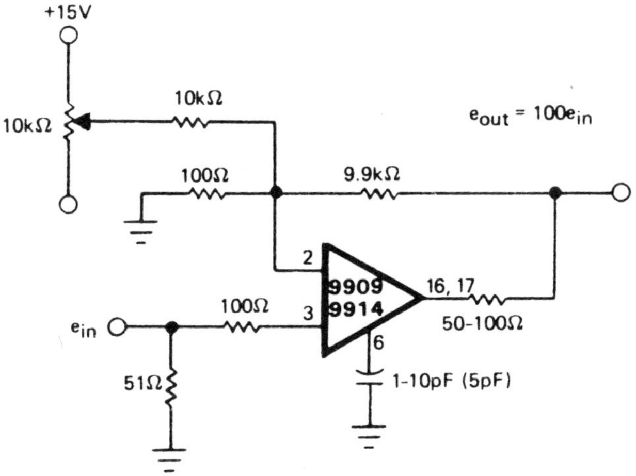

X100 non-inverting buffer amplifier with offset (OE).

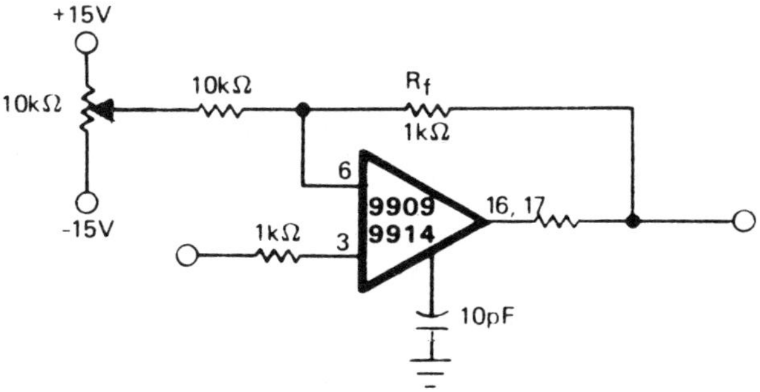

Voltage follower with offset (OE).

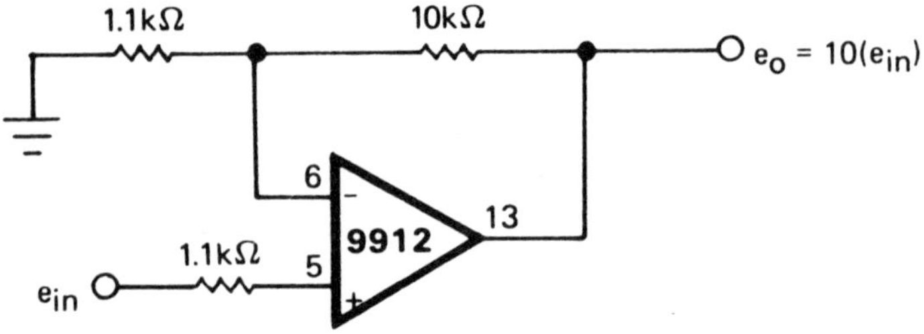

X10 non-inverting buffer (OE).

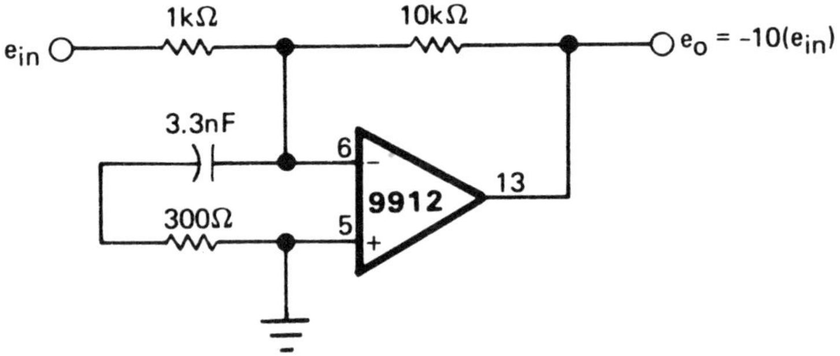

X10 inverting buffer amplifier (OE).

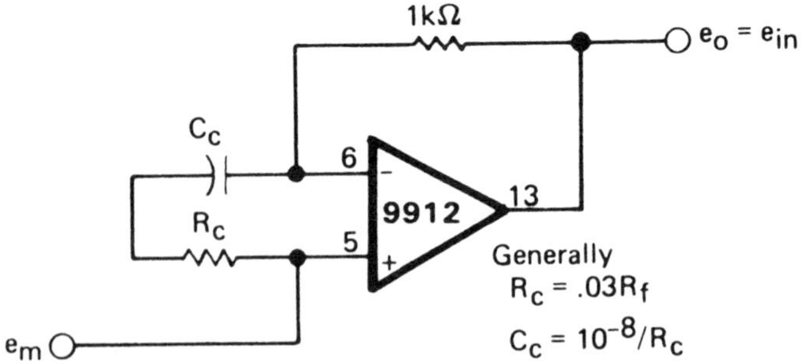

Voltage follower (OE).

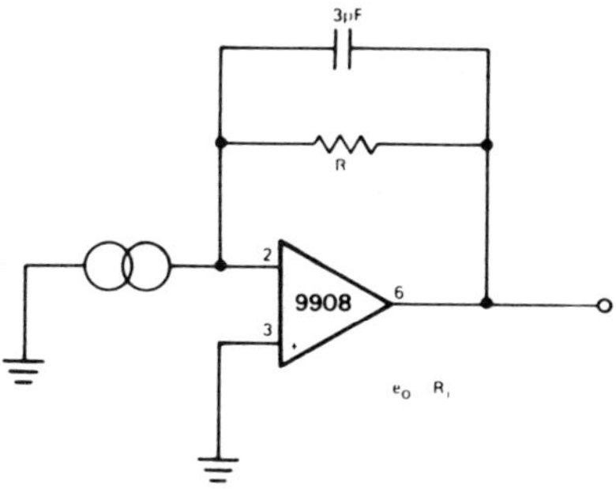

Transimpedance amplifier (OE).

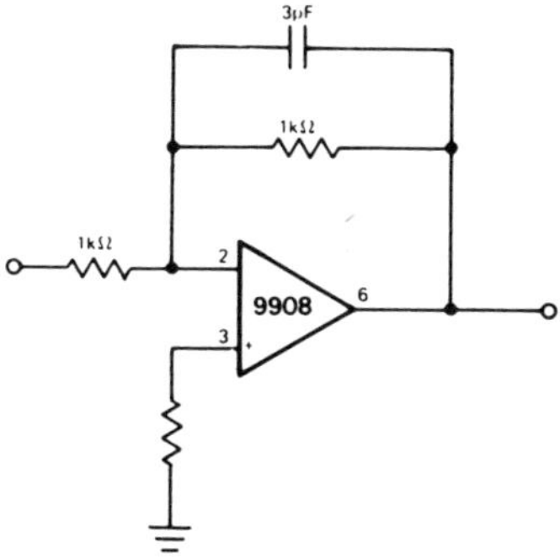

Unity gain inverter (OE).

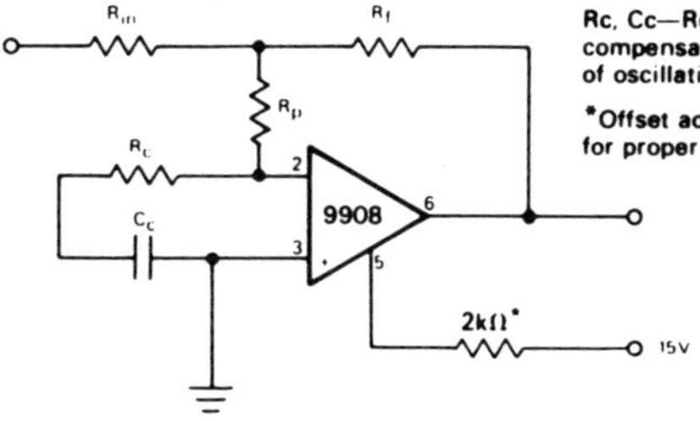

Rp—Parasitic Oscillation Supression

Rc, Cc—Resistive and capacitive compensation for the frequency of oscillation

*Offset adjustment needed for proper operation.

Standard operational amplifier set up (OE).

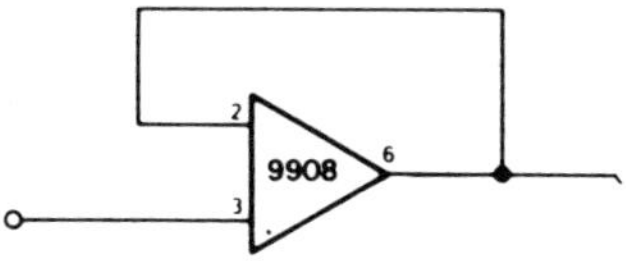

Follower (OE).

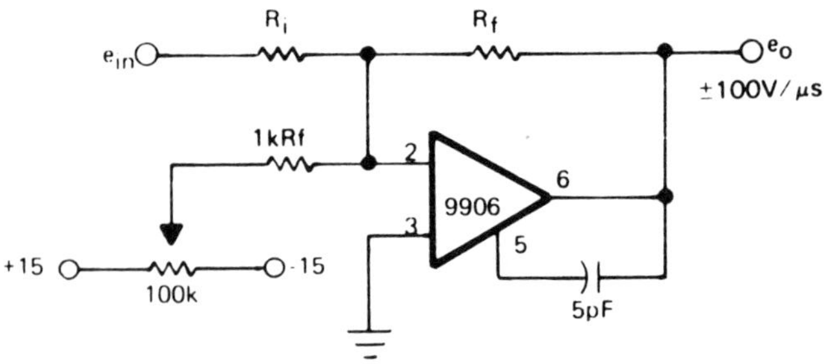

Offset voltage correction and Miller compensation (OE).

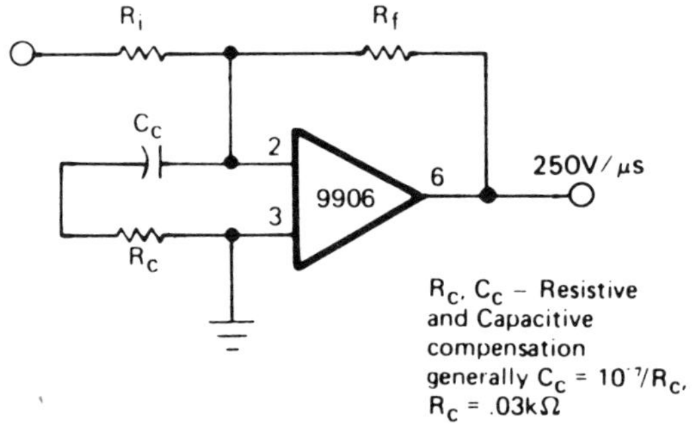

X10 inverter (OE).

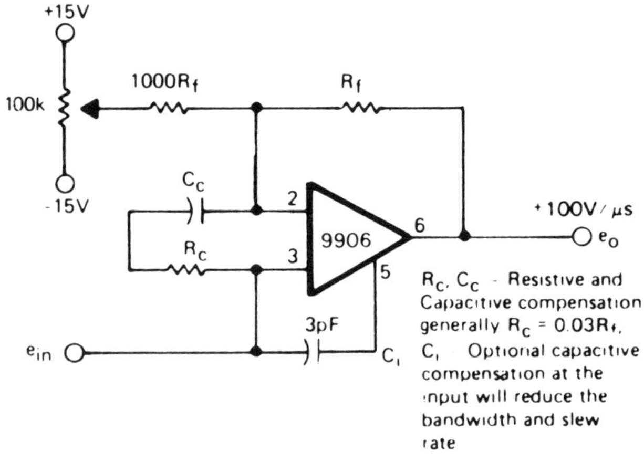

Voltage follower with offset correction (OE).

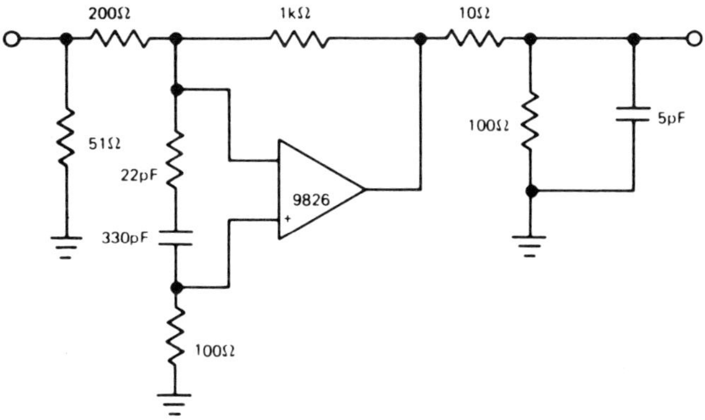

For wide band high frequency operation (OE).

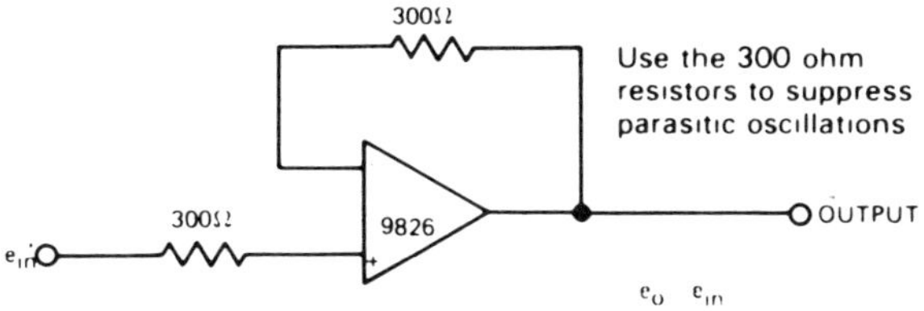

Voltage follower (OE).

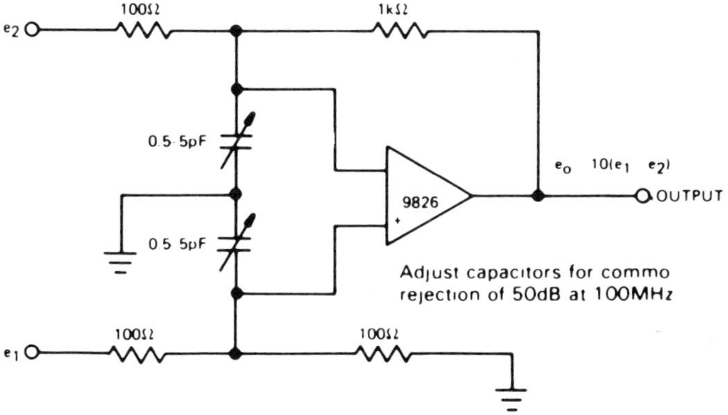

Differential amplifier (OE).

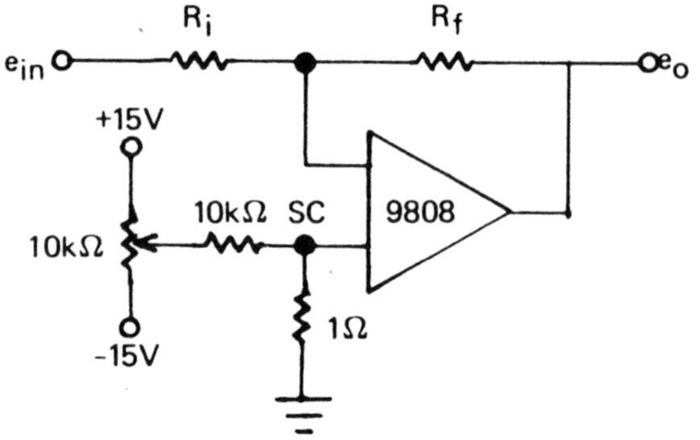

Offset voltage adjustment (OE).

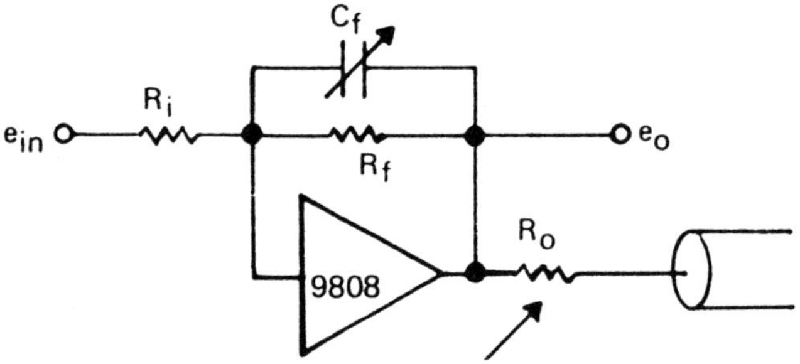

Cable driver (OE).

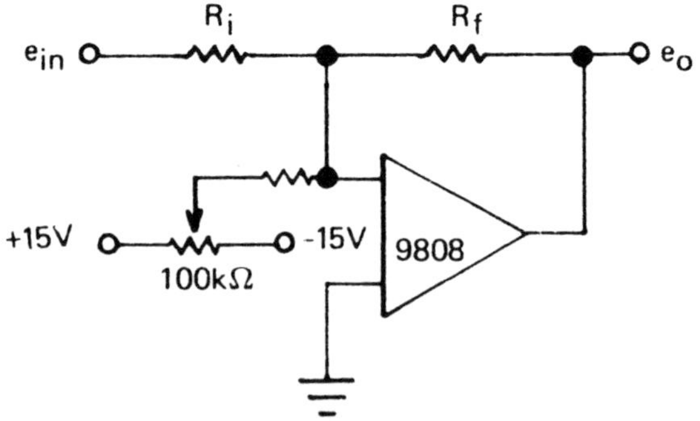

Alternative offset voltage adjustment (OE).

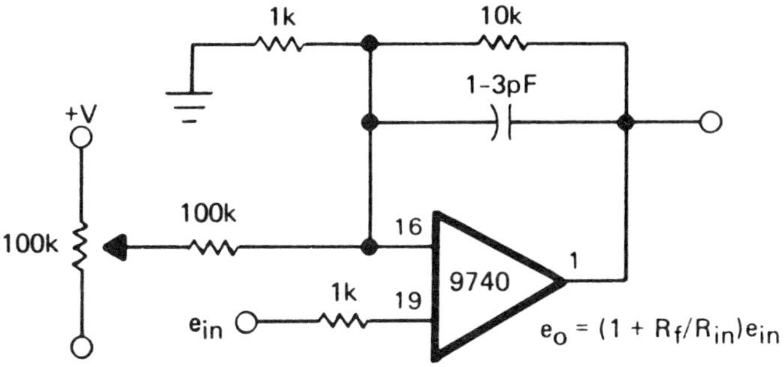

X10 non-inverting buffer amplifier with offset (OE).

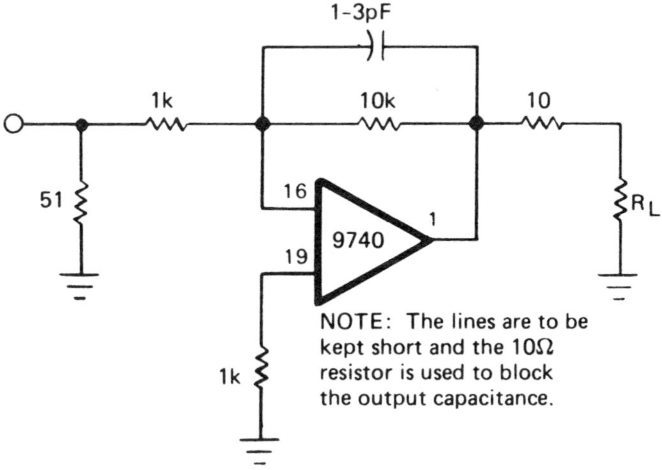

X10 buffer amplifier (OE).

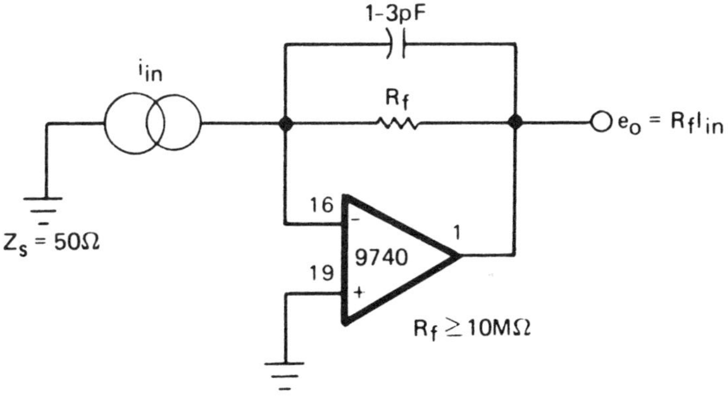

Current gain (OE).

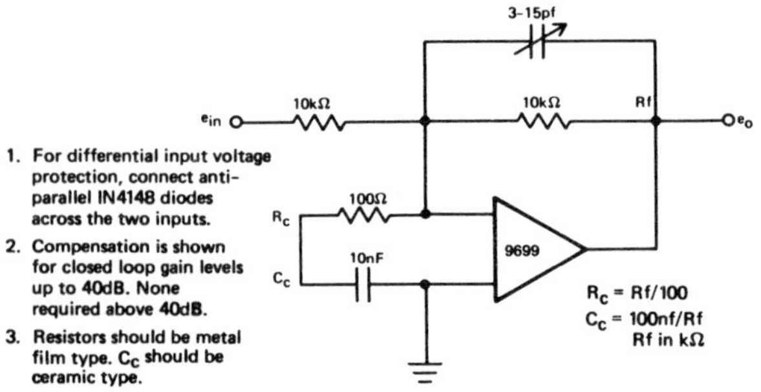

Unity gain inverter (OE).

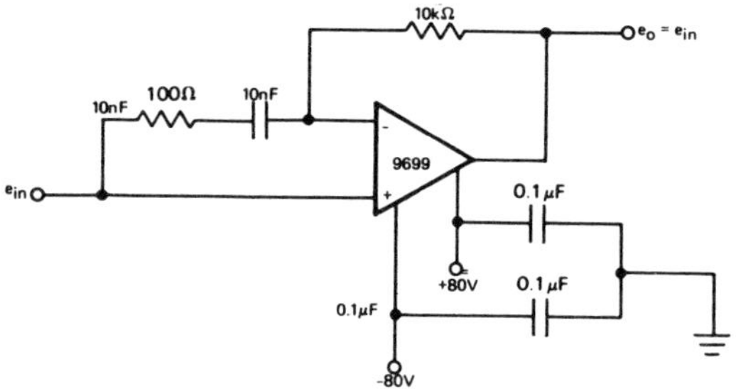

Voltage follower (OE).

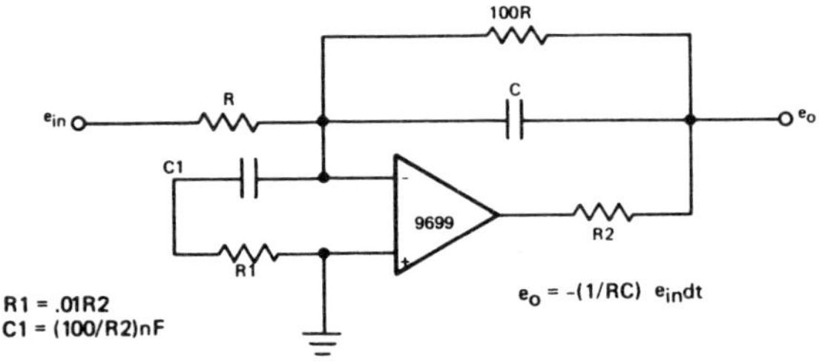

Integrator (OE).

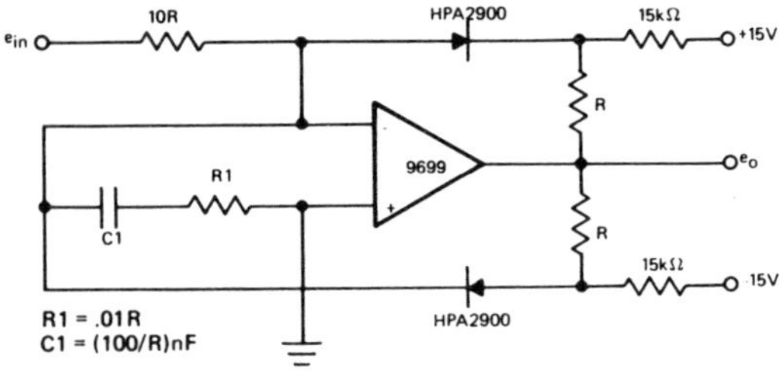

Zero crossing detector (OE).

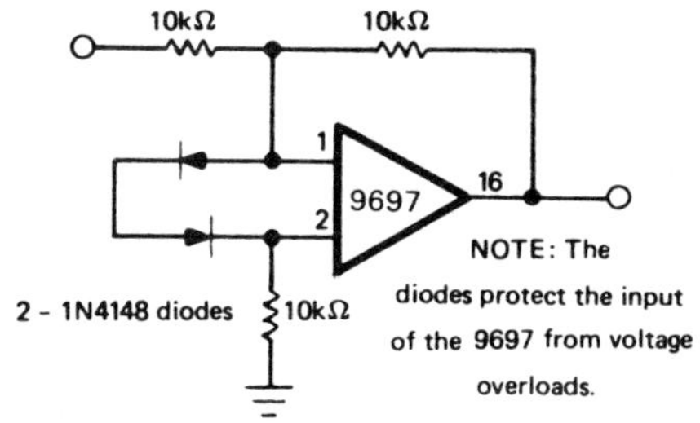

Unity gain buffer with input protection (OE).

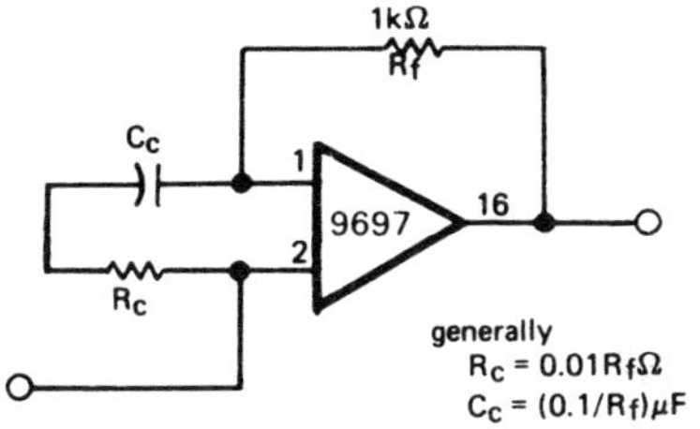

Voltage follower (OE).

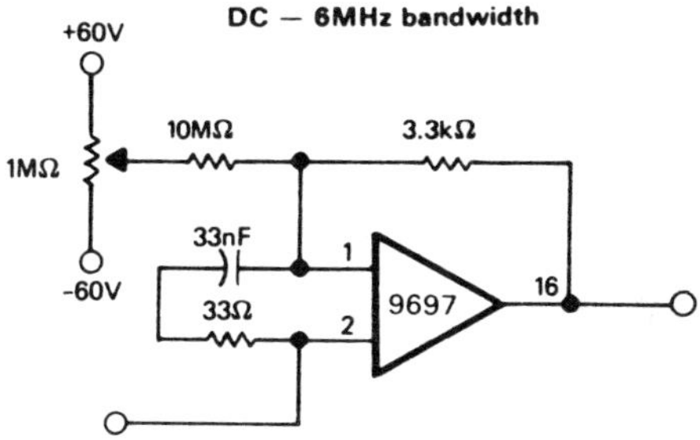

Precision unity gain voltage follower (OE).

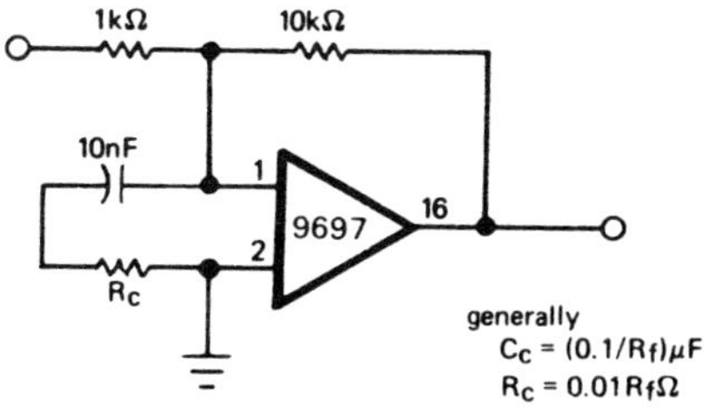

X10 inverting buffer amplifier (OE).

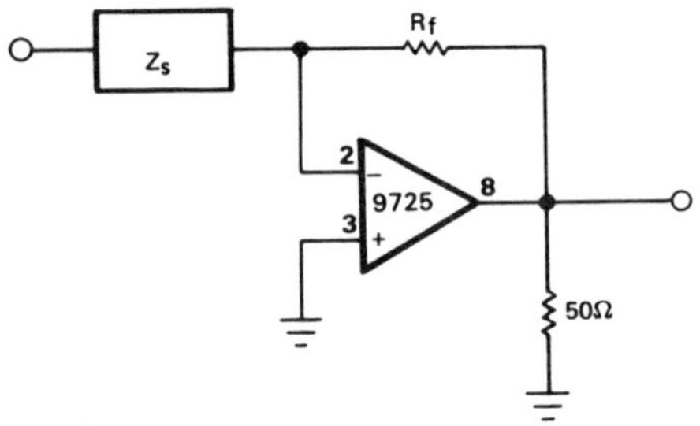

Current booster (OE).

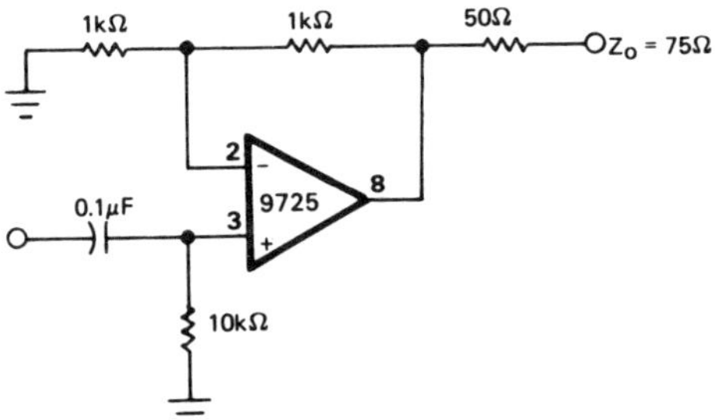

25MHz video line driver (OE).

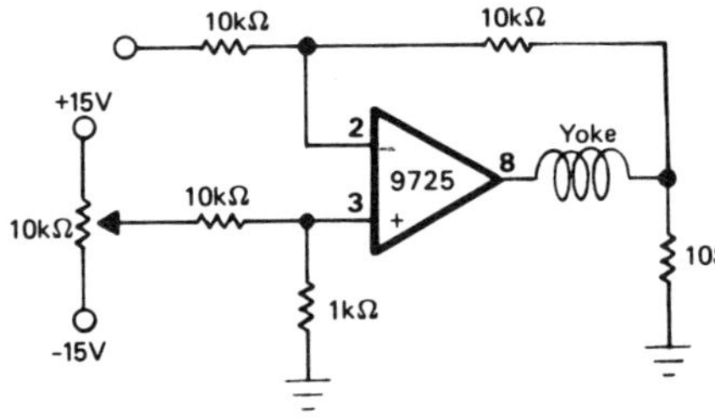

Wide band deflection amplifier (OE).

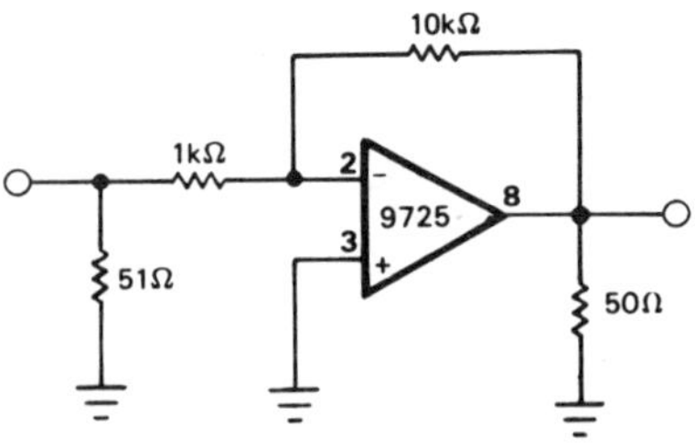

X10 amplifier (OE).

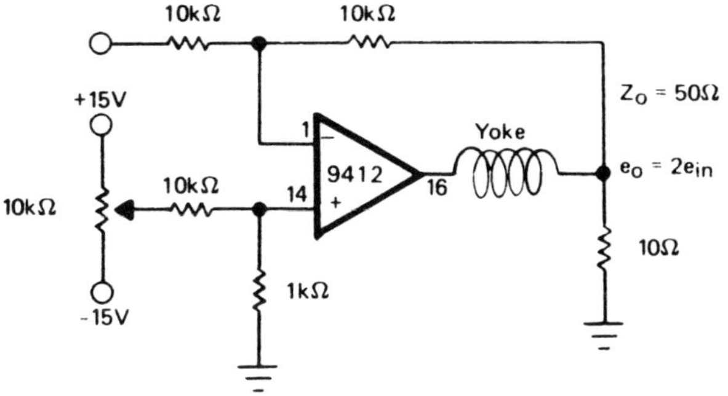

Wide band deflection amplifier (OE).

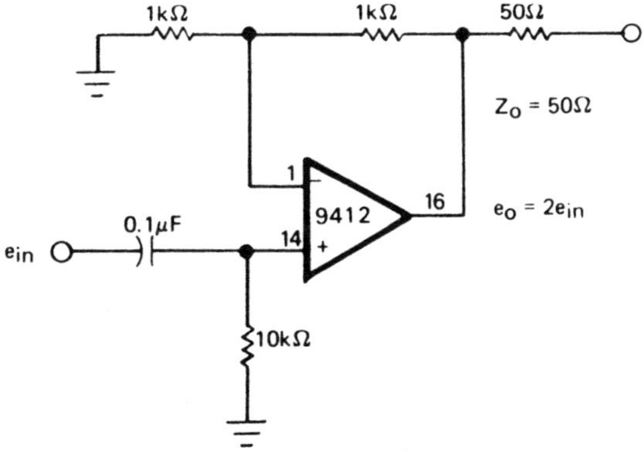

25MHz video line driver (OE).

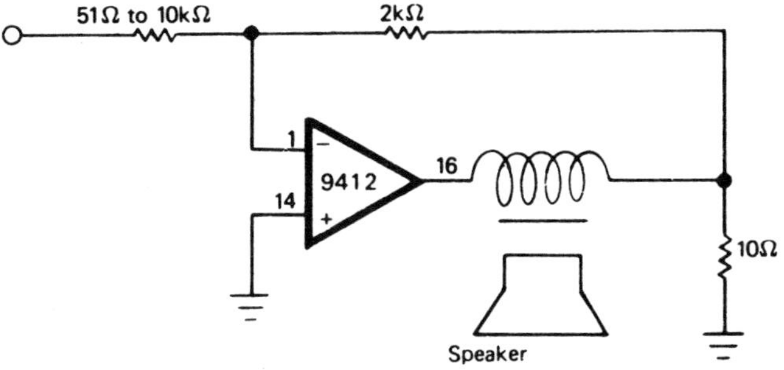

Two-watt hi-fi audio amplifier (OE).

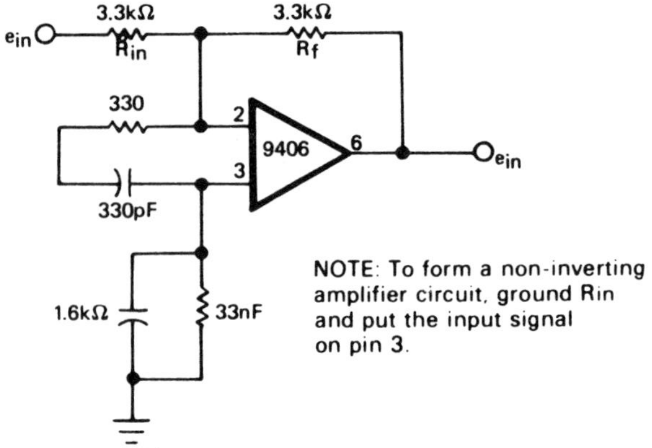

Unity gain inverter (OE).

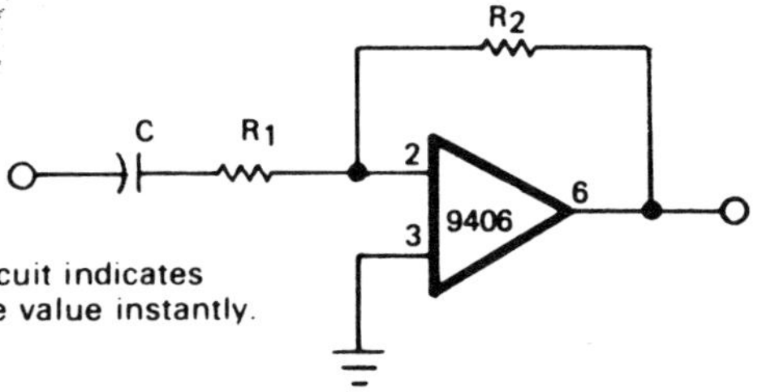

Derivative circuit (OE).

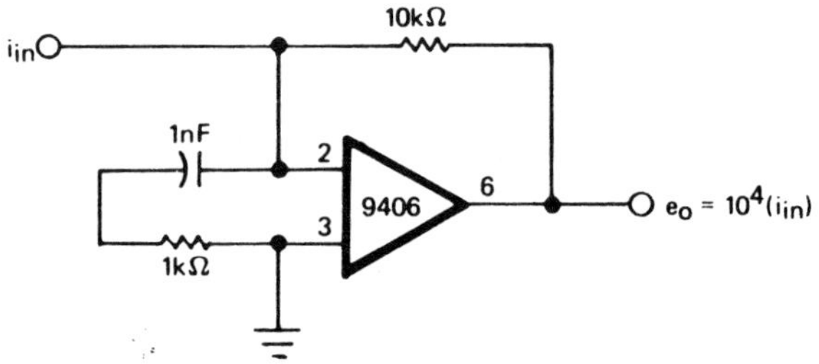

Isolation of pin 2 from signal current source with 10-100 ohms may be necessary for stable operation.

Current-to-voltage converter dc-20MHz (OE).

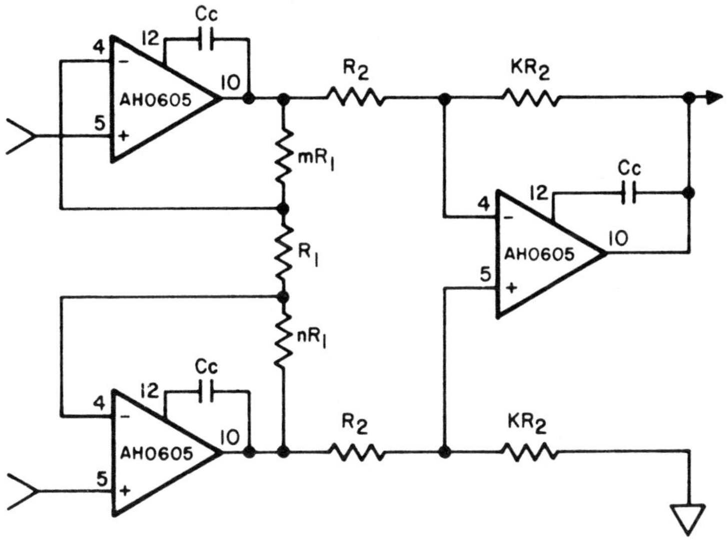

Instrumentation amplifier (OE).

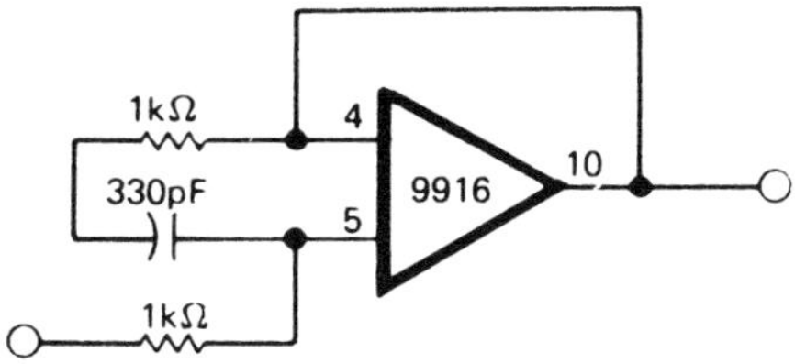

Voltage follower with components (OE).

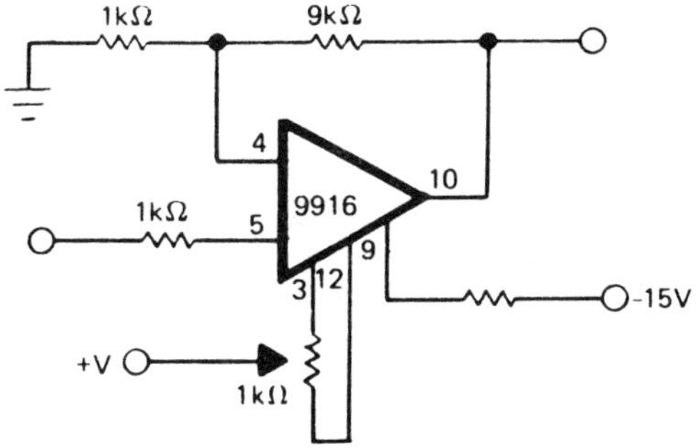

X10 non-inverting buffer amplifier with offset (OE).

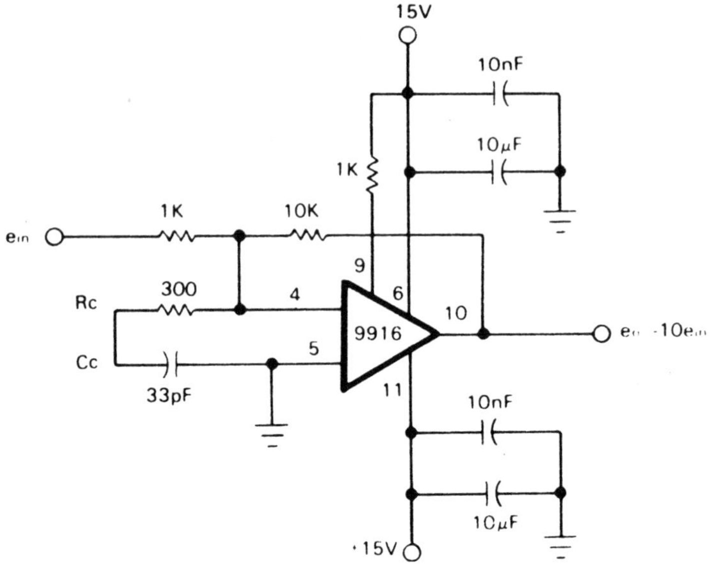

X10 inverting buffer amplifier with compensation (OE).

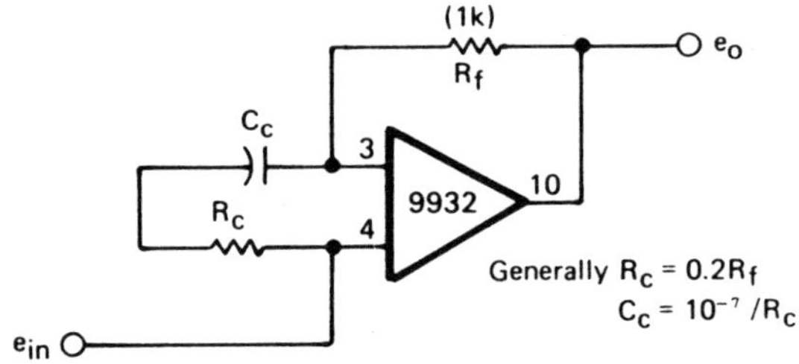

Voltage follower (OE).

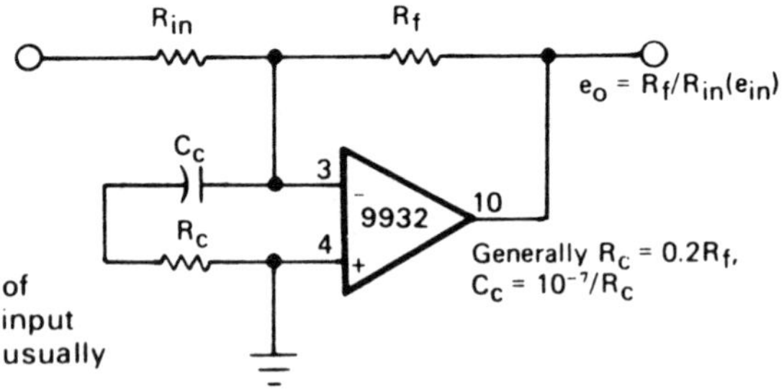

NOTE: For gains of 3 and above the input compensation is usually not required.

Video amplifier (OE).

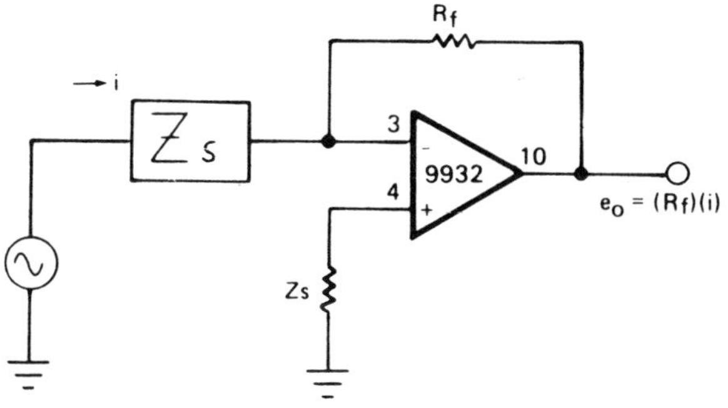

Transimpedance amplification (OE).

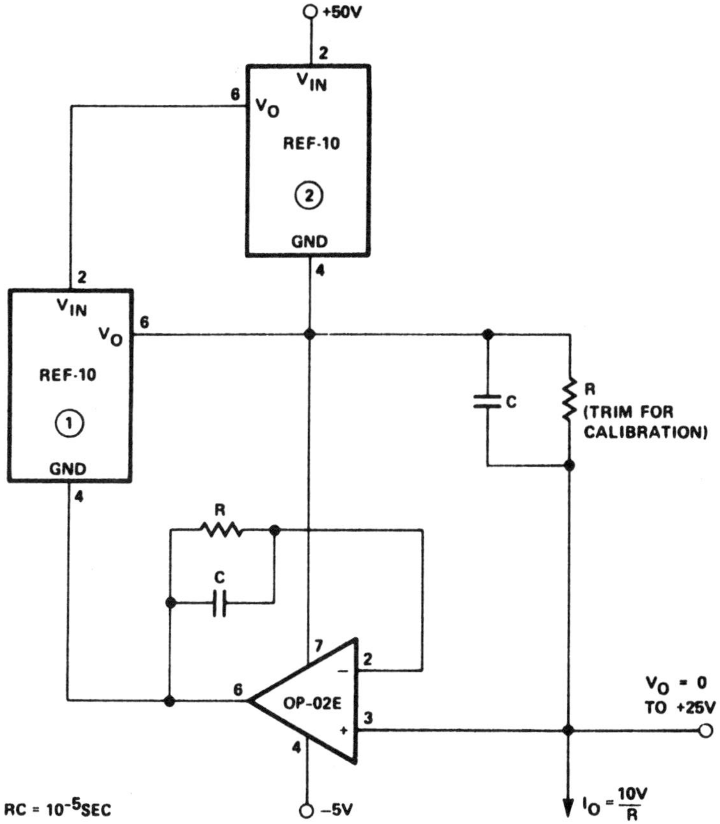

Precision current source (PM).

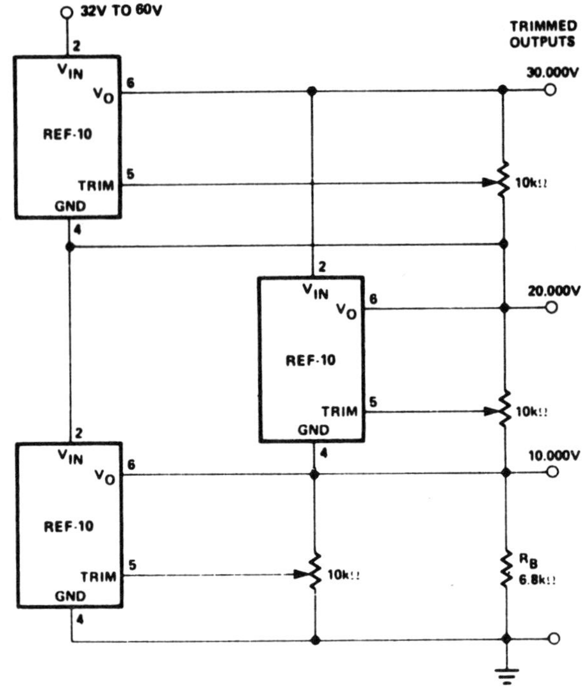

Reference stack with excellent line regulation (PM).

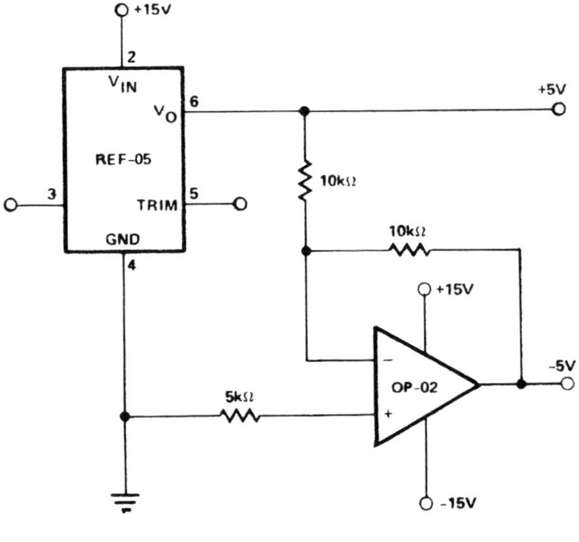

±5V reference (PM).

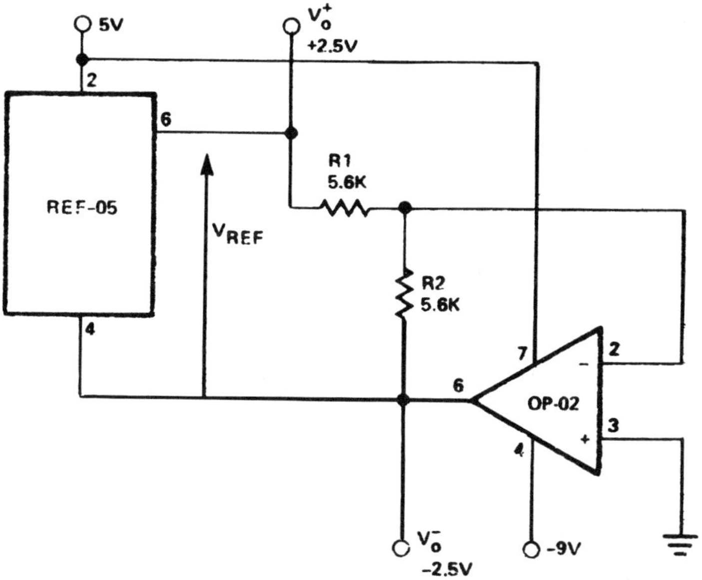

$$V_o^+ = \frac{R_1}{R_1 + R_2}\left(V_{REF}\right), \quad V_o^- = \frac{R_2}{R_1 + R_2}\left(V_{REF}\right)$$

±2.5V reference (PM).

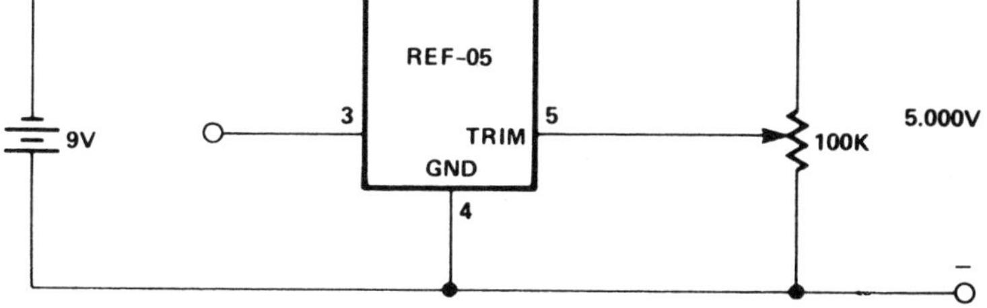

Precision calibration standard (PM).

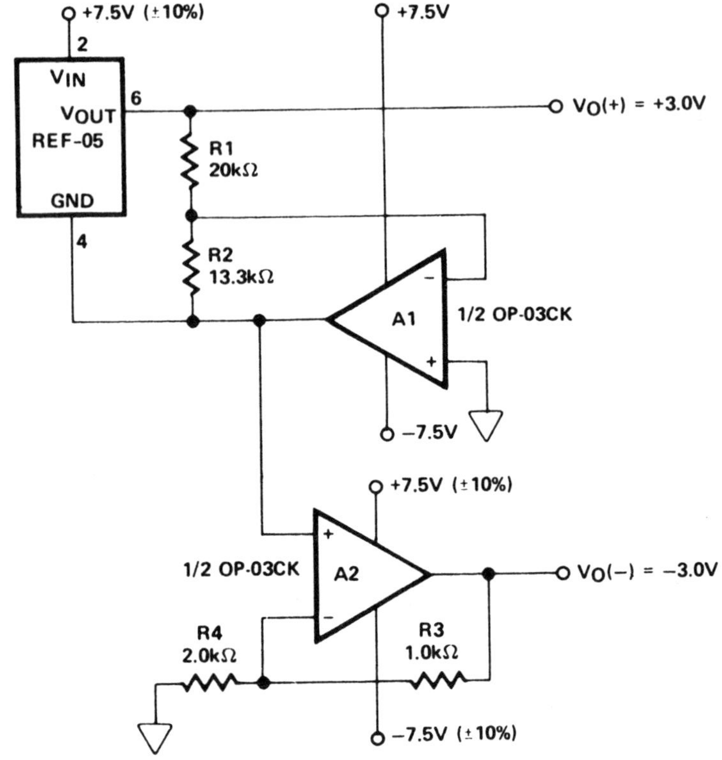

±3V reference (PM).

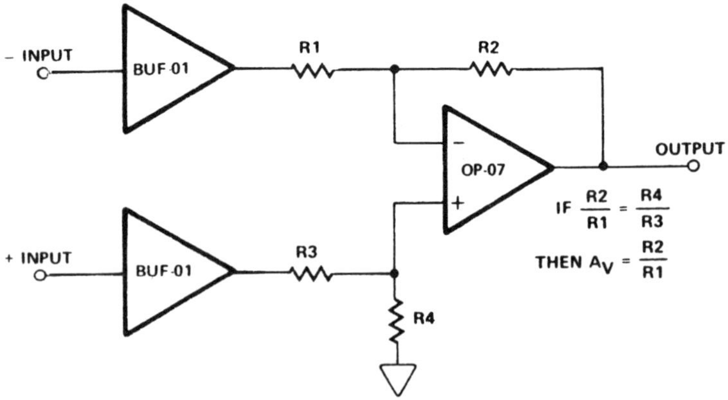

High-impedance differential amplifier (PM).

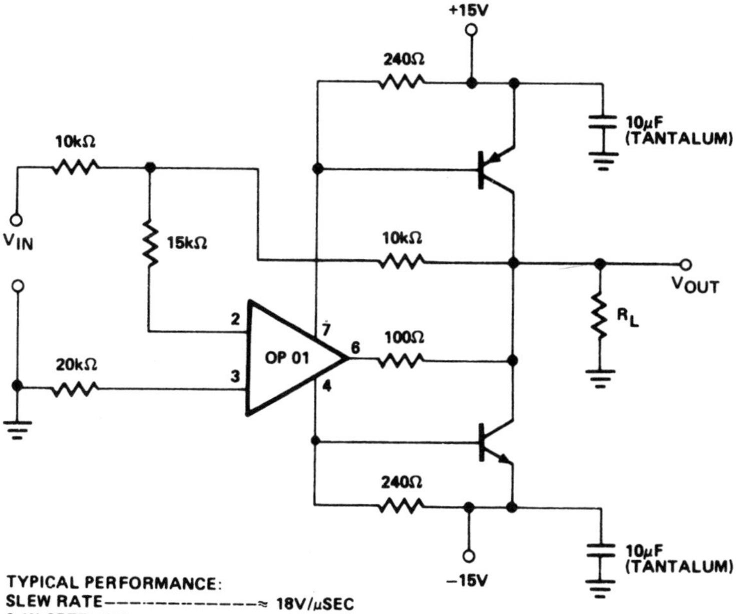

TYPICAL PERFORMANCE:
SLEW RATE—————————≈ 18V/μSEC
0.1% SETTLING ——————4μSEC (R_L = 500)
QUIESCENT SUPPLY CURRENT———1.5mA

Precision power booster circuit (PM).

Chapter 4

Battery Chargers

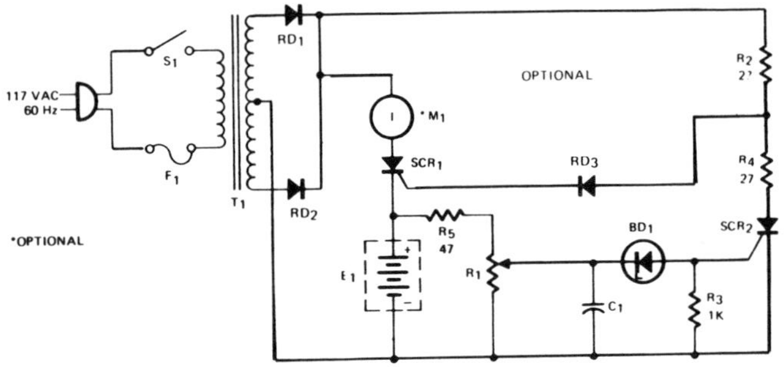

T_1	120 Volt to 24 Volt center-tapped (UTC-FT10, or equivalent)
$RD_{1,2}$	Rectifier WEP4085
RD_3	Rectifier WEP157
R_1	500 Ohm, 2 Watt Potentiometer
R_2, R_4	25 Ohm, 5 Watt Resistor WEP 24-3037
R_3	1K Ohm, 1/2 Watt Resistor WEP 22-2096
R_5	47 Ohm, 1 Watt Resistor WEP 22-3064
M_1	0-10 amp DC Ammeter
E_1	12 Volt Battery
SCR_1	SCR WEP6271
SCR_2	SCR WEP6241
BD_1	1 Watt, 8.2 Volt Zener Diode WEP1108
C_1	100 μF, 25 Volt Electrolytic Capacitor WEP AX100-25
S_1	Toggle Switch WEP TS101
F_1	Fuse, 1.5 Amp WEP FG1½ -2
	Socket WEP8050
	Heat Exchanger WEP8021
	PC Board

SCR battery charger (W).

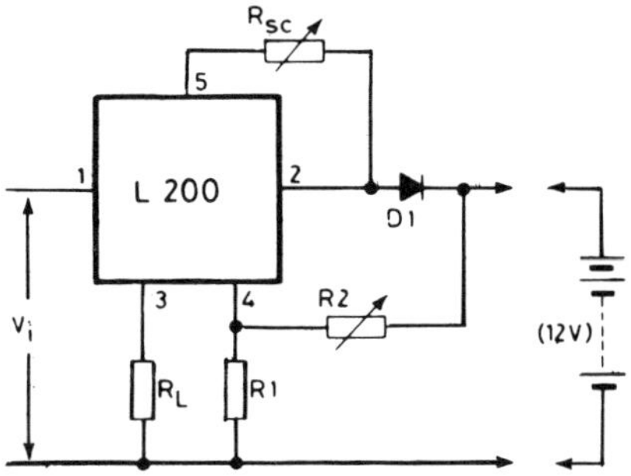

Constant current battery charger (SGS).

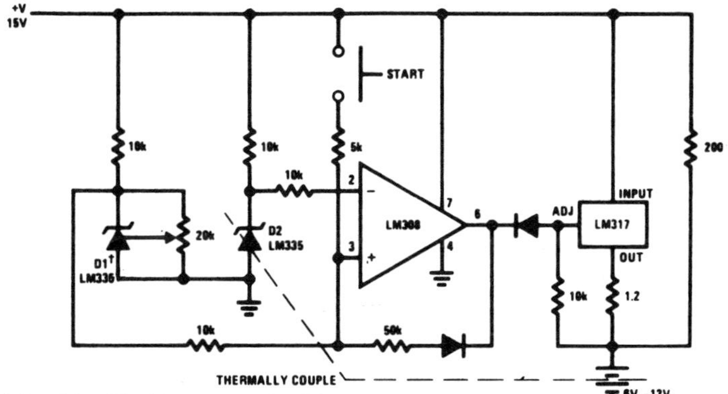

† Adjust D1 to 50 mV greater V_Z than D2.
Charge terminates on 5°C temperature rise. Couple D2 to battery.

Fast charger for nickel-cadmium batteries (NS).

Chapter 5

Filters

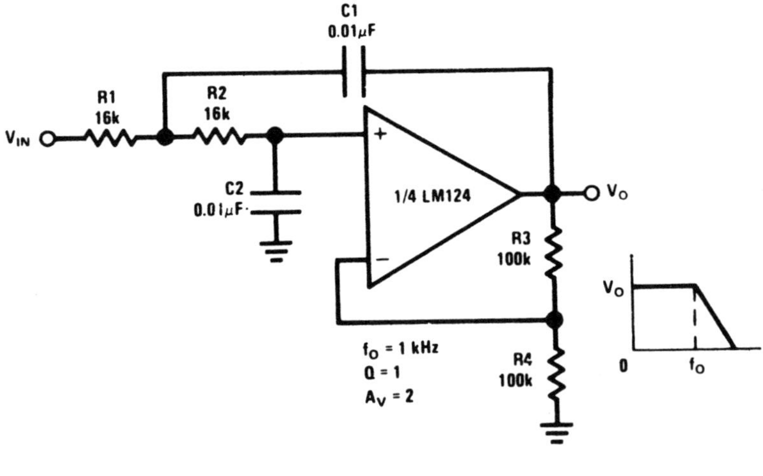

Dc coupled low-pass RC active filter (NS).

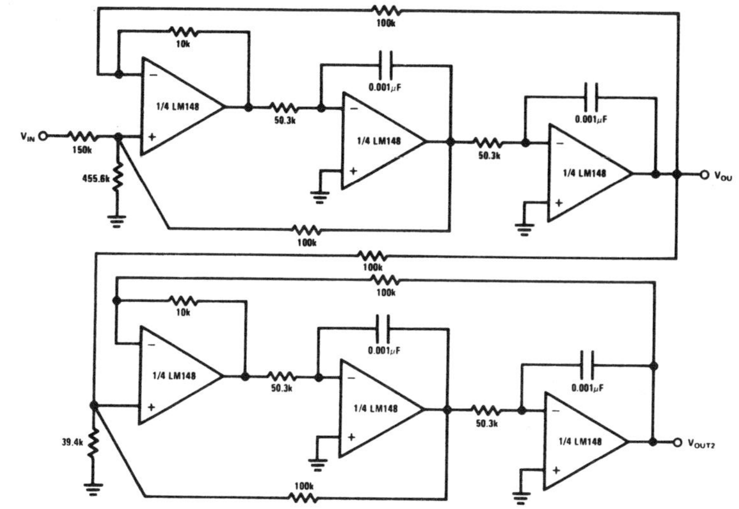

Use general equations, and tune each section separately

$Q_{1st SECTION} = 0.541$, $Q_{2nd SECTION} = 1.306$

The response should have 0 dB peaking

A 1 kHz 4 pole Butterworth (NS).

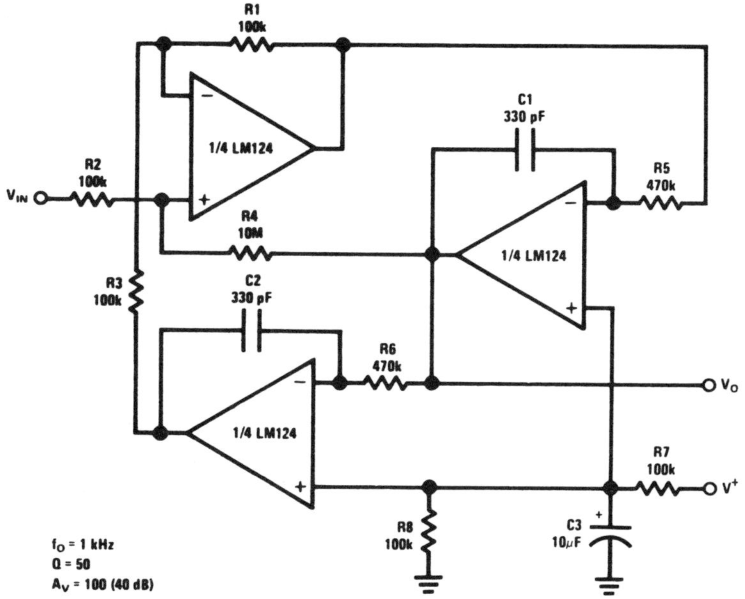

"BI-QUAD" RC active bandpass filter (NS).

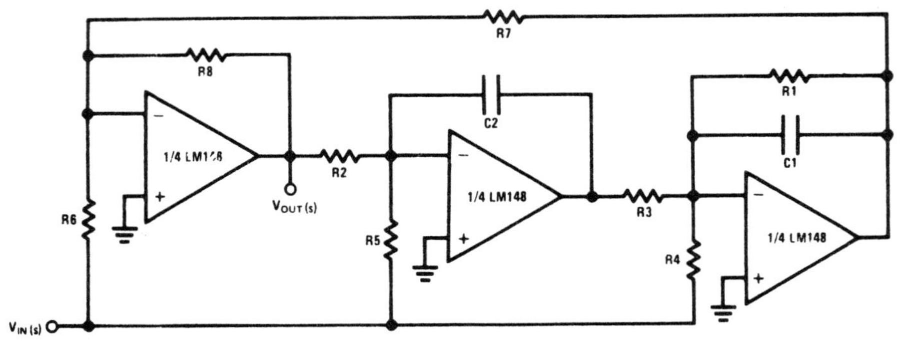

$$Q = \sqrt{\frac{R8}{R7}} \times \frac{R1C1}{\sqrt{R3C2R2C1}}, \quad f_o = \frac{1}{2\pi}\sqrt{\frac{R8}{R7}} \times \frac{1}{\sqrt{R2R3C1C2}}, \quad f_{NOTCH} = \frac{1}{2\pi}\sqrt{\frac{R6}{R3R5R7C1C2}}$$

Necessary condition for notch: $\frac{1}{R6} = \frac{R1}{R4R7}$

Ex: f_{NOTCH} = 3 kHz, Q = 5, R1 = 270k, R2 = R3 = 20k, R4 = 27k, R5 = 20k, R6 = R8 = 10k, R7 = 100k, C1 = C2 = 0.001μF

Better noise performance than the state-space approach

A 3 amplifier bi-quad notch filter (NS).

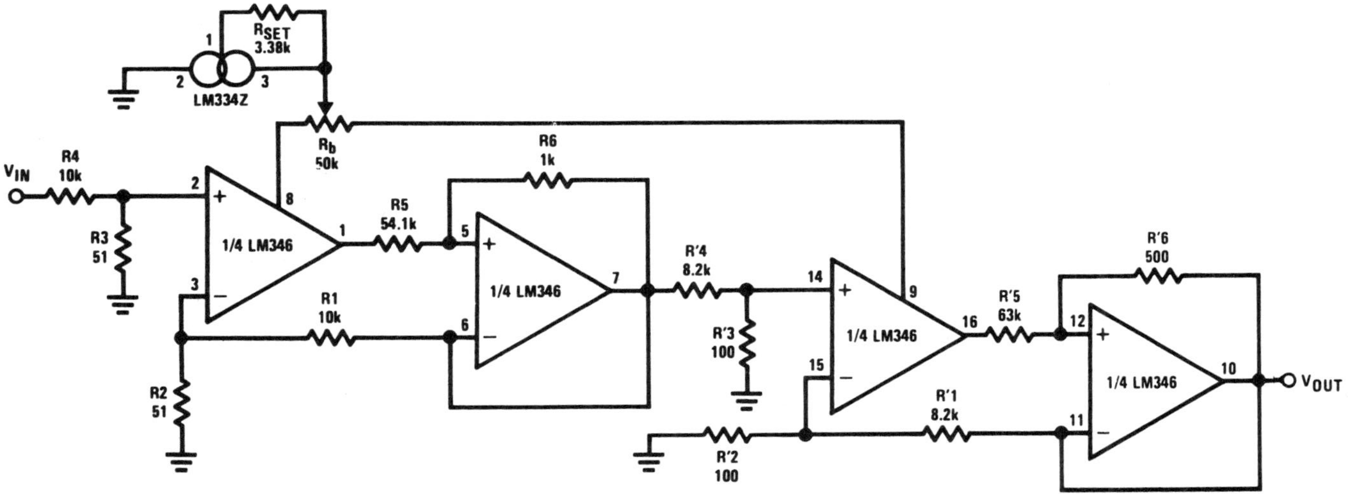

Ex: $f_c = 20$ kHz, H_o (gain of the filter) = 1, $Q_{o1} = 0.541$, $Q_{o2} = 1.306$.

- Since for this filter the GBW product of all 4 amplifiers has been designed to be the same (~1 MHz) only one current source can be used to bias the circuit. Fine tuning can be further accomplished through R_b.

 A 4th order Butterworth lowpass capacitorless filter (NS).

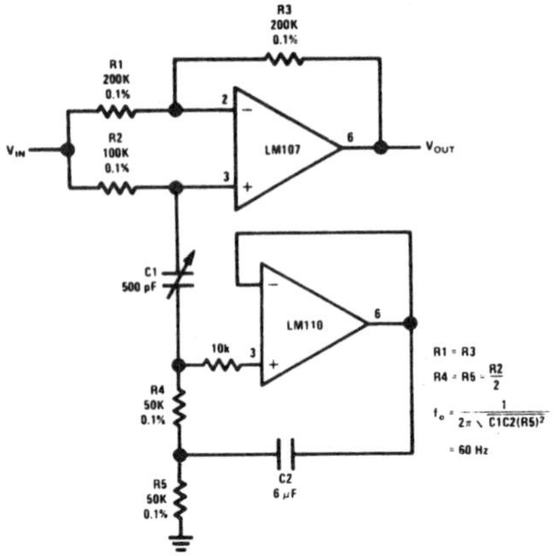

Tunable notch filter (NS).

High Q notch filter (NS).

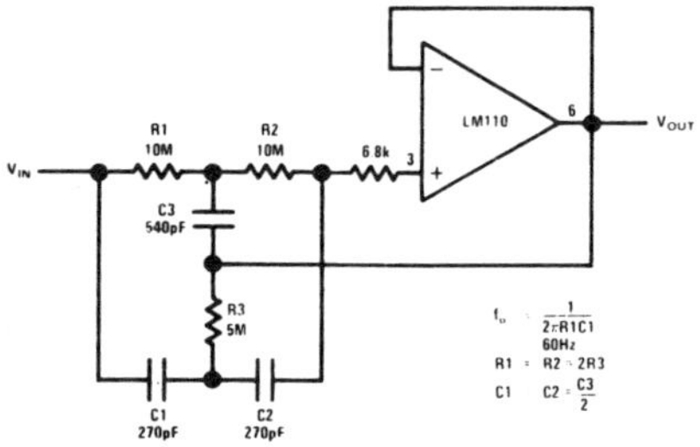

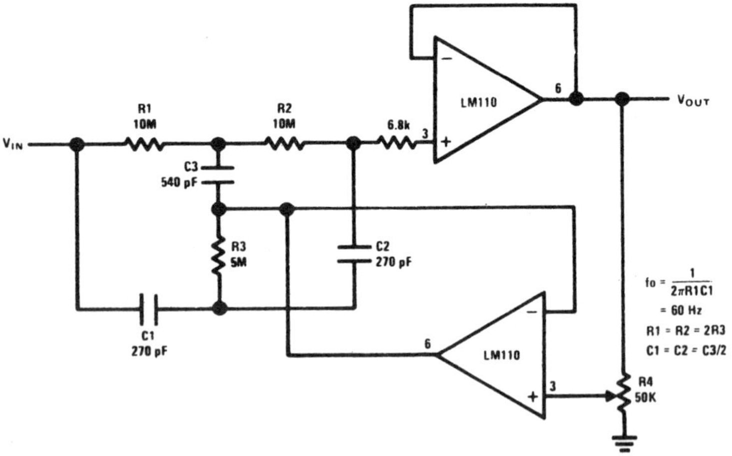

Adjustable Q notch filter (NS).

Lowpass active filter (NS).

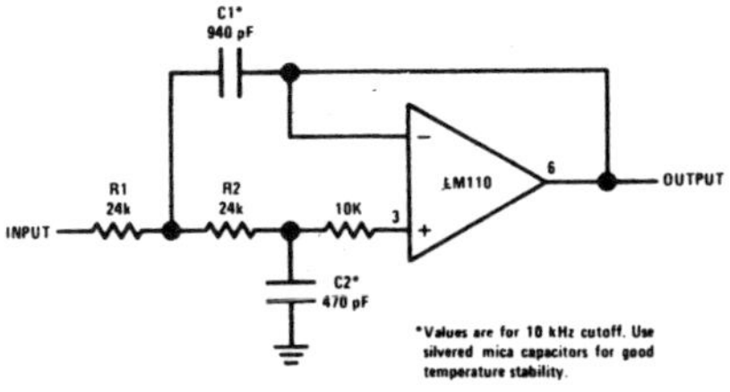

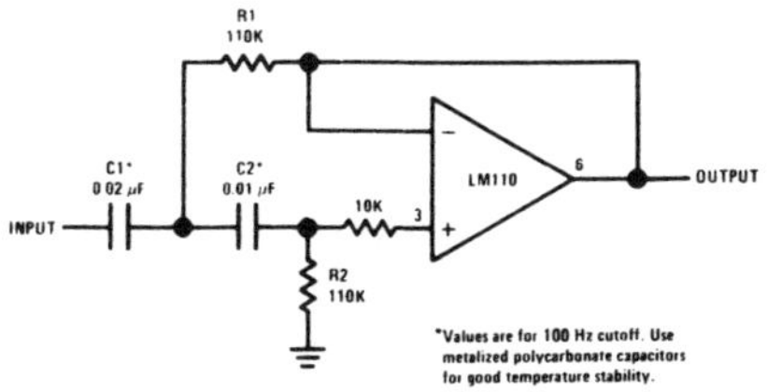

Highpass active filter (NS).

Bandpass filter (NS).

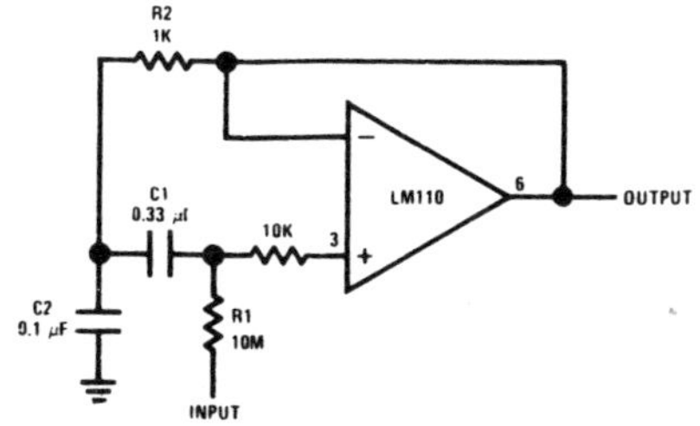

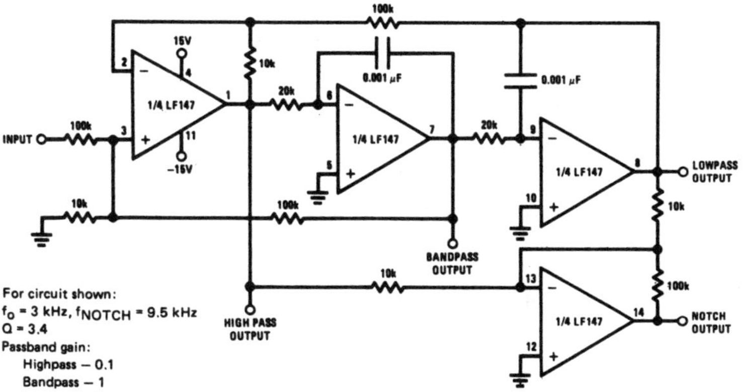

For circuit shown:
$f_o = 3$ kHz, $f_{NOTCH} = 9.5$ kHz
$Q = 3.4$
Passband gain:
- Highpass – 0.1
- Bandpass – 1
- Lowpass – 1
- Notch – 10

- $f_o \times Q \leq 200$ kHz
- 10V peak sinusoidal output swing without slew limiting to 200 kHz
- See LM148 data sheet for design equations

Universal state variable filter (NS).

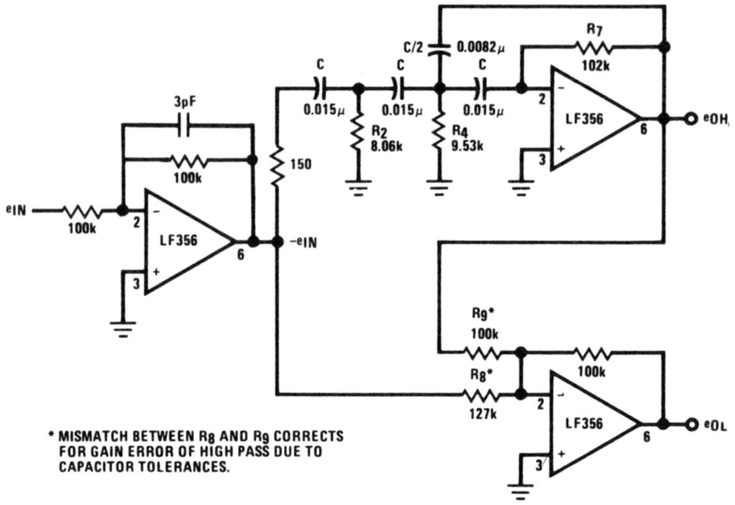

Asymmetrical 3rd order Butterworth active crossover network (NS).

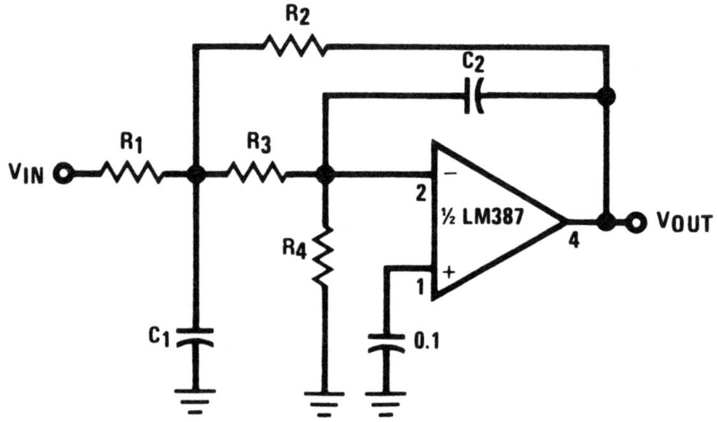

LM387 lowpass active filter (NS).

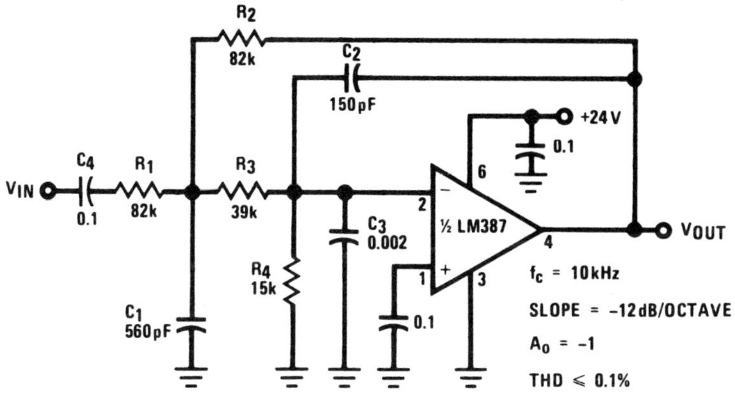

Scratch filter using LM387 (NS).

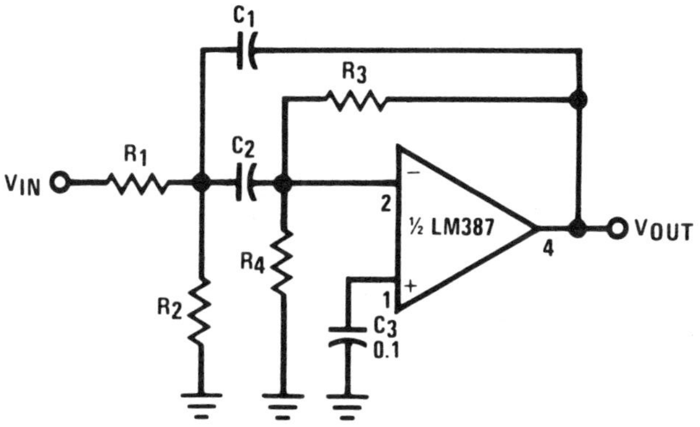

LM387 bandpass active filter (NS).

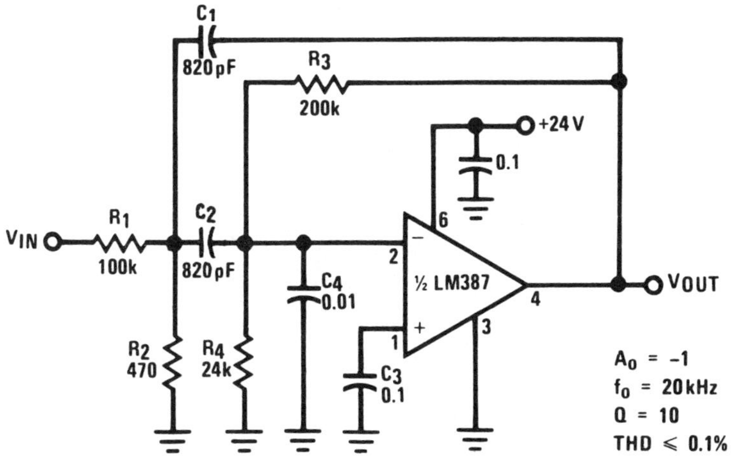

20 kHz bandpass active filter (NS).

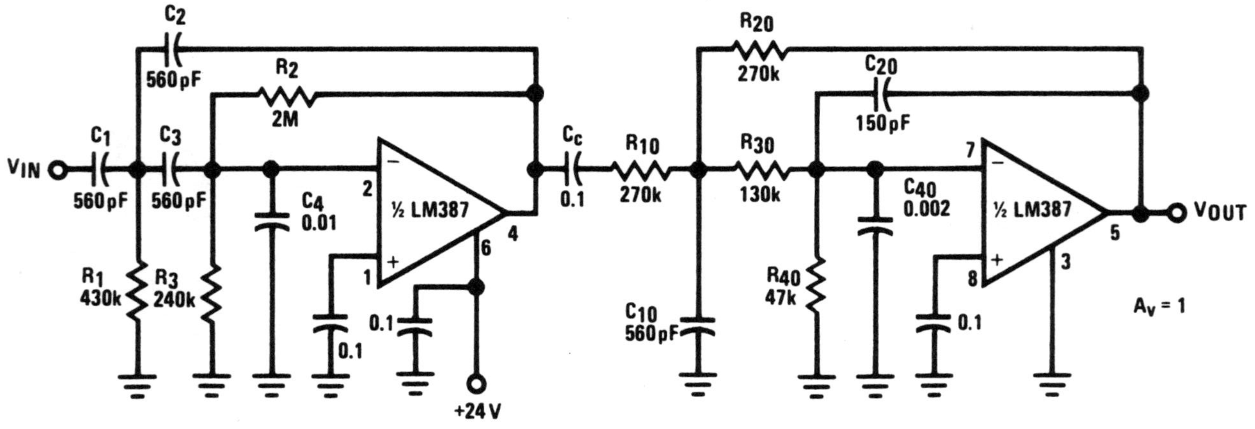

Speech filter (300 Hz-3kHz bandpass) (NS).

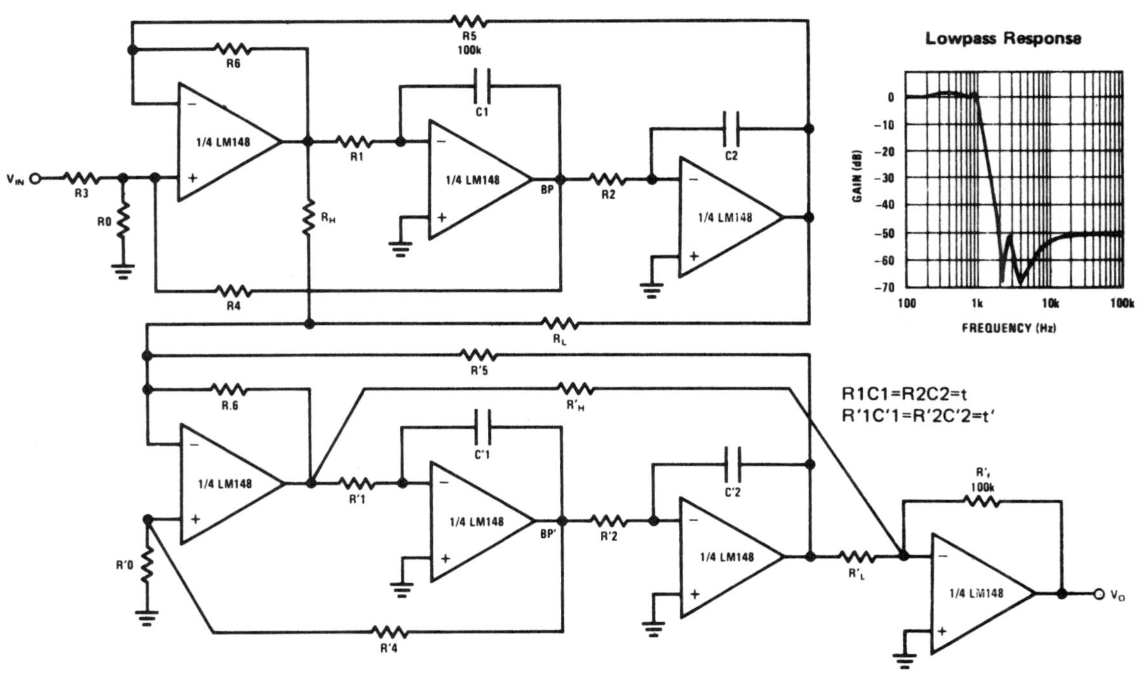

95 A 4th-order 1 kHz elliptic filter (4 poles, 4 zeros) (NS).

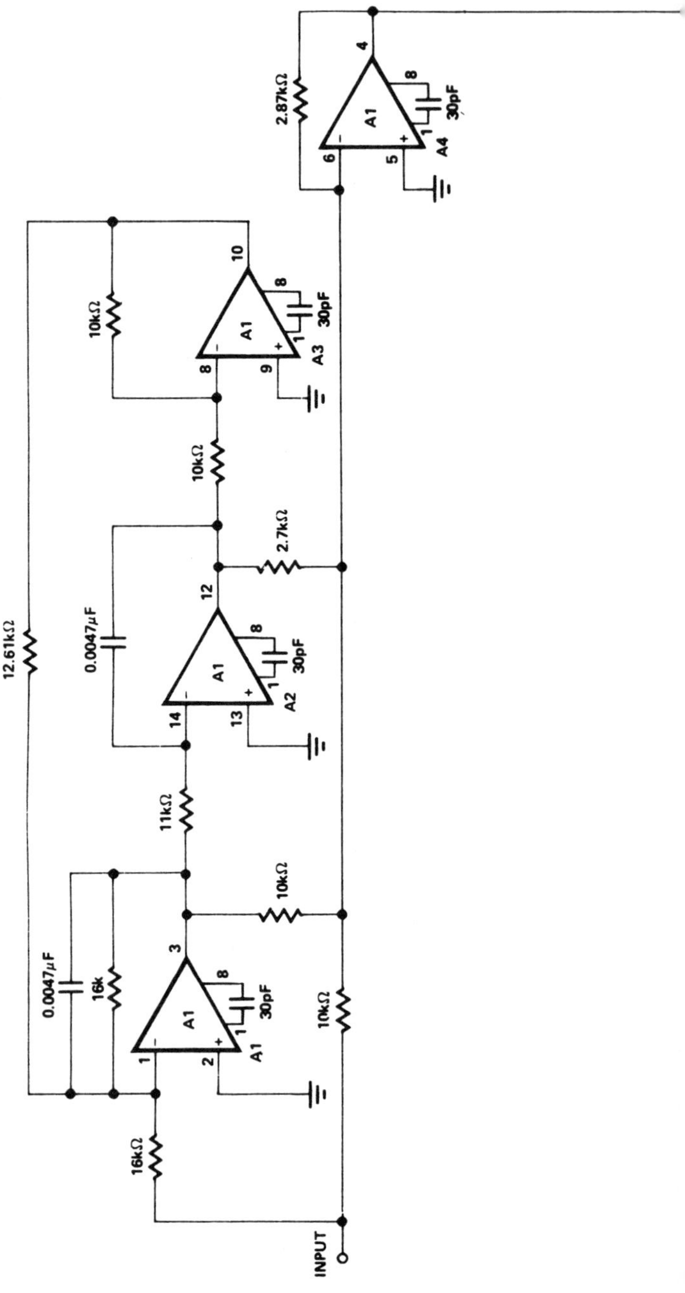
INPUT
16kΩ
0.0047μF
16k
A1
30pF
10kΩ
11kΩ
12.61kΩ
A2
2.7kΩ
A3
2.87kΩ
A4

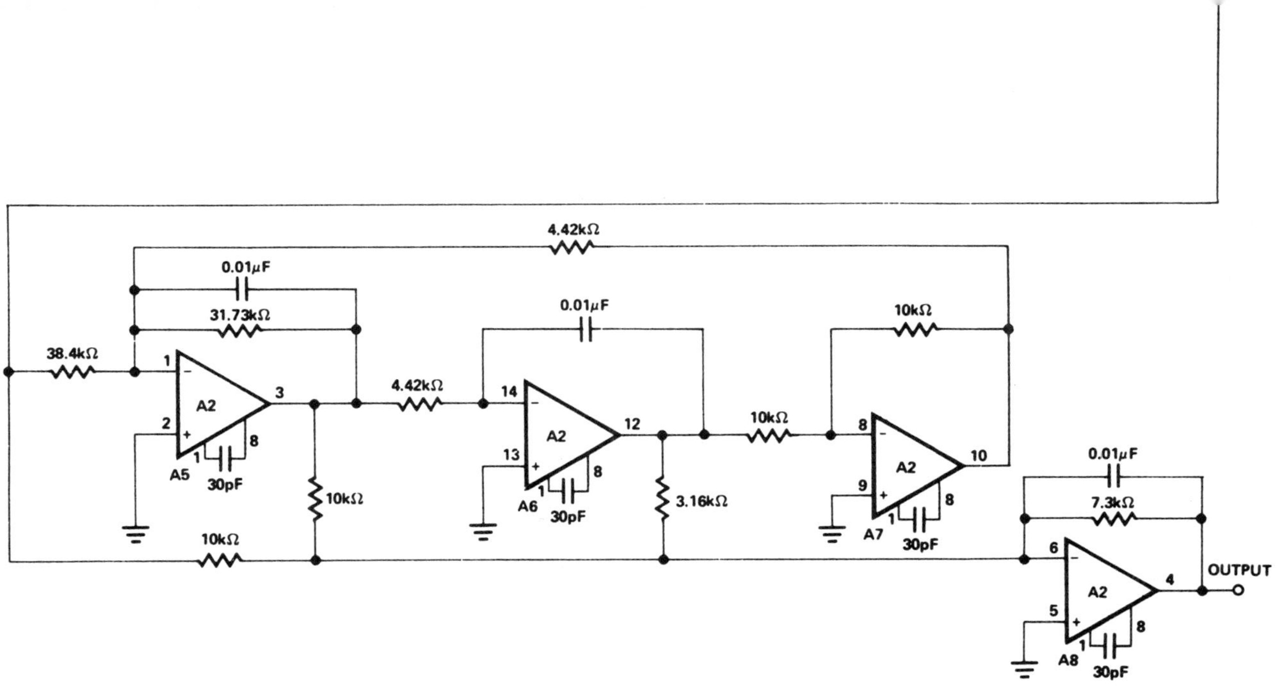

5-pole active filter (PM).

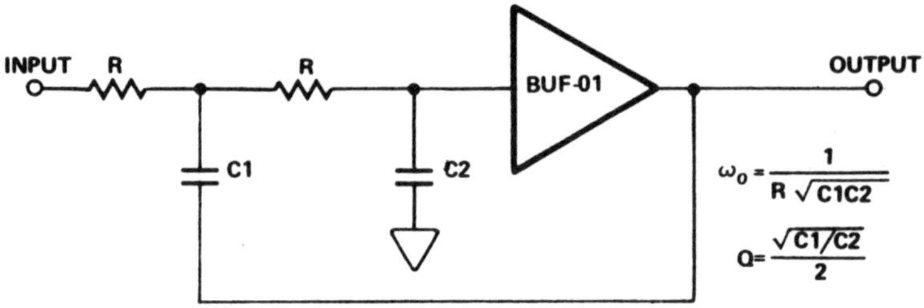

Second-order lowpass filter (PM).

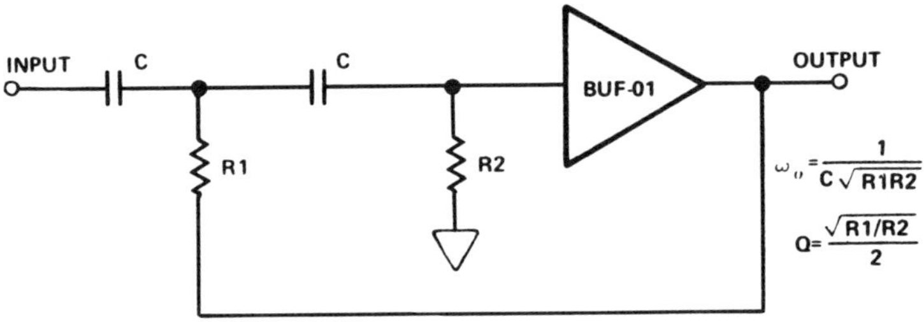

Second-order highpass filter (PM).

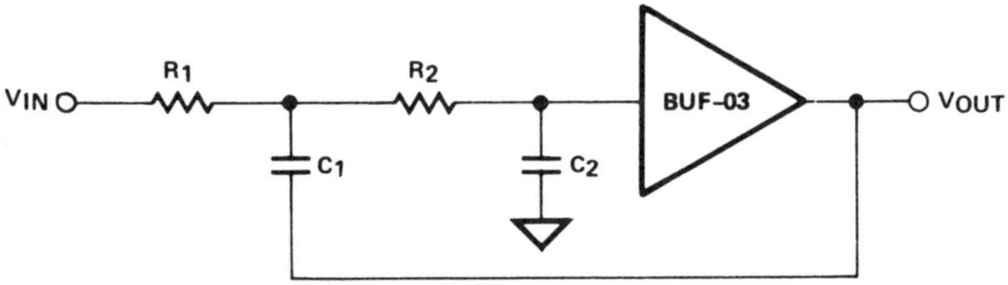

$$\omega_o = \left(\frac{1}{R_1R_2C_1C_2}\right)^{1/2}$$

IF $R_1 = R_2 = R$, THEN

$$Q = \frac{(C_1/C_2)^{1/2}}{2}$$

f_o Hz	R Ω	C_1 pF	C_2 pF	Q
1.0MEG	1.02K	220	110	0.71

Lowpass filter (high frequency) (PM).

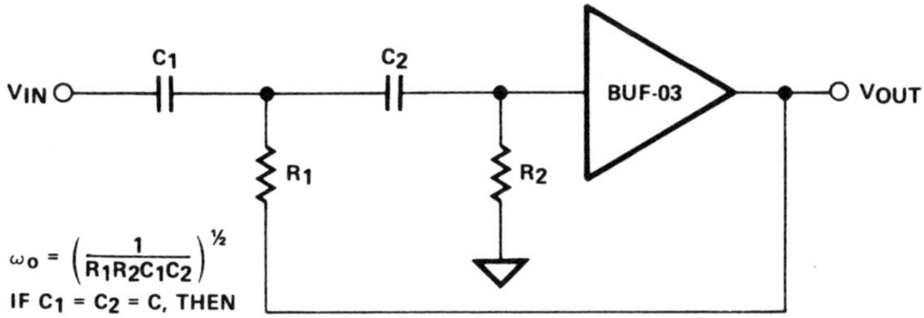

$$\omega_o = \left(\frac{1}{R_1R_2C_1C_2}\right)^{1/2}$$

IF $C_1 = C_2 = C$, THEN

$$Q = \frac{(R_1/R_2)^{1/2}}{2}$$

f_o Hz	C pF	R_1 Ω	R_2 Ω	Q
500K	220	2.05K	1.02K	0.71

Highpass filter (high frequency) (PM).

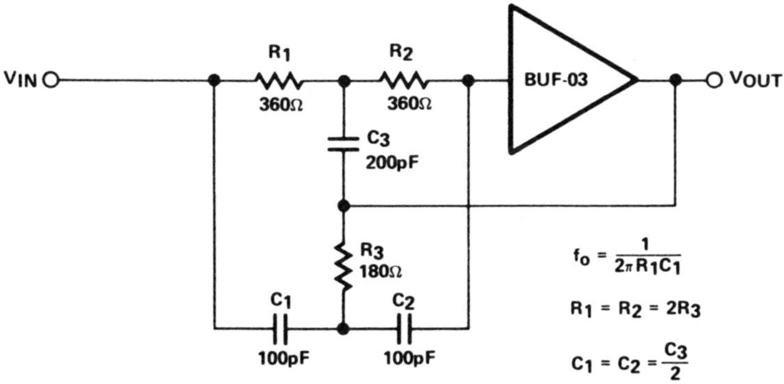

$$f_o = \frac{1}{2\pi R_1C_1}$$

$$R_1 = R_2 = 2R_3$$

$$C_1 = C_2 = \frac{C_3}{2}$$

Notch filter at 4.5MHz (PM).

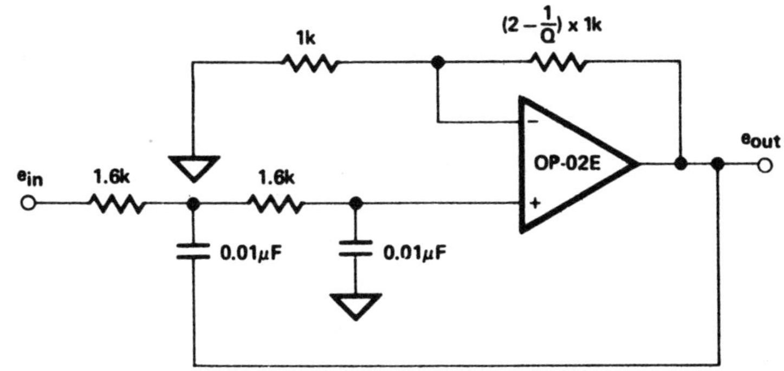

Equal component Sallen-Key filter (PM).

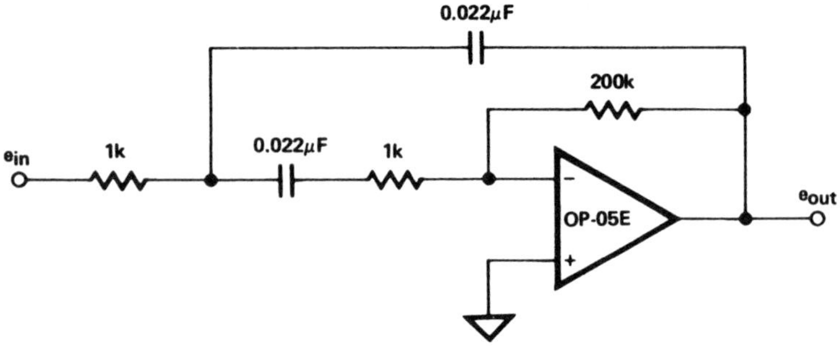

500Hz Sallen-Key bandpass filter (PM).

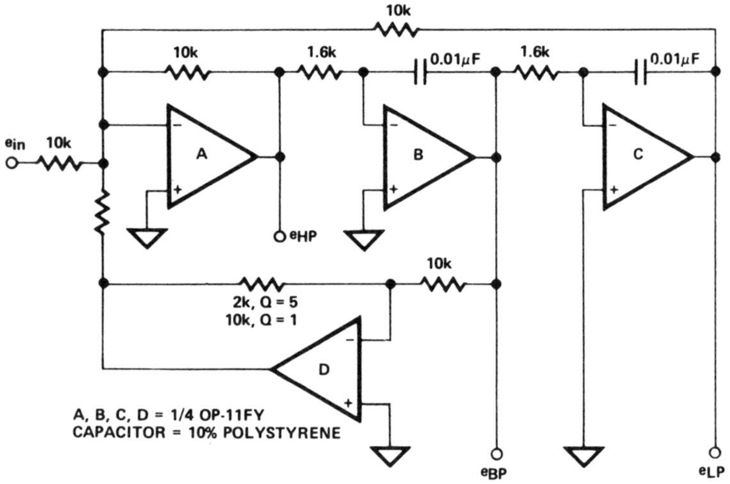

Practical state-variable 10kHz active filter (PM).

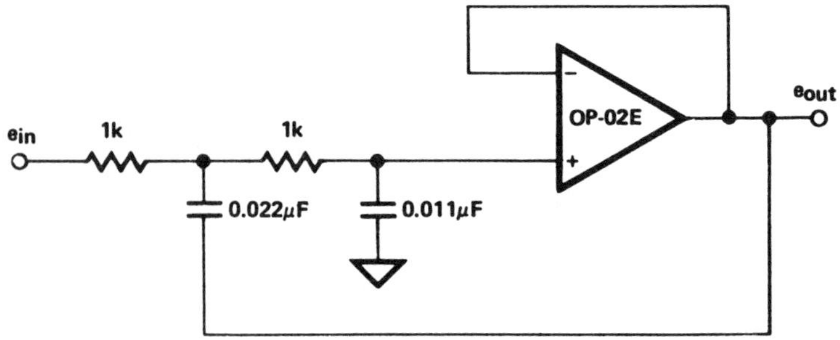

10kHz Sallen-Key filter (PM).

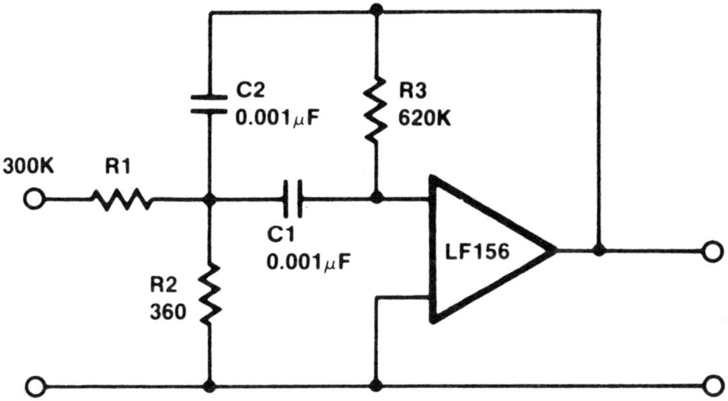

10kHz bandpass filter (S).

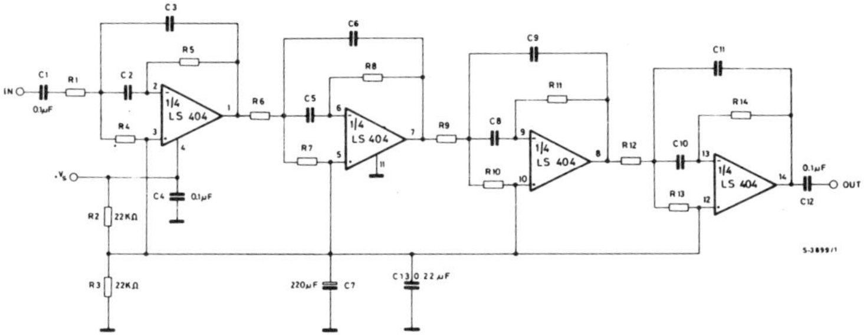

f_c = 1.180Hz; A = 1; $C_2 = C_3 = C_5 = C_6 = C_8 = C_9 = C_{10} = C_{11}$ = 3.300 pF;
$R_1 = R_6 = R_9 = R_{12}$ = 160 KΩ; $R_5 = R_8 = R_{11} = R_{14}$ = 330KΩ; $R_4 = R_7 = R_{10} = R_{13}$ = 5.3KΩ

Multiple feedback 8-pole bandpass filter (SGS).

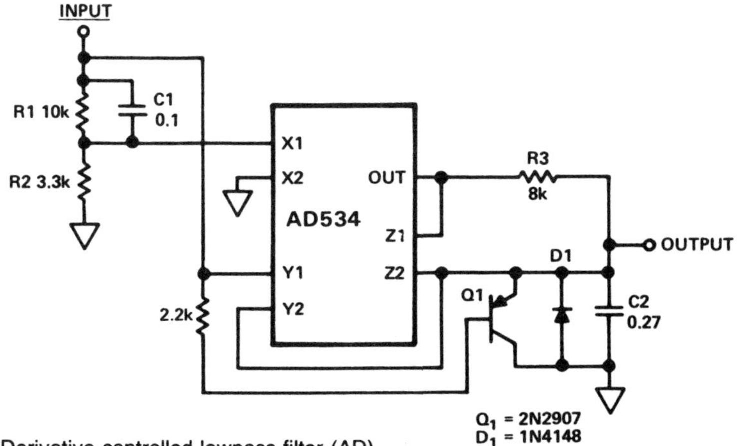

Derivative-controlled lowpass filter (AD).

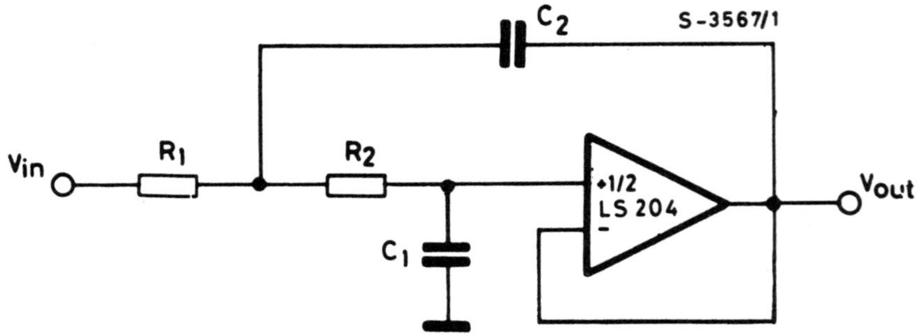

$$\frac{V_o}{V_i} = \frac{1}{1 + 2\,\xi\,\frac{S}{\omega_c} + \frac{S^2}{\omega_c^{\,2}}}$$

where:

$\omega_c = 2\pi\ f_c$ with f_c = cutoff frequency

ξ = damping factor.

Design of 2nd order active lowpass filter (Sallen and Key configuration unity gain op-amp) (SGS).

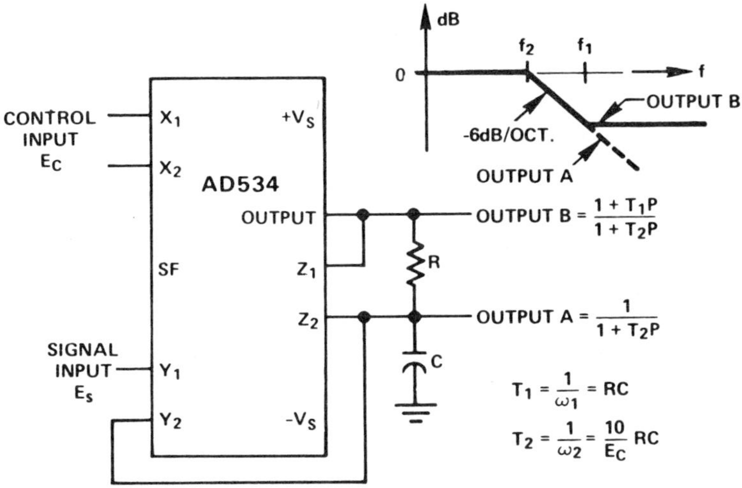

Voltage-controlled lowpass filter (AD).

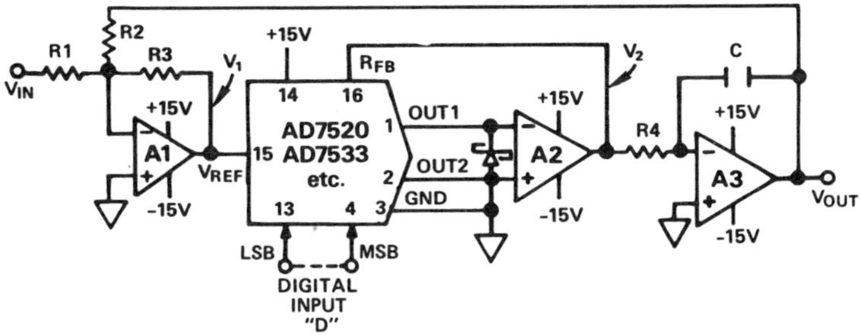

Digitally programmable lowpass filter (2) (AD).

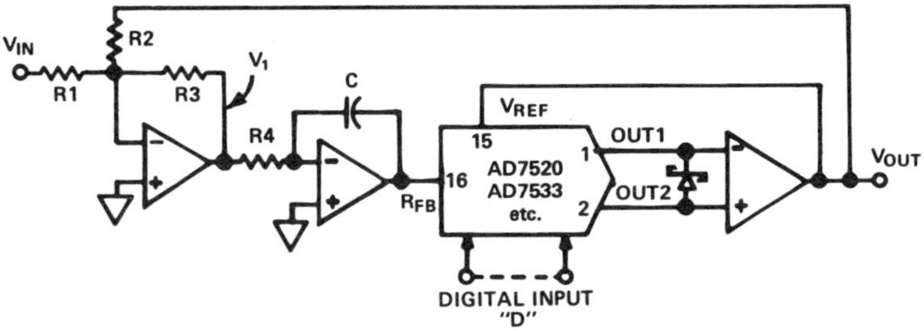

Digitally programmable lowpass filter (3) (AD).

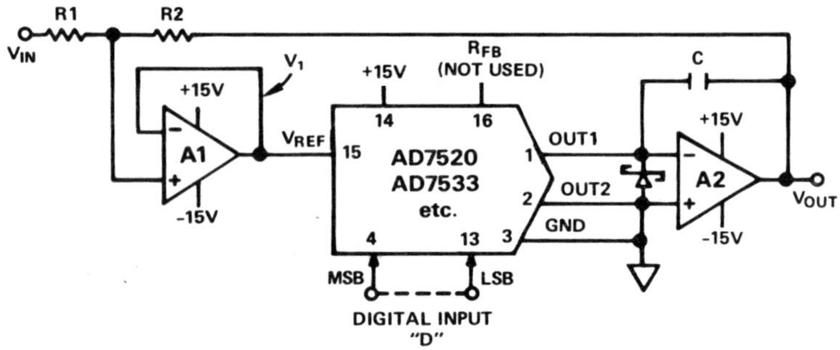

Programmable lowpass filter (1) (AD).

Chapter 6

Multivibrators, Sample and Hold, and Peak Detectors

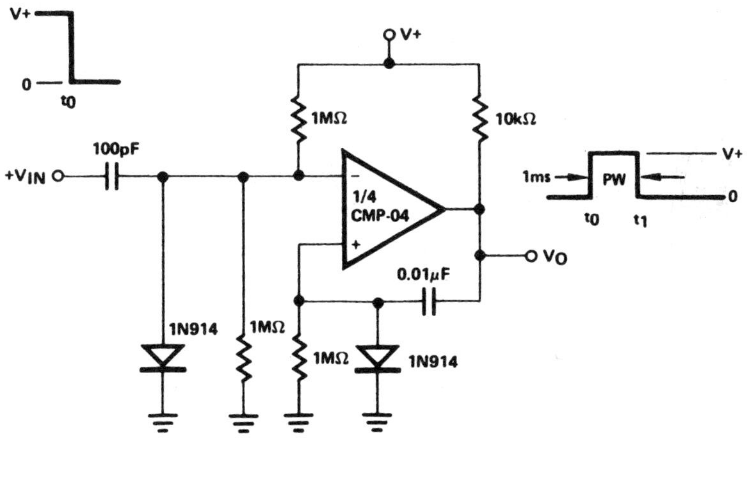

One-shot multivibrator (PM).

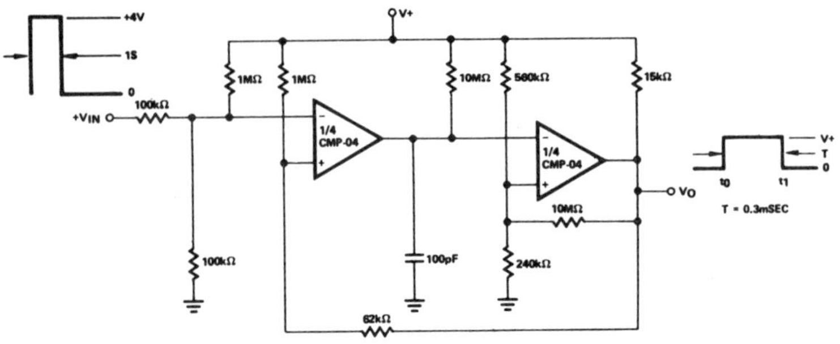

One-shot multivibrator with input lock out (PM).

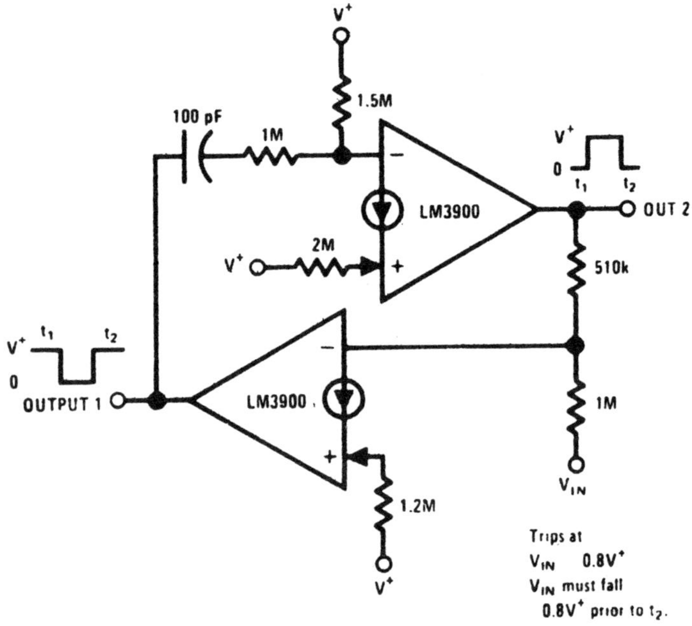

One-shot with dc input comparator (NS).

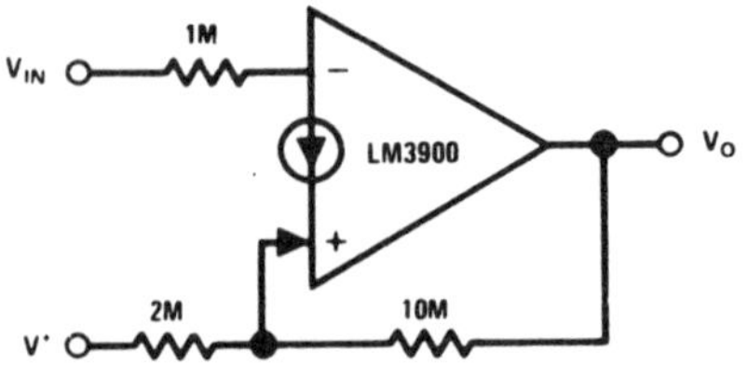

Schmitt-trigger (NS).

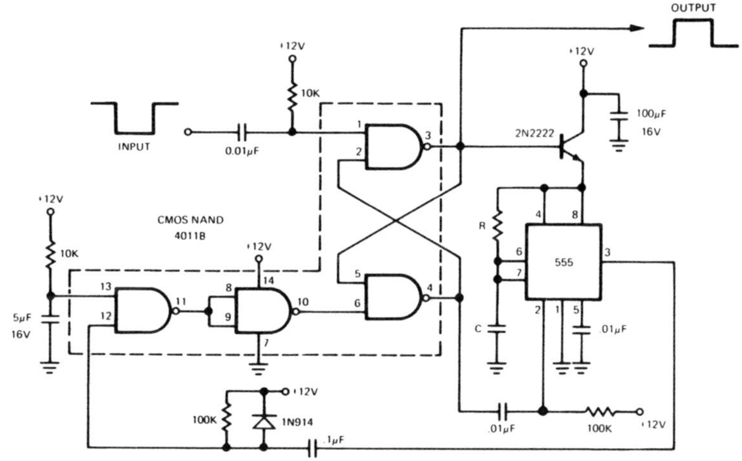

Low-power monostable (S).

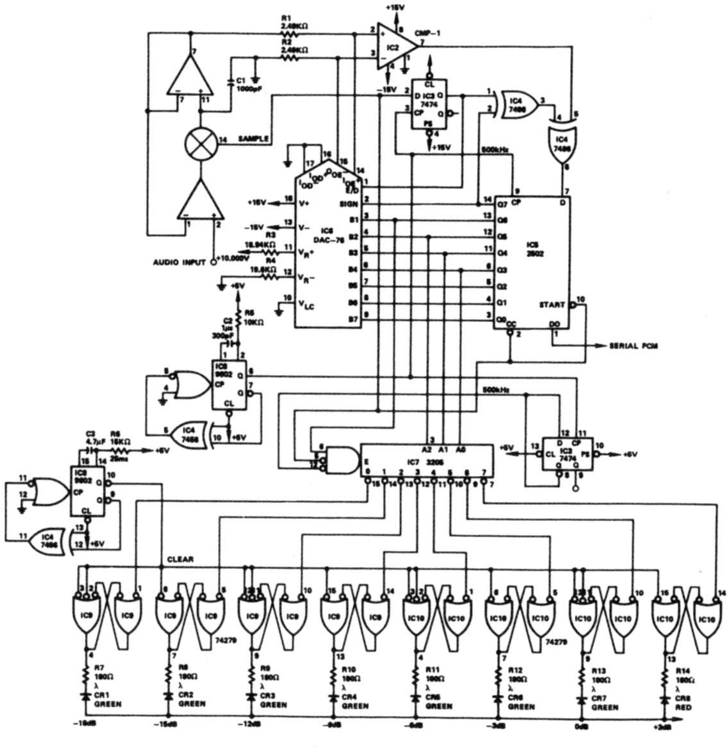

Peak-reading Vu indicator complete schematic (PM).

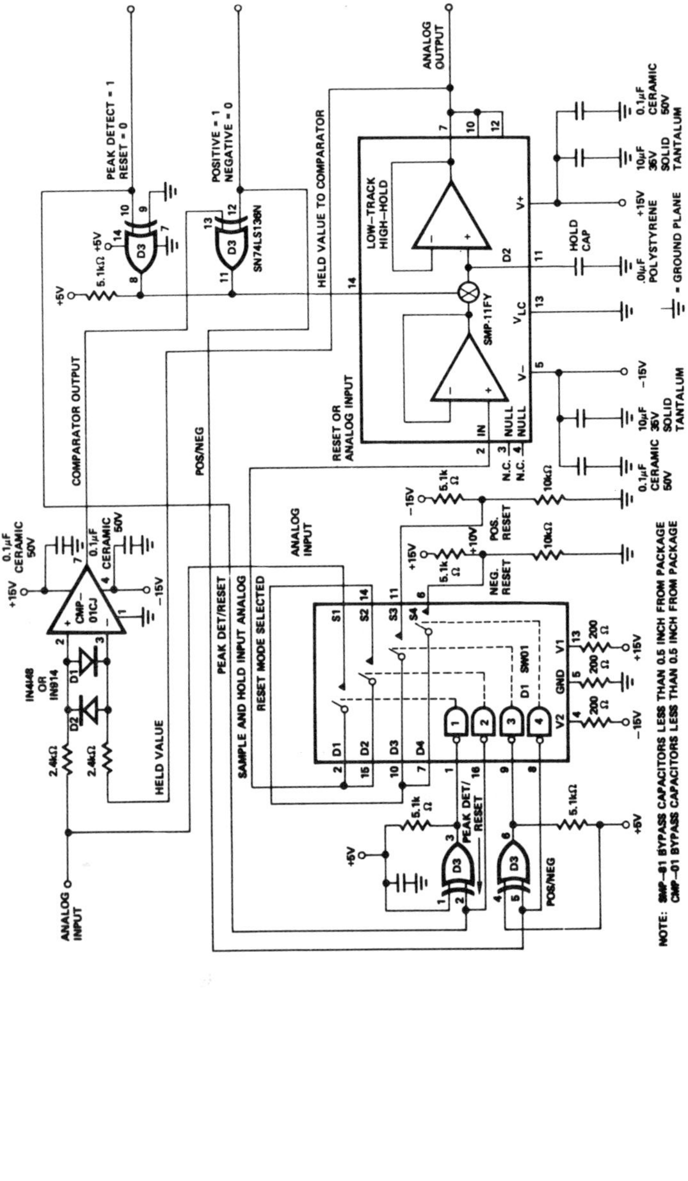

Programmable peak detector complete schematic (PM).

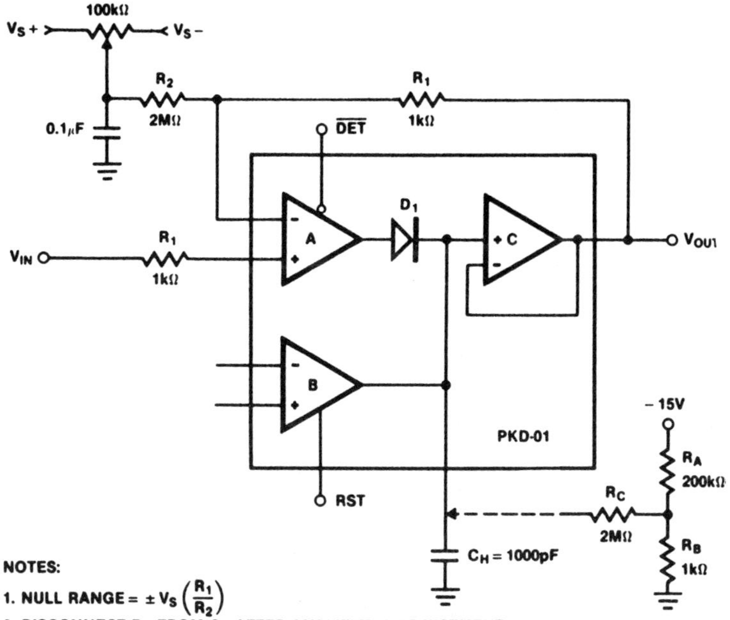

NOTES:

1. NULL RANGE = $\pm V_S \left(\frac{R_1}{R_2}\right)$
2. DISCONNECT R_C FROM C_H AFTER AMPLIFIER A ADJUSTMENT.
3. REPEAT NULL CIRCUIT FOR RESET BUFFER AMPLIFIER B IF REQUIRED. R_A, R_B AND R_C NOT NECESSARY FOR AMPLIFIER B ADJUSTMENT.

Vos null circuit for unity gain positive peak detector (PM).

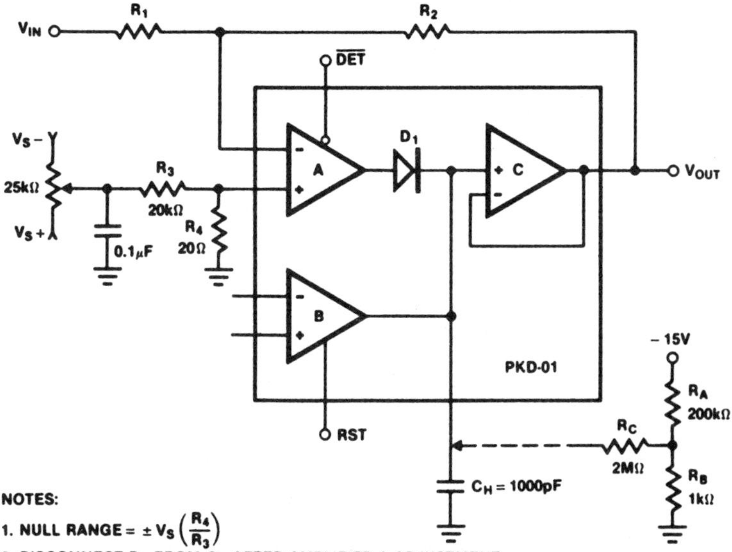

NOTES:

1. NULL RANGE = $\pm V_S \left(\frac{R_4}{R_3}\right)$
2. DISCONNECT R_C FROM C_H AFTER AMPLIFIER A ADJUSTMENT.
3. REPEAT NULL CIRCUIT FOR RESET BUFFER AMPLIFIER B IF REQUIRED.

Vos null circuit for negative peak detector (PM).

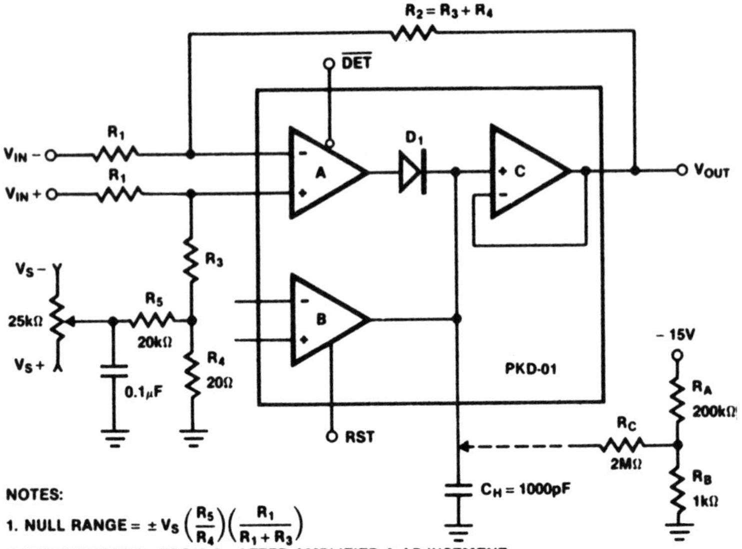

NOTES:

1. NULL RANGE = $\pm V_S \left(\frac{R_5}{R_4}\right)\left(\frac{R_1}{R_1 + R_3}\right)$
2. DISCONNECT R_C FROM C_H AFTER AMPLIFIER A ADJUSTMENT.
3. REPEAT NULL CIRCUIT FOR RESET BUFFER AMPLIFIER B IF REQUIRED.

Vos null circuit for differential peak detector (PM).

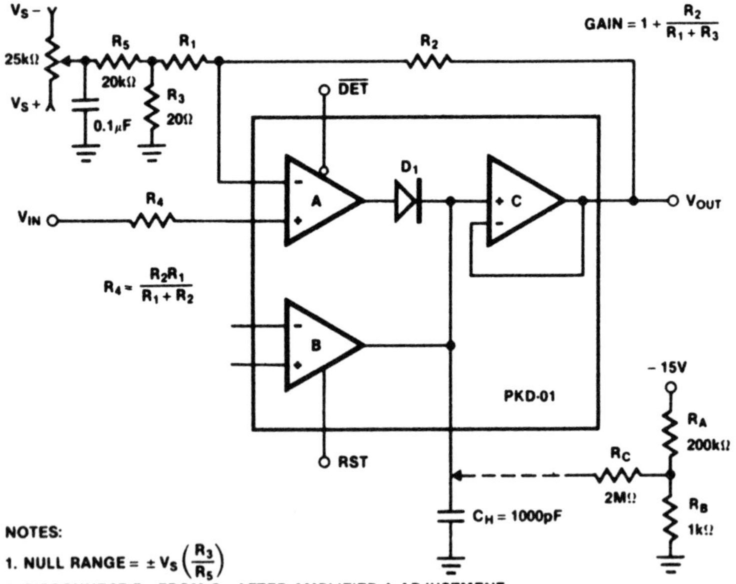

NOTES:

1. NULL RANGE = $\pm V_S \left(\frac{R_3}{R_5}\right)$
2. DISCONNECT R_C FROM C_H AFTER AMPLIFIER A ADJUSTMENT.
3. REPEAT NULL CIRCUIT FOR RESET BUFFER AMPLIFIER B IF REQUIRED.

Vos null circuit for positive peak detector with gain (PM).

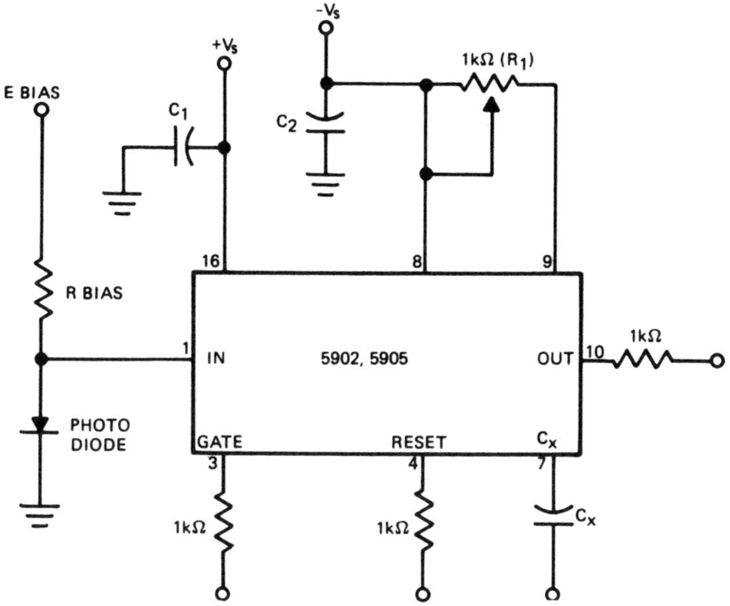

Peak light level sensing (OE).

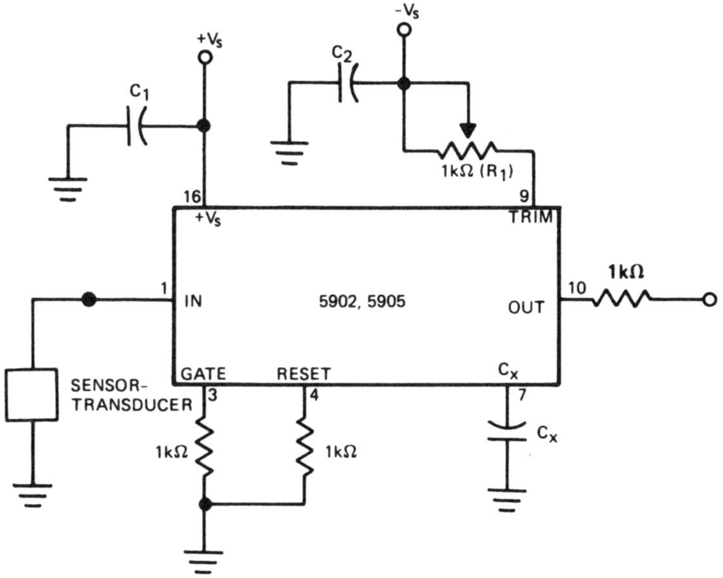

Peak thrust, shock, vibration measurement (OE).

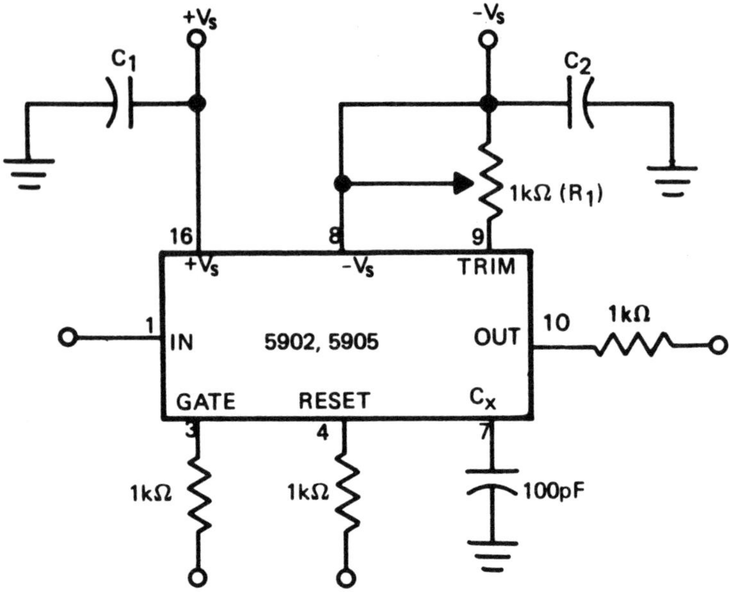

Data acquisition, high-speed peak sense mode (OE).

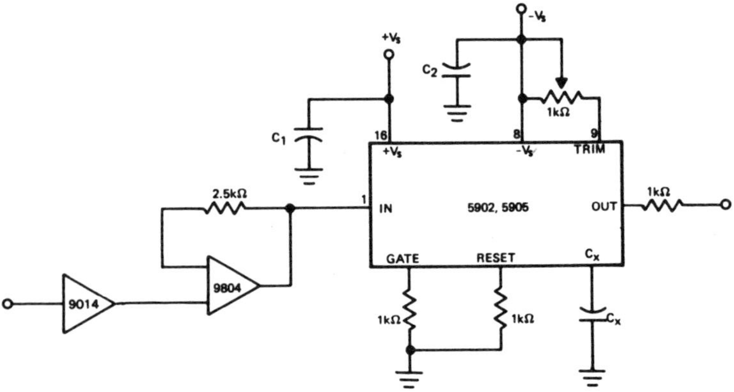

Peak measurement of absolute values (OE).

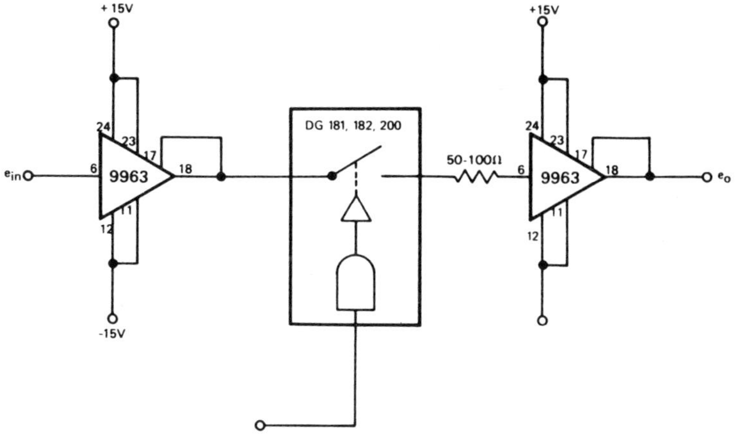

High-speed sample and hold (OE).

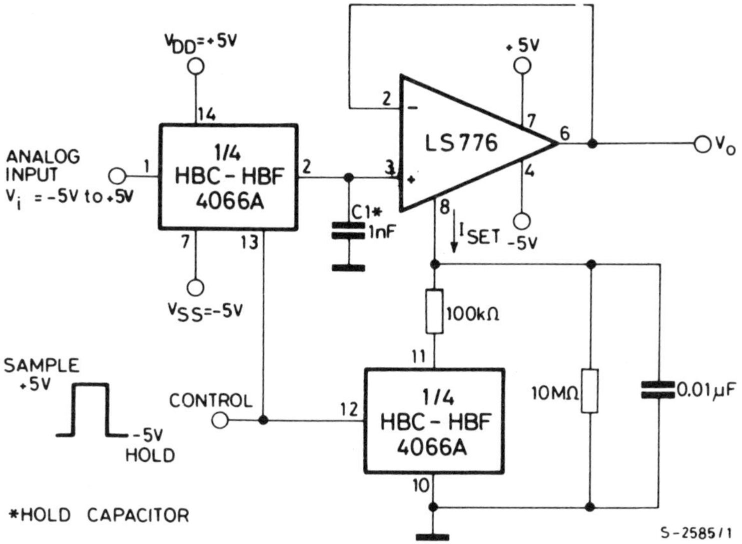

High-accuracy sample and hold (SGS).

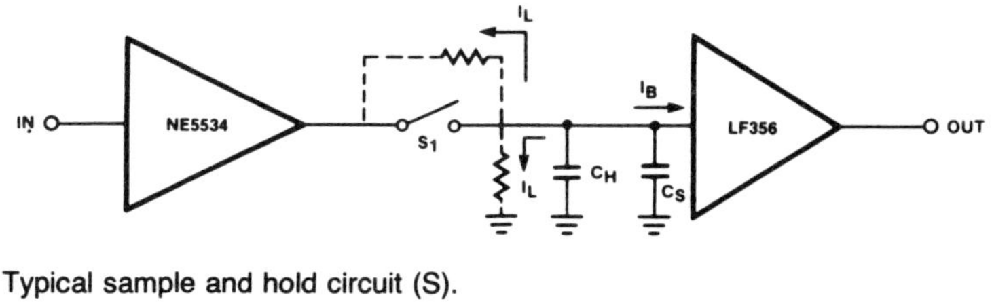

Typical sample and hold circuit (S).

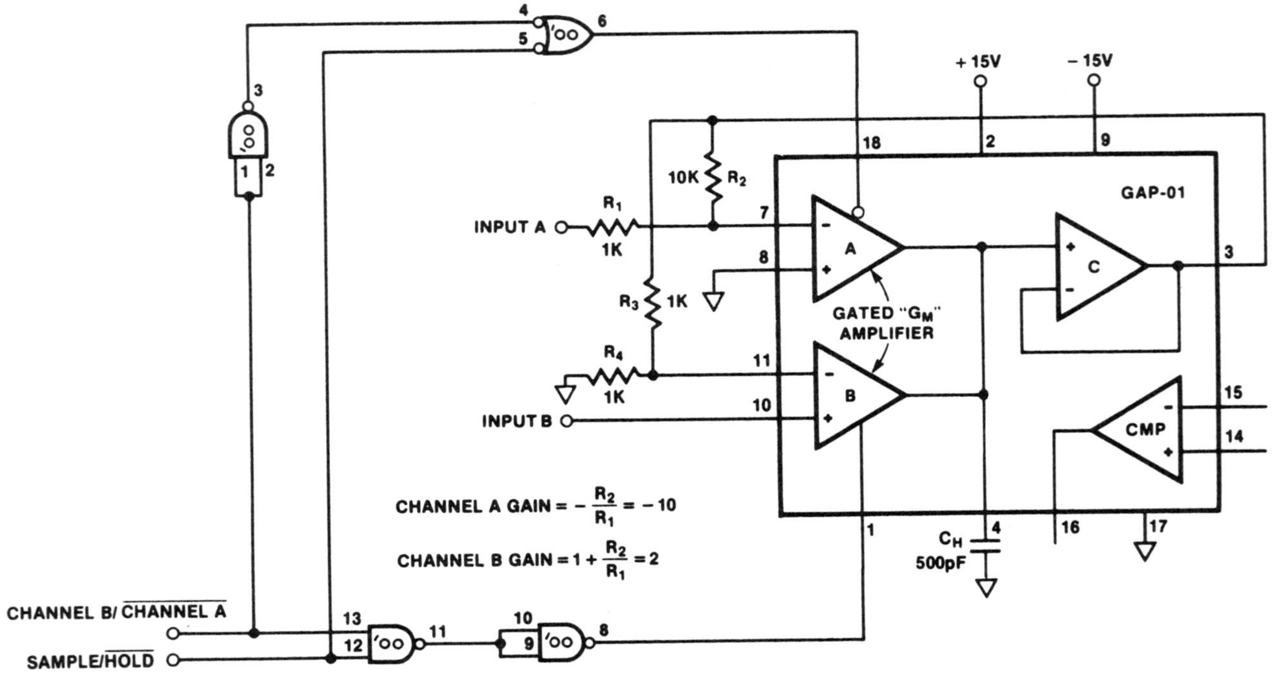

Two-channel sample/hold amplifier (PM).

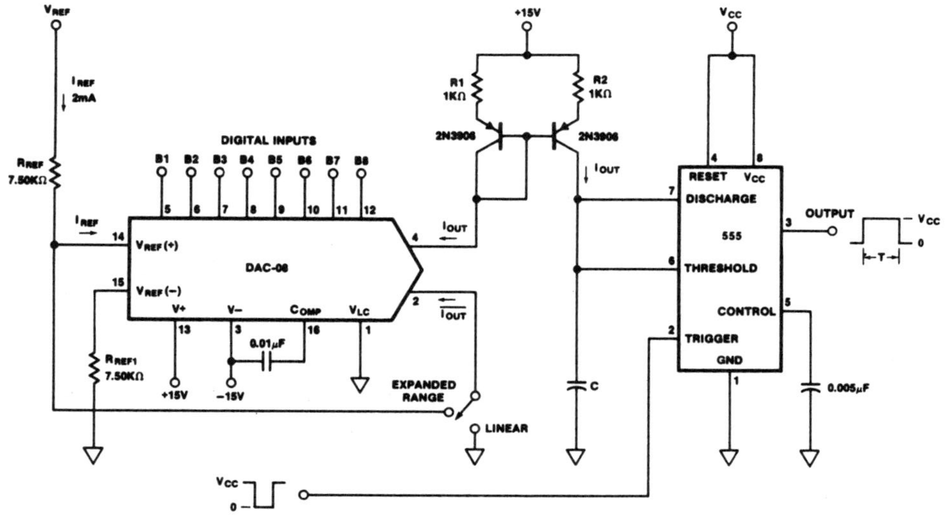

$$\text{ONE-SHOT PERIOD, } T = \frac{2}{3}\,\frac{R_{REF}C}{\{D\}}\;\frac{V_{CC}}{V_{REF}} \text{ FOR LINEAR MODE}$$

$$\text{ONE-SHOT PERIOD, } T = \frac{2}{3}\,R_{REF}C\,\frac{V_{CC}}{V_{REF}}\left[\frac{2-\{D\}}{\{D\}}\right] \text{FOR EXPANDED MODE}$$

Digitally controlled one-shot (PM).

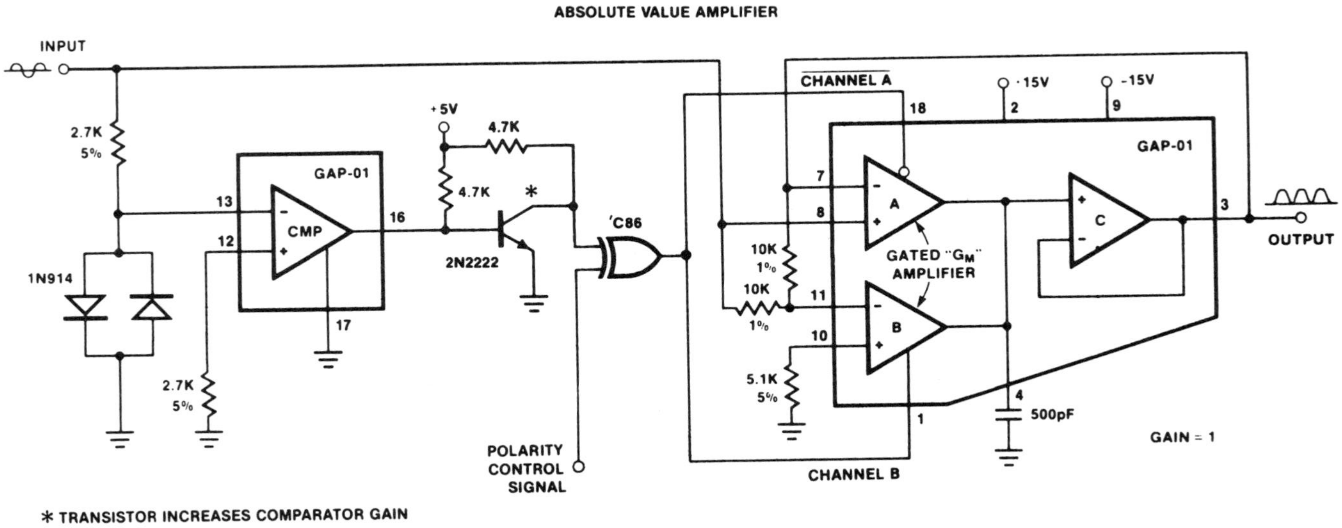

Absolute value circuit with polarity programmable output (PM).

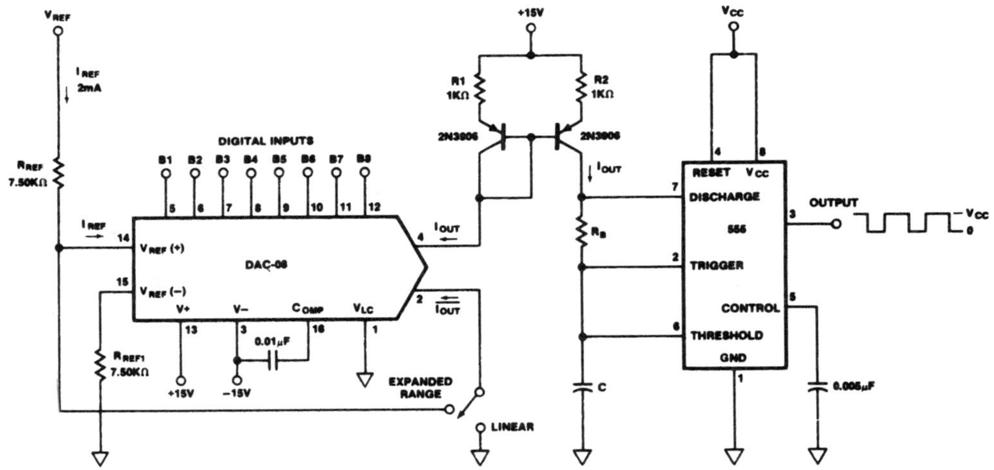

$$\text{FREQUENCY, } f = \frac{1}{\frac{1}{3}\,\frac{R_{REF}C}{\{D\}}\,\frac{V_{CC}}{V_{REF}} + 0.695\,R_B C} \quad \text{FOR LINEAR MODE}$$

$$\text{FREQUENCY, } f = \frac{1}{\frac{1}{3}\,R_{REF}C\,\frac{V_{CC}}{V_{REF}}\left[\frac{2-\{D\}}{\{D\}}\right] + 0.695\,R_B C} \quad \text{FOR EXPANDED MODE}$$

Digitally controlled astable multivibrator (PM).

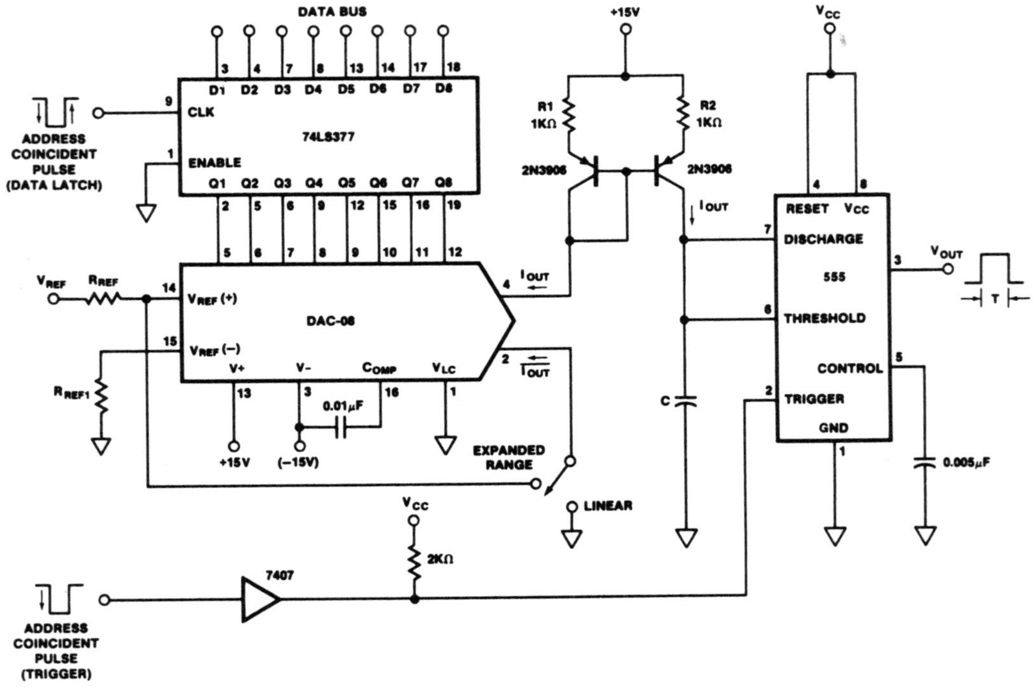

Microprocessor controlled one-shot (PM).

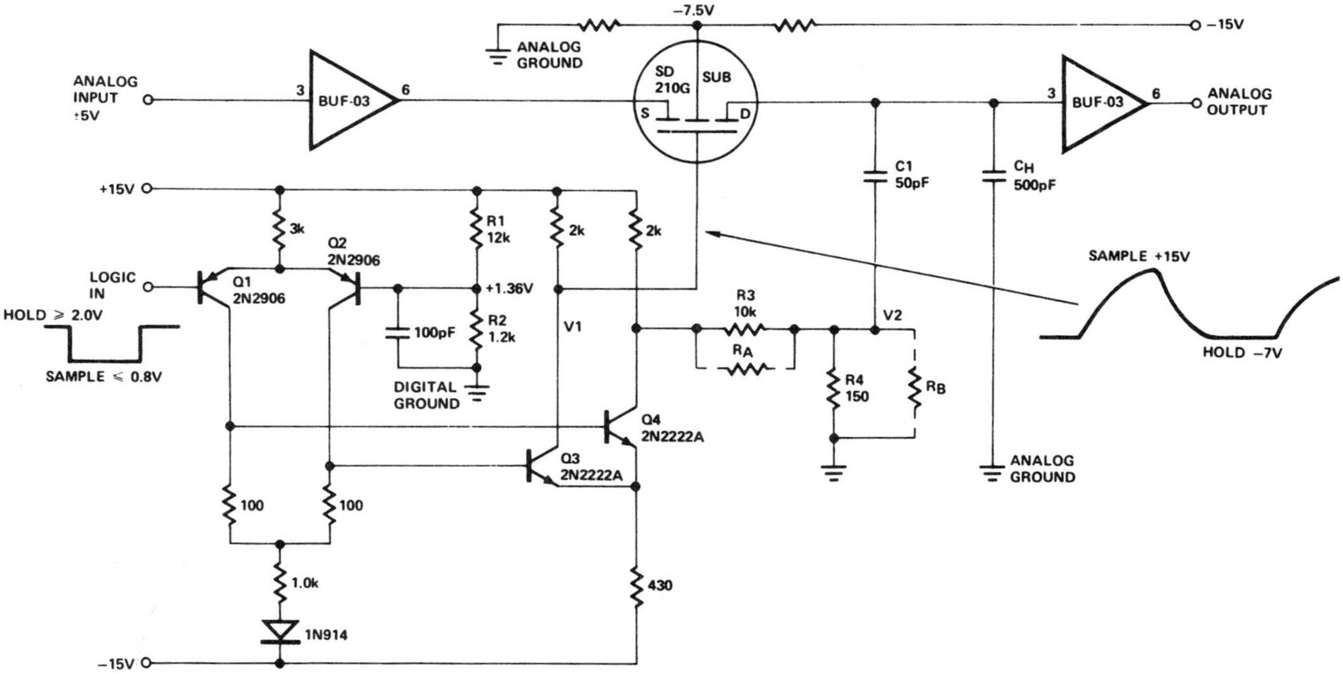

High-speed sample/hold amplifier (PM).

Chapter 7

Waveform and Function Generators

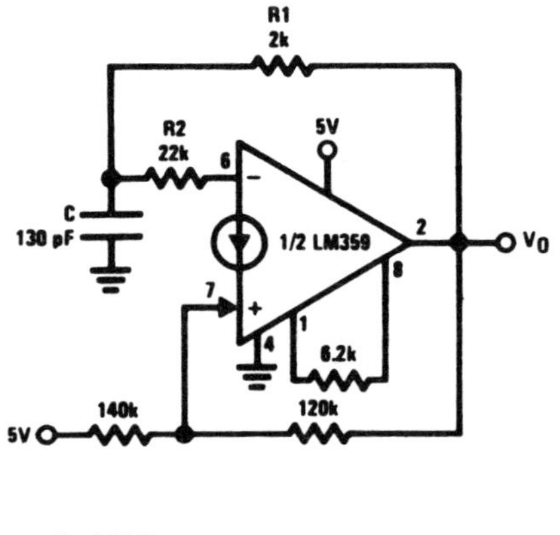

f = 1 MHz
Output is TTL compatible
Frequency is adjusted by R1 & C (R1<<R2)

Squarewave generator (NS).

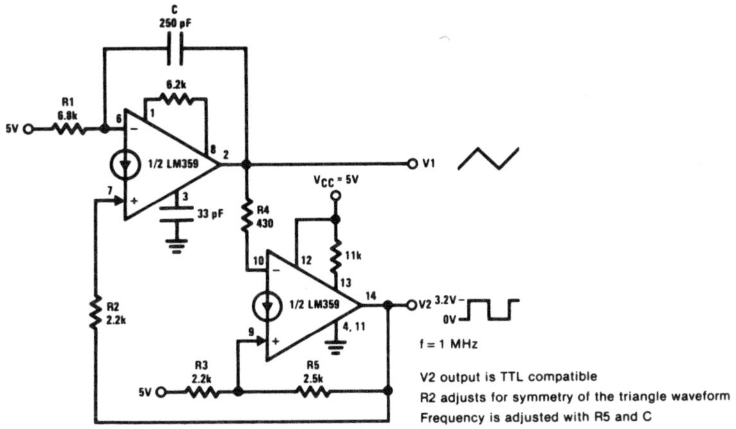

Triangle waveform generator (NS).

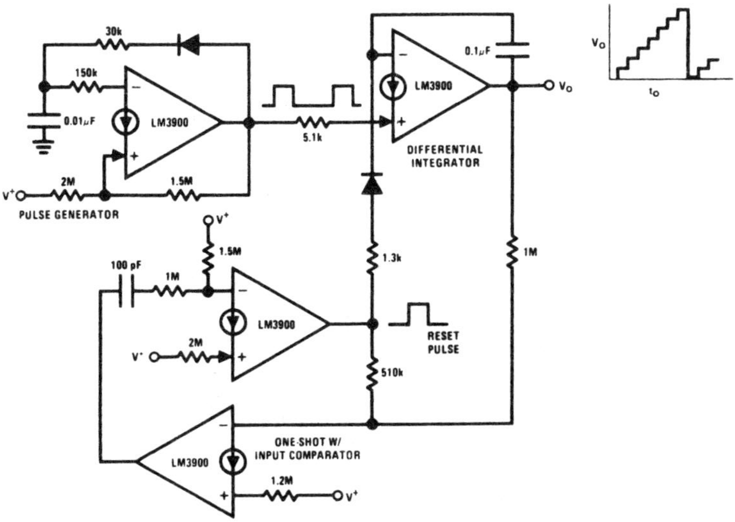

Free-running staircase generator/pulse counter (NS).

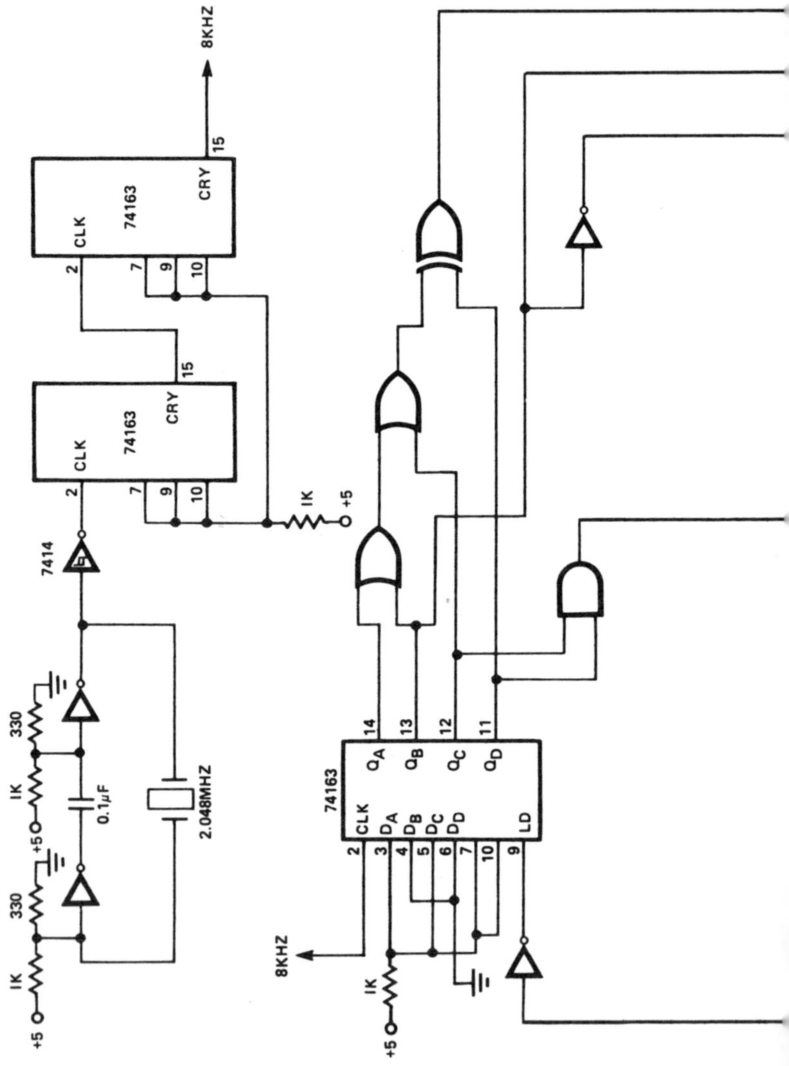

74163
CLK
CRY
8KHZ
7414
IK
+5
330
0.1µF
2.048MHZ
QA
QB
QC
QD
DA
DB
DC
DD
LD

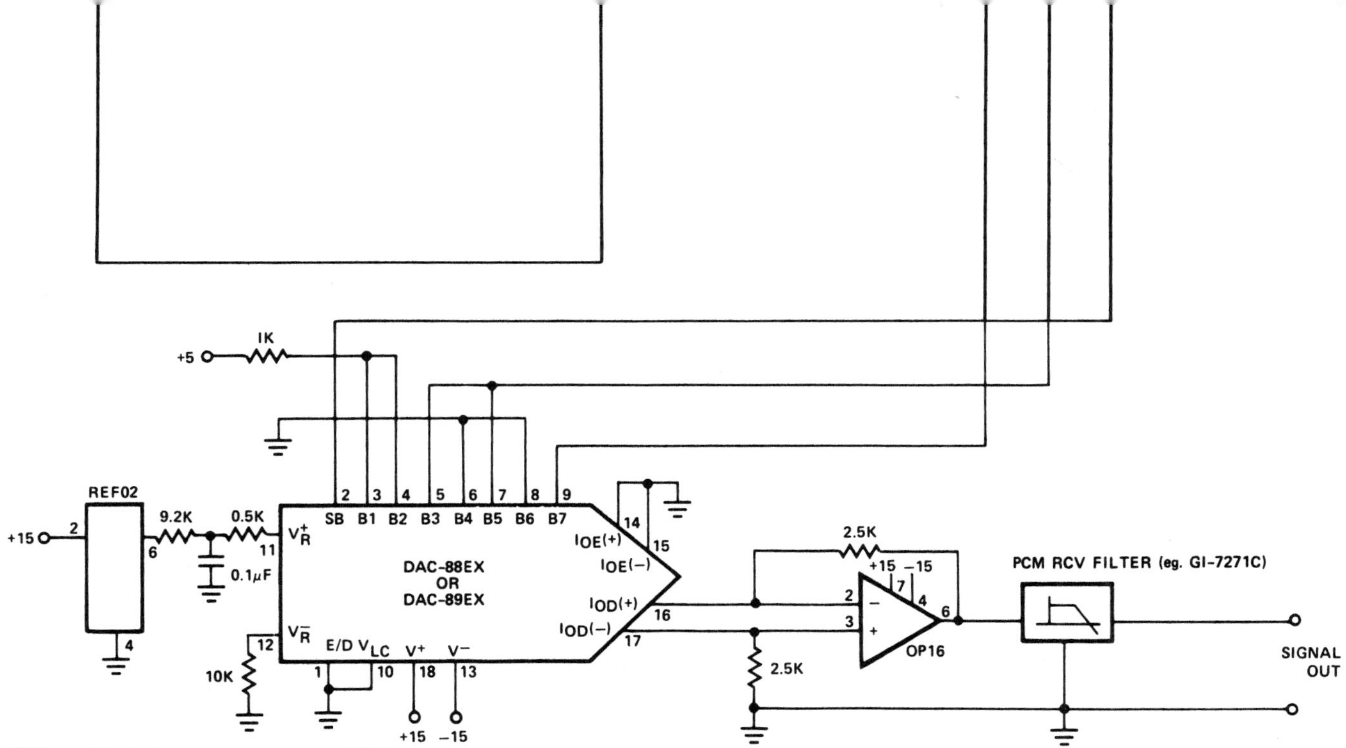

1kHz, 0dBm0 sine-wave generator (PM).

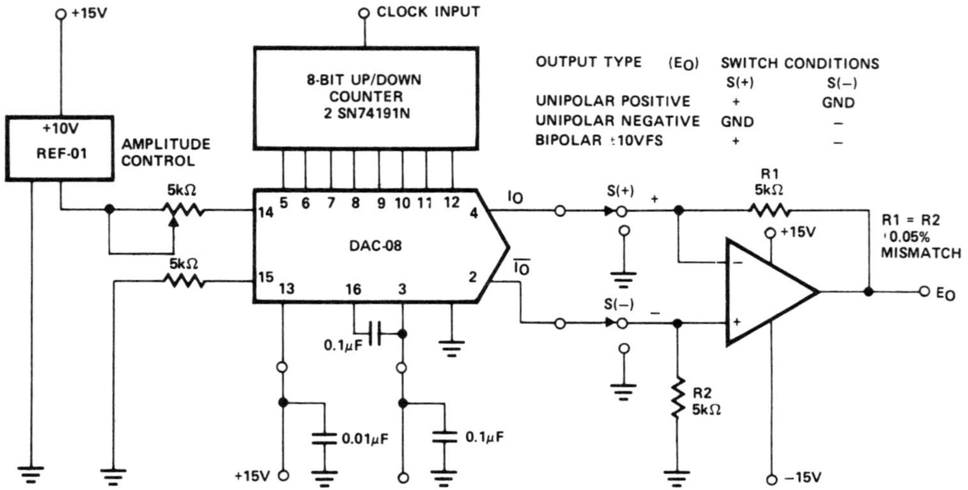

NOTES:

1. BIPOLAR OUTPUT IS SYMMETRICAL AROUND ZERO. ADJUSTABLE PEAK-TO-PEAK AMPLITUDE.
2. FOR TRIANGLE WAVE, COUNT UP TO FULL REVERSE AND COUNT DOWN.
3. FOR POSITIVE-GOING SAWTOOTH, COUNT UP TO FULL CLEAR, REPEAT.
4. FOR NEGATIVE-GOING SAWTOOTH, COUNT DOWN. CLEAR REPEAT.
5. FOR OTHER WAVEFORMS, USE A ROM PROGRAMMED WITH THE DESIRED FUNCTION.
6. 85nSEC SETTLING TIME PERMITS WAVEFORM PERIOD OF 25.6μSEC OR 39kHz REPETITION RATE IN THE UNIPOLAR POSITIVE MODE.

High-speed waveform generator (PM).

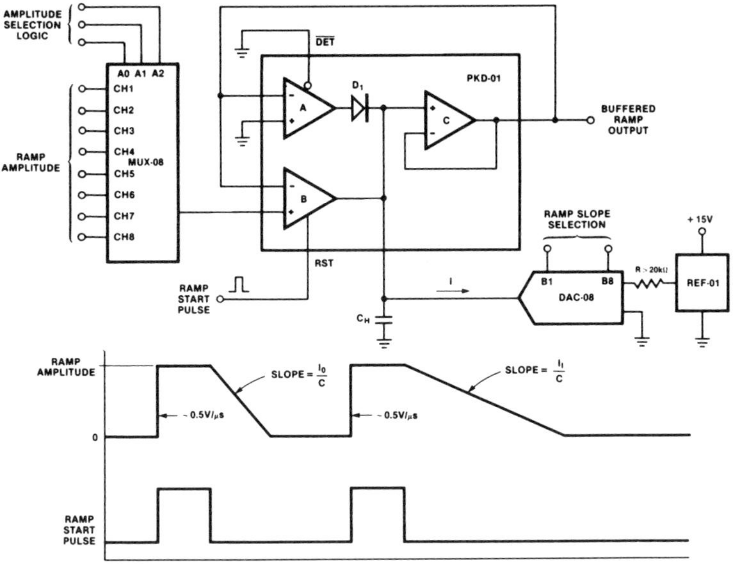

NOTES: 1. NEGATIVE SLOPE OF RAMP IS SET BY DAC-08 OUTPUT CURRENT.
2. DAC-08 IS DIGITALLY CONTROLLED CURRENT GENERATOR. THE MAXIMUM FULL SCALE CURRENT MUST BE LESS THAN 0.5mA.

Programmable low frequency ramp generator (PM).

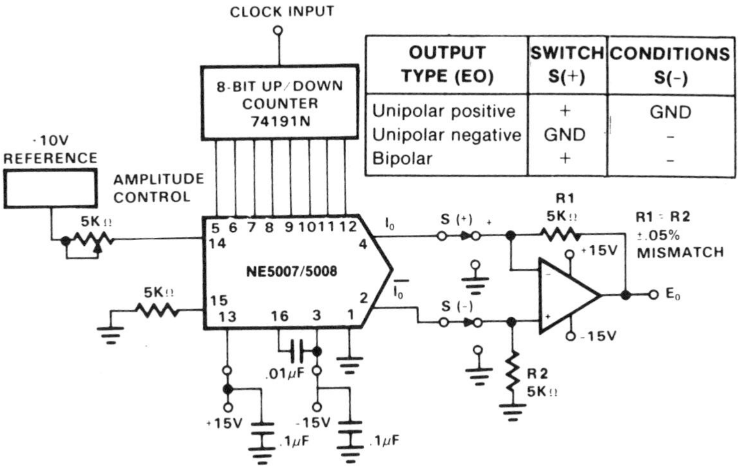

OUTPUT TYPE (EO)	SWITCH S(+)	CONDITIONS S(-)
Unipolar positive	+	GND
Unipolar negative	GND	-
Bipolar	+	-

NOTES

1. Bipolar output is symmetrical around zero, adjustable peak to peak amplitude.
2. For triangle wave, count up to full, reverse and count down.
3. For positive-going sawtooth, count up to full, clear, repeat.
4. For negative-going sawtooth, count down, clear, repeat.
5. For other waveforms, use a ROM programmed with the desired function.

High-speed waveform generator (S).

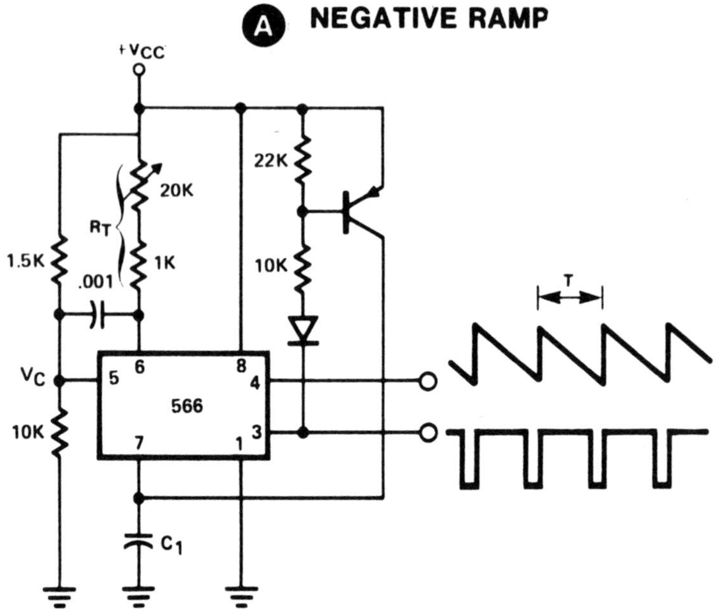

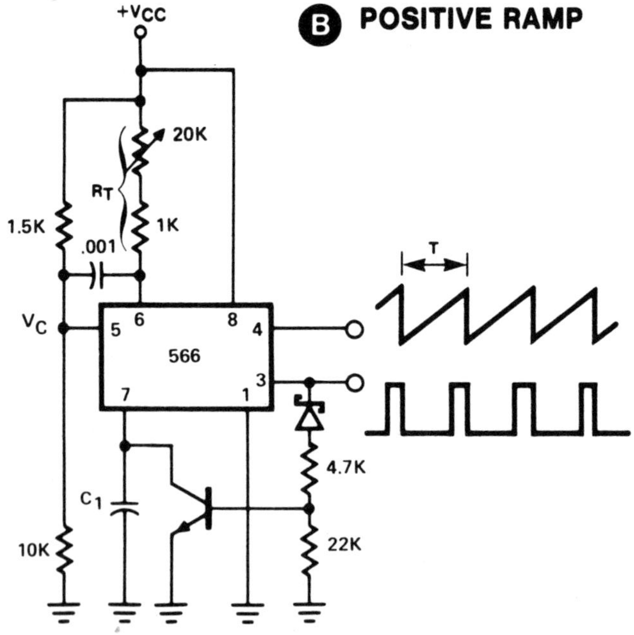

Ramp generators (S).

A POSITIVE SAWTOOTH

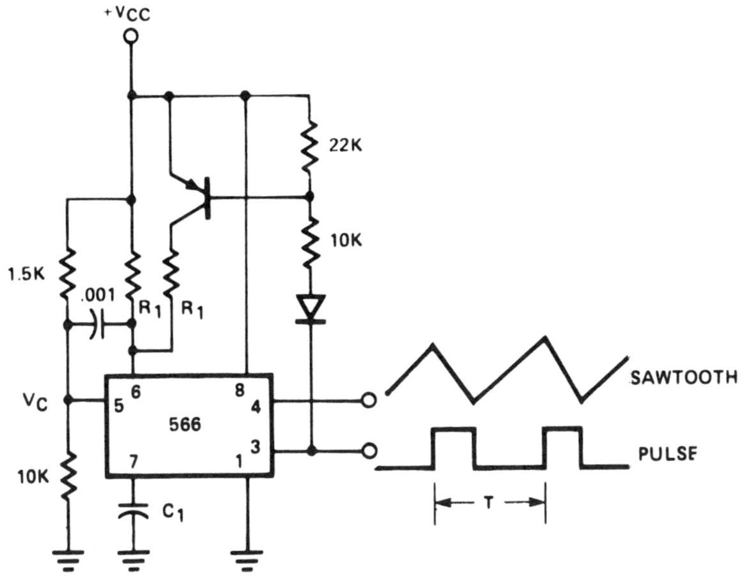

B NEGATIVE SAWTOOTH

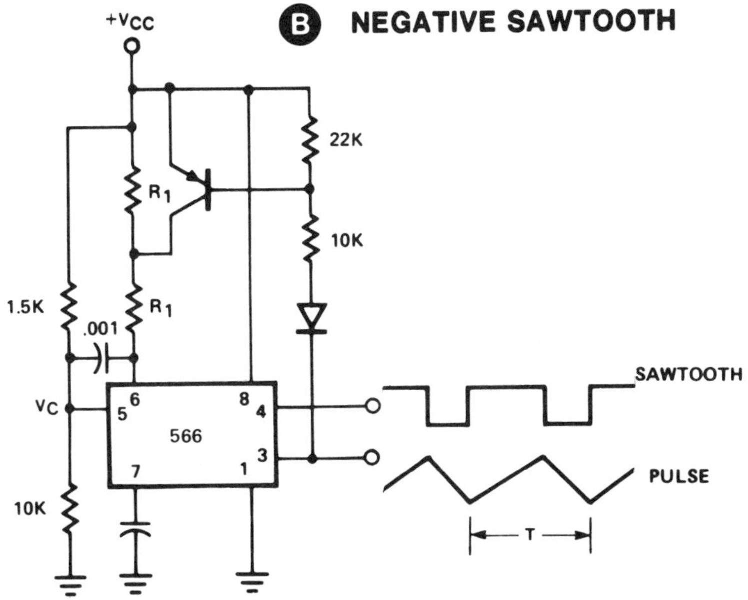

Sawtooth and pulse generators (S).

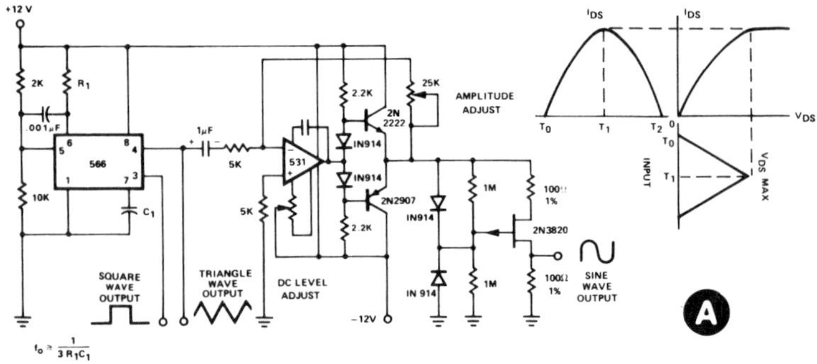

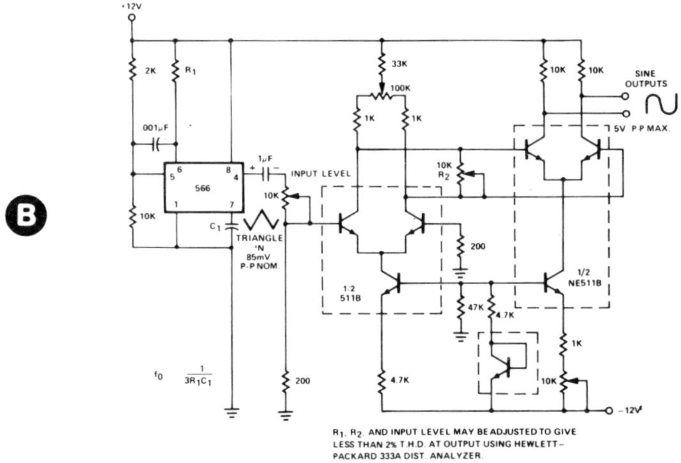

Triangle-to-sine converters (S).

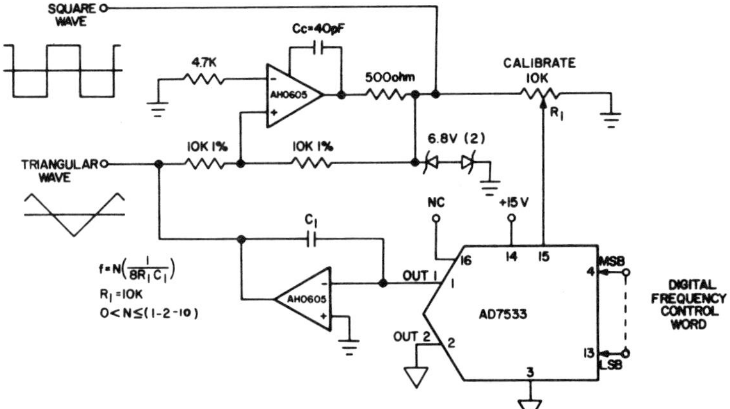

Digitally controllable function generator (OE).

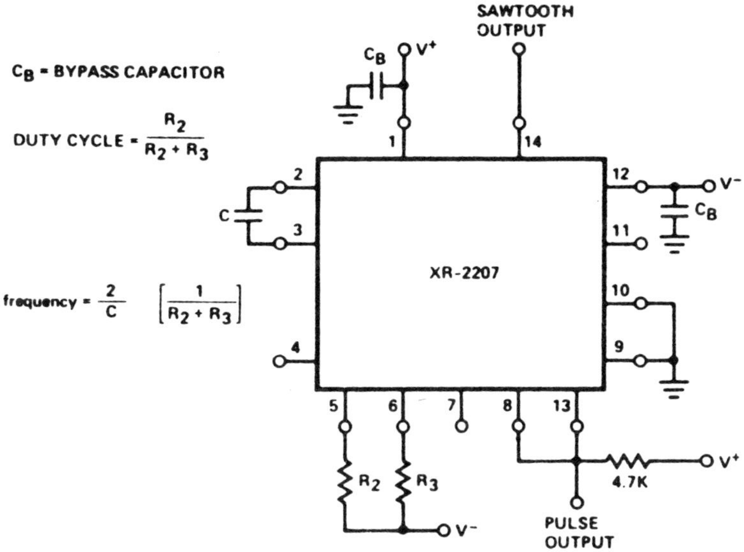

Sawtooth and pulse outputs (E).

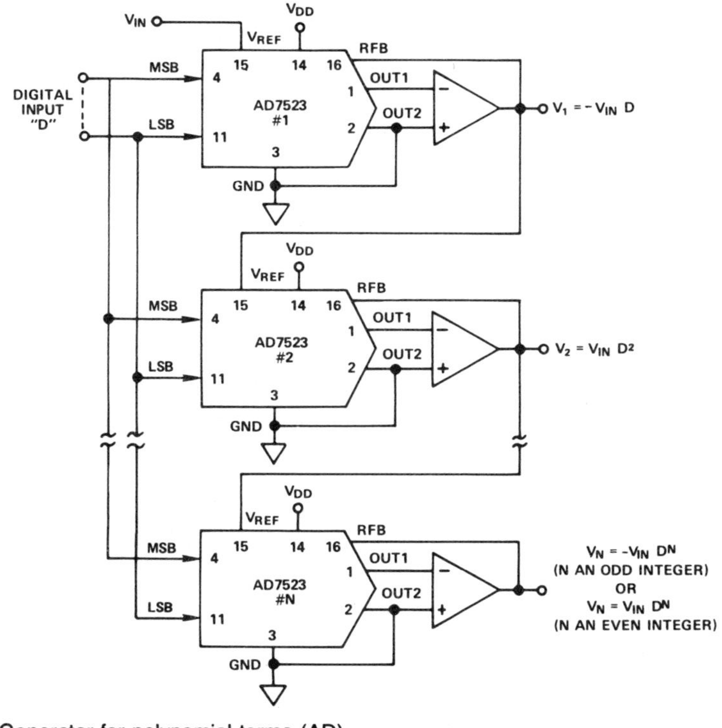

Generator for polynomial terms (AD).

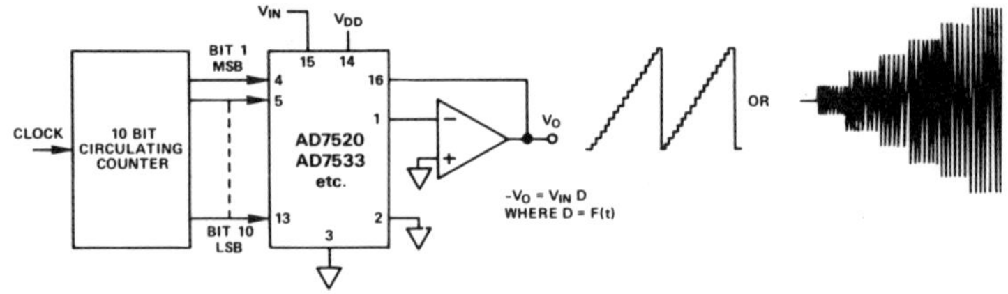

Sawtooth generator (AD).

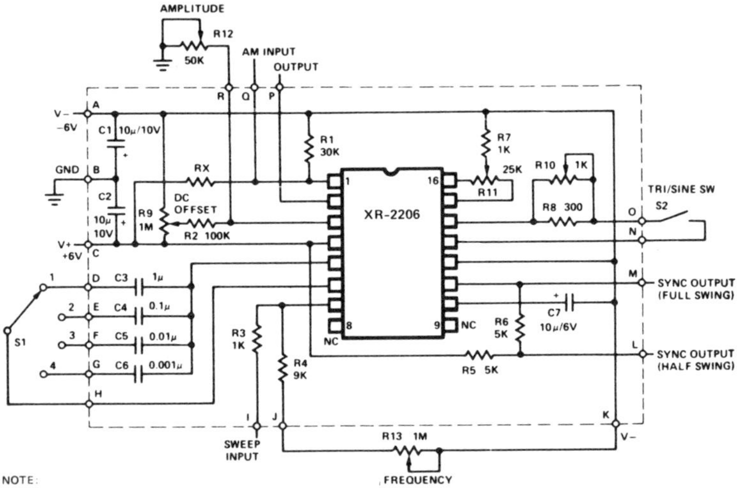

NOTE:

1. For single supply operation lift GND connection keeping R12 across terminals R and B intact, and connect terminal A to GND.

2. For maximum output, R_X may be open. R_X = 68 KΩ is recommended for external amplitude modulation.

Circuit connection diagram for function generator (E).

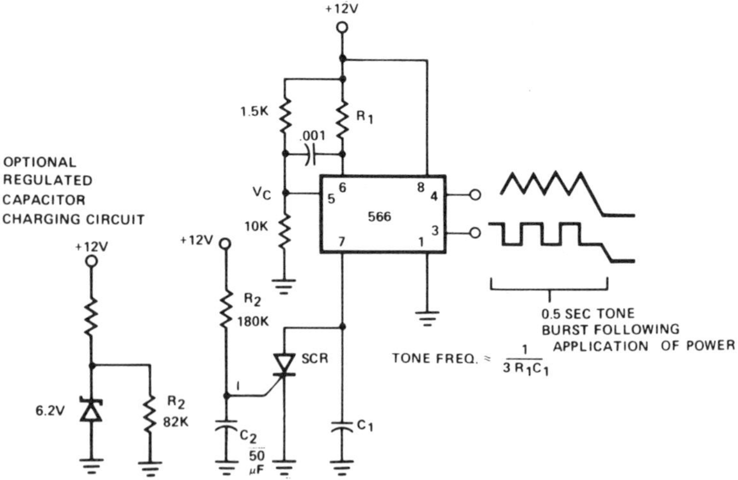

Single-burst tone generator (S).

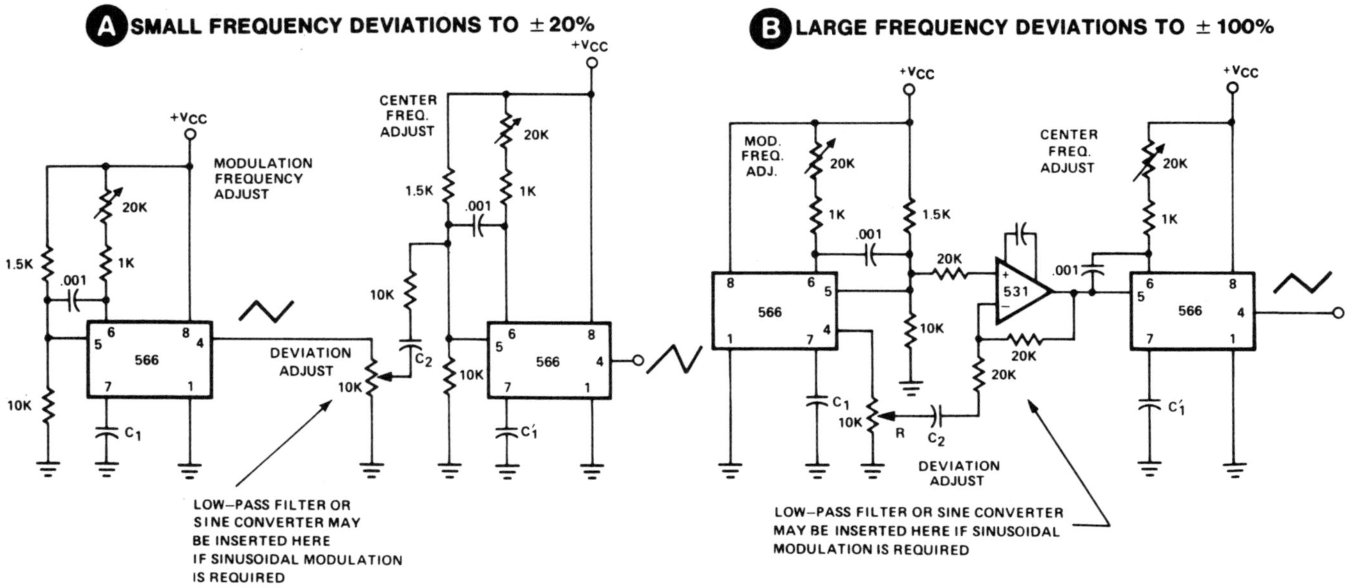

Frequency modulated generator (S).

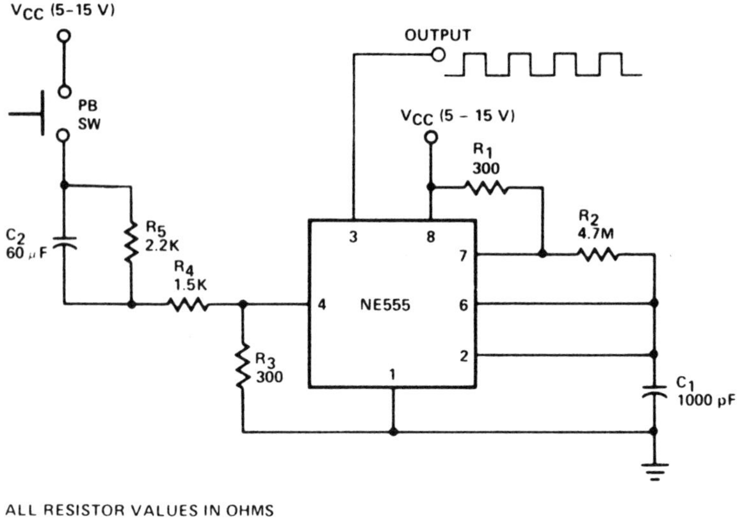

Squarewave tone burst generator (S).

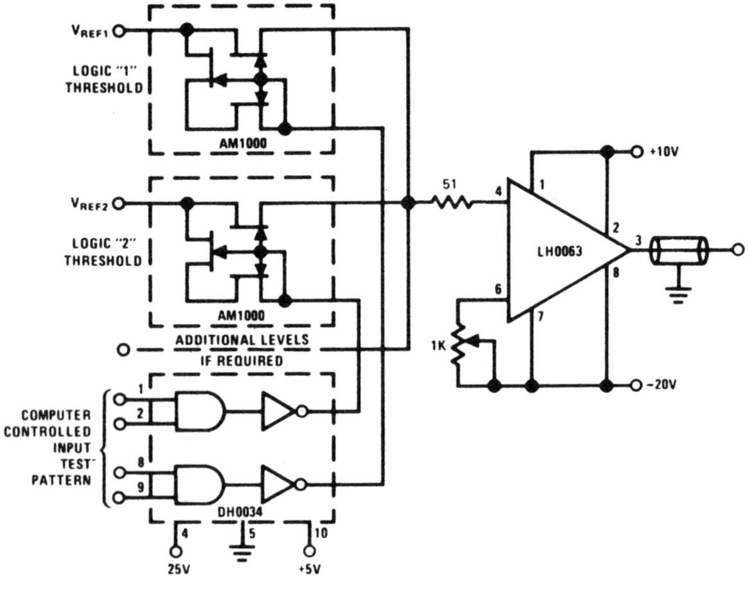

High-speed automatic test equipment forcing function generator (NS).

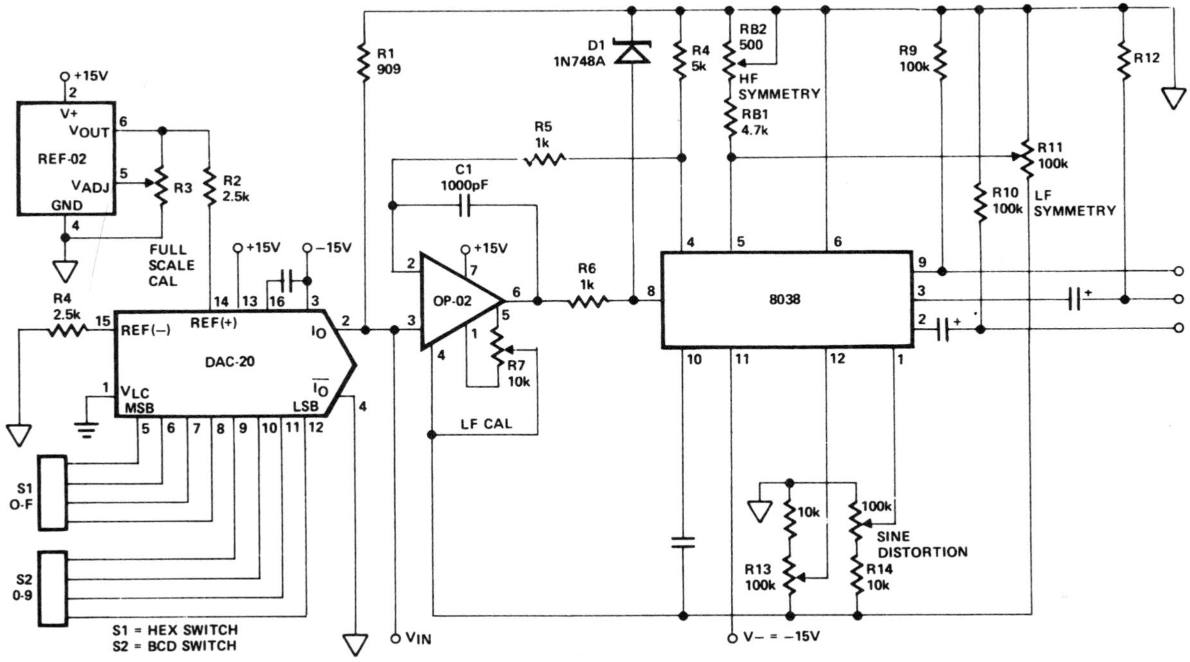

Complete circuit diagram for programmed function generator (PM).

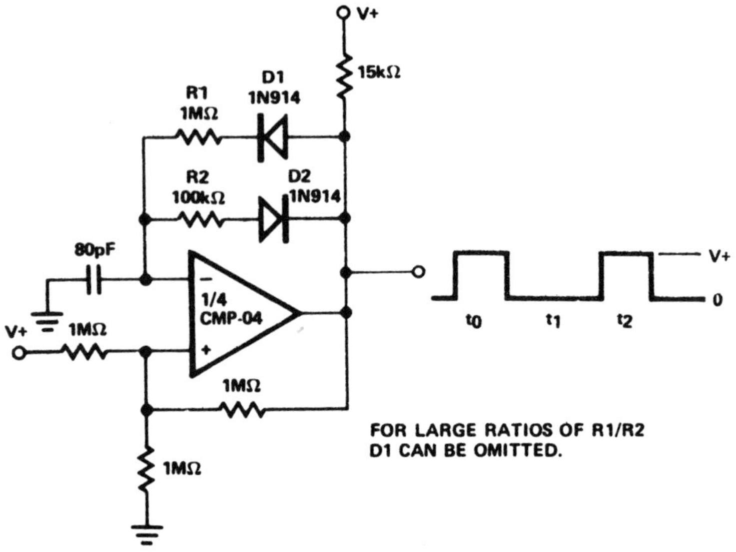

Pulse generator (PM).

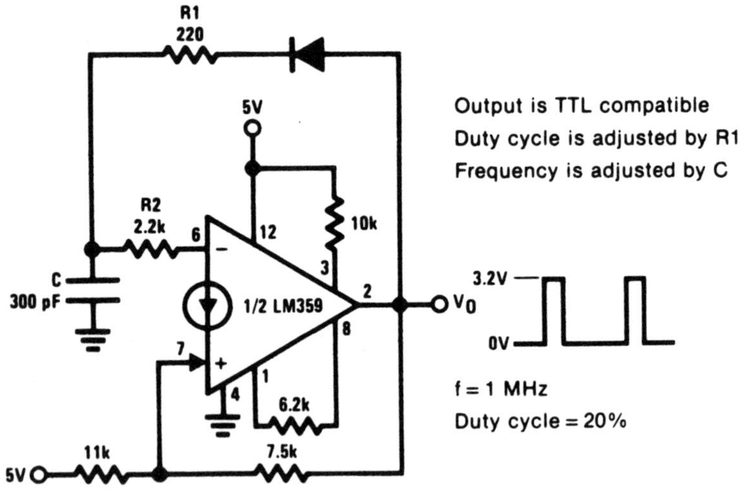

Pulse generator (NS).

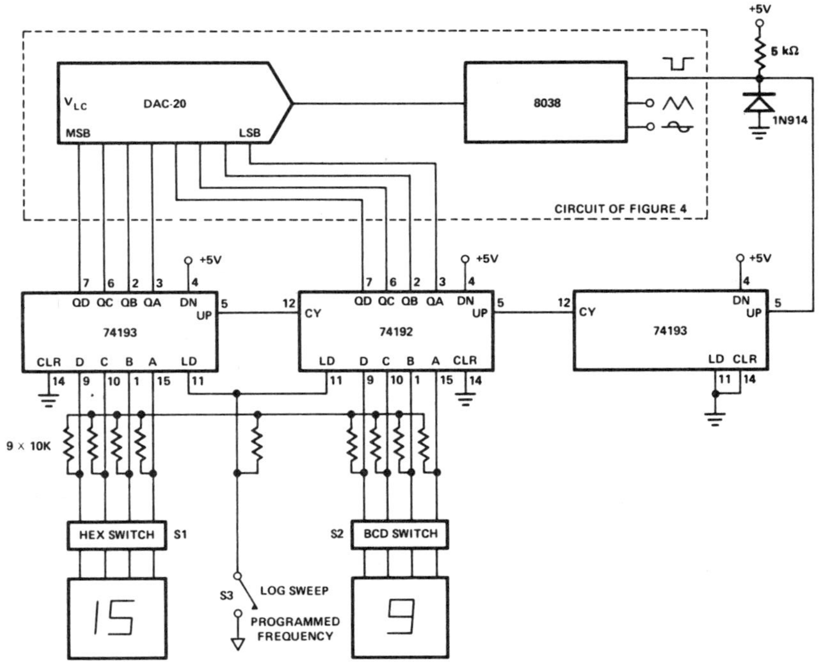

Log sweep modification of programmable function generator (PM).

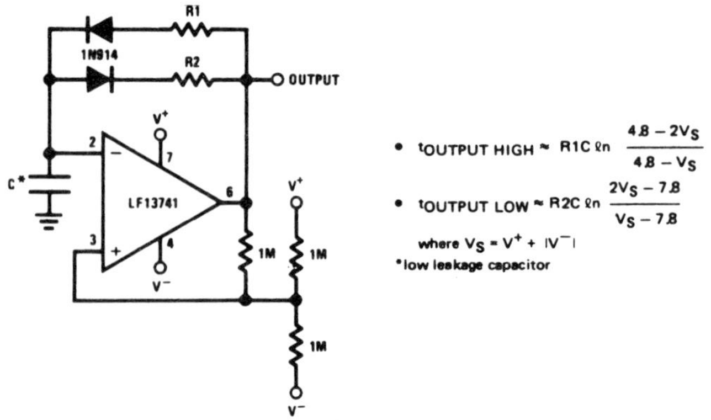

Ultra-low (or high) duty cycle pulse generator (NS).

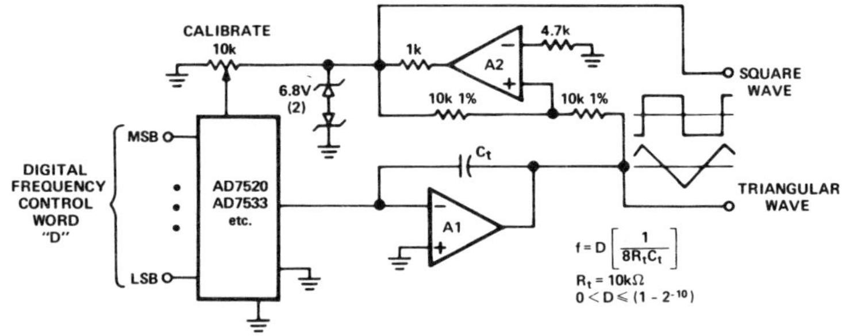

Programmable function generator; p-p amplitude of both waveforms is ≅ 15V (AD).

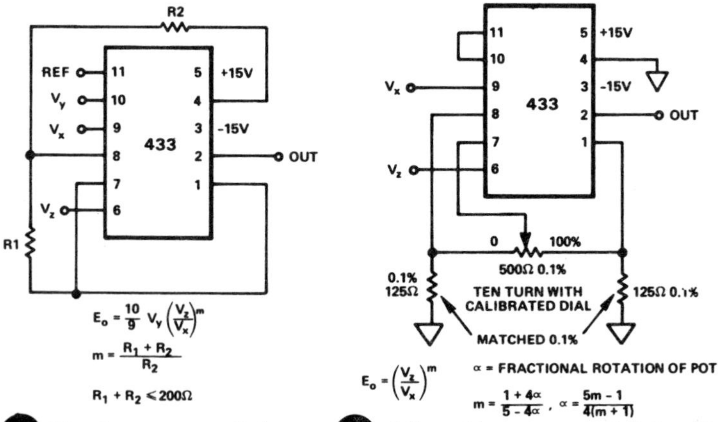

A *Fixed exponent, m>1* **B** *Adjustable exponent, 1/5<m<5*

The 433 as an arbitrary fixed power generator or adjustable power/root generator (AD).

Chapter 8

Clock Circuits

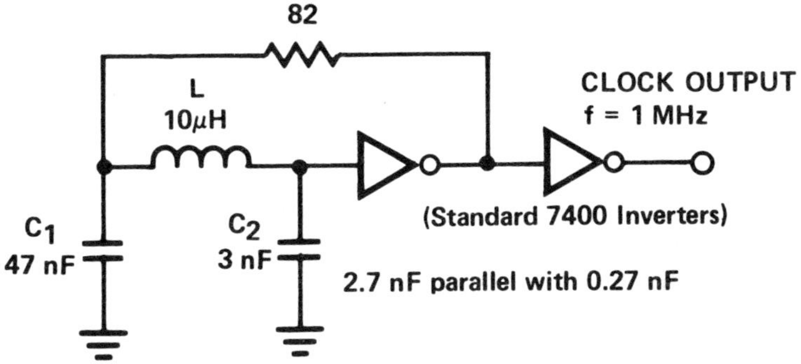

LC clock generator (S).

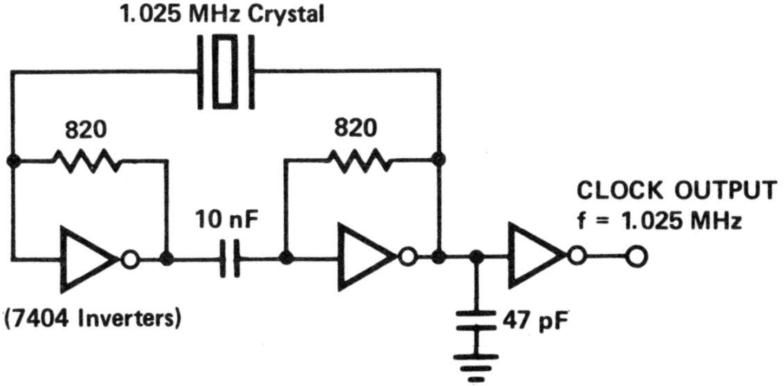

Clock generator using a non-TV standard crystal (S).

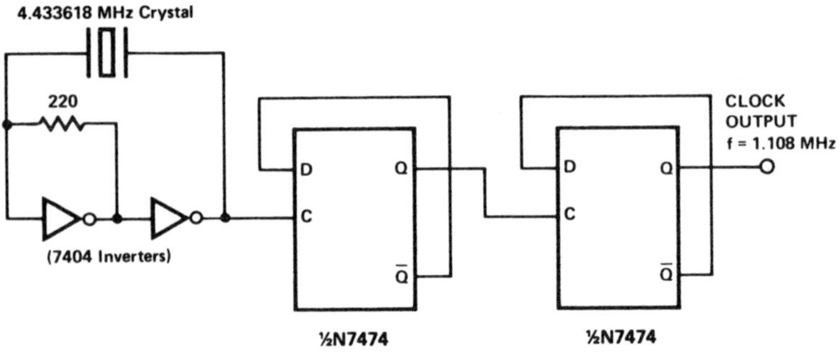

Low-cost color TV crystal clock generator (S).

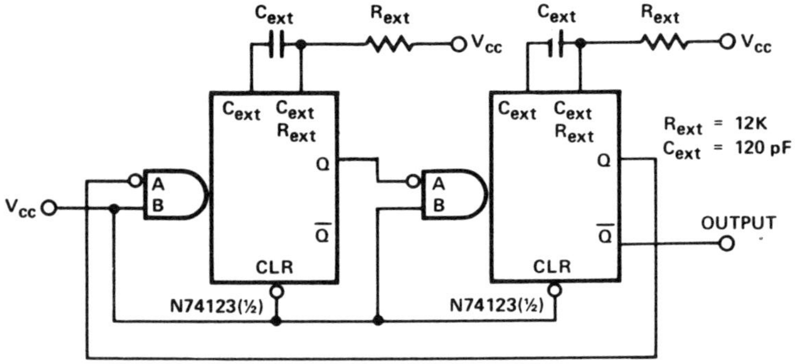

RC clock generator with monostable circuit N74123 (S).

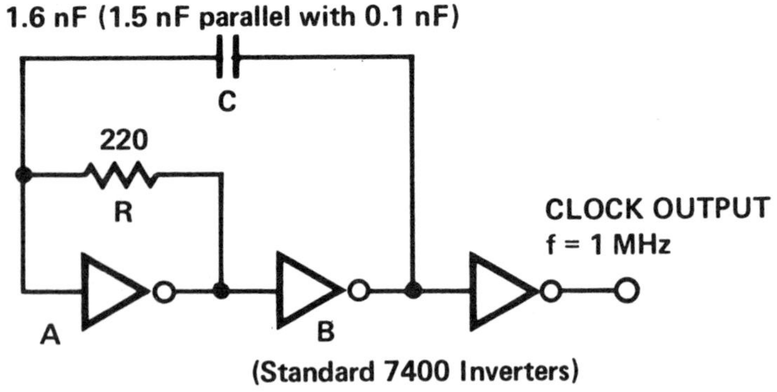

RC clock generator (S).

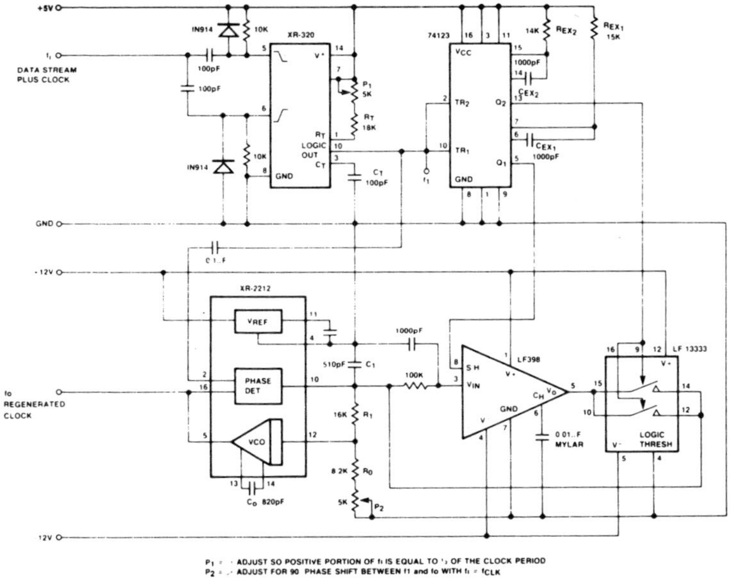

Complete clock regenerator (E).

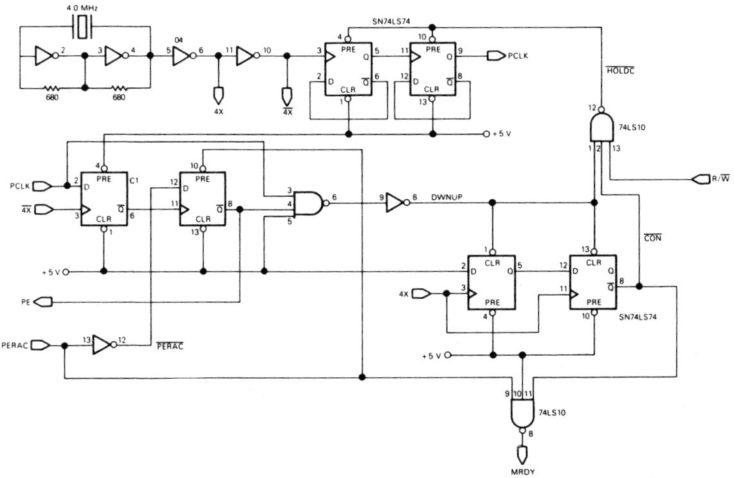

MC6809 half-speed peripheral clock circuit (M).

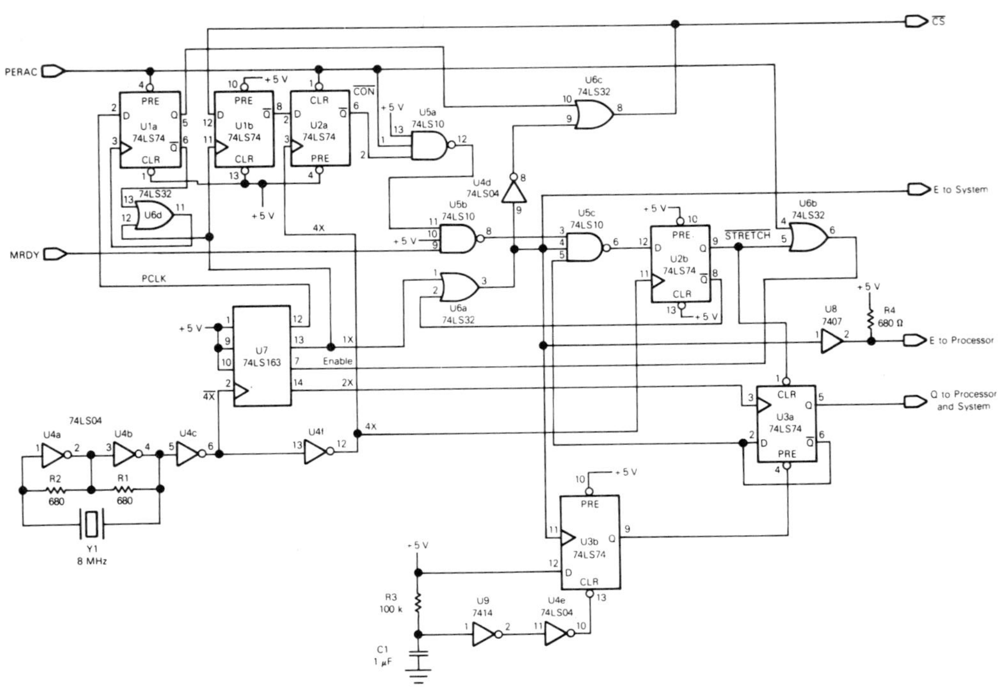

Clock generator circuit for MC6809E (M).

Chapter 9

Analog to Digital and Digital to Analog

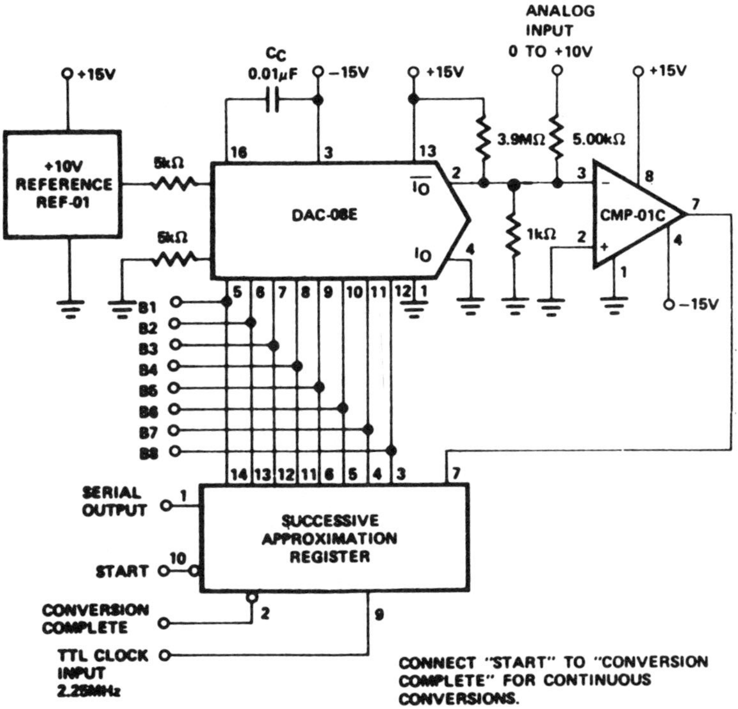

3 IC low cost A/D converter (PM).

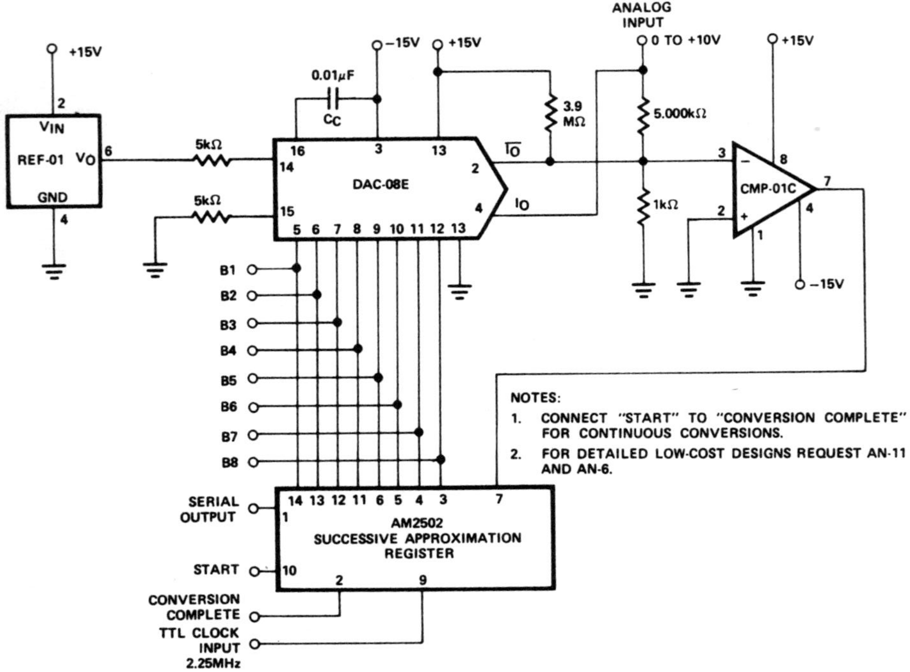

Four IC low-cost A/D converter (PM).

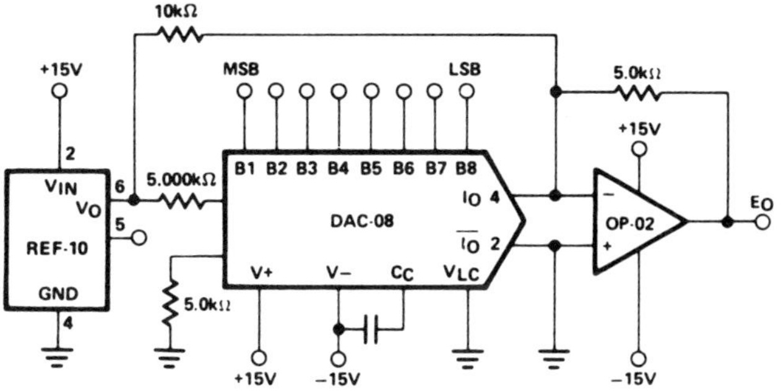

	B1	B2	B3	B4	B5	B6	B7	B8	E
POS FULL SCALE −1 LSB	1	1	1	1	1	1	1	1	+4.960
ZERO SCALE	1	0	0	0	0	0	0	0	0.000
NEG FULL SCALE +1 LSB	0	0	0	0	0	0	0	1	−4.960
NEG FULL SCALE	0	0	0	0	0	0	0	0	−5.000

D/A converter reference (PM).

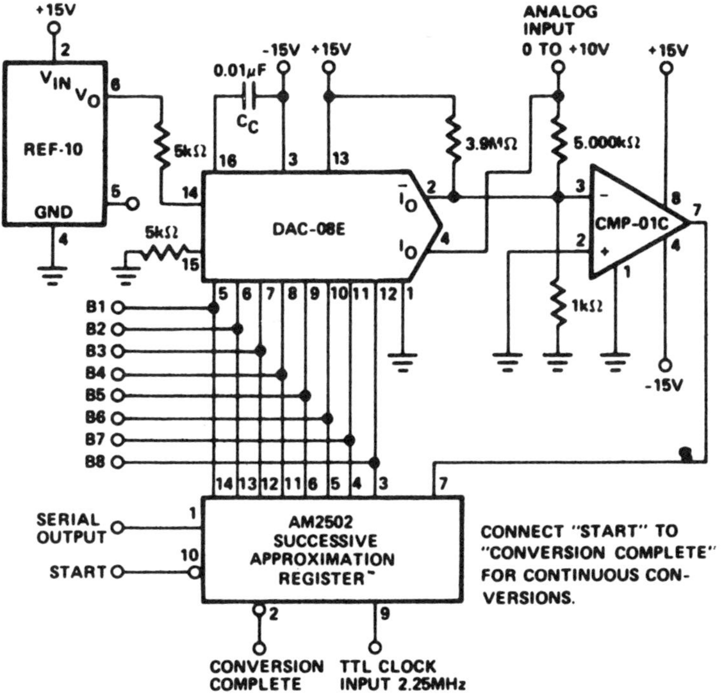

A/D converter reference (PM).

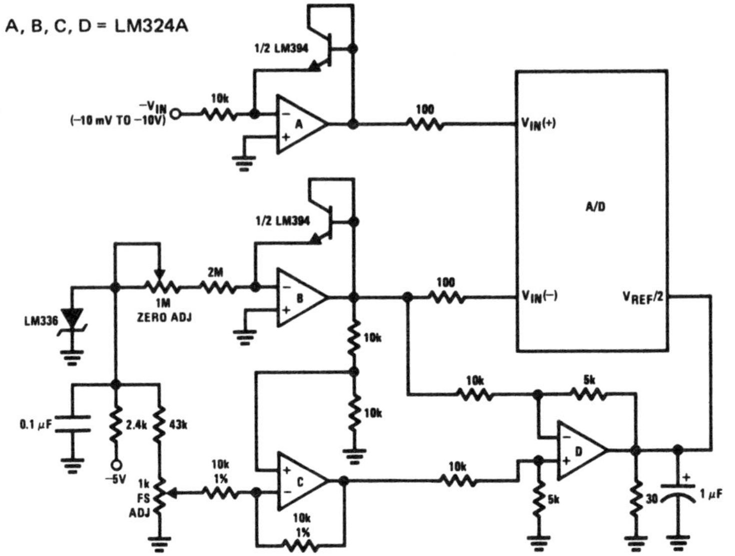

3-decade logarithmic A/D converter (NS).

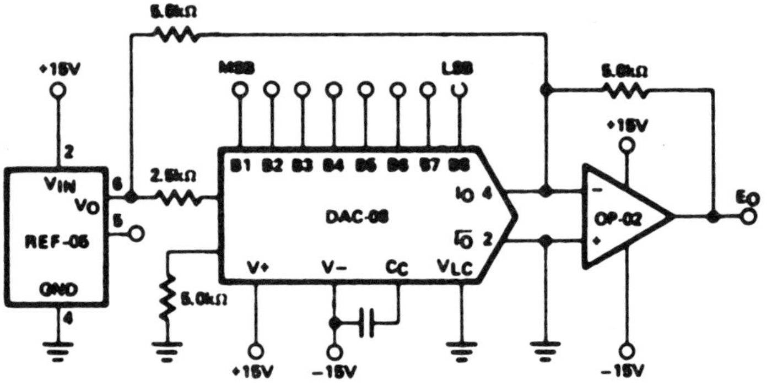

POS FULL SCALE −1 LSB	B1	B2	B3	B4	B5	B6	B7	B8	E
ZERO SCALE	1	1	1	1	1	1	1	1	+4.960
NEG FULL SCALE +1 LSB	1	0	0	0	0	0	0	0	0.000
NEG FULL SCALE	0	0	0	0	0	0	0	1	−4.960
	0	0	0	0	0	0	0	0	−5.000

D/A converter reference (PM).

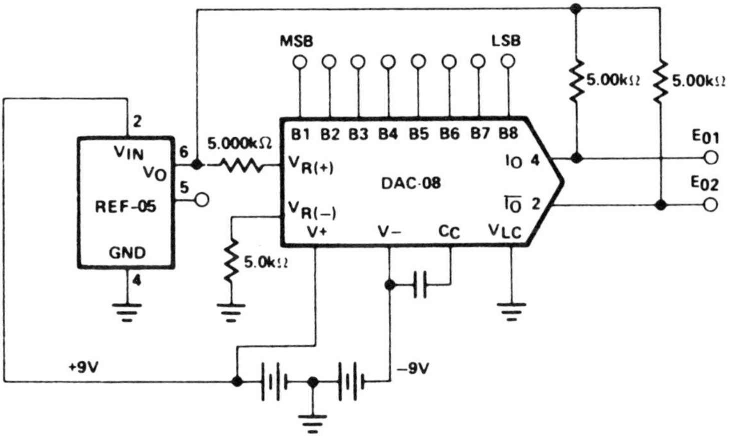

Battery-operated D/A converter reference (PM).

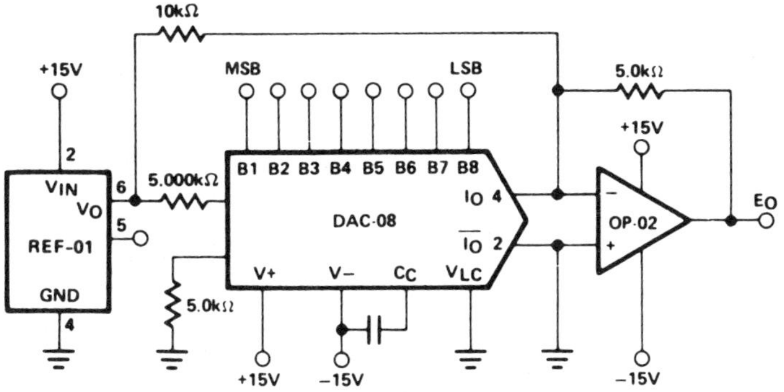

	B1	B2	B3	B4	B5	B6	B7	B8	E
POS FULL SCALE −1 LSB	1	1	1	1	1	1	1	1	+4.960
ZERO SCALE	1	0	0	0	0	0	0	0	0.000
NEG FULL SCALE +1 LSB	0	0	0	0	0	0	0	1	−4.960
NEG FULL SCALE	0	0	0	0	0	0	0	0	−5.000

D/A converter reference (PM).

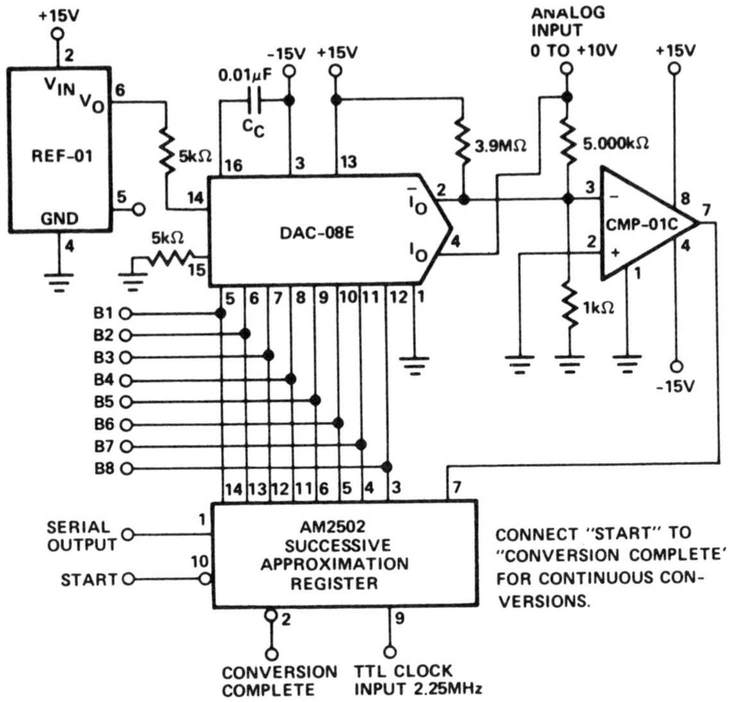

A/D converter reference (PM).

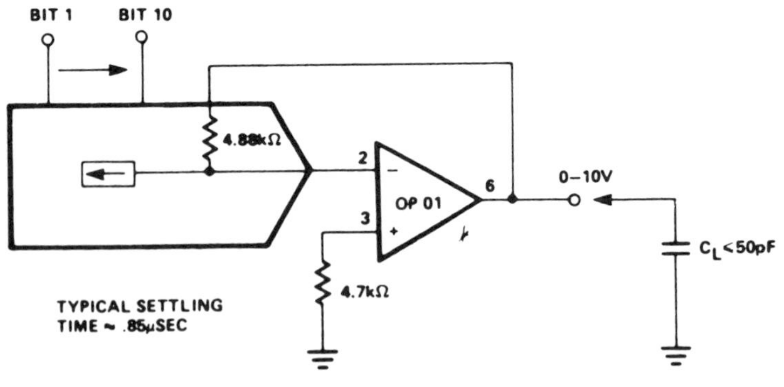

Fast voltage output D/A converter (PM).

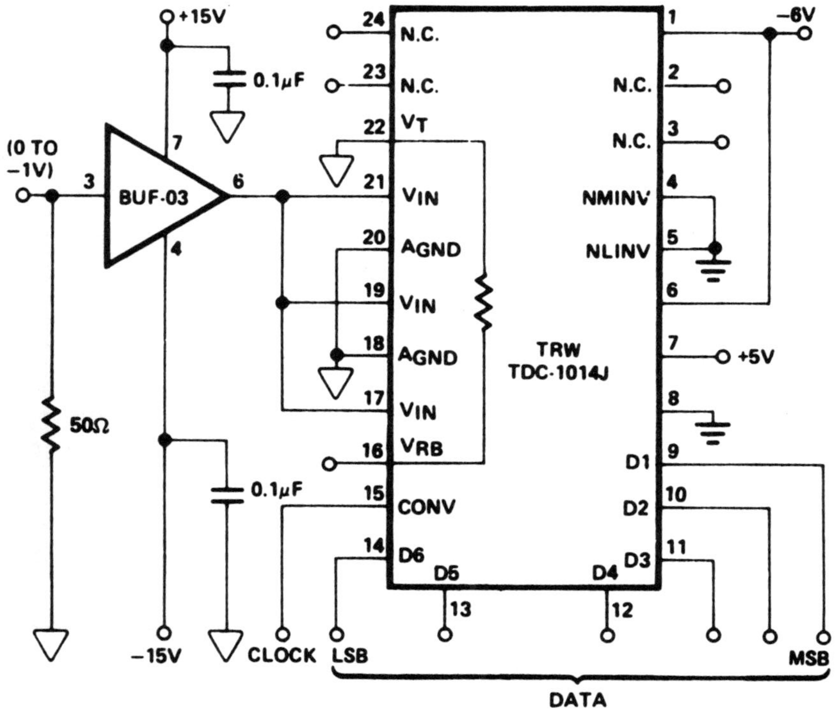

High speed 6-bit A/D buffer (PM).

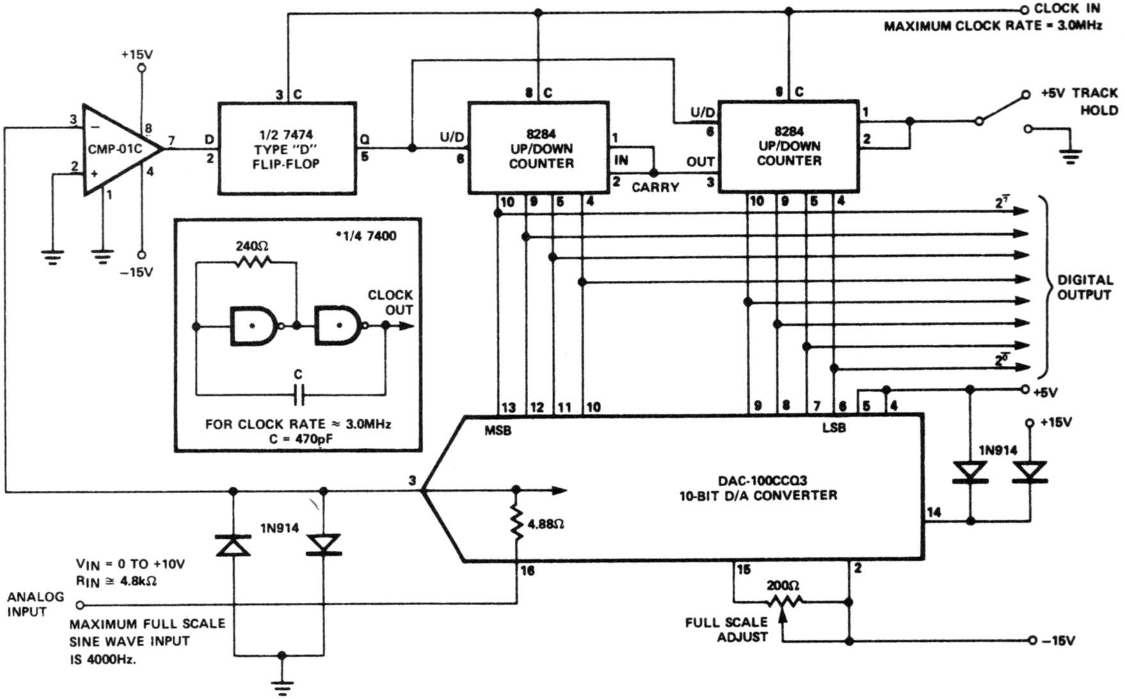

8-bit tracking A/D converter (PM).

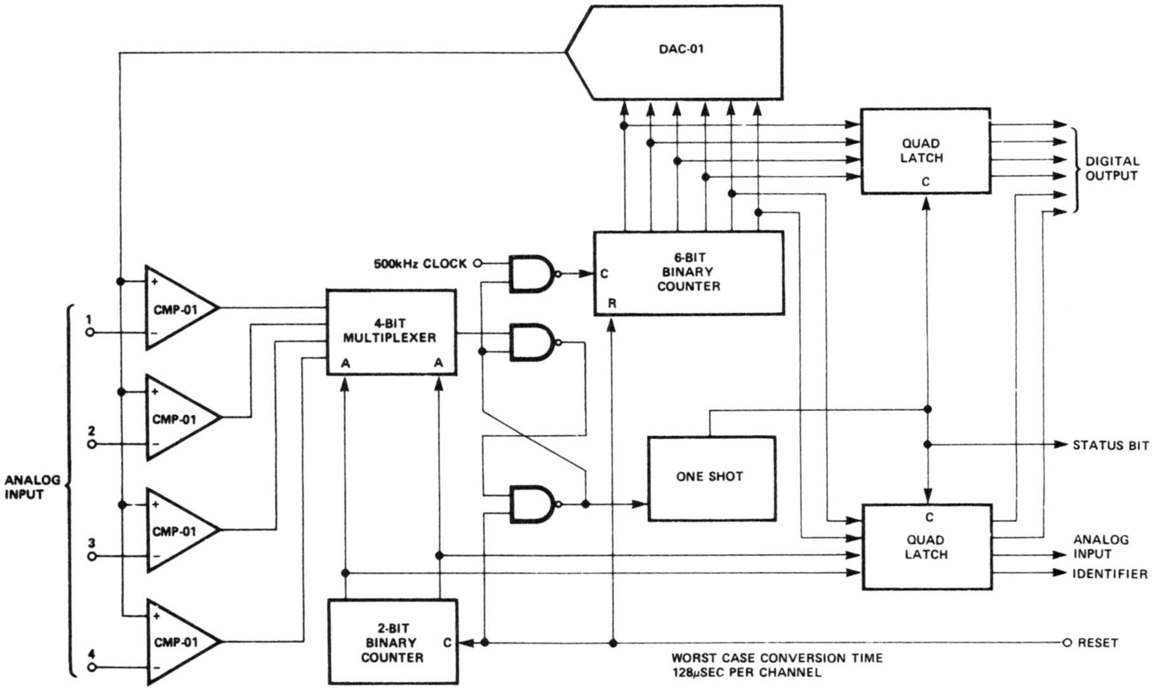

4-channel digitally multiplexed ramped A/D converter (PM).

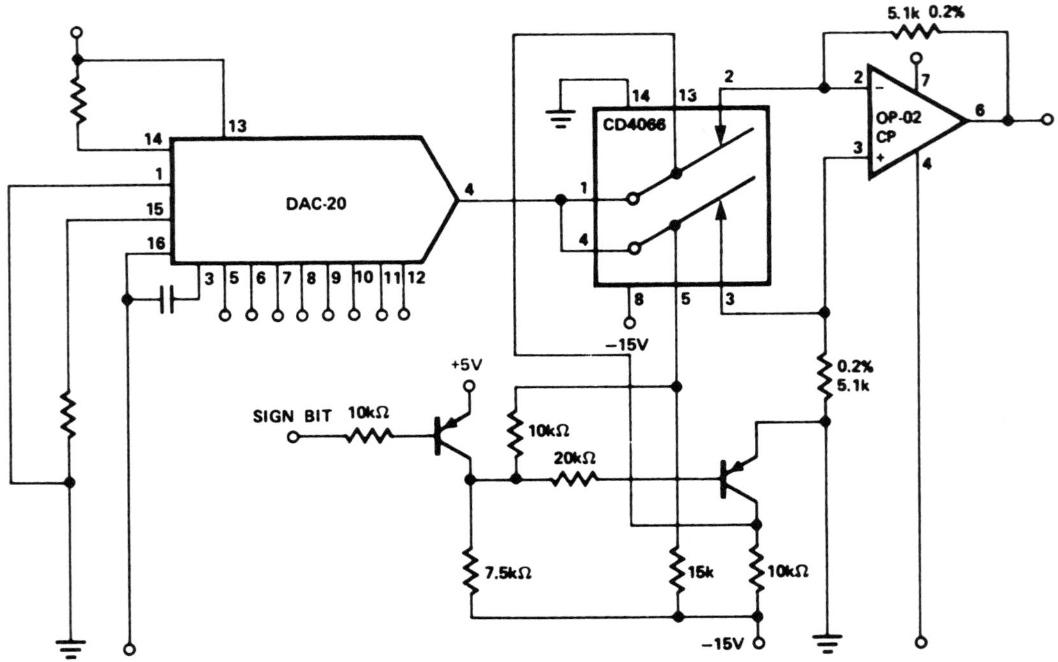

Sign-magnitude BCD DAC (PM).

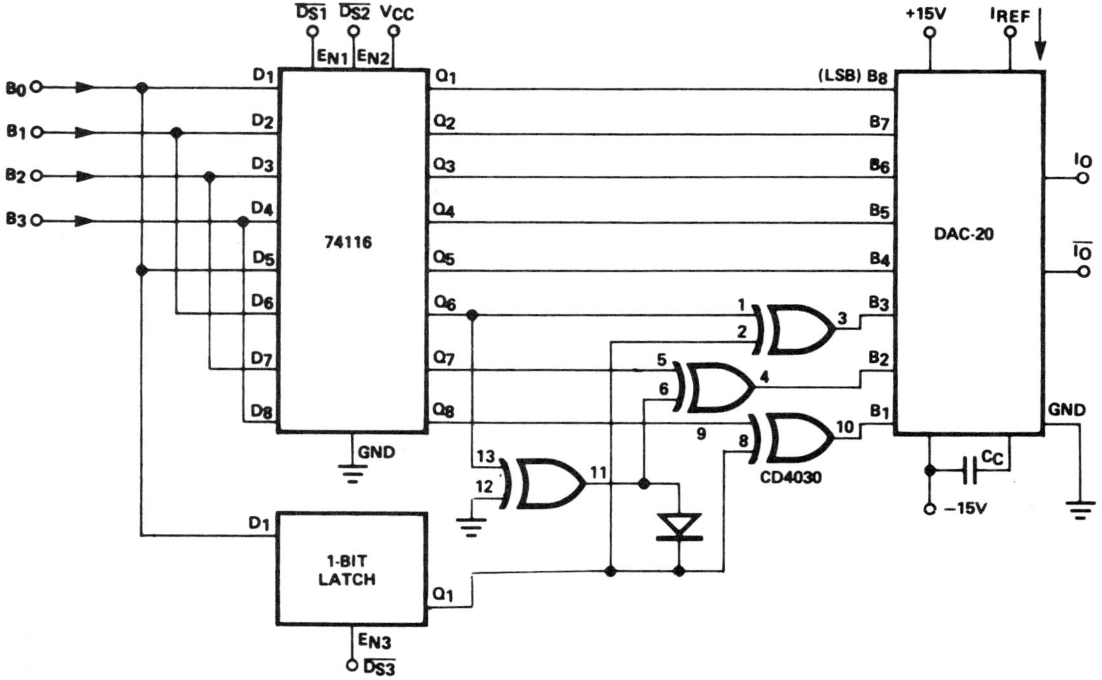

"2-½" digit operation (159 count) (PM).

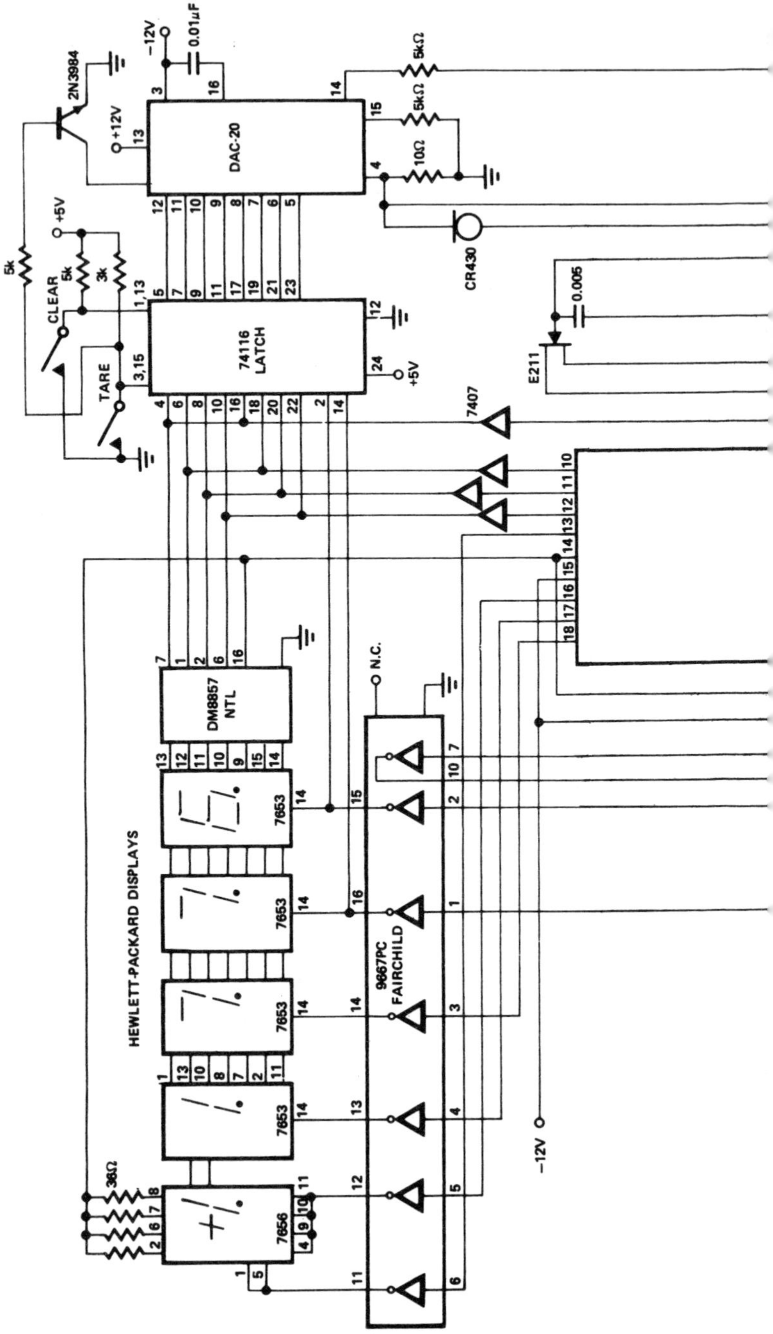

2N3984
−12V
0.01μF
+12V
DAC-20
5kΩ
5kΩ
10Ω
CR430
+5V
5k
5k
3k
CLEAR
TARE
74116
LATCH
+5V
0.005
E211
7407
DM8857
NTL
N.C.
HEWLETT-PACKARD DISPLAYS
7653
7653
7653
7653
7656
36Ω
9667PC
FAIRCHILD
−12V

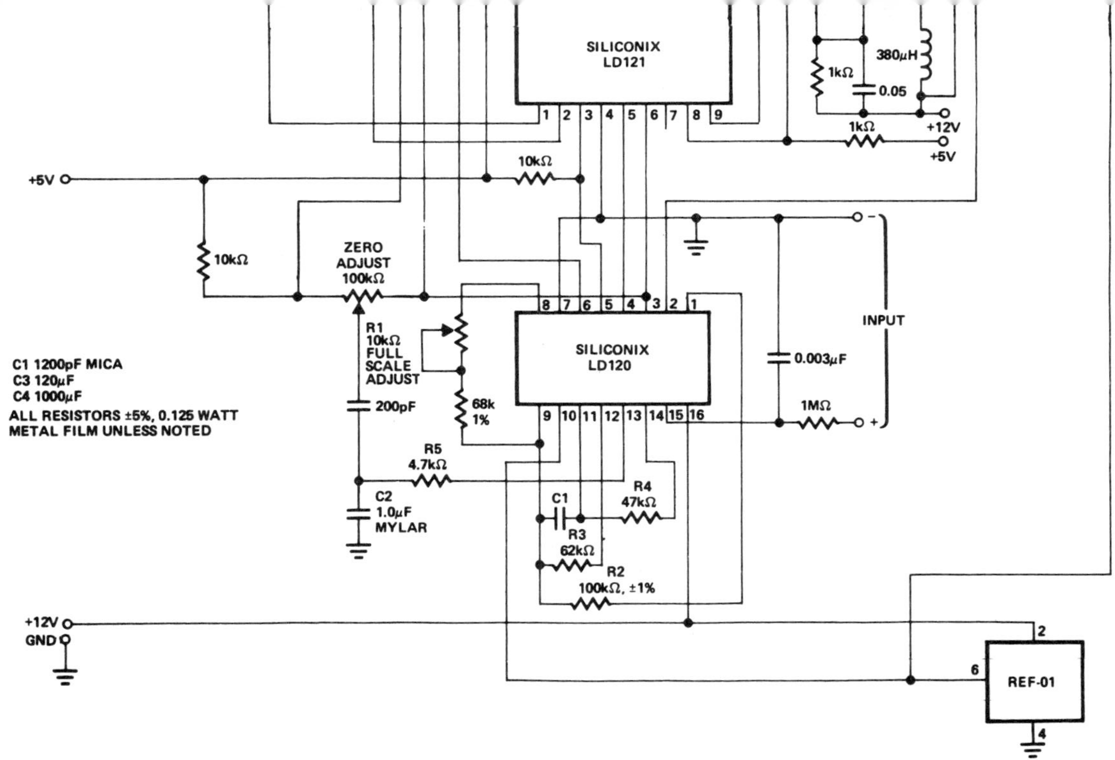

Automatic nulling DPM (PM).

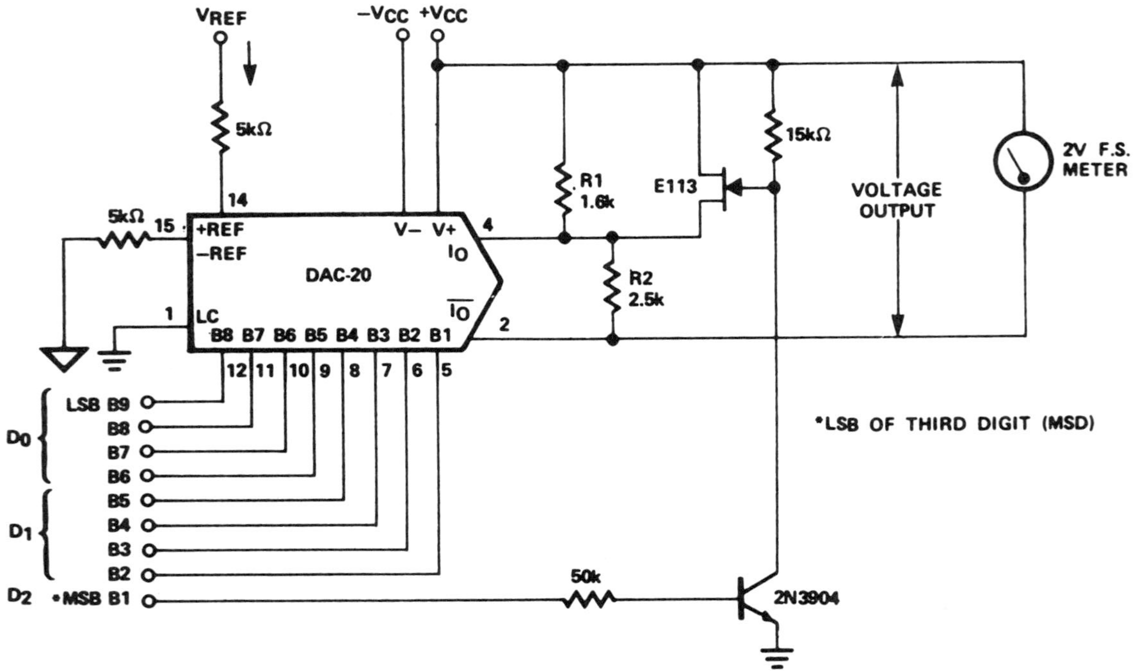

"2-½" digit operation (199 count) (PM).

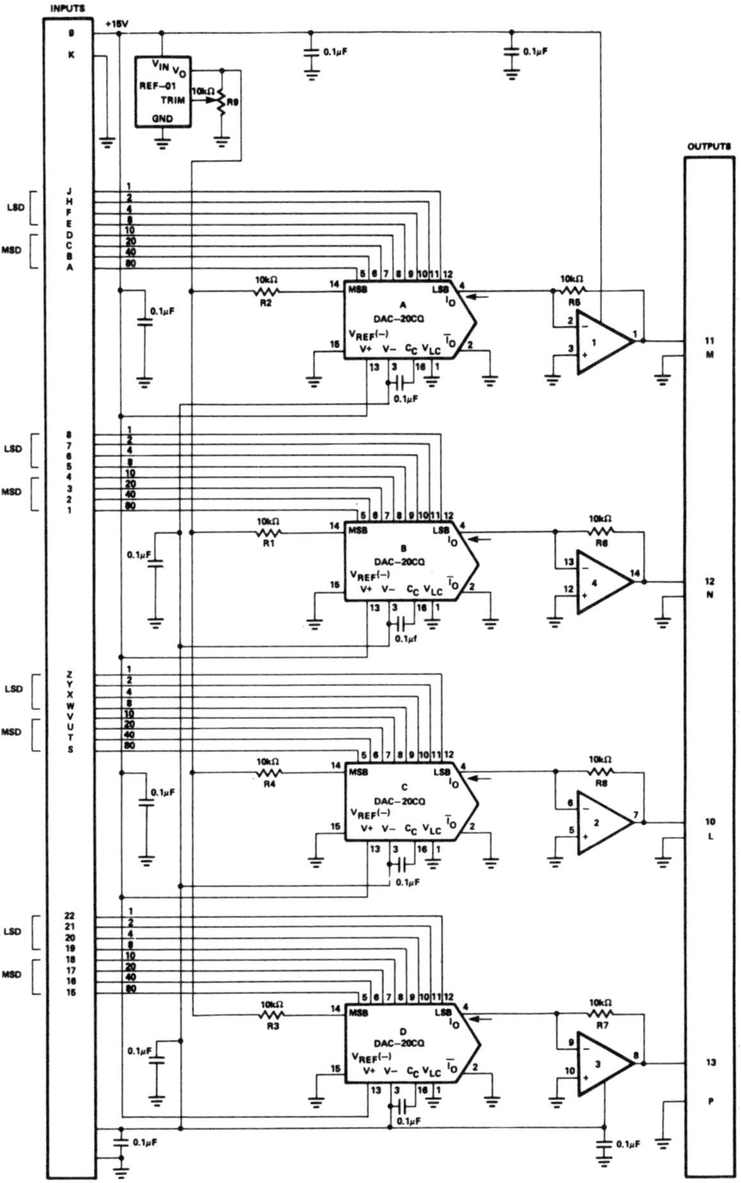

Four-channel BCD DAC complete schematic (PM).

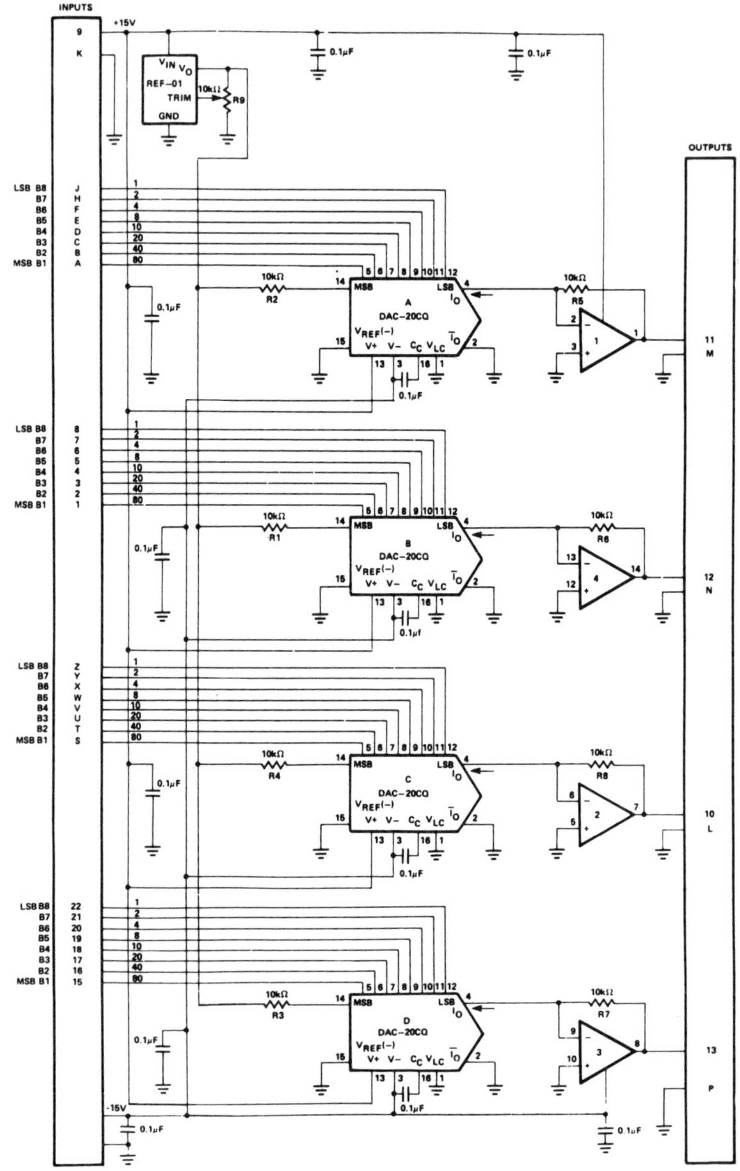

Four-channel binary DAC complete schematic (PM).

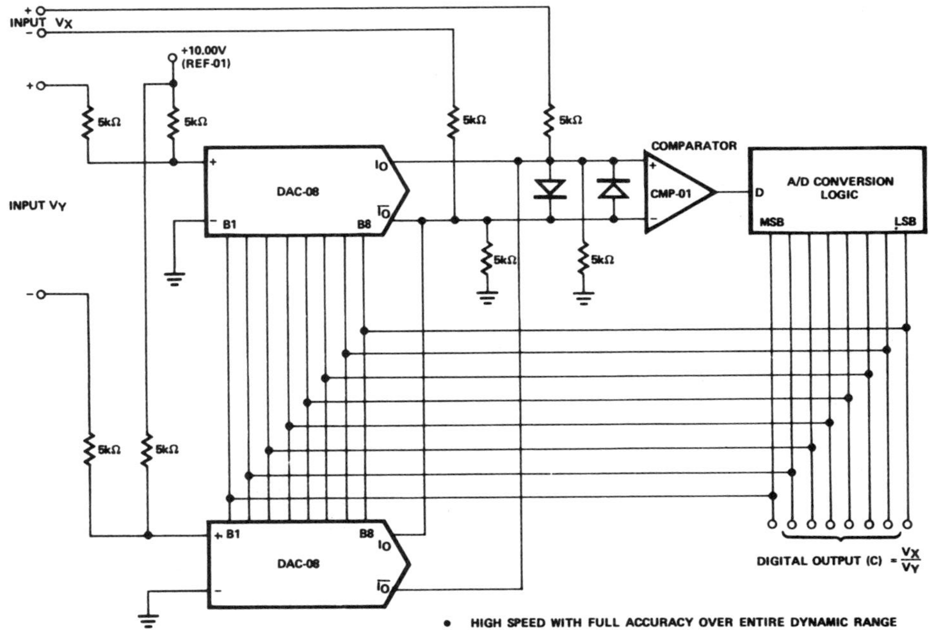

- HIGH SPEED WITH FULL ACCURACY OVER ENTIRE DYNAMIC RANGE
- ±5V DIFFERENTIAL INPUTS WITH ±5V COMMON MODE RANGE
- DIODES ARE 1N4148'S

Four-quadrant ratiometric A/D conversion basic connections (PM).

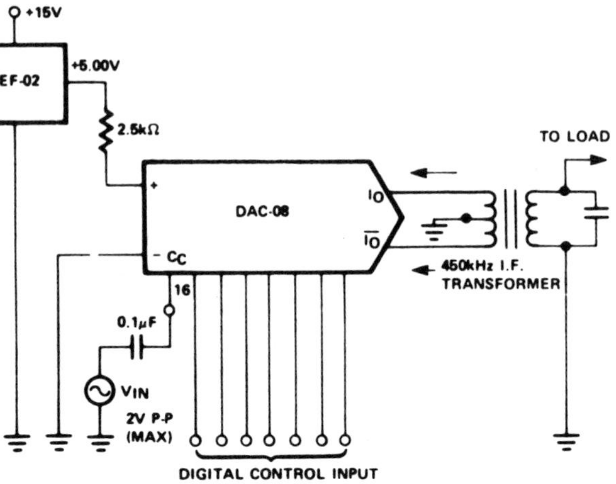

High input impedance ac-coupled multiplication (I.F. amplifier/digital attenuator) (PM).

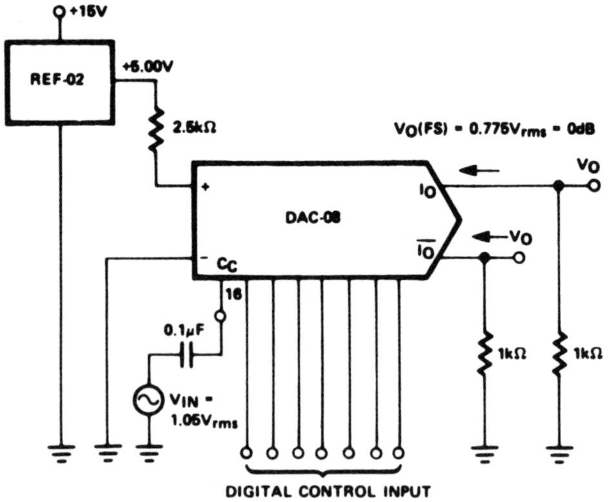

High input impedance ac-coupled multiplication (audio frequency digital attenuator) (PM).

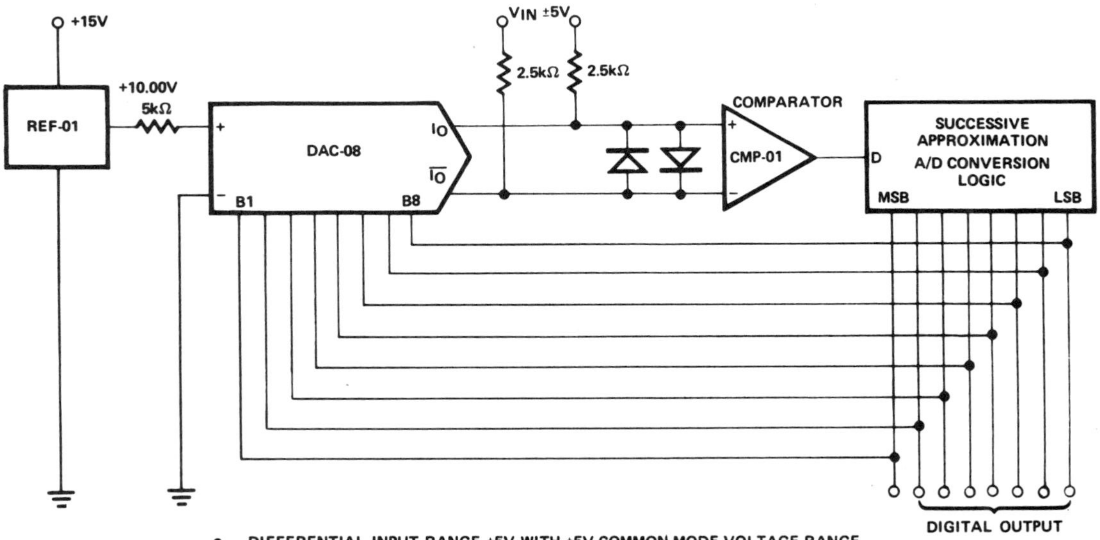

- DIFFERENTIAL INPUT RANGE ±5V WITH ±5V COMMON MODE VOLTAGE RANGE
- USE 5kΩ INPUT RESISTORS FOR DIFFERENTIAL INPUT RANGE OF ±10V WITH 0V COMMON MODE VOLTAGE RANGE
- USE OP-04 OR OP-14 DUAL MATCHED OP AMPS CONNECTED AS VOLTAGE FOLLOWERS IF HIGH INPUT IMPEDANCE IS REQUIRED
- DIODES ARE 1N4148'S

Differential input A/D conversion basic connections (PM).

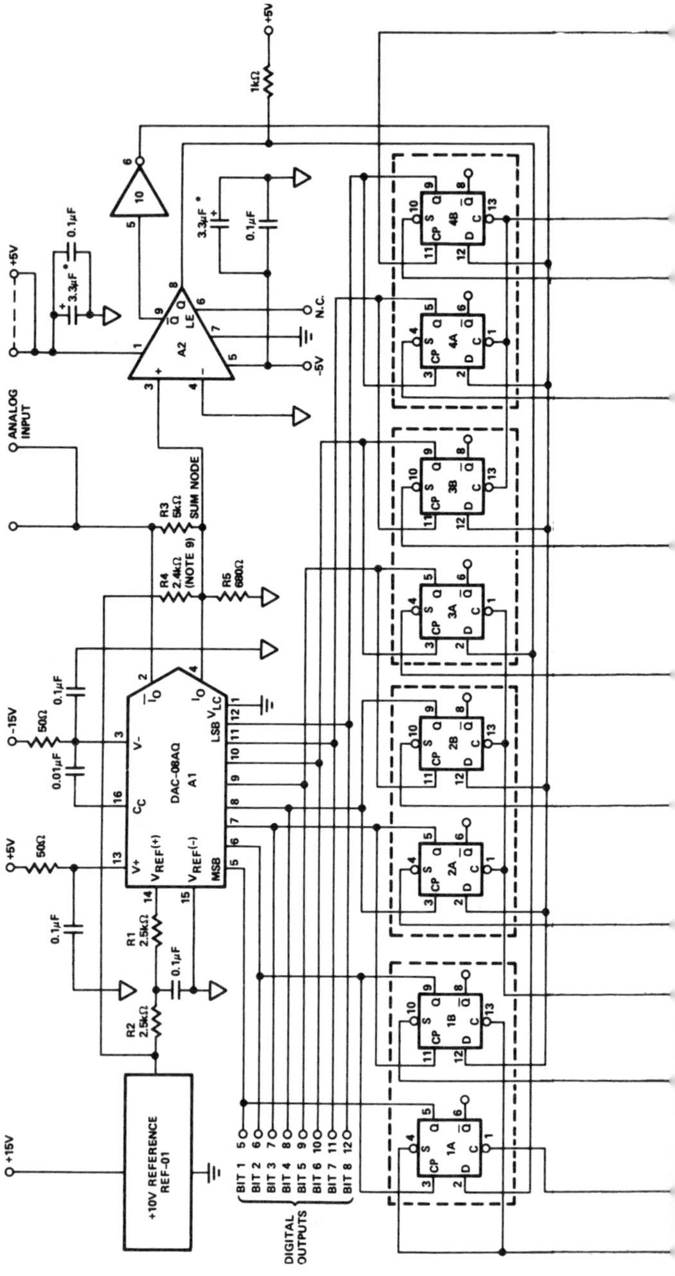
+10V REFERENCE
REF-01
DAC-08AQ
A1
A2
ANALOG INPUT
SUM NODE
DIGITAL OUTPUTS
BIT 1
BIT 2
BIT 3
BIT 4
BIT 5
BIT 6
BIT 7
BIT 8

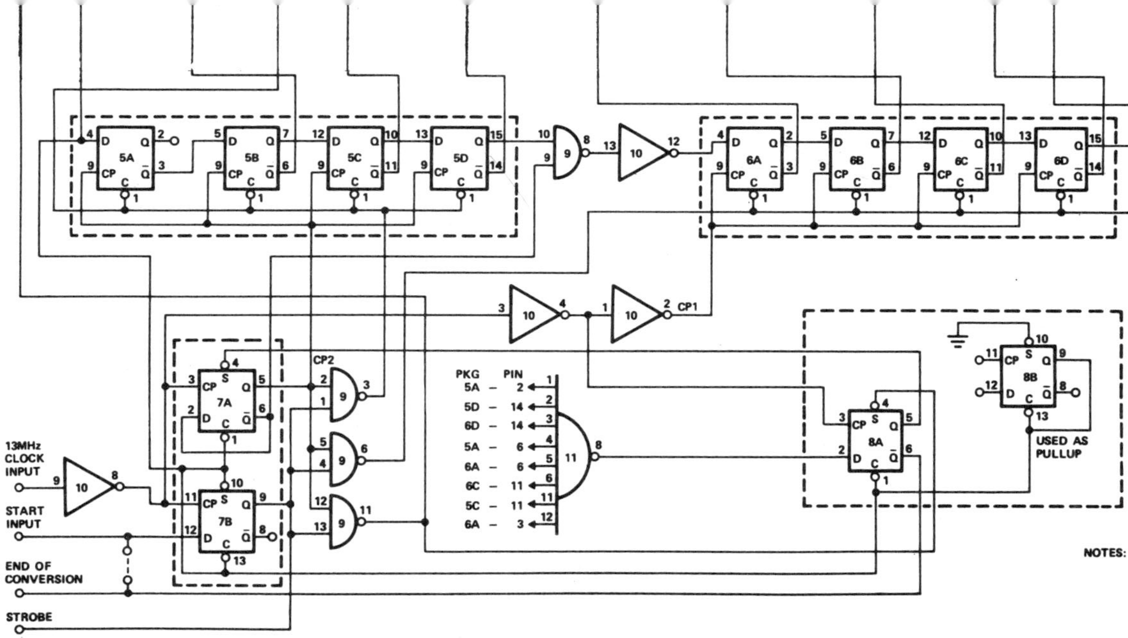

NOTES:
1. 1, 2, 3, 4, 7, 8. SN74S74N
2. 5,6, SN74S175N
3. 9 SN74S00N
4. 10 SN74S04N
5. 11 SN74S30N
6. A1 DAC-08AQ
7. A2 AM 686HC
8. *OPTIONAL BYPASS CAPACITORS
9. OPTIONAL SUM NODE BIASING RESISTOR PROVIDES +1/2 LSB (4μA)

Complete schematic 8-bit, 1μs A/D (PM).

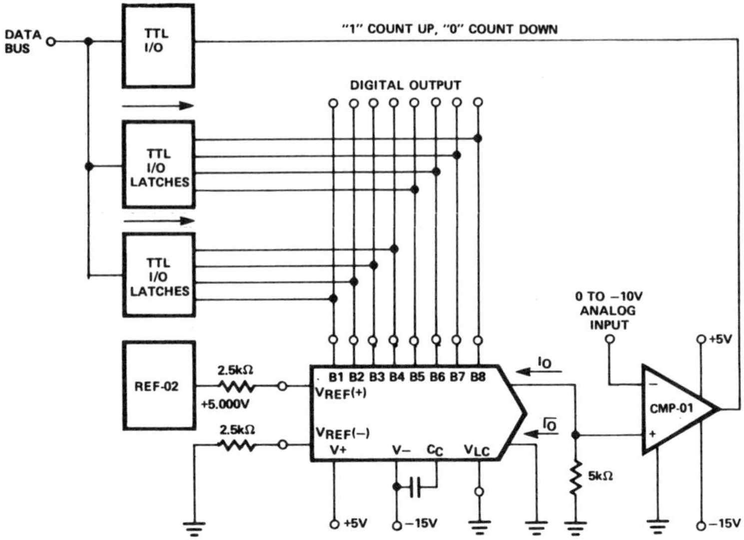

- USEFUL FOR ENCODING DC INPUTS
- ELIMINATES UP/DOWN COUNTERS
- ANSWER CONTINUOUSLY AVAILABLE
- LOW POWER CONSUMPTION
- HI-Z INPUT
- RAM OUTPUTS MAY BE USED IN PLACE OF TTL I/O

Microprocessor controlled tracking A/D converter (PM).

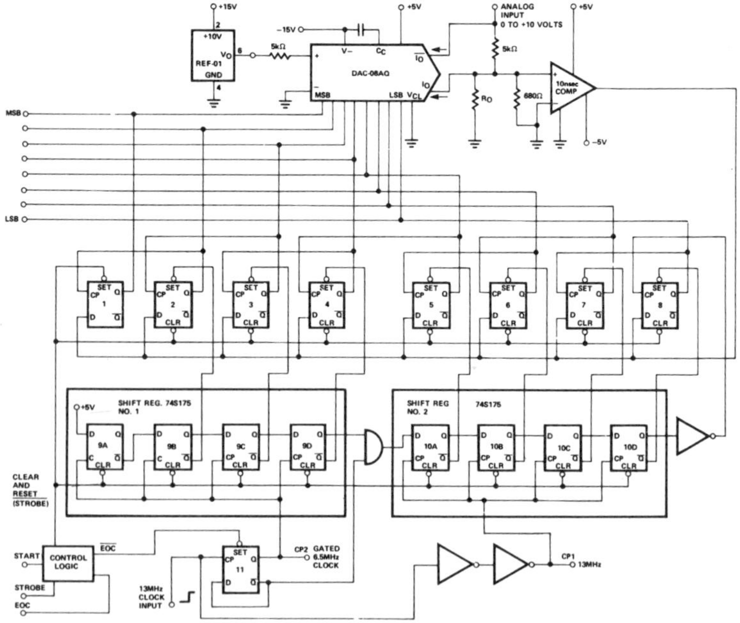

Simplified schematic 1μs A/D (PM).

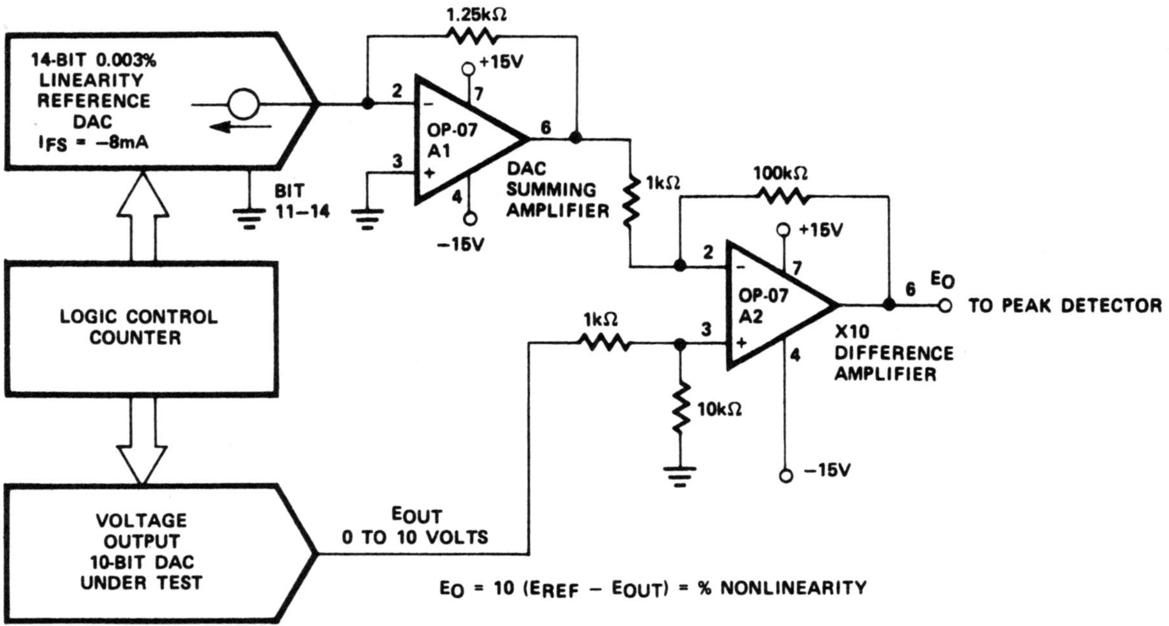

D/A converter test system (PM).

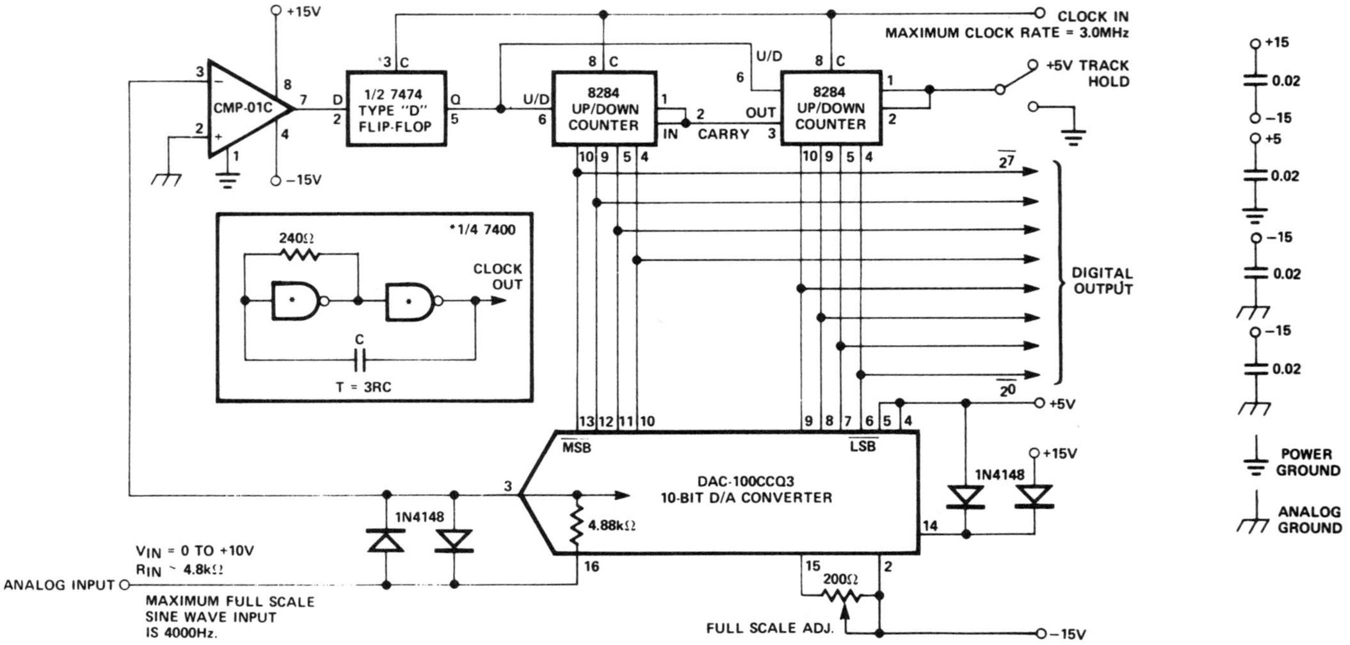

8-bit tracking A/D converter (PM).

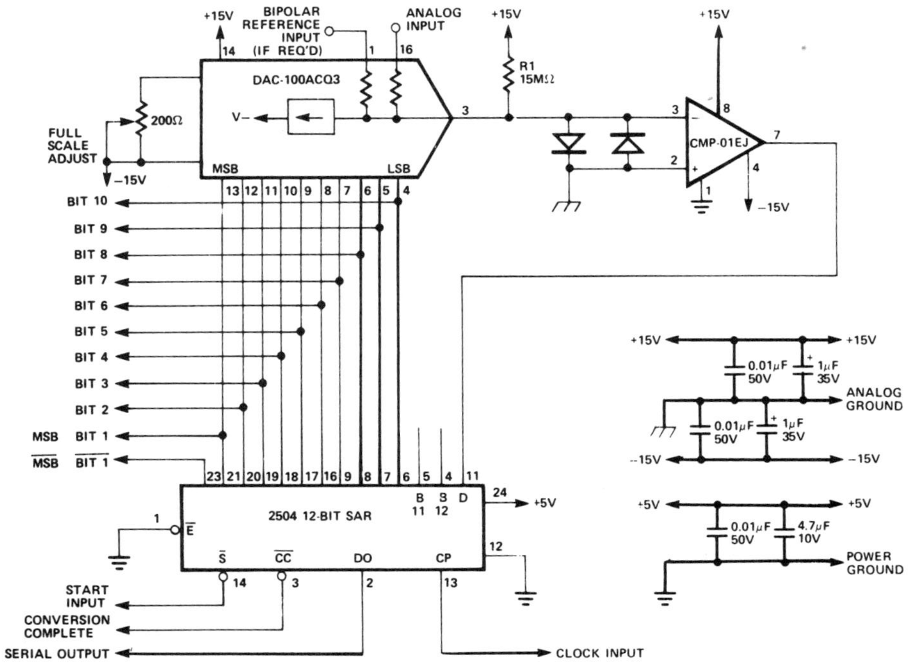

10-bit A/D schematic (PM).

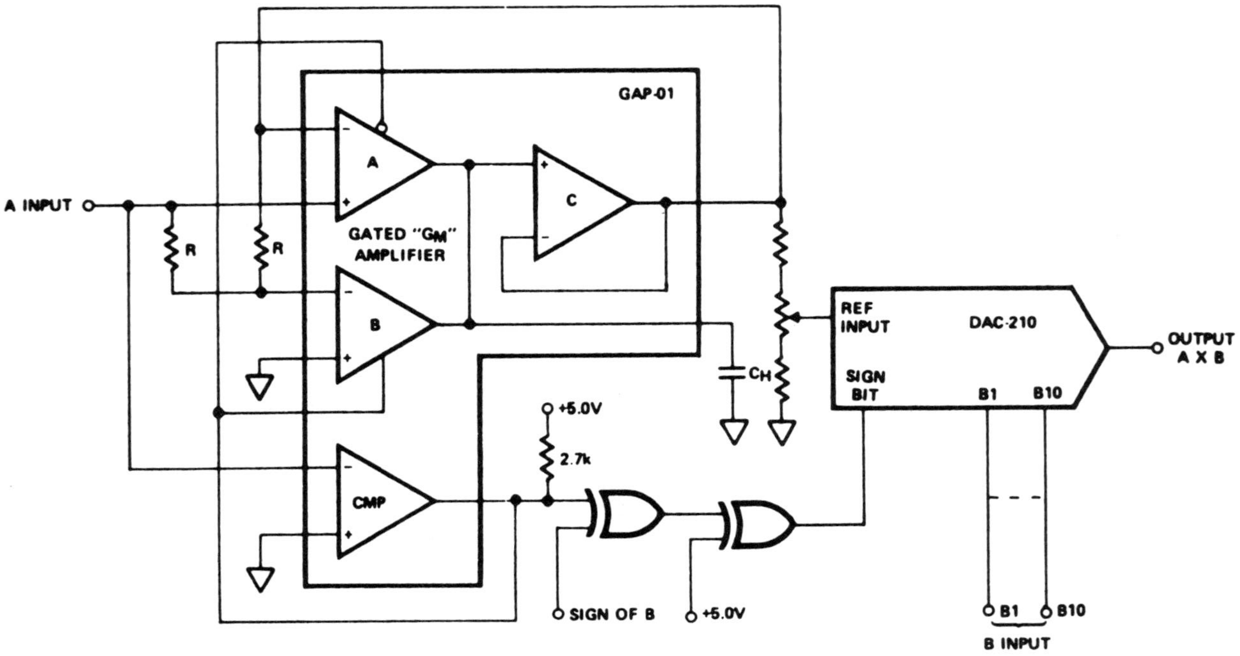

Four quadrant multiplying DAC (PM).

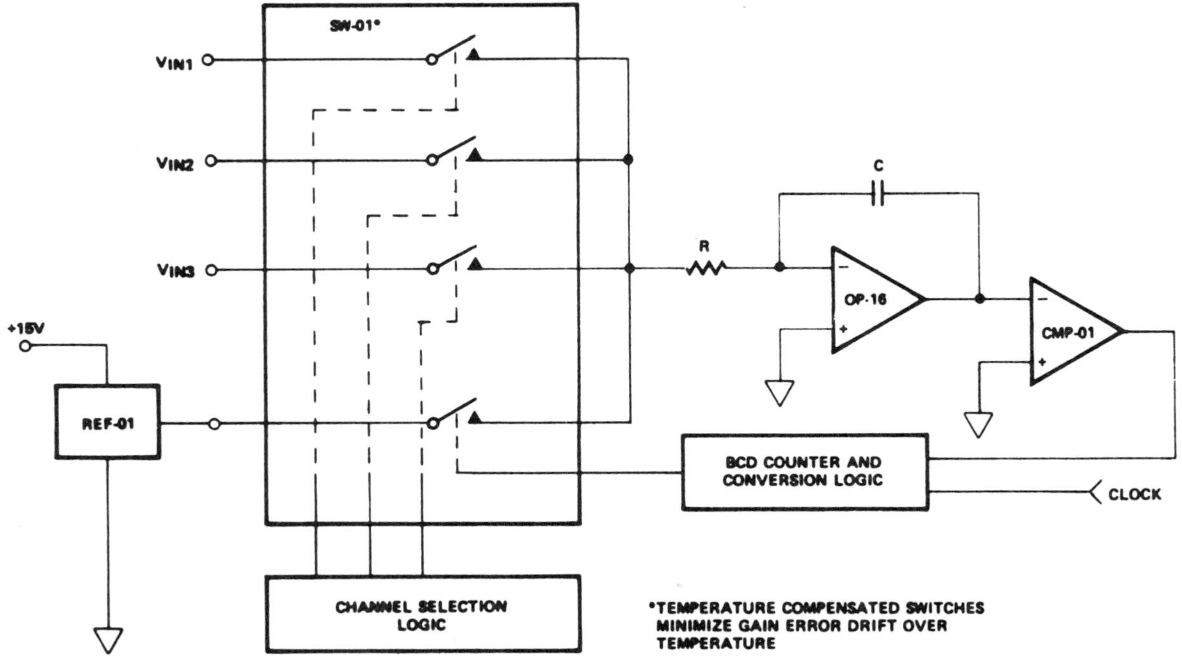

Dual slope A/D conversion (PM).

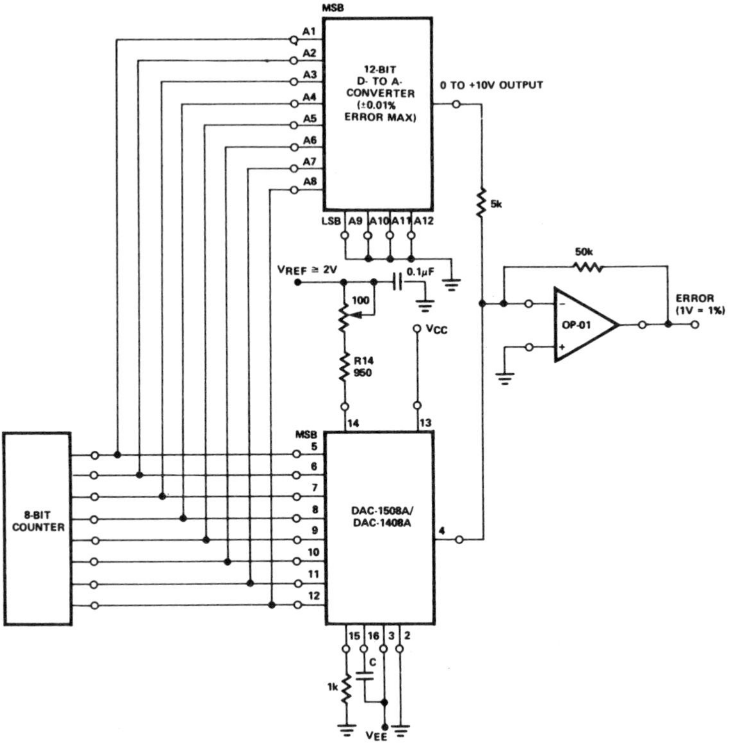

Relative accuracy test circuit (PM).

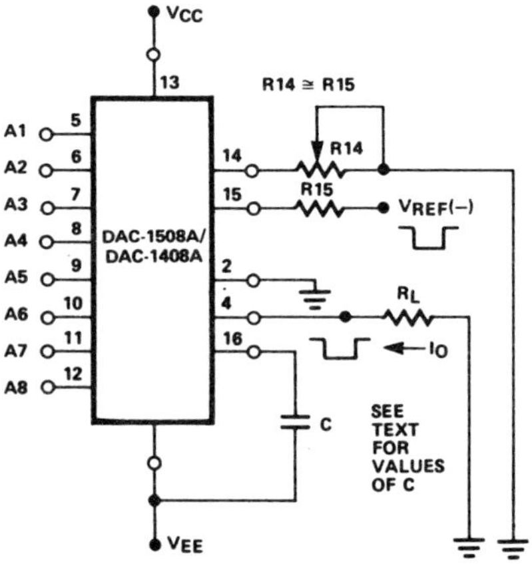

Use with negative VREF (PM).

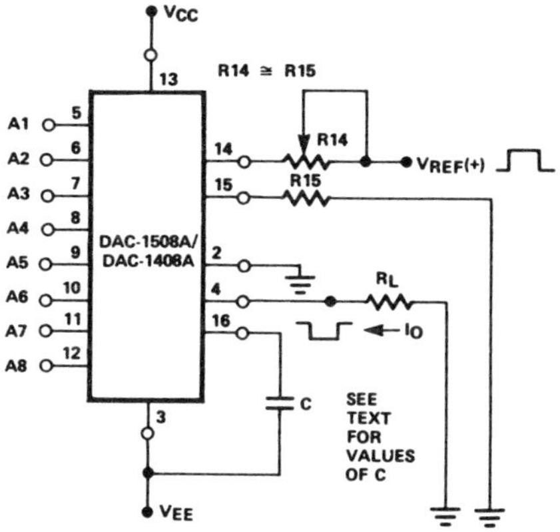

Use with positive VREF (PM).

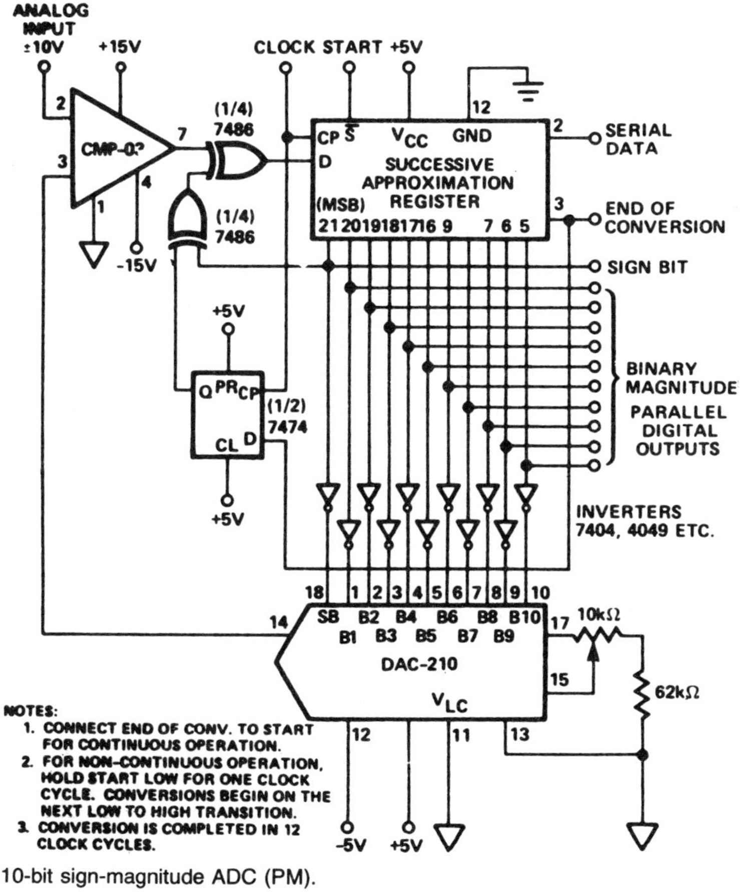

10-bit sign-magnitude ADC (PM).

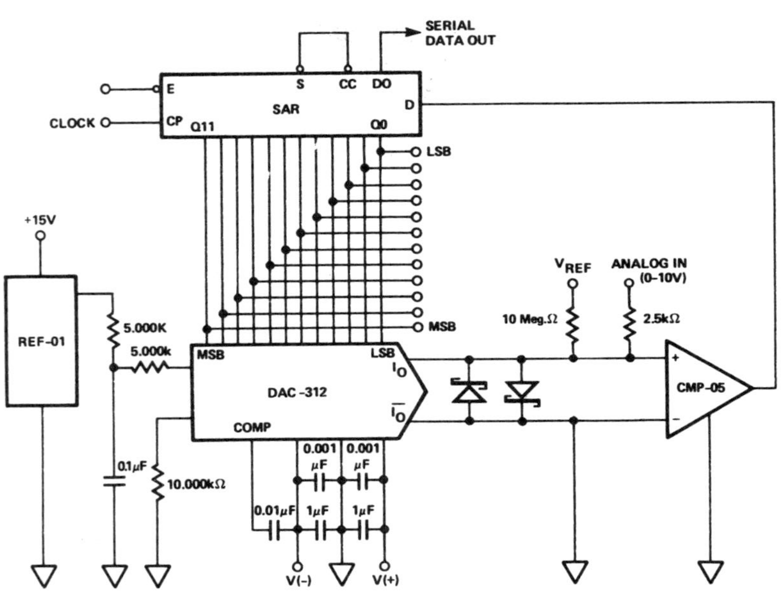

NOTE:
DEVICE(S) CONNECTED TO ANALOG INPUT MUST BE CAPABLE OF SOURCING 4.0mA.
A BUFFER (eg. BUF-02) MAY BE REQUIRED.

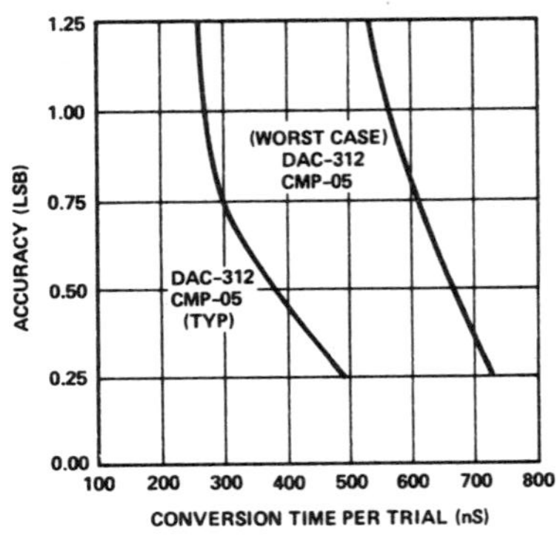

CONVERSION TIME (nS)	TYP	WORST CASE
SAR	33	55
CMP-05	92	125
TOTAL	375nS	680nS
× 13	4.9μS	8.8μS

12-bit fast A/D converter (PM).

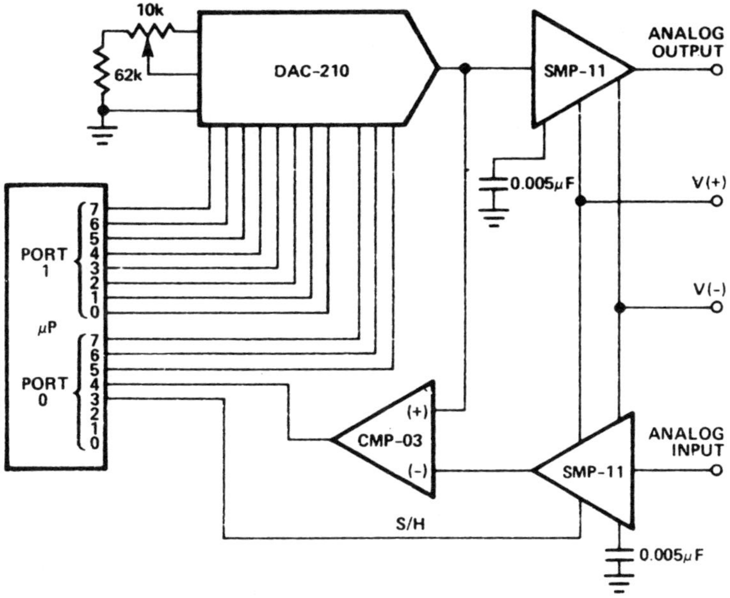

Time multiplexed A/D converter (PM).

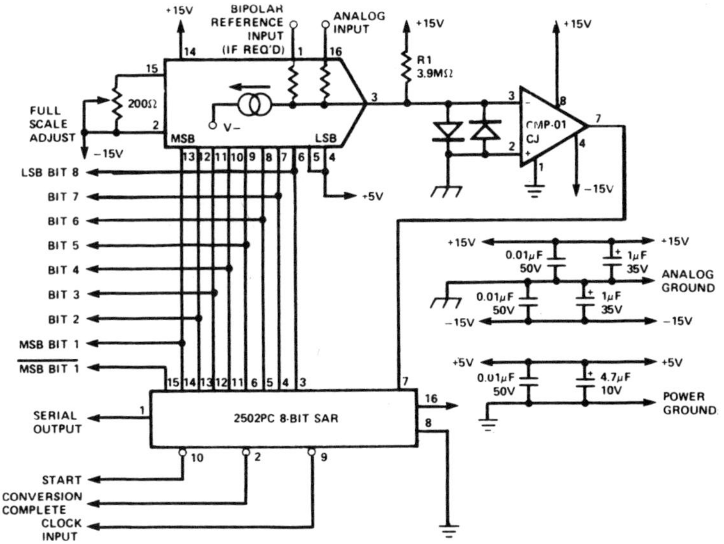

NOTE
FOR A COMPLETE TREATMENT OF SA ADC'S CONSTRUCTED WITH THE DAC-100, REFER TO AN-11, "A LOW COST, EASY-TO-BUILD SUCCESSIVE APPROXIMATION A/D CONVERTER"

Successive approximation A/D converter (PM).

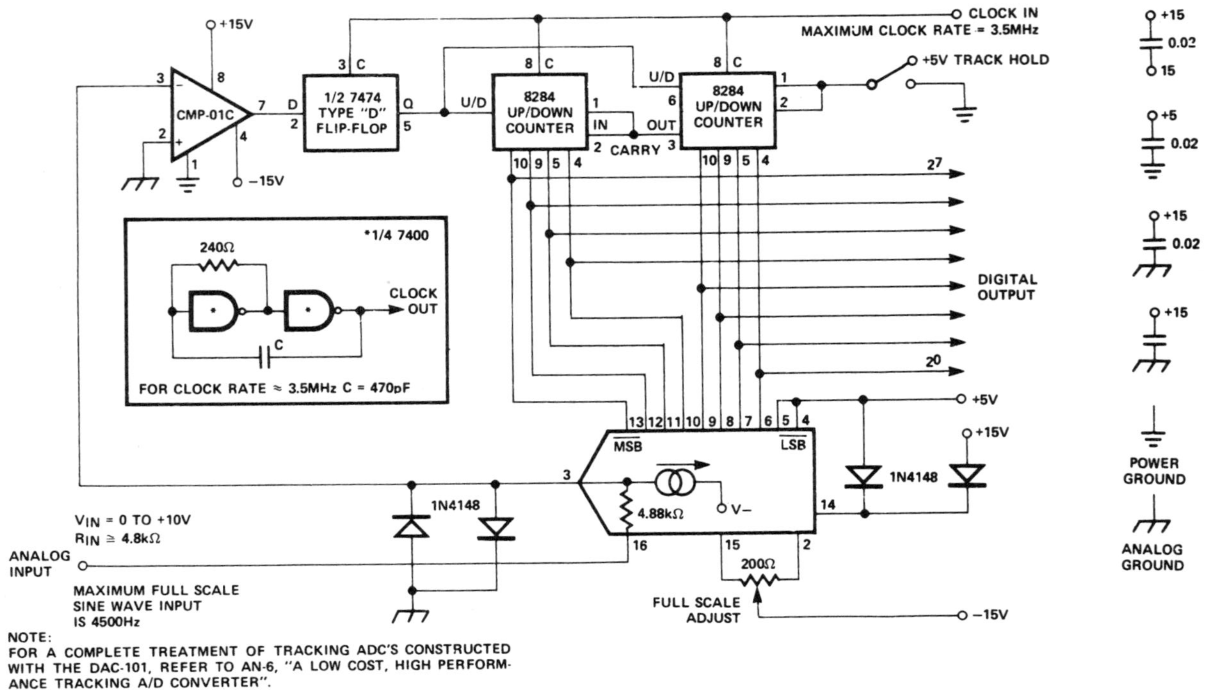

Tracking (servo-type) A/D converter (PM).

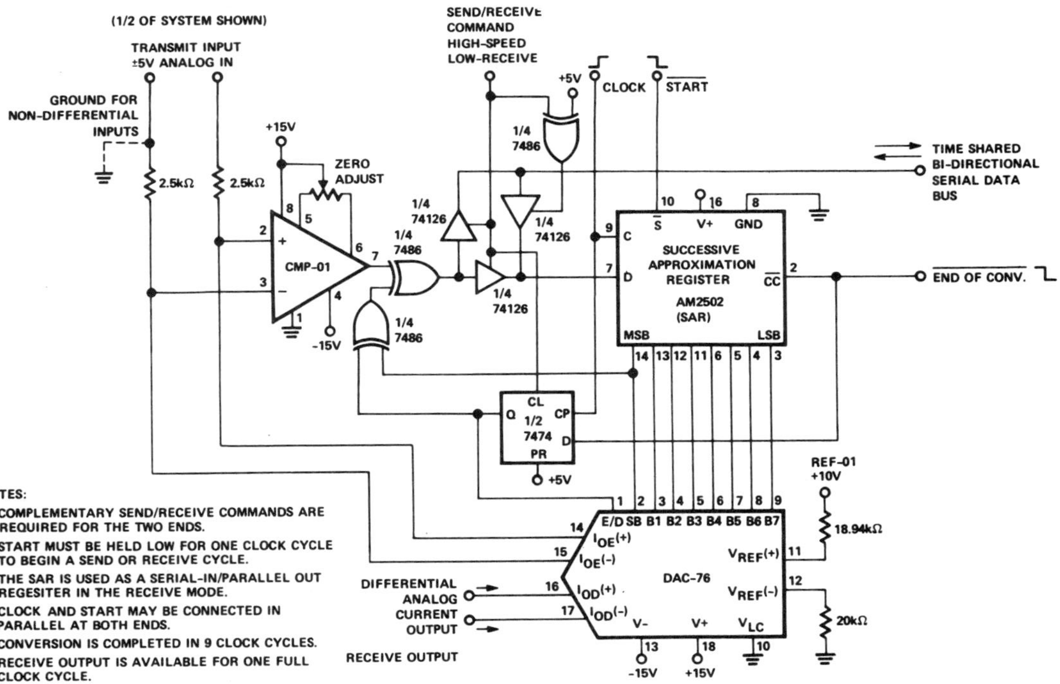

NOTES:

1. COMPLEMENTARY SEND/RECEIVE COMMANDS ARE REQUIRED FOR THE TWO ENDS.
2. START MUST BE HELD LOW FOR ONE CLOCK CYCLE TO BEGIN A SEND OR RECEIVE CYCLE.
3. THE SAR IS USED AS A SERIAL-IN/PARALLEL OUT REGESITER IN THE RECEIVE MODE.
4. CLOCK AND START MAY BE CONNECTED IN PARALLEL AT BOTH ENDS.
5. CONVERSION IS COMPLETED IN 9 CLOCK CYCLES.
6. RECEIVE OUTPUT IS AVAILABLE FOR ONE FULL CLOCK CYCLE.

Serial data transceiving converter (PM).

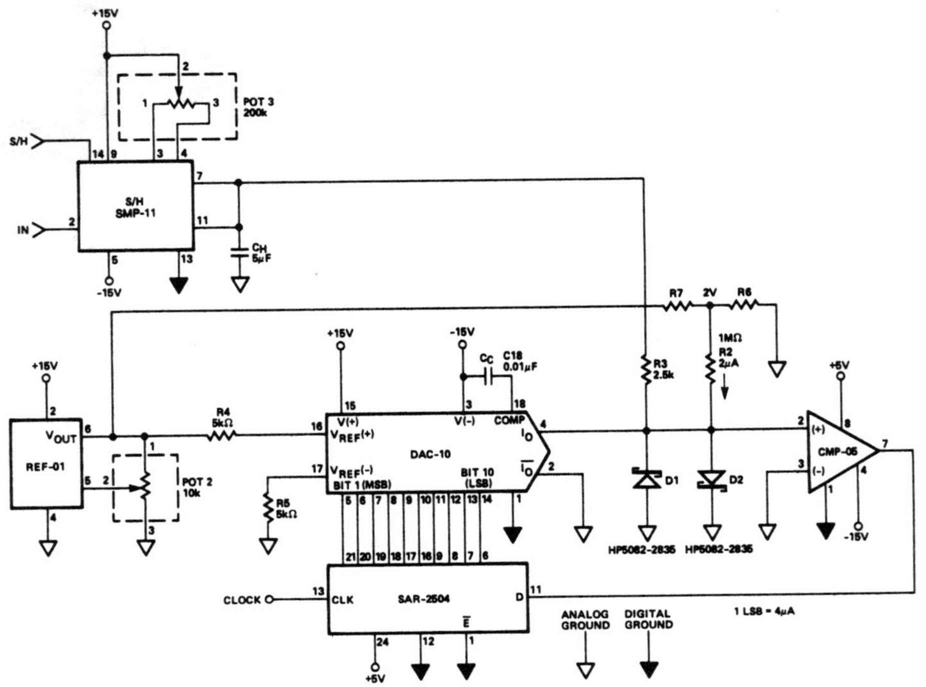

10-bit A/D converter (PM).

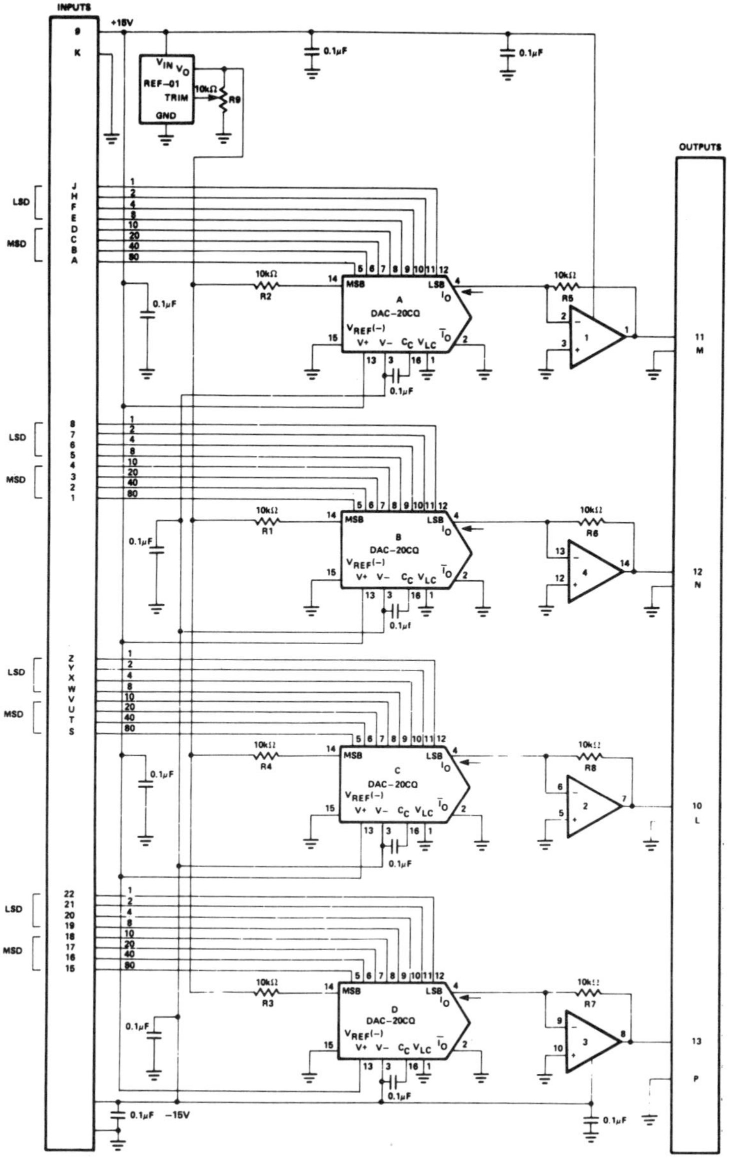

Four-channel D/A output amplifier (PM).

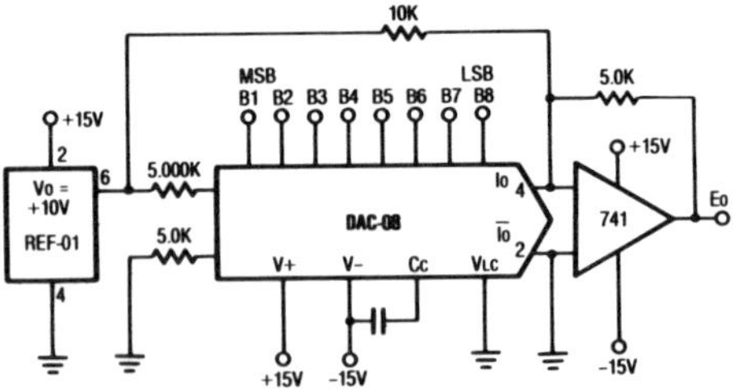

Scale	B1	B2	B3	B4	B5	B6	B7	B8	Eo
Pos Full Scale	1	1	1	1	1	1	1	1	+4.960
Zero Scale	1	0	0	0	0	0	0	0	0.00
Neg Full Scale +1 LSB	0	0	0	0	0	0	0	1	−4.960
Neg Full Scale	0	0	0	0	0	0	0	0	−5.000

Offset binary operation (R).

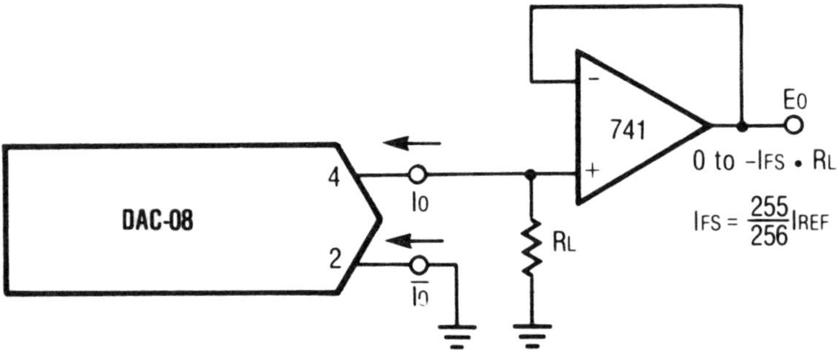

For complementary output (operation as a negative logic DAC), connect non-inverting input of Op-Amp to Īo (pin 2); connect Io (pin 4) to ground.

Positive low impedance output operation (R).

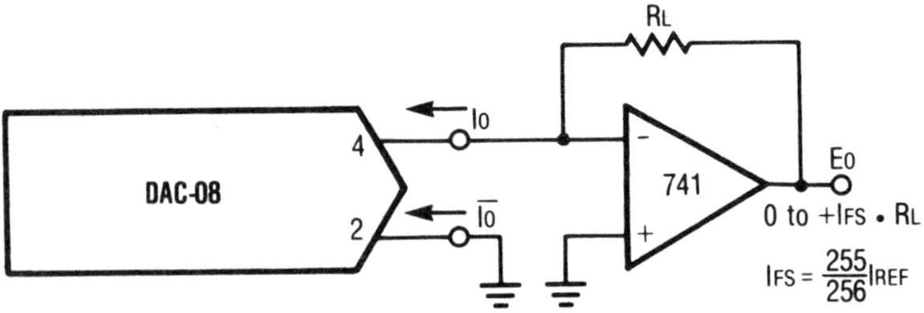

For complementary output (operation as a negative logic DAC), connect inverting input of Op-Amp to Īo (pin 2); connect Io (pin 4) to ground.

Negative low impedance output operation (R).

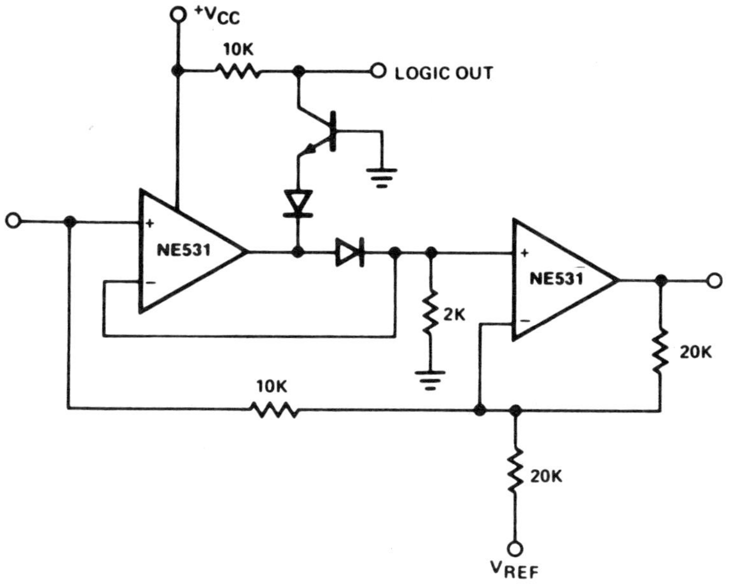

Cyclic A/D converter (S).

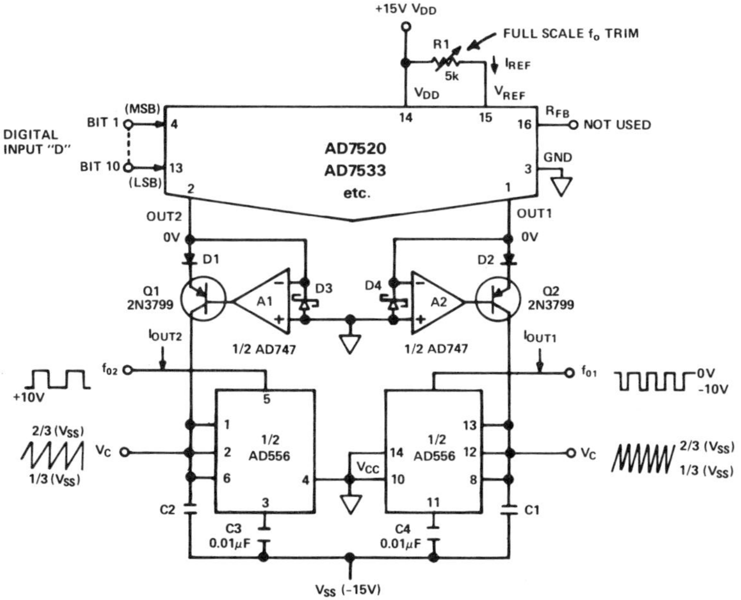

D to F converter with complementary output frequencies (AD).

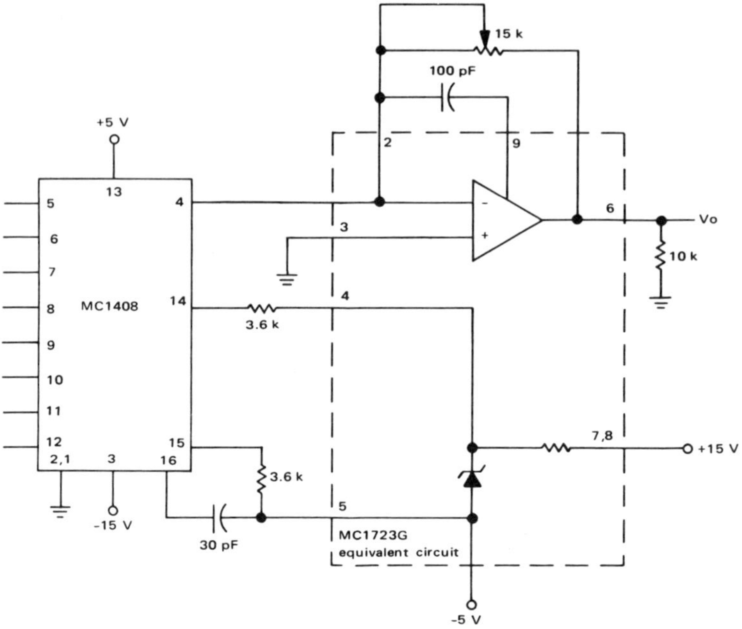

8-bit DAC with voltage regulator (M).

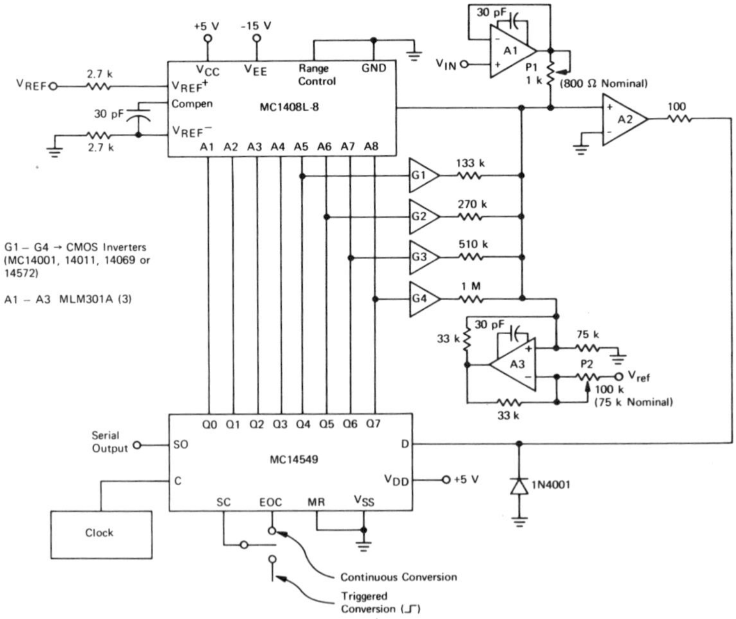

Successive approximation BCD A/D converter (M).

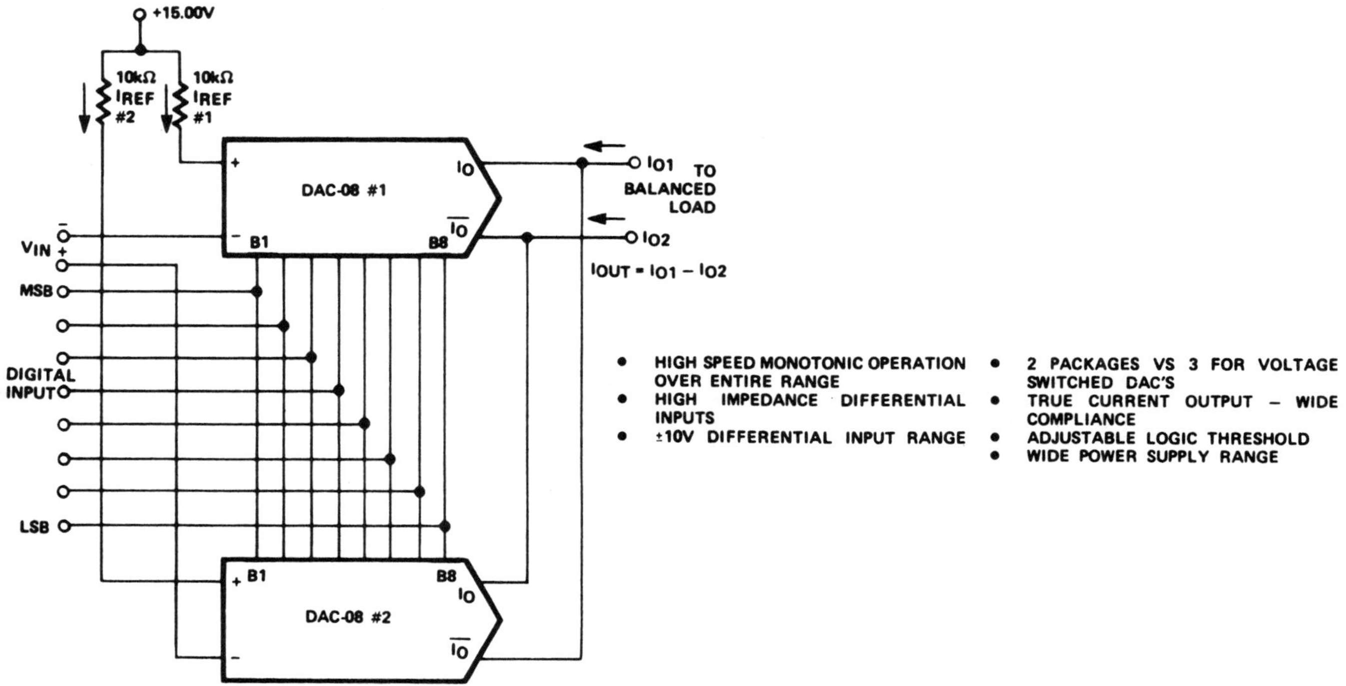

Four-quadrant multiplying DAC with impedance input (PM).

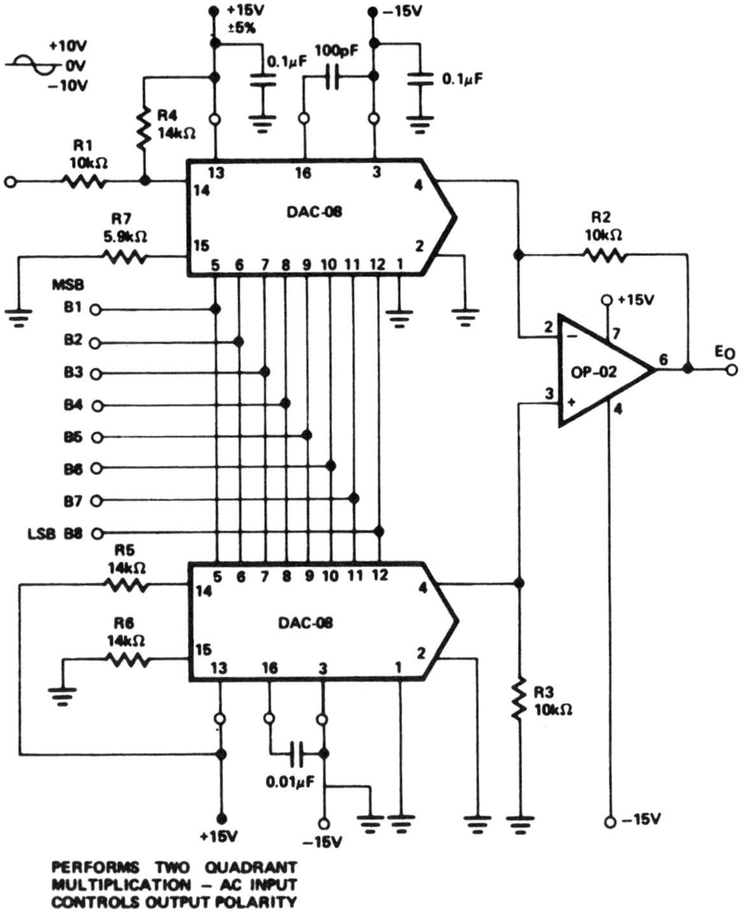

Bipolar analog two-quadrant multiplication (dc-coupled digital attenuator) (PM).

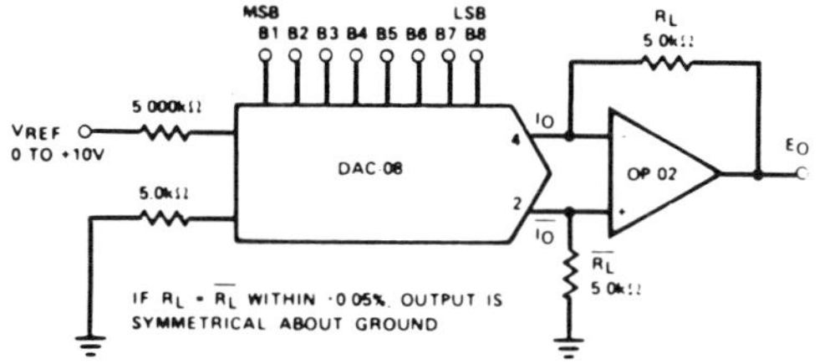

	B1	B2	B3	B4	B5	B6	B7	B8	I_O mA	$\overline{I_O}$ mA	E_O (V)
POS FULL SCALE	1	1	1	1	1	1	1	1	1.992	0.000	+9.96
POS FULL SCALE −LSB	1	1	1	1	1	1	1	0	1.984	0.008	+9.88
(+) ZERO SCALE	1	0	0	0	0	0	0	0	1.000	0.992	+0.040
(−) ZERO SCALE	0	1	1	1	1	1	1	1	0.992	1.000	−0.040
NEG FULL SCALE +LSB	0	0	0	0	0	0	0	1	0.008	1.984	−9.88
NEG FULL SCALE	0	0	0	0	0	0	0	0	0.000	1.992	−9.96

Bipolar digital two-quadrant multiplication (symmetrical offset binary) (PM).

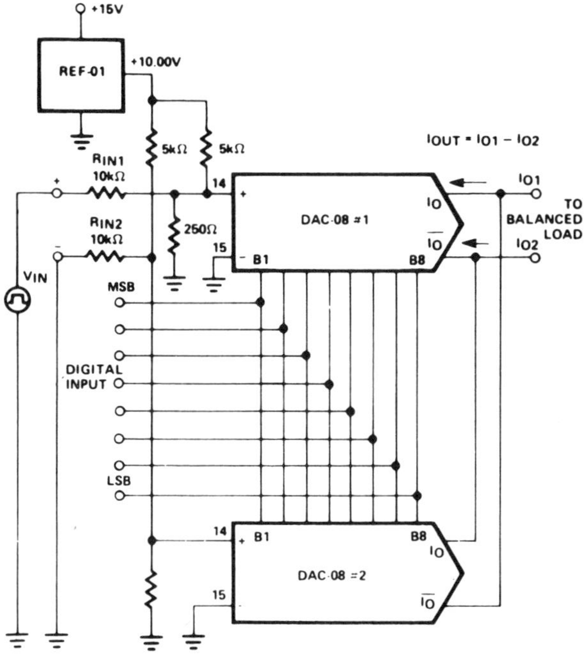

- 250Ω RESISTORS AND OMISSION OF COMPENSATION CAPACITORS ARE OPTIONAL FOR FAST PULSED REFERENCE OPERATION.
- INPUT DIFFERENTIAL AND COMMON MODE RANGES ARE EXTENDABLE BY INCREASING 10kΩ RESISTORS. EXAMPLE: 100kΩ FOR 100V.
- HIGH SPEED MULTIPLYING CONNECTION WITH MONOTONIC OPERATION OVER ENTIRE RANGE.
- NOT NECESSARY TO MODULATE BOTH DAC'S WITH THE REFERENCE INPUT.

Four-quadrant multiplying DAC with extendable input range and highest speed (PM).

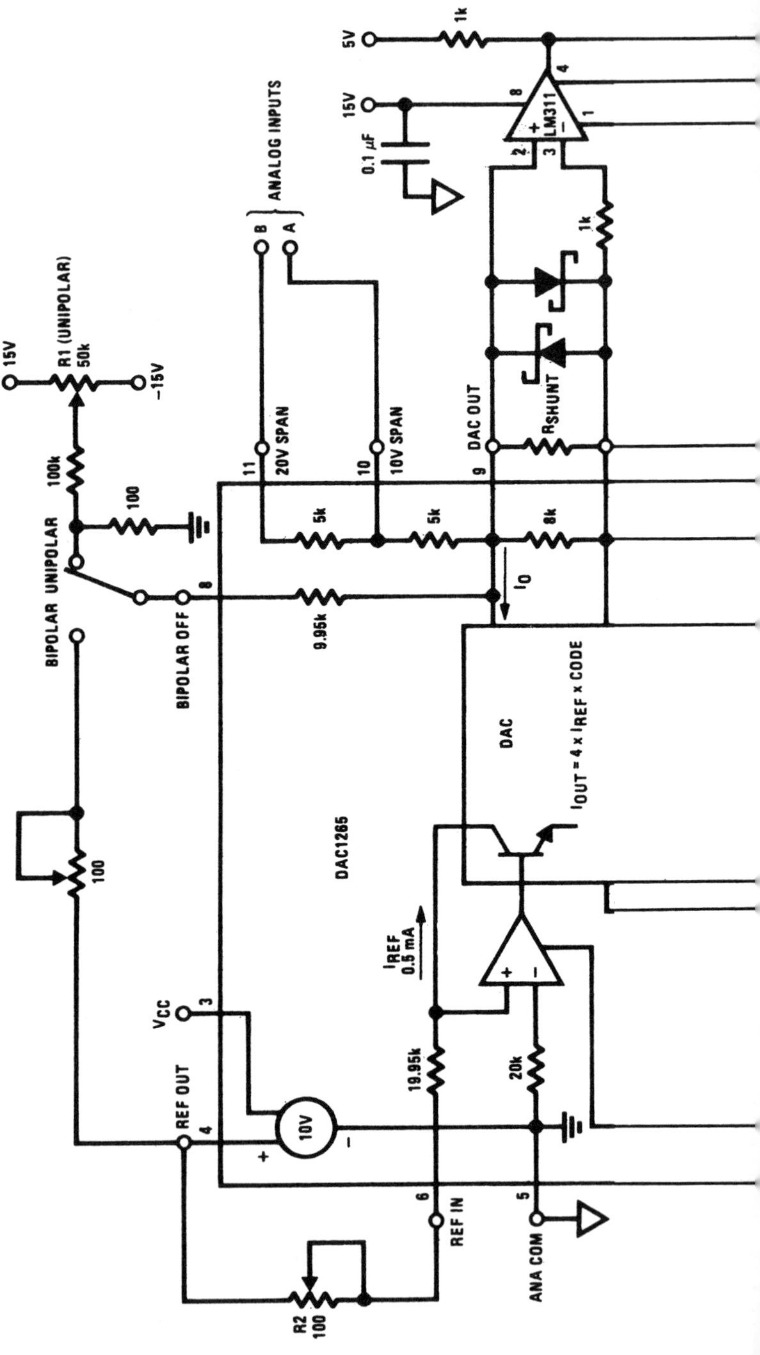

5V
1k
15V
0.1 μF
LM311
ANALOG INPUTS
B
A
15V
R1 (UNIPOLAR)
50k
−15V
100k
100
BIPOLAR
UNIPOLAR
BIPOLAR OFF
20V SPAN
10V SPAN
DAC OUT
R_{SHUNT}
5k
5k
8k
9.95k
I_O
DAC
$I_{OUT} = 4 \times I_{REF} \times CODE$
DAC1265
I_{REF} 0.5 mA
VCC
REF OUT
10V
19.95k
20k
REF IN
ANA COM
R2
100

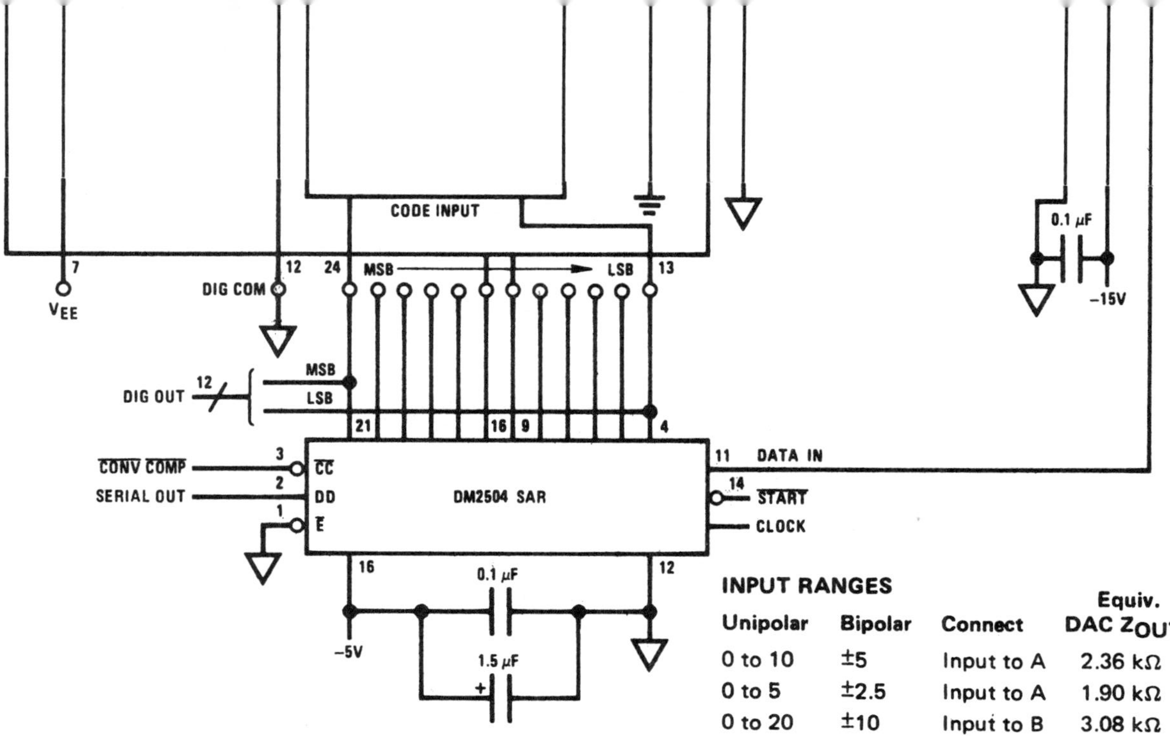

INPUT RANGES

Unipolar	Bipolar	Connect	Equiv. DAC Z_{OUT}
0 to 10	±5	Input to A	2.36 kΩ
0 to 5	±2.5	Input to A	1.90 kΩ
0 to 20	±10	Input to B	3.08 kΩ
		B to DAC OUT	

Fast precision analog-to-digital converter (NS).

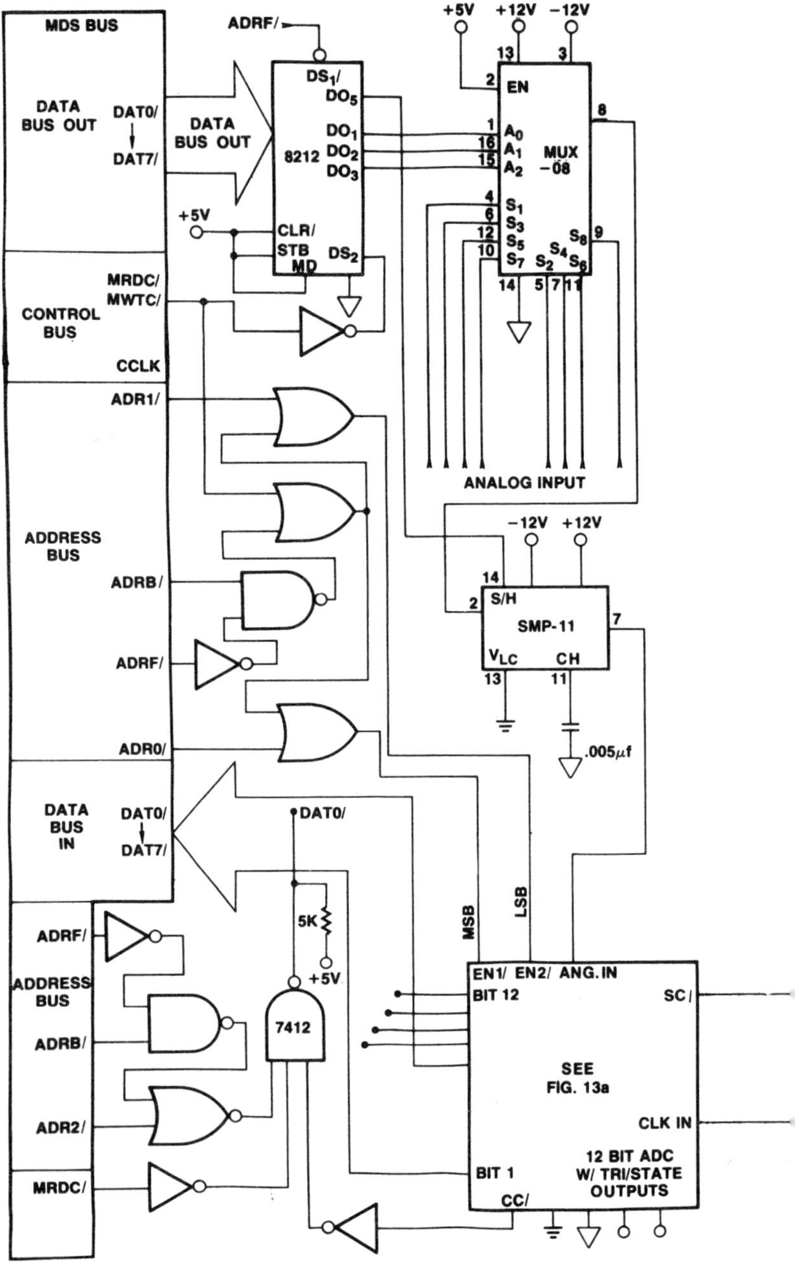

Complete, 12-bit, 8-channel high-speed data acquisition system (PM).

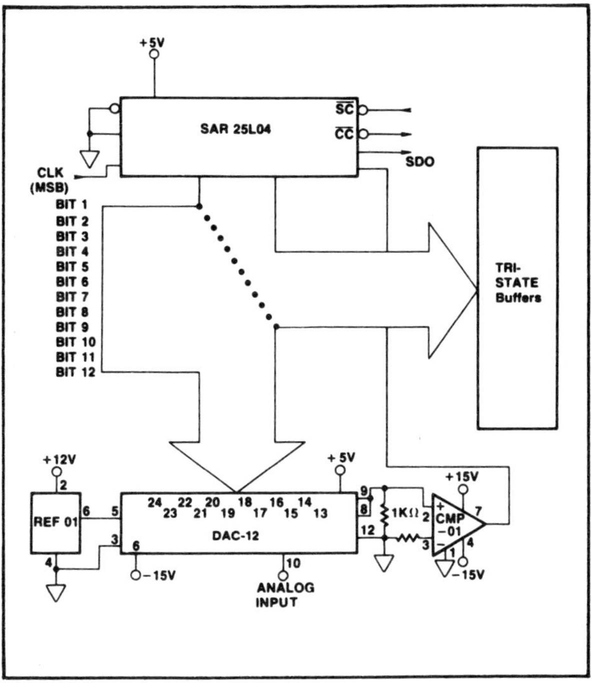
+5V
SAR 25L04
SC
CC
SDO
CLK
(MSB)
BIT 1
BIT 2
BIT 3
BIT 4
BIT 5
BIT 6
BIT 7
BIT 8
BIT 9
BIT 10
BIT 11
BIT 12
TRI-
STATE
Buffers
+12V
REF 01
DAC-12
ANALOG
INPUT
−15V
+5V
1KΩ
CMP
−01
+15V
−15V

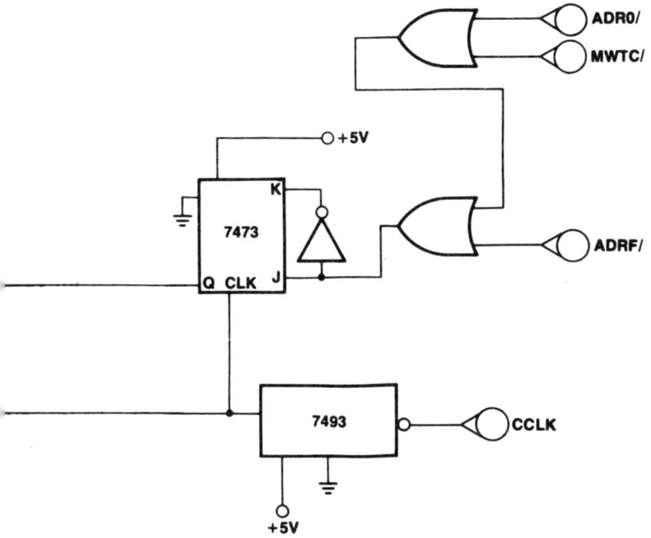
ADR0/
MWTC/
+5V
7473
K
J
Q CLK
ADRF/
7493
CCLK
+5V

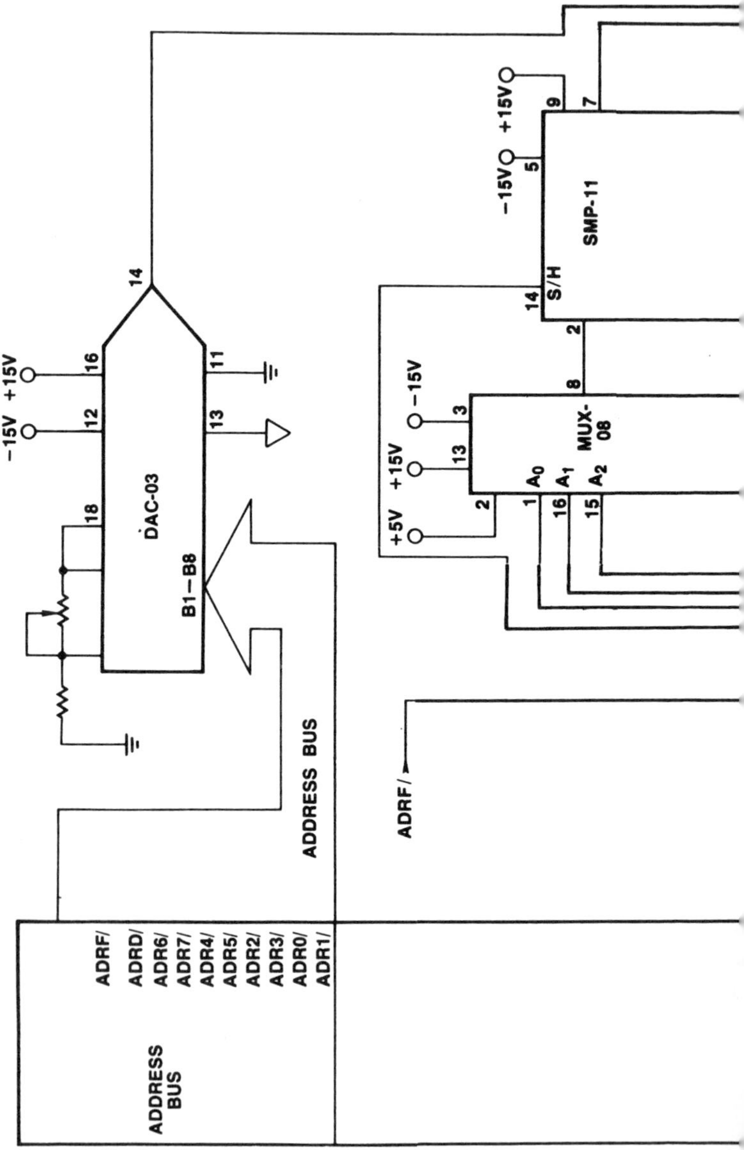
DAC-03
B1—B8
−15V +15V
ADDRESS BUS
ADRF/
SMP-11
S/H
MUX-08
+5V
ADDRESS BUS
ADRF/
ADRD/
ADR6/
ADR7/
ADR4/
ADR5/
ADR2/
ADR3/
ADR0/
ADR1/

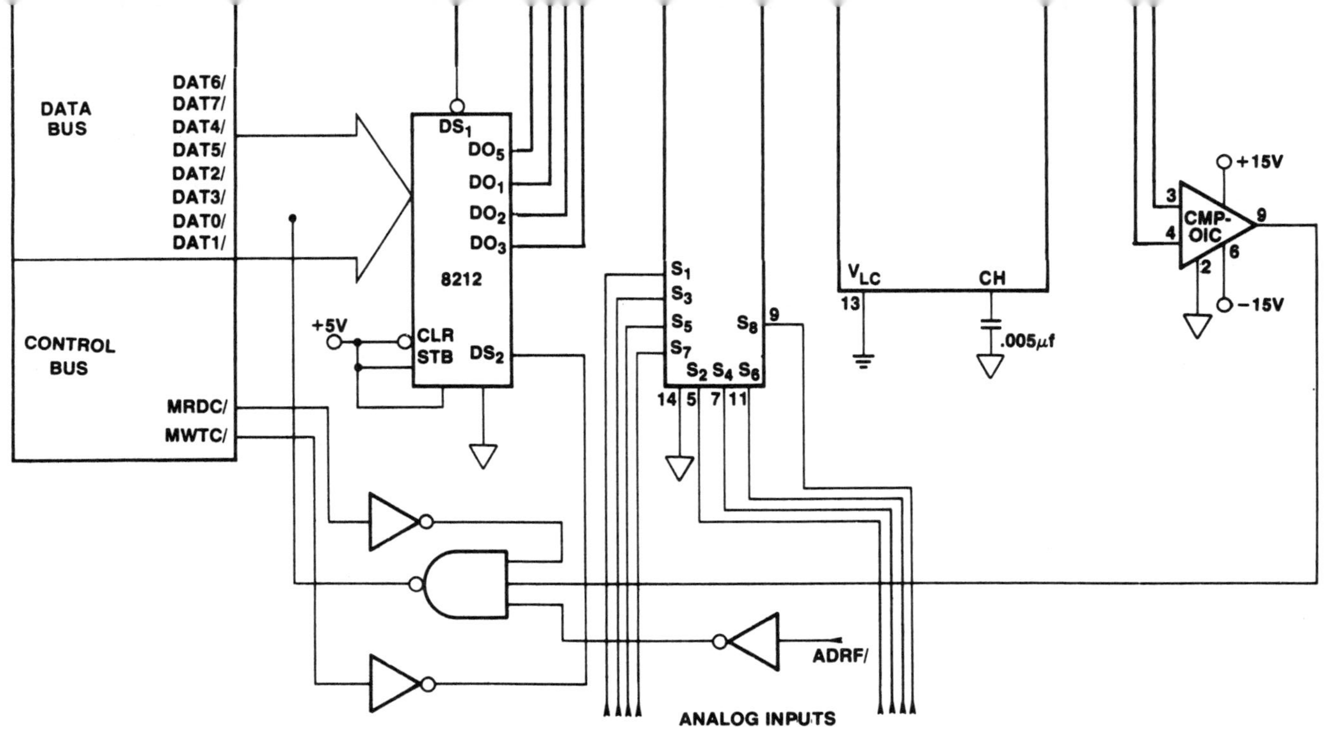

8-bit, 8-channel data acquisition system with software ADC (PM).

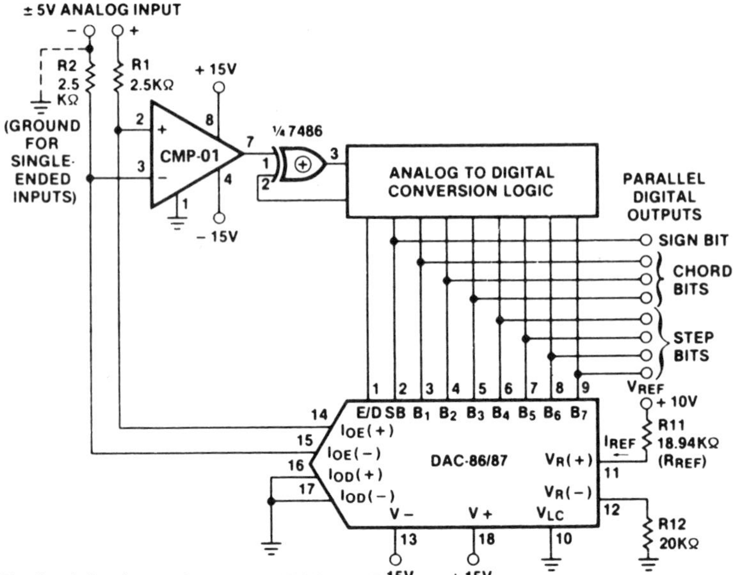

Basic eight-channel encoder (PM).

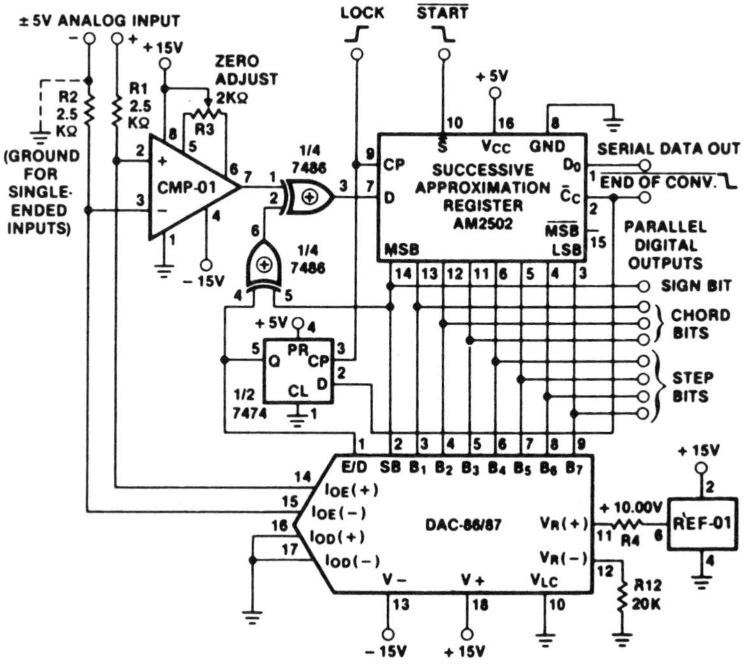

Eight-channel encoder using LSI SAR (PM).

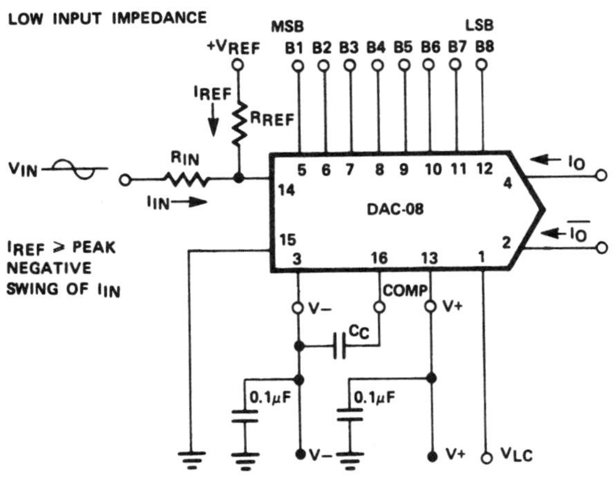

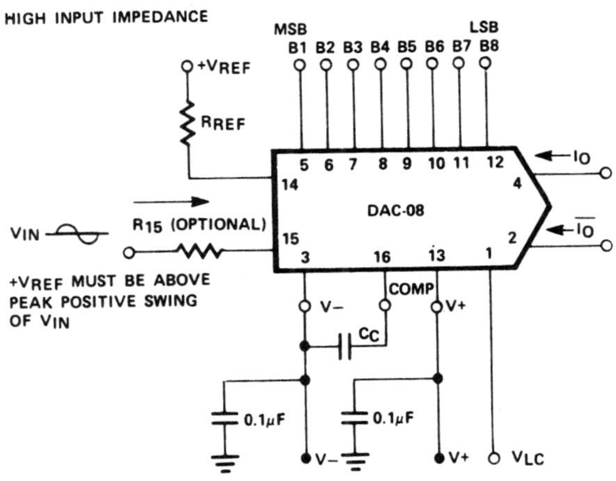

Bipolar reference connections (PM).

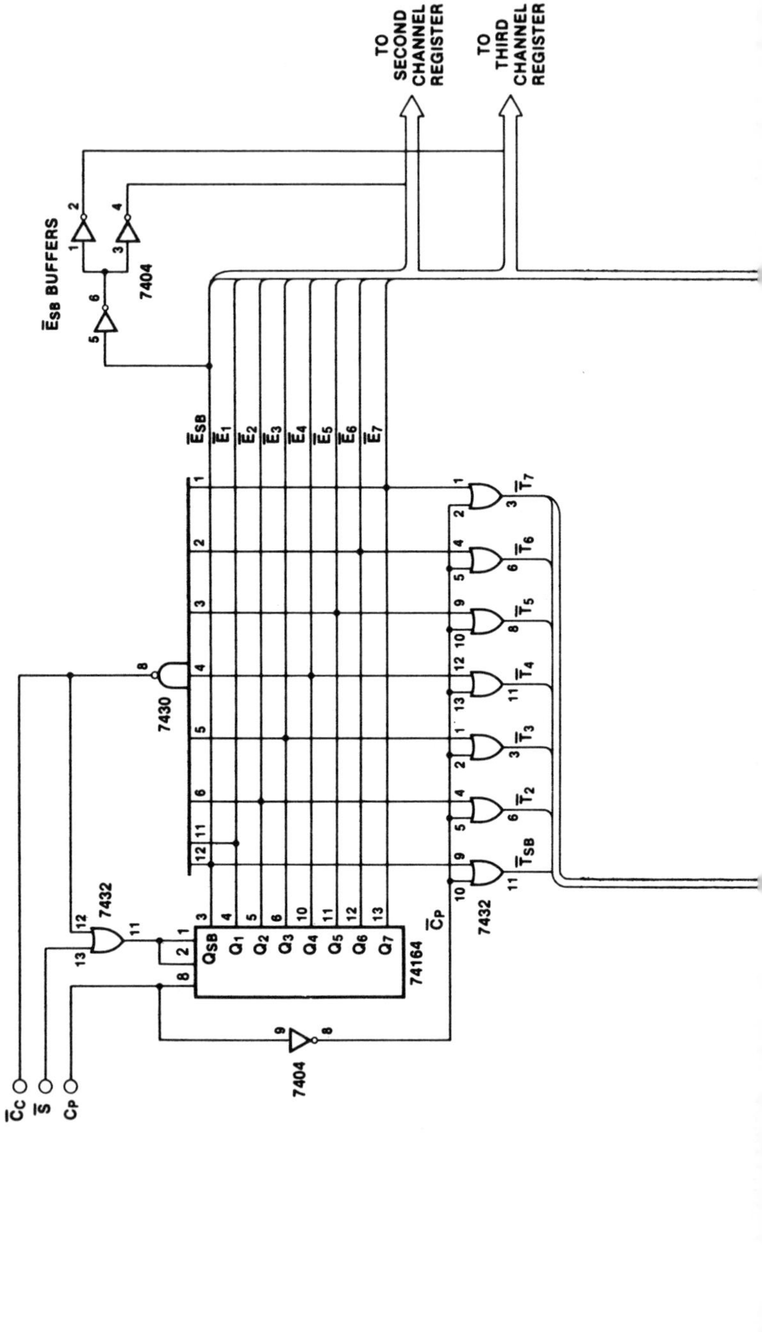

TO SECOND CHANNEL REGISTER
TO THIRD CHANNEL REGISTER
$\overline{E}_{SB}$ BUFFERS
7404
7430
7432
74164
$\overline{C}_C$
$\overline{S}$
C_P
$\overline{C}_P$
Q_{SB} Q_1 Q_2 Q_3 Q_4 Q_5 Q_6 Q_7
$\overline{E}_{SB}$ $\overline{E}_1$ $\overline{E}_2$ $\overline{E}_3$ $\overline{E}_4$ $\overline{E}_5$ $\overline{E}_6$ $\overline{E}_7$
$\overline{T}_7$ $\overline{T}_6$ $\overline{T}_5$ $\overline{T}_4$ $\overline{T}_3$ $\overline{T}_2$ $\overline{T}_{SB}$

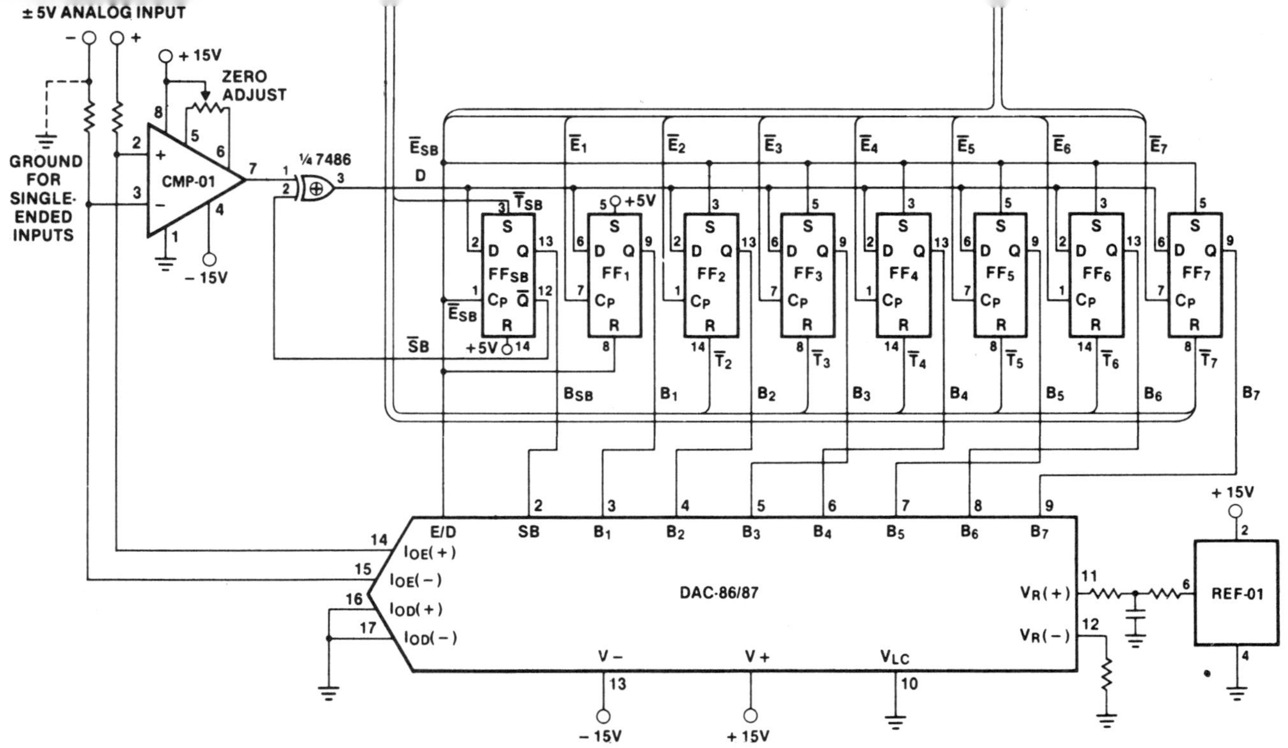

Improved eight-channel encoder complete schematic (PM).

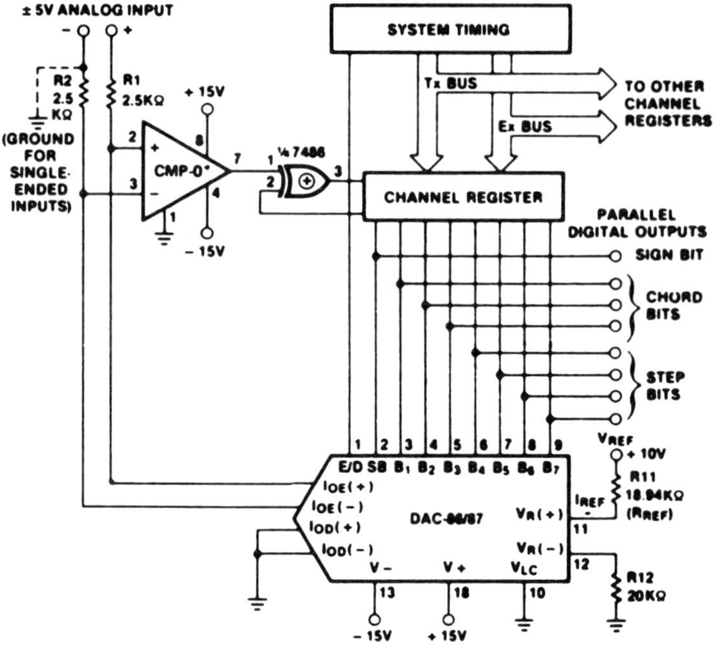

Eight-channel encoder with improved design SAR (PM).

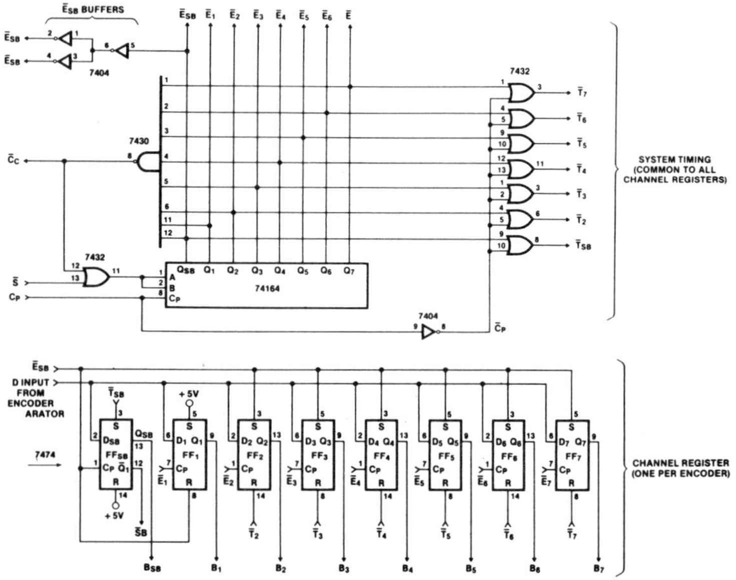

Successive approximation register (PM).

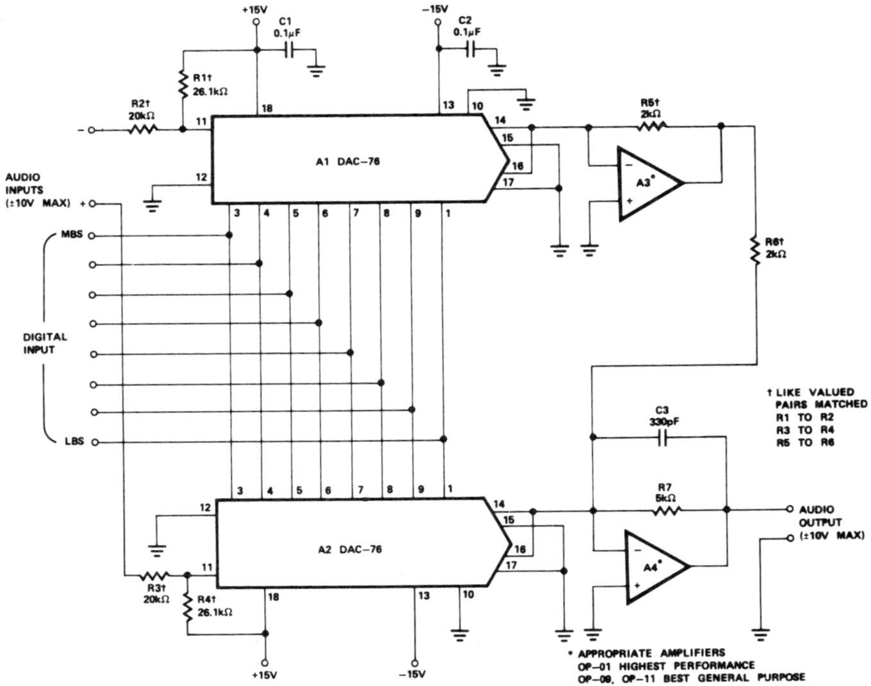

"Clickless" attenuator/amplifier (PM).

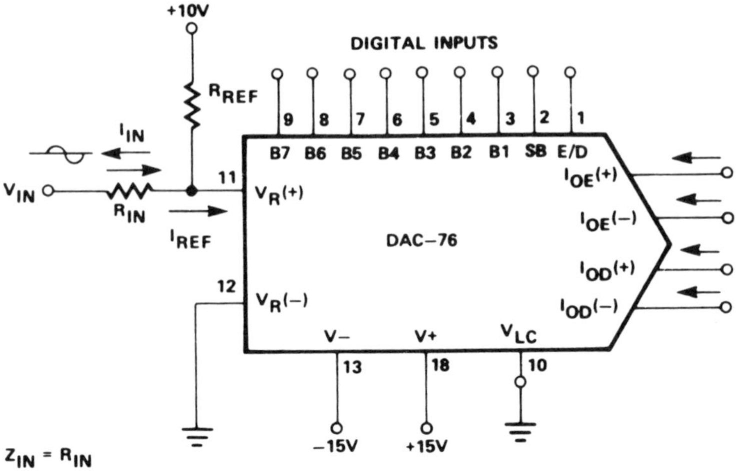

$Z_{IN} = R_{IN}$

$$I_{REF} = \frac{V_{IN}}{R_{IN}} + \frac{V_{REF}}{R_{REF}}$$

$I_{FS} \approx 4\,I_{REF}$

Low input impedance connection (PM).

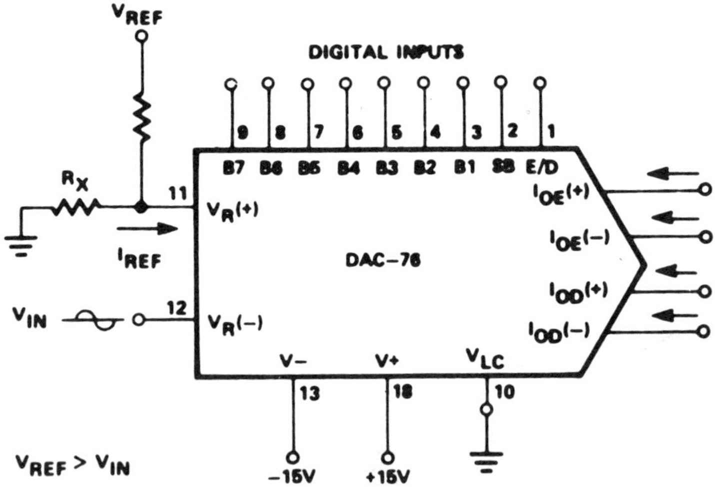

$V_{REF} > V_{IN}$

$$I_{FS} = \frac{V_{IN}}{R_{REF} \parallel RX} + \frac{V_{REF}}{R_{REF}}$$

$I_{FS} \approx 4\,I_{REF}$

High input impedance connection (PM).

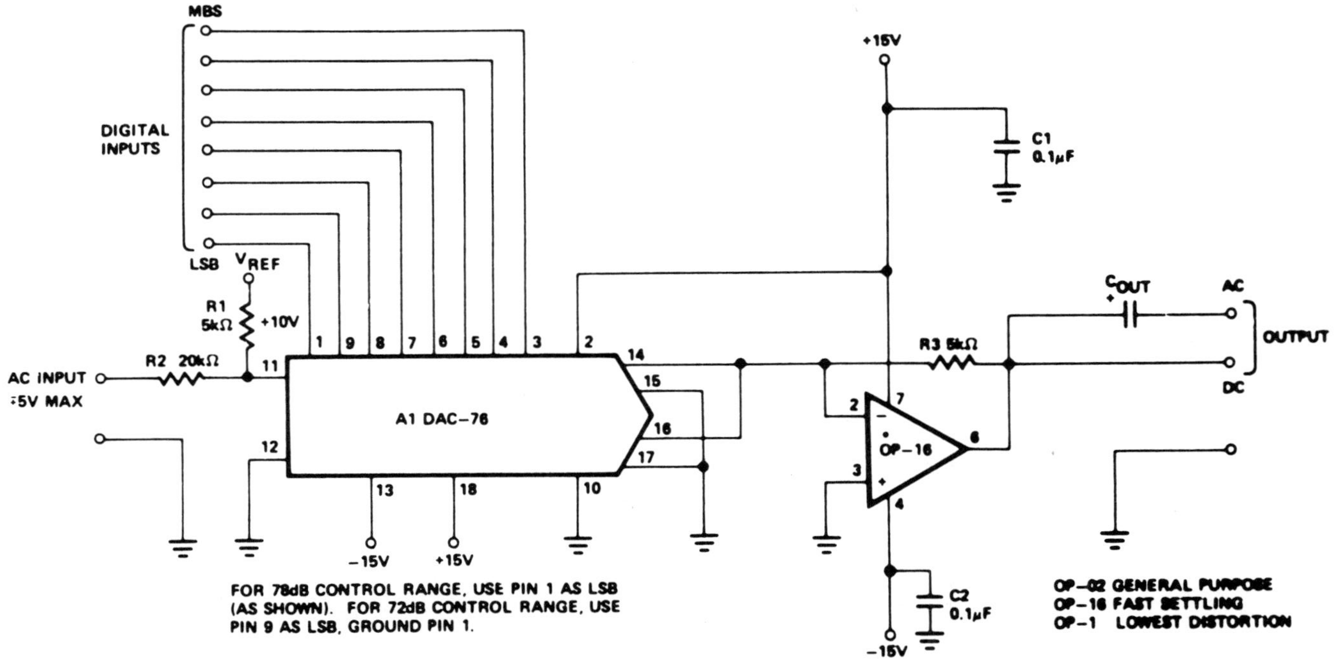

Basic exponential control characteristic two quadrant multiplier (PM).

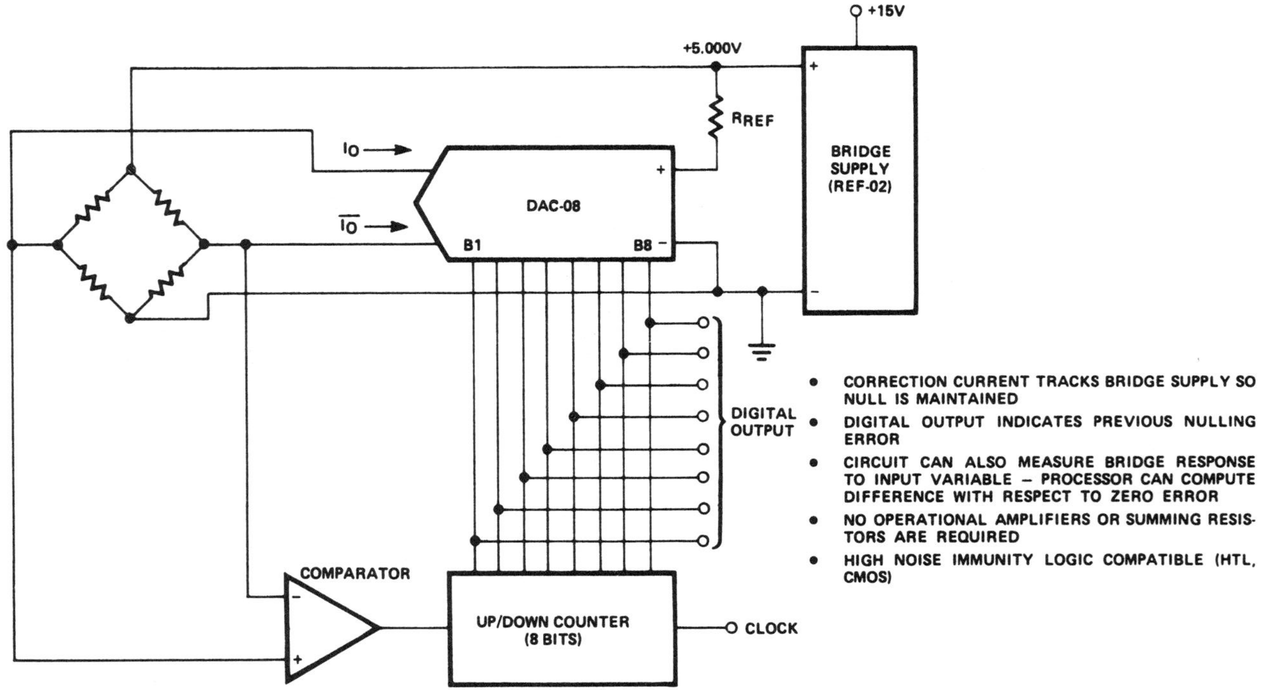

Bridge transducer null (PM).

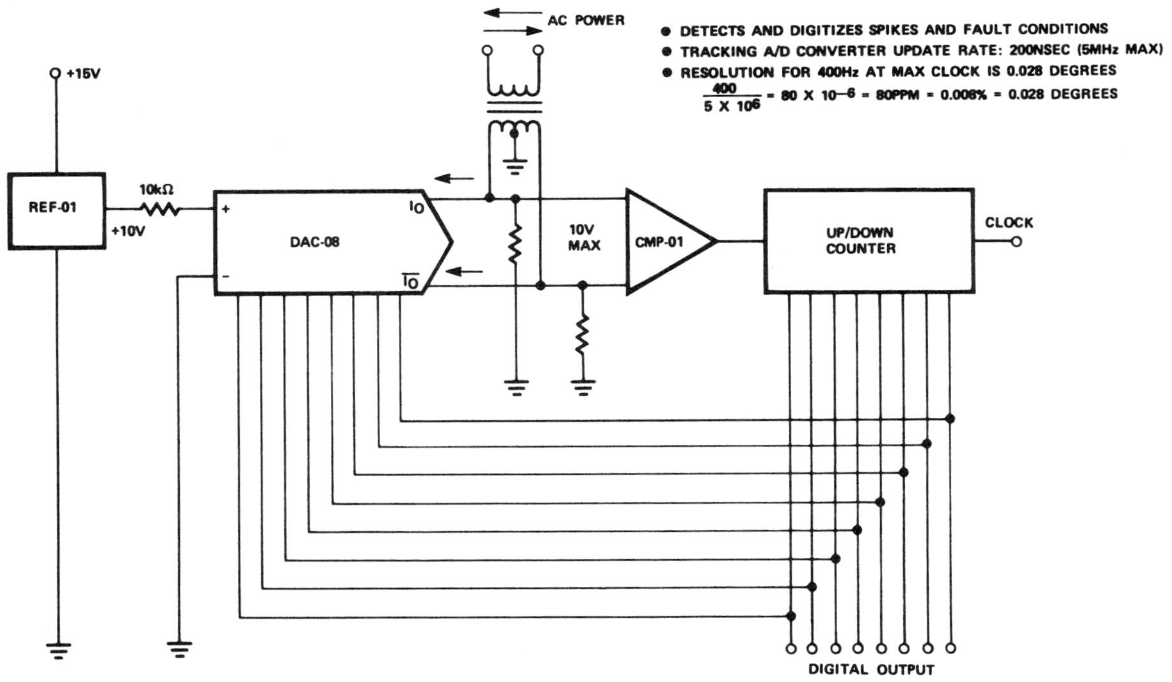

Power fault monitor and detector (PM).

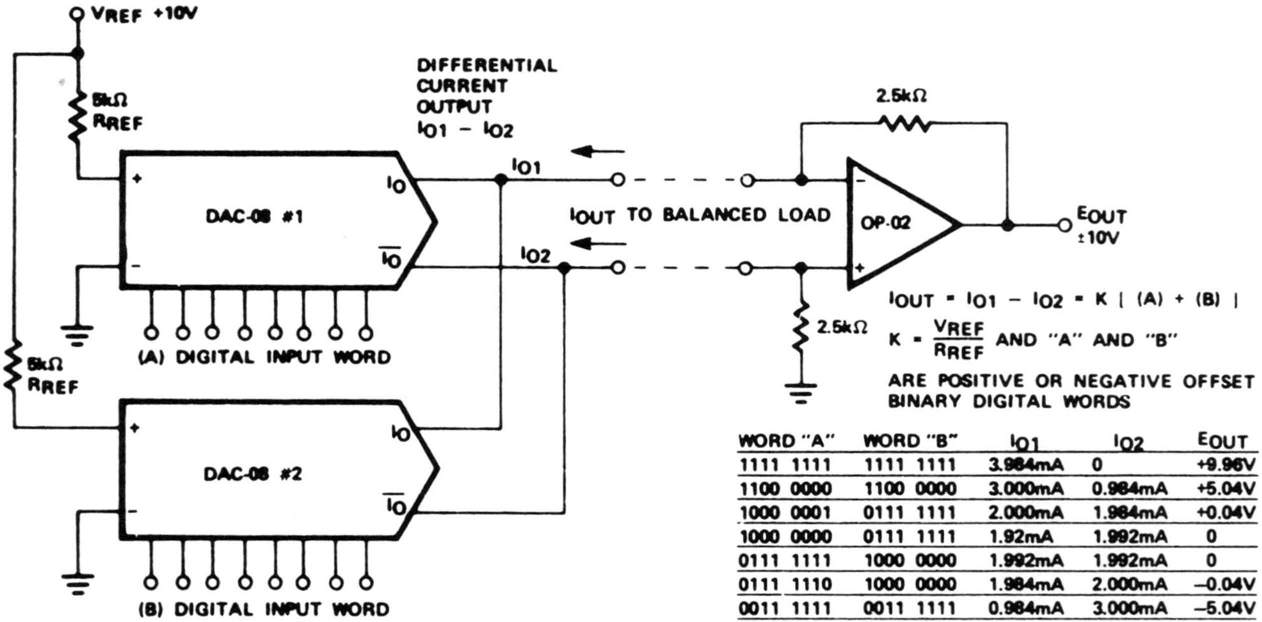

WORD "A"	WORD "B"	I_{O1}	I_{O2}	E_{OUT}
1111 1111	1111 1111	3.984mA	0	+9.96V
1100 0000	1100 0000	3.000mA	0.984mA	+5.04V
1000 0001	0111 1111	2.000mA	1.984mA	+0.04V
1000 0000	0111 1111	1.92mA	1.992mA	0
0111 1111	1000 0000	1.992mA	1.992mA	0
0111 1110	1000 0000	1.984mA	2.000mA	−0.04V
0011 1111	0011 1111	0.984mA	3.000mA	−5.04V
0000 0000	0000 0000	0	3.984mA	−9.96V

Four-quadrant algebraic digital computation (PM).

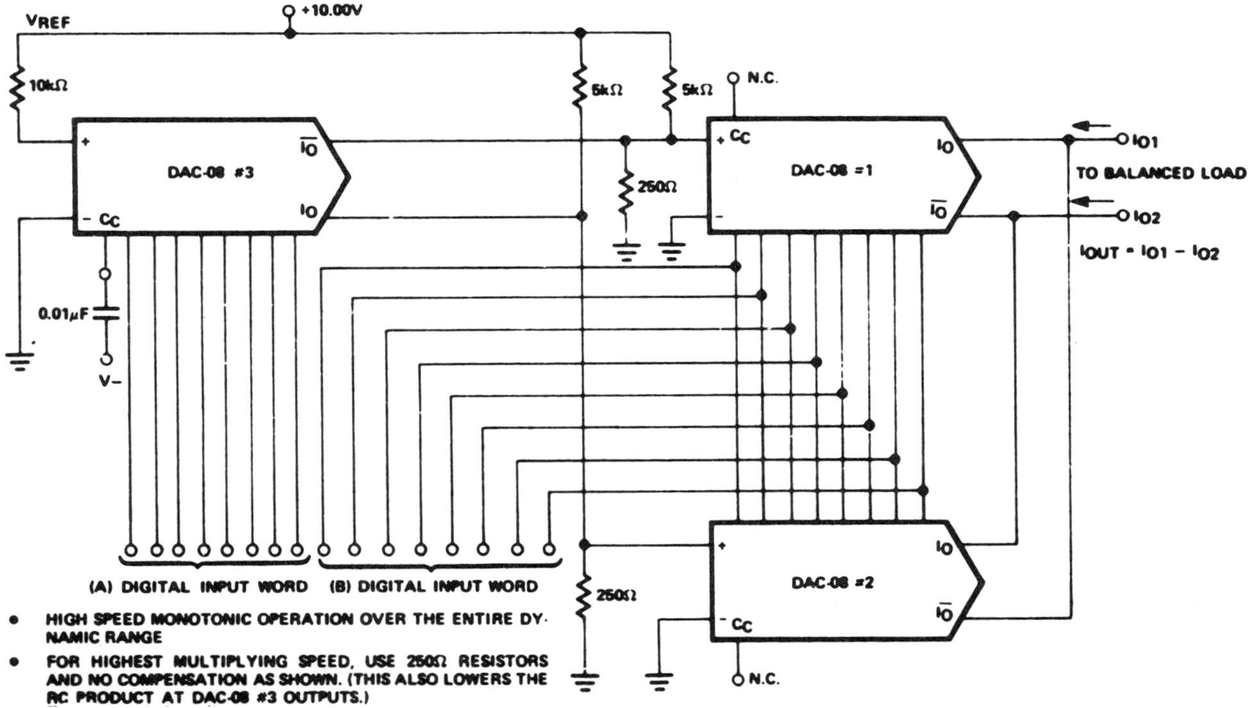

Four-quadrant 8-bit × 8-bit digital multiplier (PM).

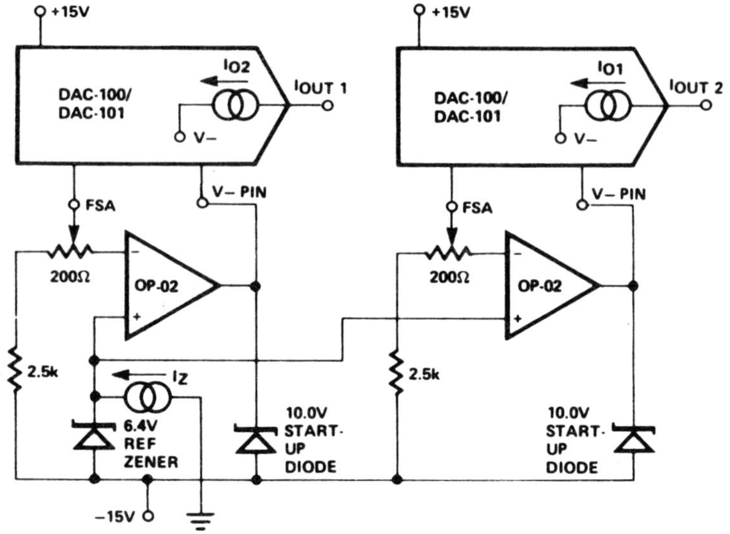

External reference connection (PM).

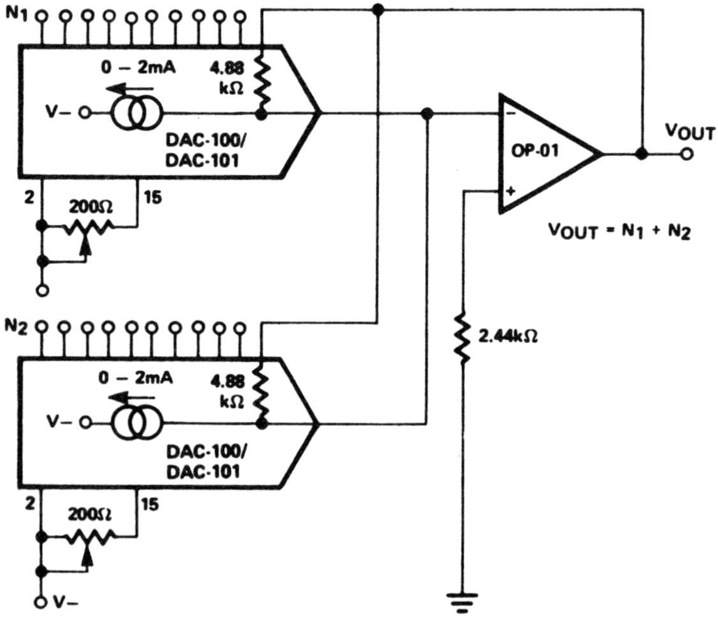

Analog sum of two digital numbers (PM).

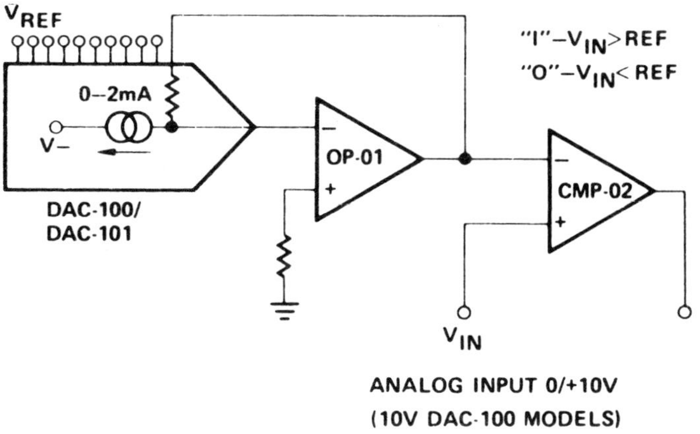

Digitally programmed level detector (PM).

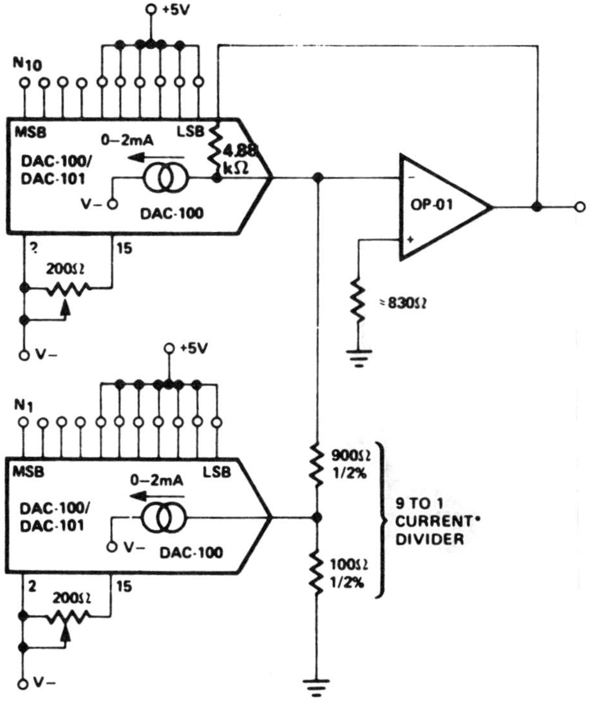

Binary-coded-decimal D/A conversion (PM).

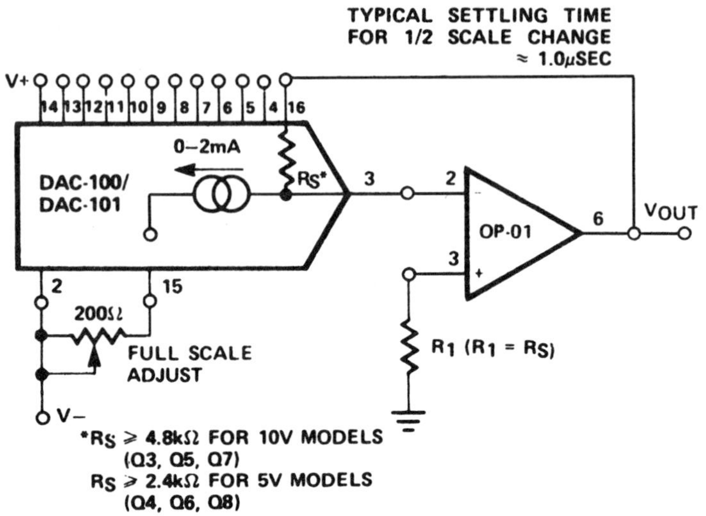

Basic unipolar voltage output circuit (PM).

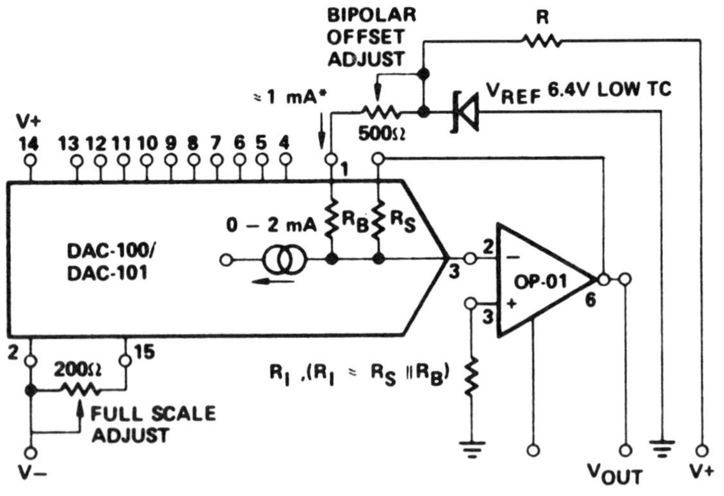

Basic bipolar voltage output circuit (PM).

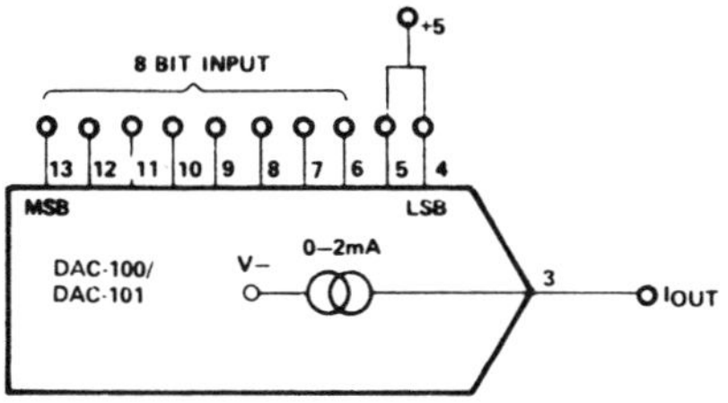

Reduced resolution application (PM).

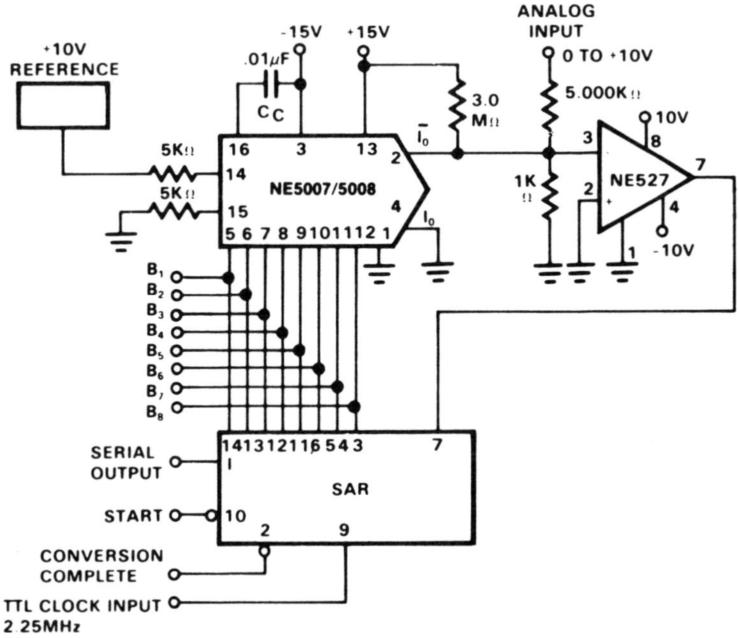

NOTE

Connect "start" to "conversion complete" for continuous conversions.

3 IC low cost A-to-D converter (S).

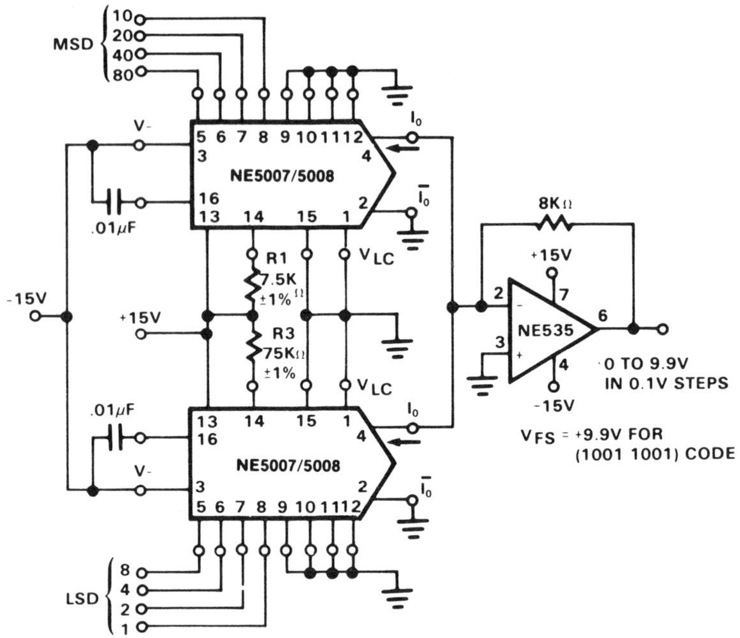

NOTE

Output is directly proportional to positive power supply.

Low cost 2-digit BCD DAC (S).

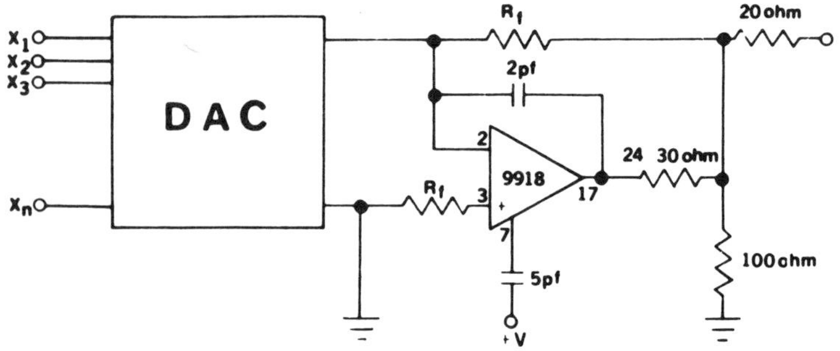

High speed digital-to-analog converter w/fast setting (OE).

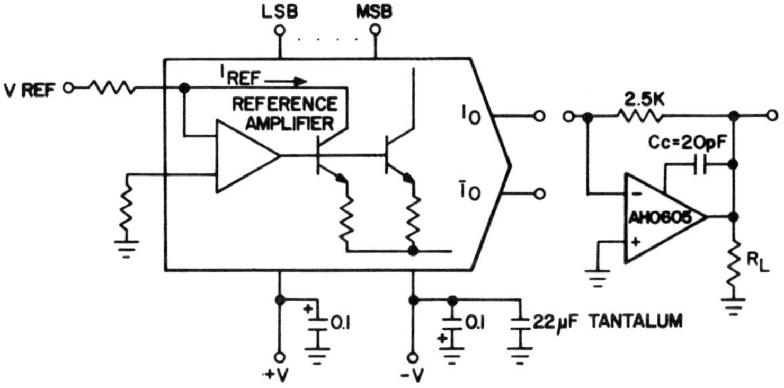

Current to voltage conversion for a high speed multiplying DAC (OE).

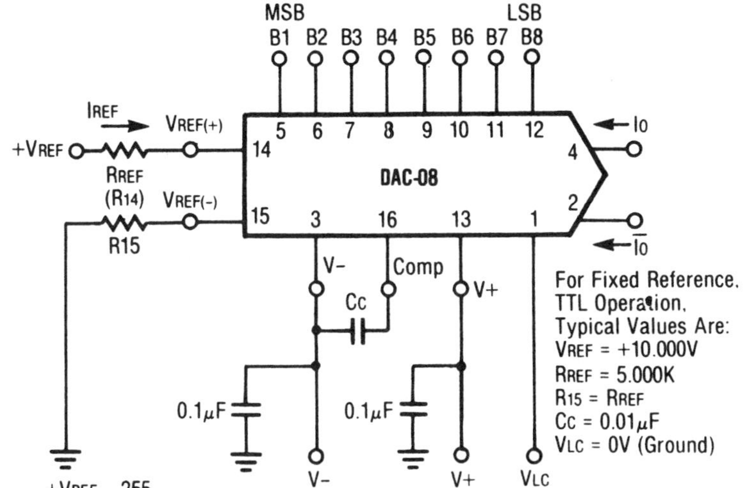

$$I_{FS} = \frac{+V_{REF}}{R_{REF}} \times \frac{255}{256}$$

$I_O + \overline{I_O} = I_{FS}$ For All Logic States

Basic positive reference operation (R).

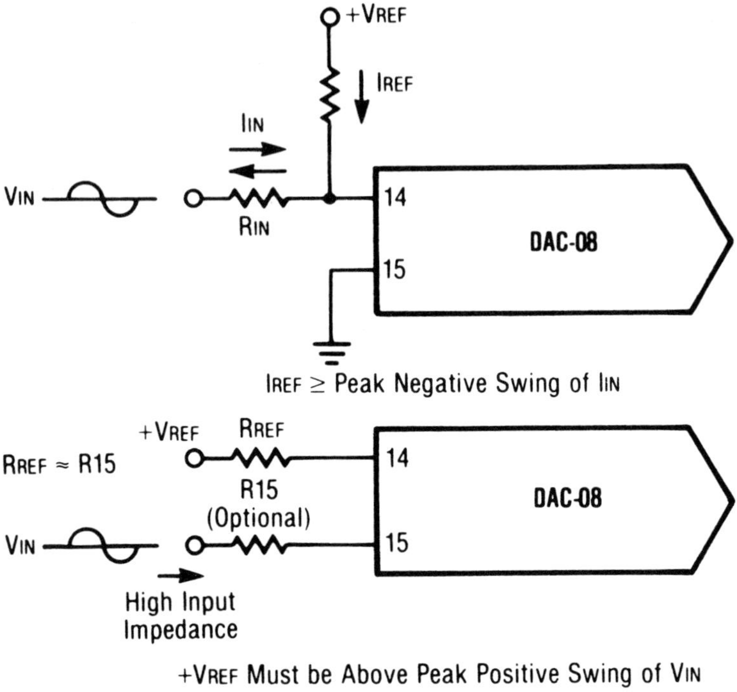

Accommodating bipolar references (R).

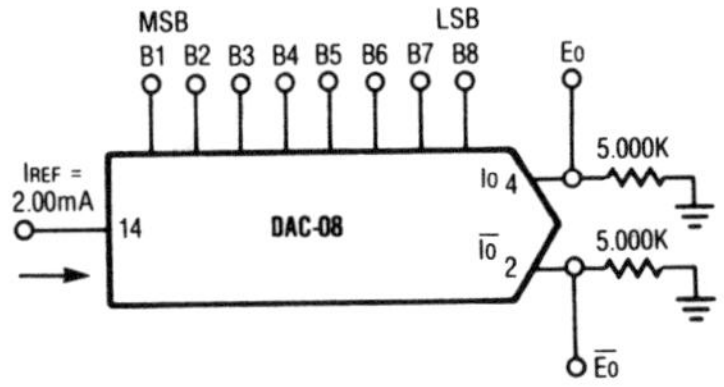

Scale	B1 B2 B3 B4 B5 B6 B7 B8	IomA	$\overline{\text{IomA}}$	Eo	$\overline{\text{Eo}}$
Full Scale	1 1 1 1 1 1 1 1	1.992	.000	-9.960	-.000
Half Scale +LSB	1 0 0 0 0 0 0 1	1.008	.984	-5.040	-4.920
Half Scale	1 0 0 0 0 0 0 0	1.000	.992	-5.000	-4.960
Half Scale -LSB	0 1 1 1 1 1 1 1	.992	1.000	-4.960	-5.000
Zero Scale +LSB	0 0 0 0 0 0 0 1	.008	1.984	-0.040	-9.920
Zero Scale	0 0 0 0 0 0 0 0	.000	1.992	.000	-9.960

Basic unipolar negative operation (R).

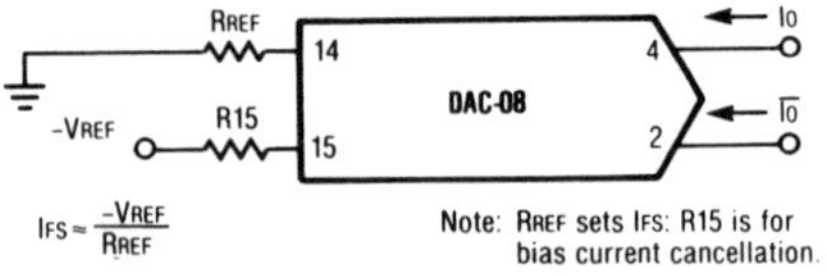

Basic negative reference operation (R).

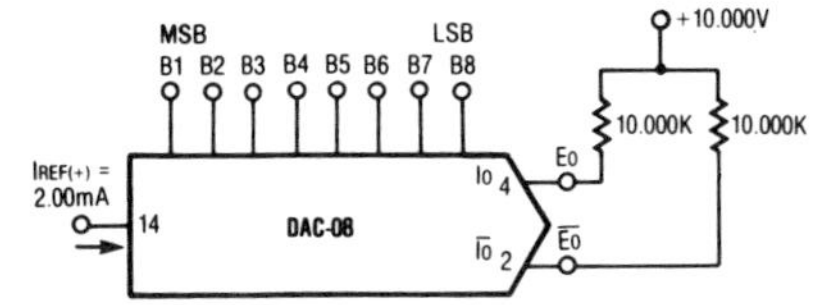

Scale	B1 B2 B3 B4 B5 B6 B7 B8	Eo	$\overline{\text{Eo}}$
Pos Full Scale	1 1 1 1 1 1 1 1	-9.920	+10.000
Pos Full Scale -LSB	1 1 1 1 1' 1 1 0	-9.840	+9.920
Zero Scale +LSB	1 0 0 0 0 0 0 1	-0.080	+0.160
Zero Scale	1 0 0 0 0 0 0 0	0.000	+0.080
Zero Scale -LSB	0 1 1 1 1 1 1 1	+0.080	0.000
Neg Full Scale +LSB	0 0 0 0 0 0 0 1	+9.920	-9.840
Neg Full Scale	0 0 0 0 0 0 0 0	+10.000	-9.920

Basic bipolar output operation (R).

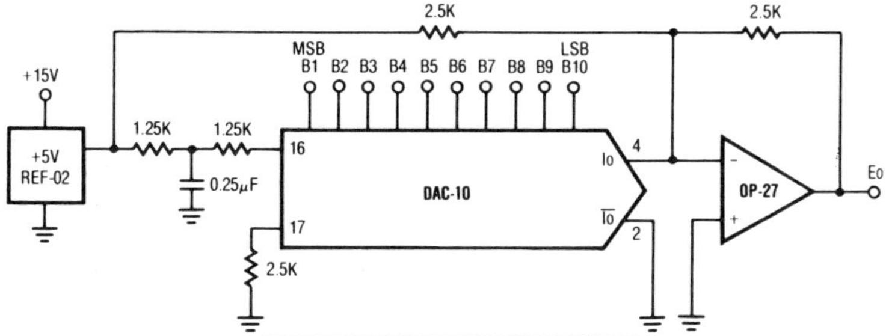

	B1 B2 B3 B4 B5 B6 B7 B8 B9 B10	Eo
Pos Full Scale	1 1 1 1 1 1 1 1 1 1	+4.990
Zero Scale	1 0 0 0 0 0 0 0 0 0	0.00
Neg Full Scale +1 LSB	0 0 0 0 0 0 0 0 0 1	-4.990
Neg Full Scale	0 0 0 0 0 0 0 0 0 0	-5.000

Offset binary operation (R).

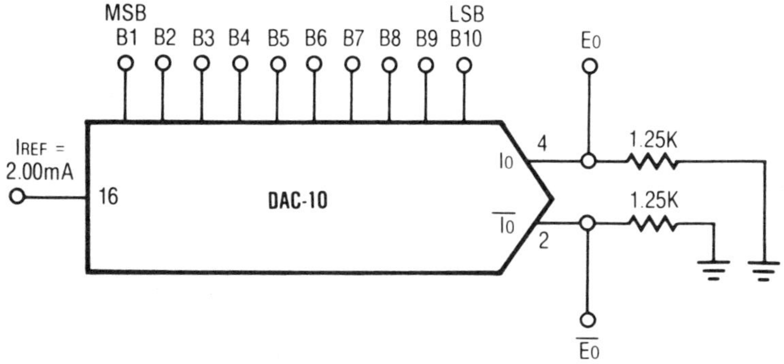

	B1 B2 B3 B4 B5 B6 B7 B8 B9 B10	IomA	$\overline{\text{IomA}}$	Eo	$\overline{\text{Eo}}$
Full Scale	1 1 1 1 1 1 1 1 1 1	3.996	0.000	-4.995	-0.000
Half Scale +LSB	1 0 0 0 0 0 0 0 0 1	2.004	1.992	-2.505	-2.490
Half Scale	1 0 0 0 0 0 0 0 0 0	2.000	1.996	-2.500	-2.495
Half Scale -LSB	0 1 1 1 1 1 1 1 1 1	1.996	2.000	-2.495	-2.500
Zero Scale +LSB	0 0 0 0 0 0 0 0 0 1	0.004	3.992	-0.005	-4.990
Zero Scale	0 0 0 0 0 0 0 0 0 0	0.000	3.996	-0.000	-4.995

Basic unipolar negative operation (R).

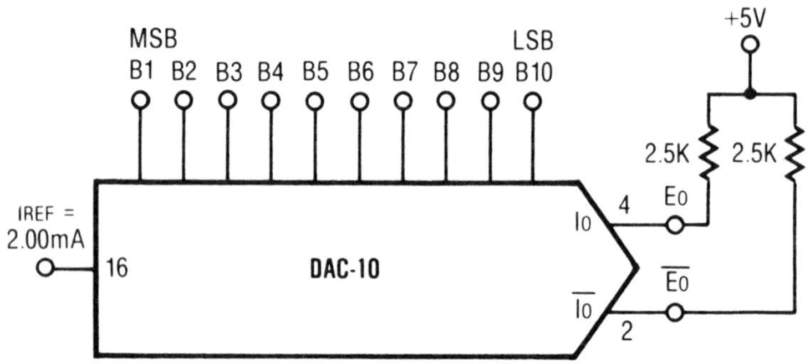

	B1 B2 B3 B4 B5 B6 B7 B8 B9 B10	Eo	$\overline{Eo}$
Pos Full Scale	1 1 1 1 1 1 1 1 1 1	-4.990	+5.000
Pos Full Scale -LSB	1 1 1 1 1 1 1 1 1 0	-4.980	+4.990
Zero Scale +LSB	1 0 0 0 0 0 0 0 0 1	-0.010	+0.020
Zero Scale	1 0 0 0 0 0 0 0 0 0	0.000	+0.010
Zero Scale -LSB	0 1 1 1 1 1 1 1 1 1	+0.010	0.000
Neg Full Scale +LSB	0 0 0 0 0 0 0 0 0 1	+4.990	-4.980
Neg Full Scale	0 0 0 0 0 0 0 0 0 0	+5.000	-4.990

Basic bipolar output operation (R).

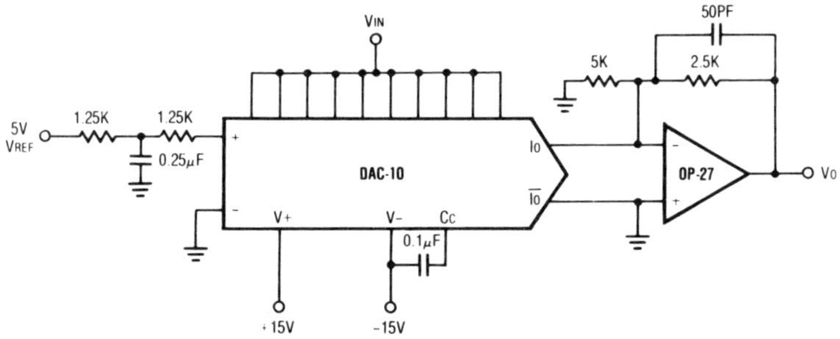

Settling time using OP-27 (R).

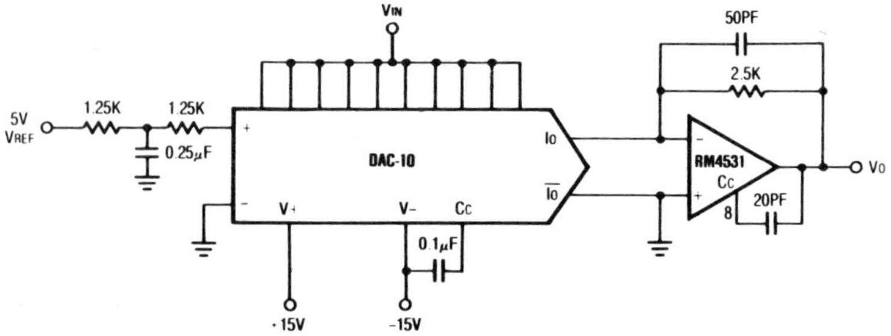

Settling time using RM4531 (R).

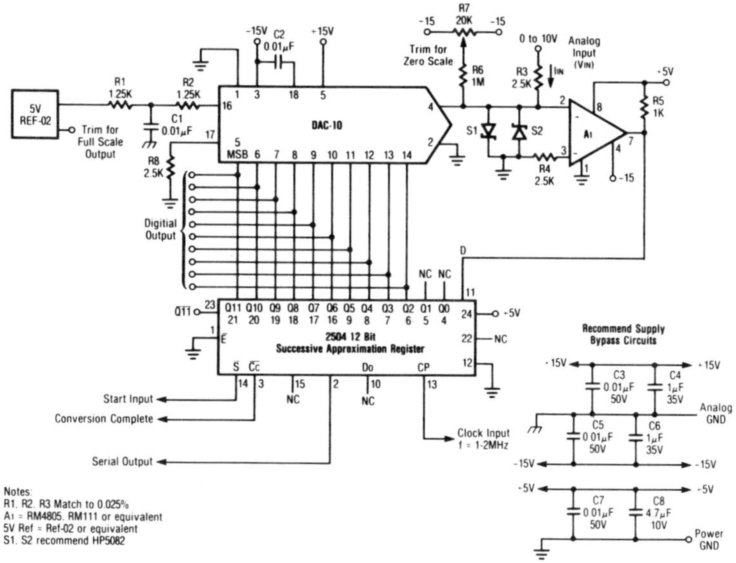

10-bit successive approximation A/D converter (R).

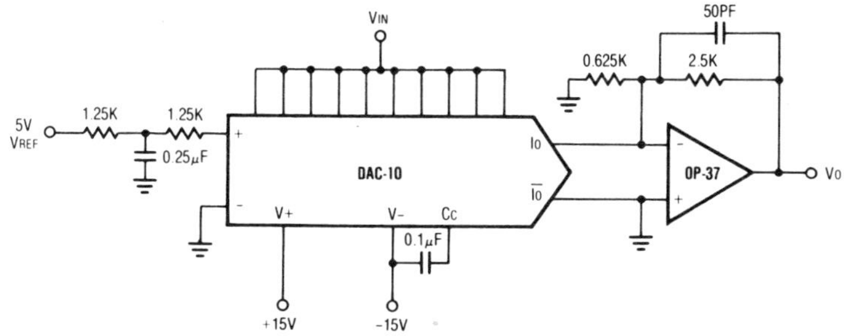

Settling time using OP-37 (R).

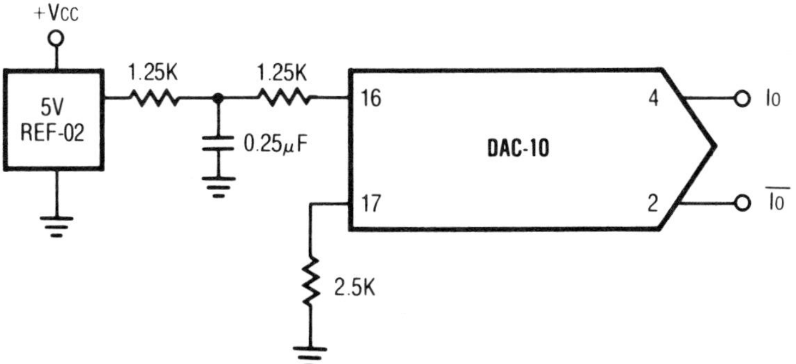

Input reference noise limiting filter (R).

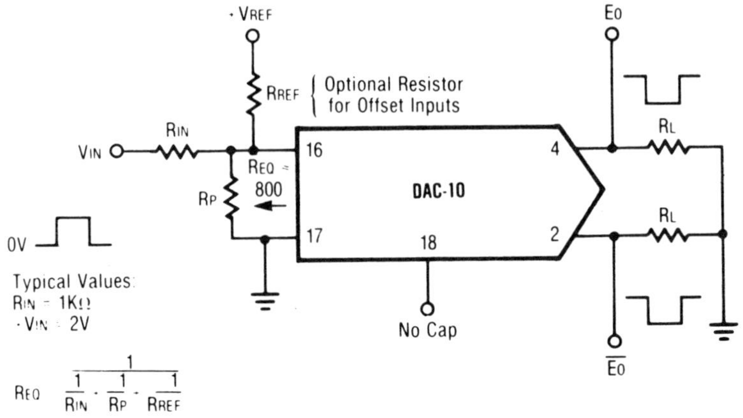

Pulsed reference operation (R).

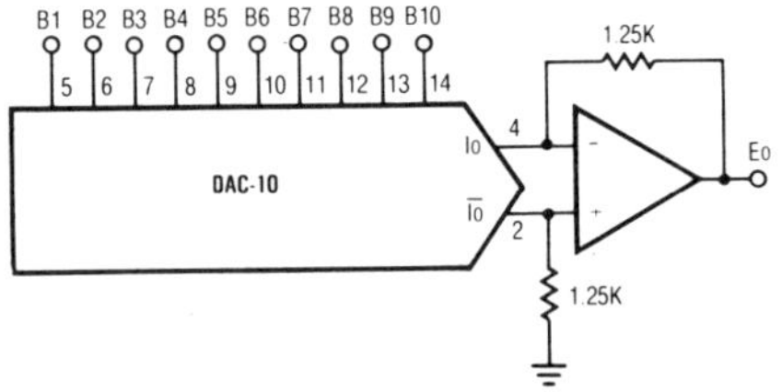

	B1	B2	B3	B4	B5	B6	B7	B8	B9	B10	E0
Pos Full Scale	1	1	1	1	1	1	1	1	1	1	+4.995
Pos Full Scale -LSB	1	1	1	1	1	1	1	1	1	0	+4.985
(+) Zero Scale	1	0	0	0	0	0	0	0	0	0	+0.005
(-) Zero Scale	0	1	1	1	1	1	1	1	1	1	-0.005
Neg Full Scale +LSB	0	0	0	0	0	0	0	0	0	1	-4.985
Neg Full Scale	0	0	0	0	0	0	0	0	0	0	-4.995

Bipolar operation (R).

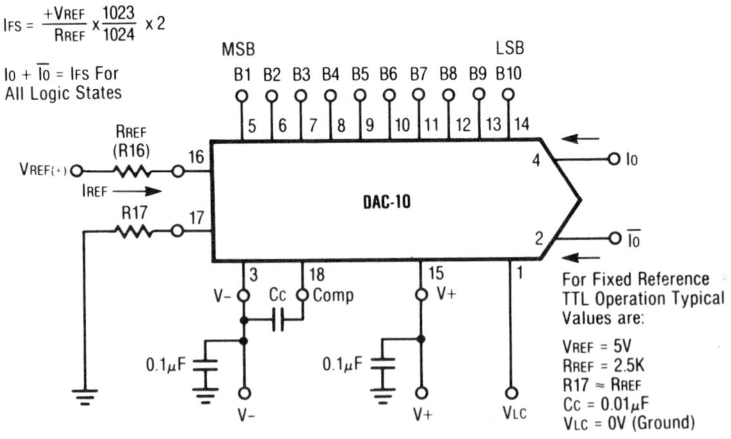

Positive reference operation (R).

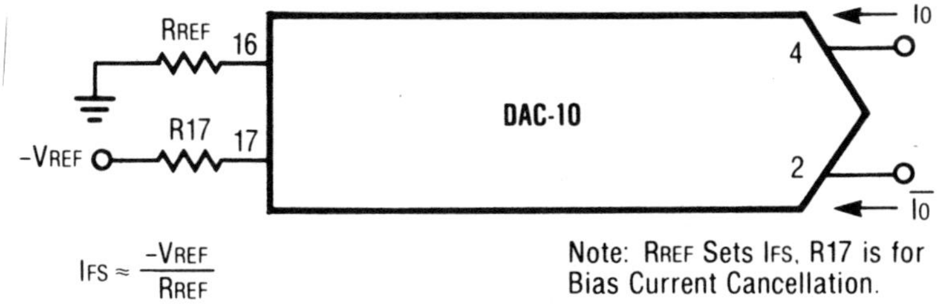

Negative reference operation (R).

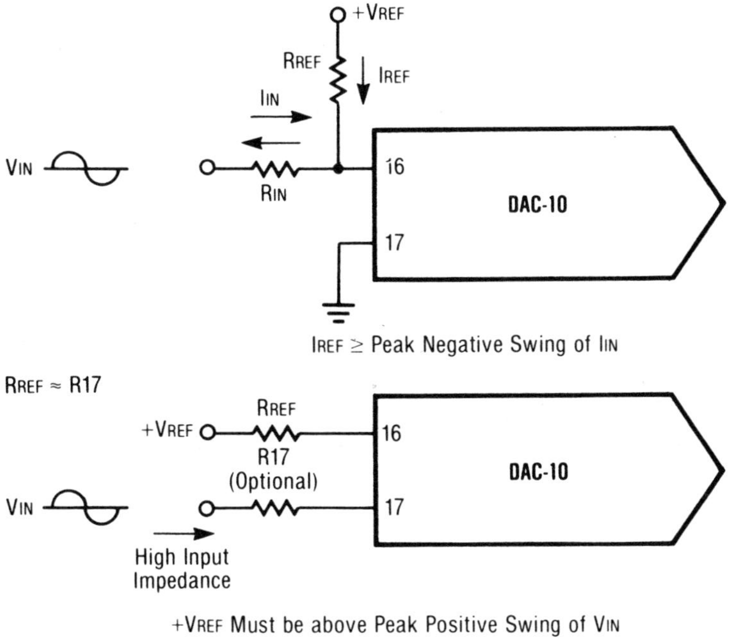

Providing offsets to accommodate bipolar references (R).

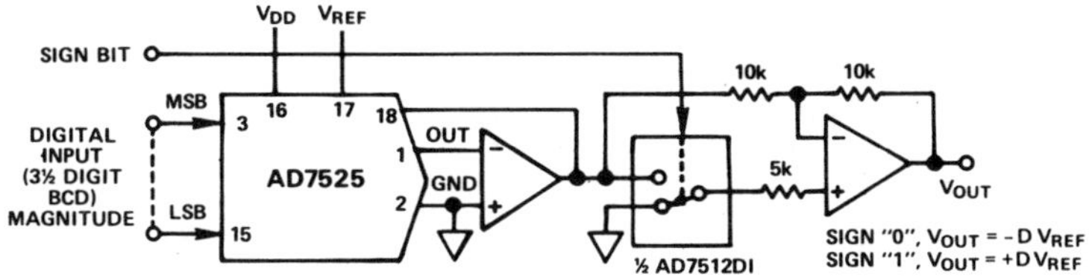

Sign/magnitude DAC (AD).

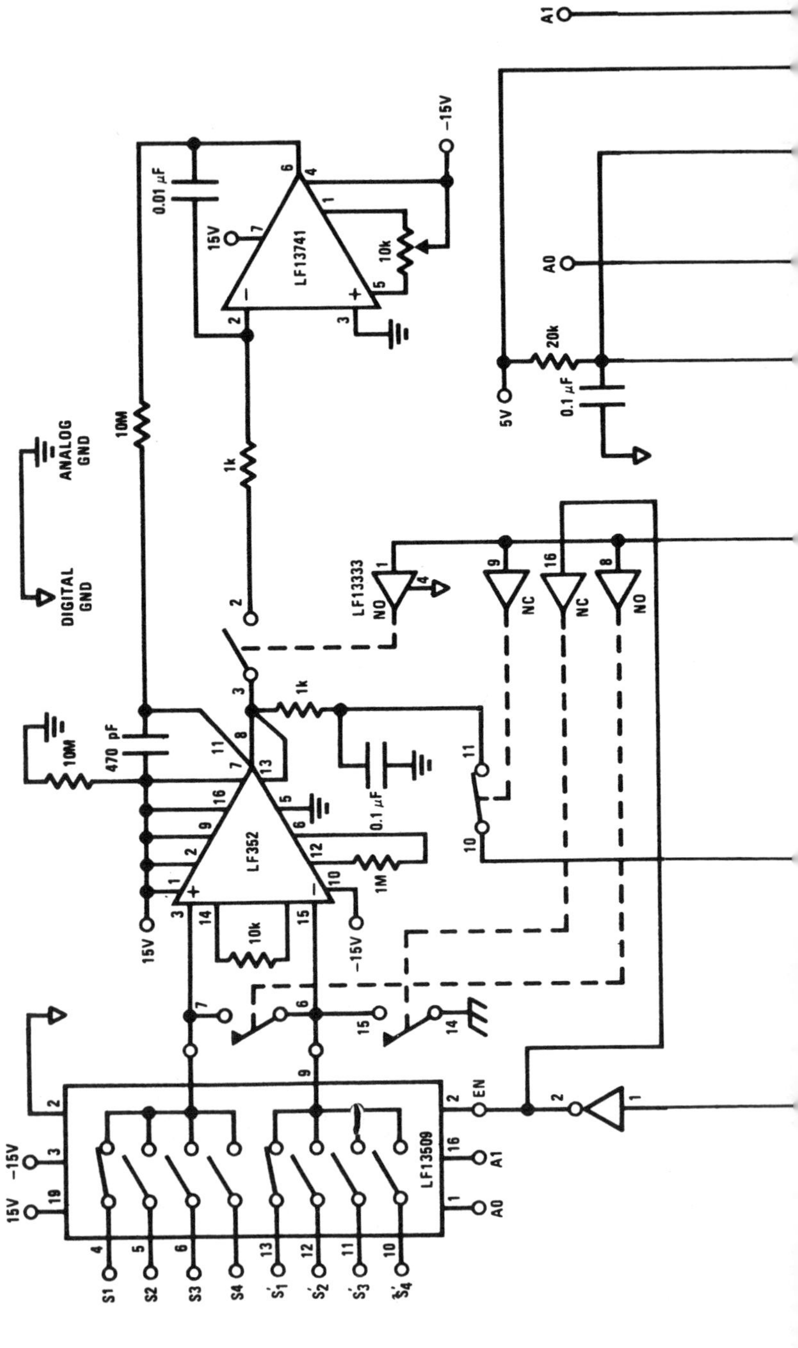
ANALOG GND
DIGITAL GND
10M
0.01 μF
−15V
15V
LF13741
10k
1k
20k
0.1 μF
5V
A0
A1
LF13333
NO
NC
470 pF
LF352
1M
LF13509
EN
S1
S2
S3
S4
S′1
S′2
S′3
S′4

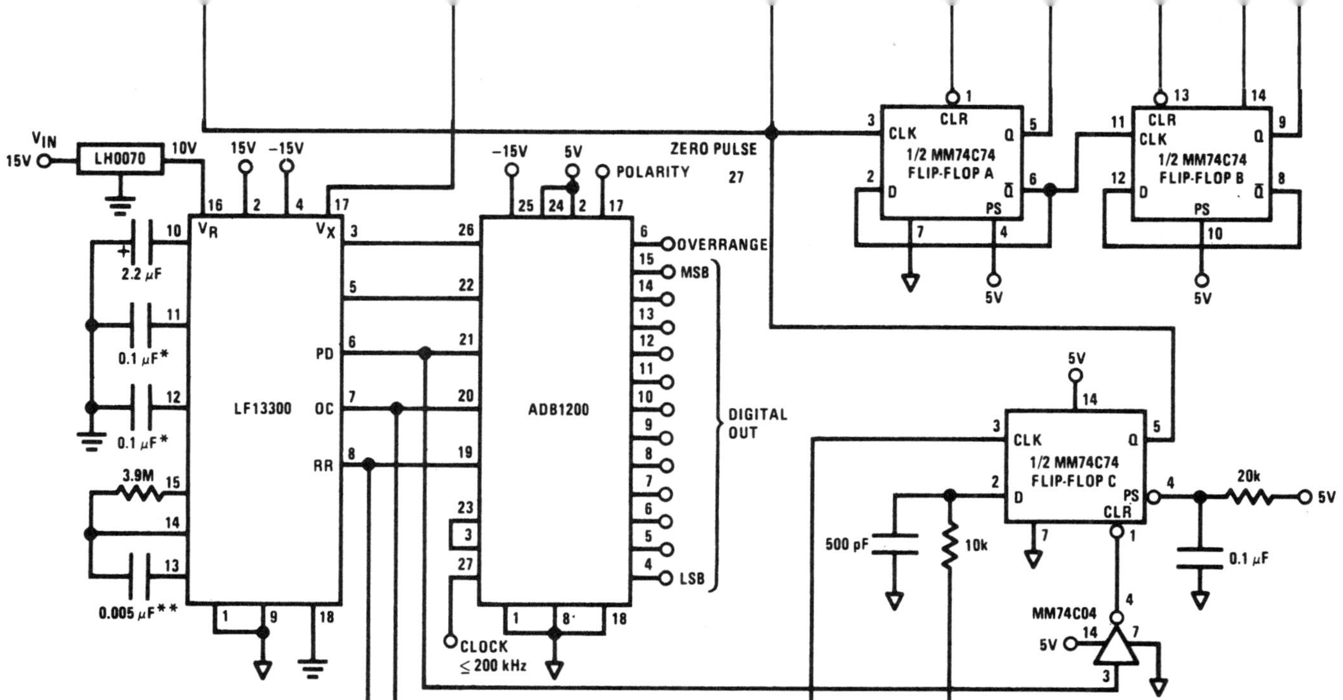

*Low leakage mylar

**Polypropylene

$f_{CLOCK\ MAX}$ = 200 kHz

4-channel differential multiplexer with autozeroed instrumentation amplifier and 12-bit A/D converter (NS).

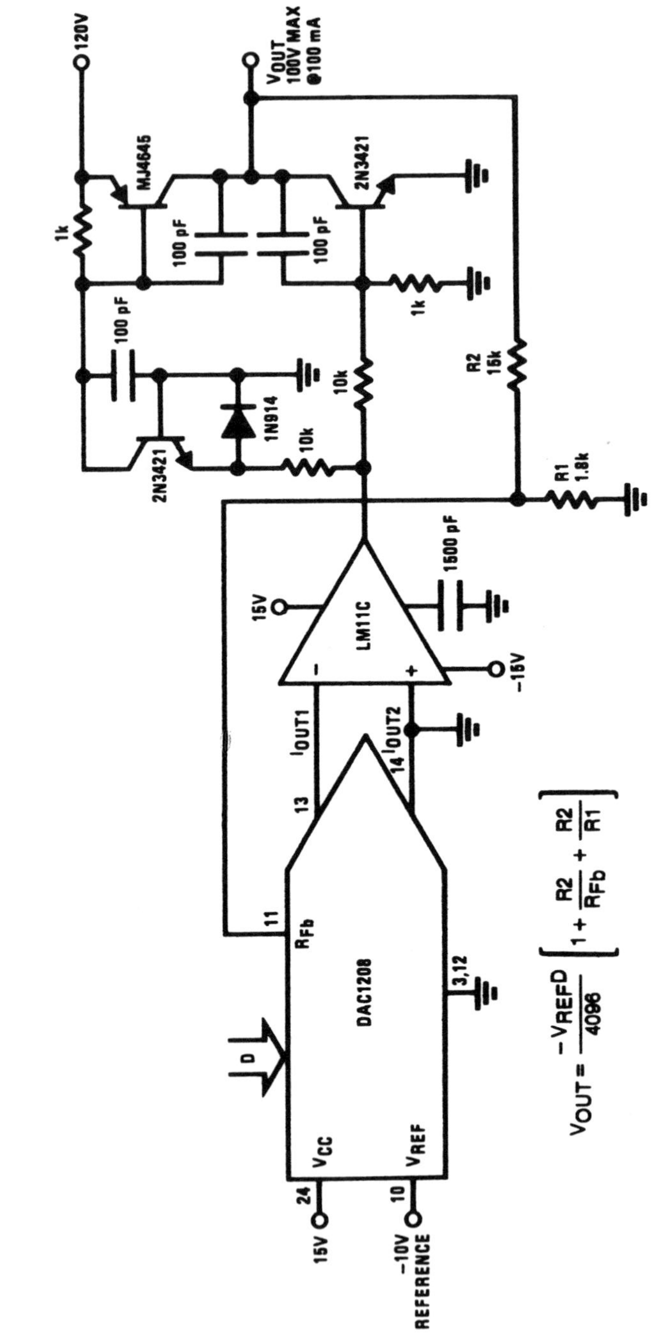

High voltage, power DAC (NS).

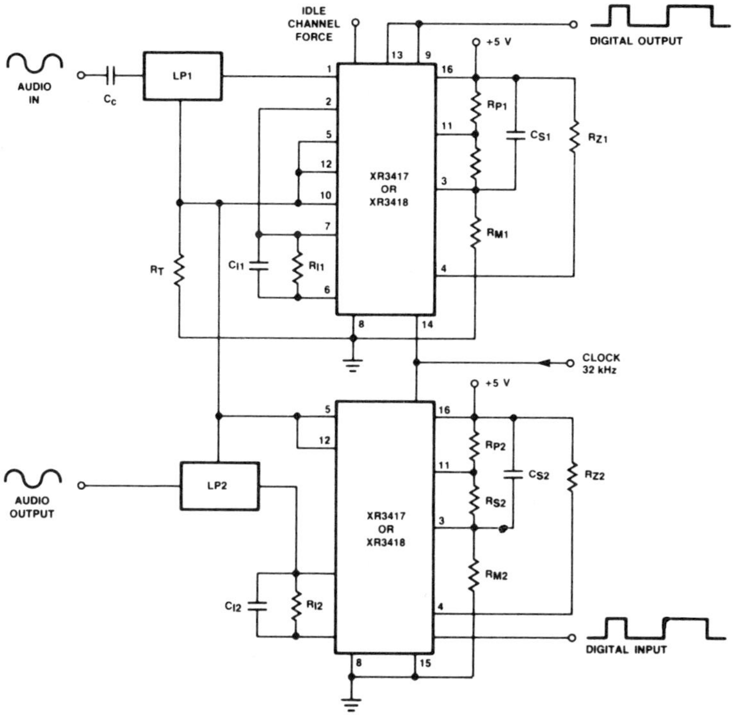

LP1 and LP2 — Lowpass Filter (Design for System Requirements)

Full duplex voice codec (E).

Chapter 10

Audio and Video Circuits

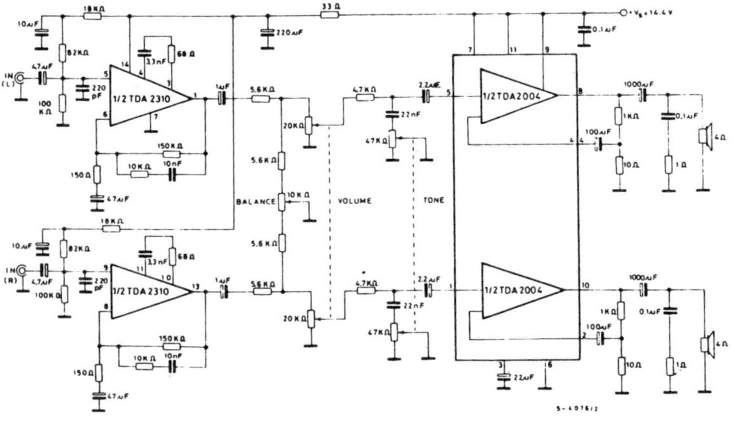

High quality, 10 + 10W cassette player (SGS).

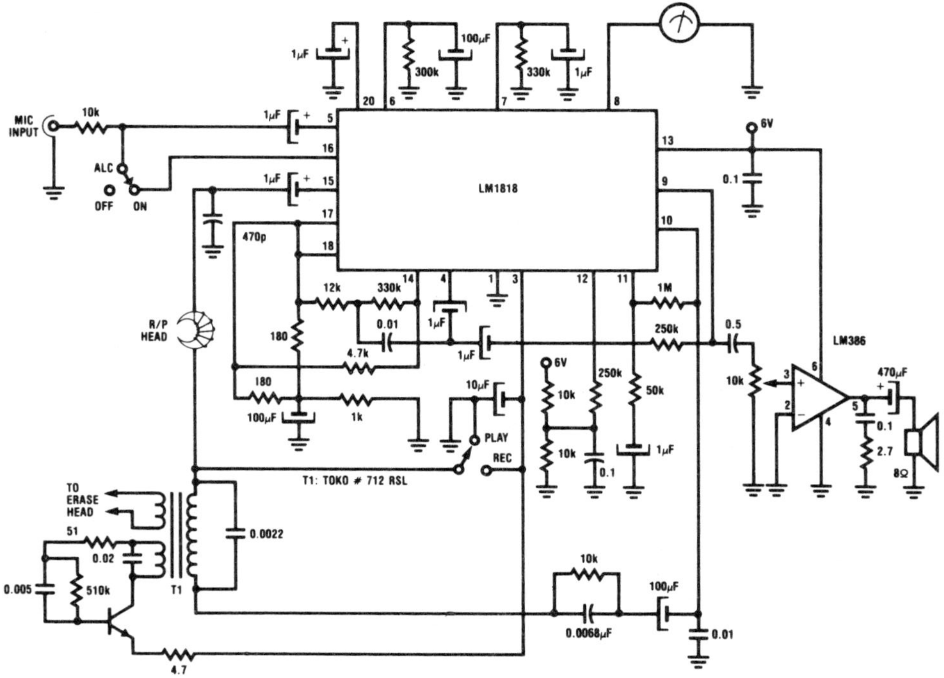

Monaural cassette recorder (NS).

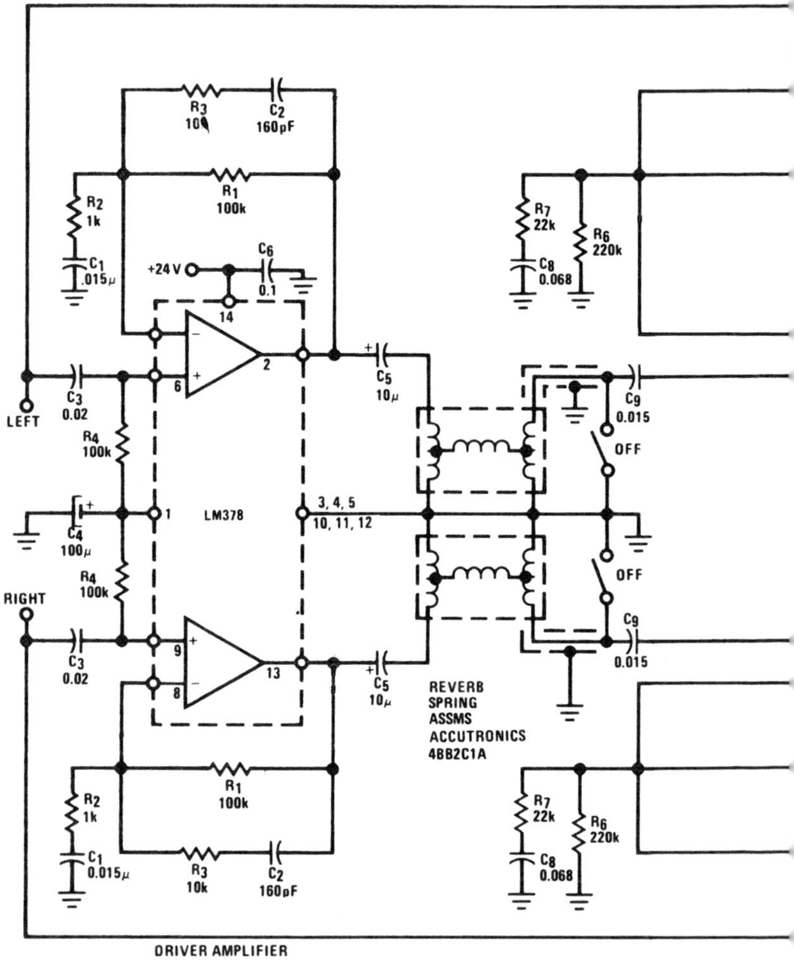

Stereo reverb system (NS).

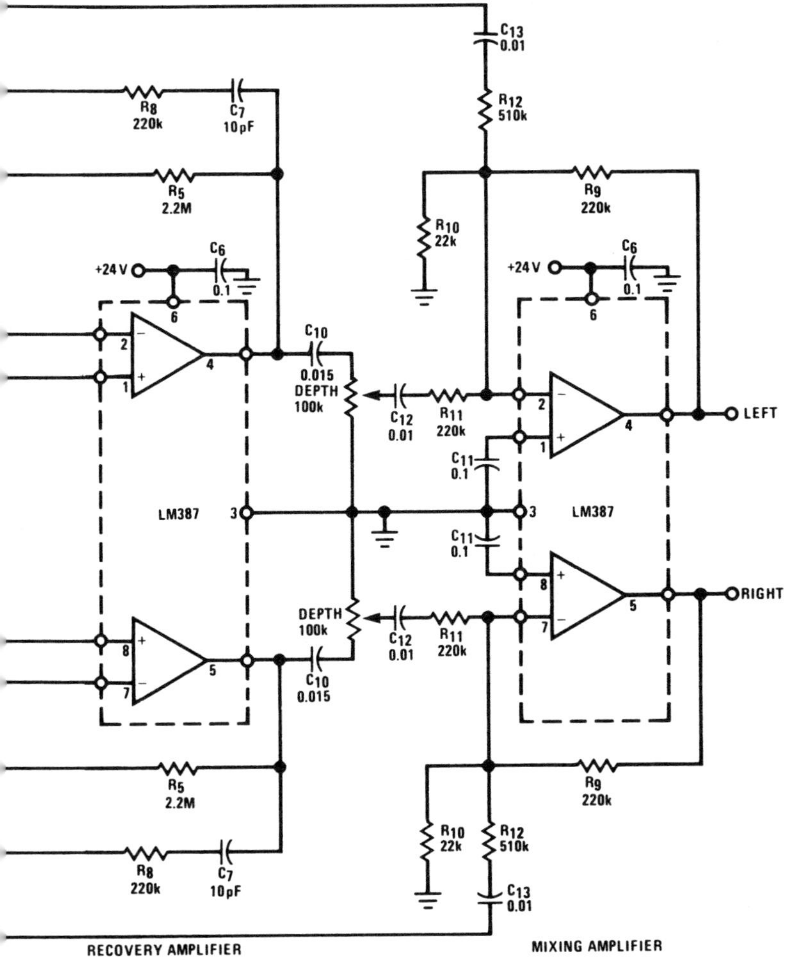
C13
0.01
R8
220k
C7
10pF
R12
510k
R5
2.2M
R9
220k
R10
22k
C6
+24 V
0.1
C6
+24 V
0.1
6
6
C10
2
0.015
4
DEPTH
100k
1
C12
0.01
R11
220k
2
4
LEFT
1
C11
0.1
LM387
3
3
LM387
C11
0.1
8
RIGHT
DEPTH
100k
C12
0.01
R11
220k
5
7
8
5
C10
0.015
7
R5
2.2M
R9
220k
R10
22k
R12
510k
R8
220k
C7
10pF
C13
0.01
RECOVERY AMPLIFIER
MIXING AMPLIFIER

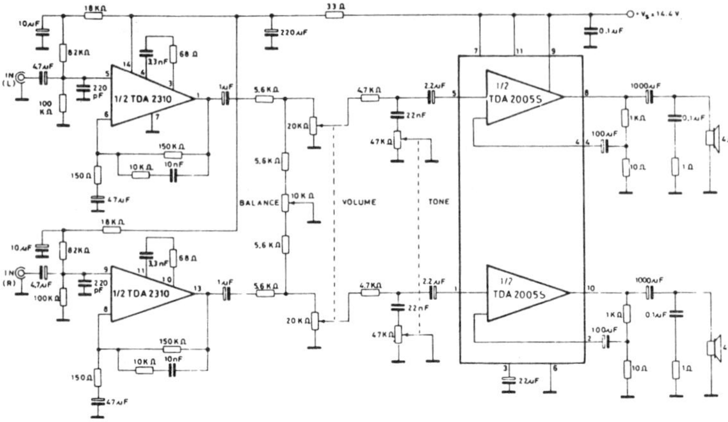

High quality, 10 + 10W car radio tape player (SGS).

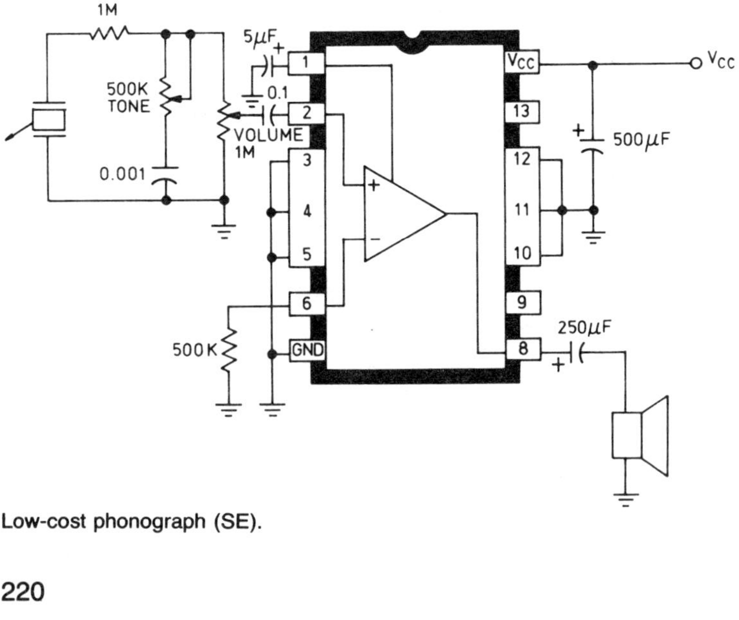

Low-cost phonograph (SE).

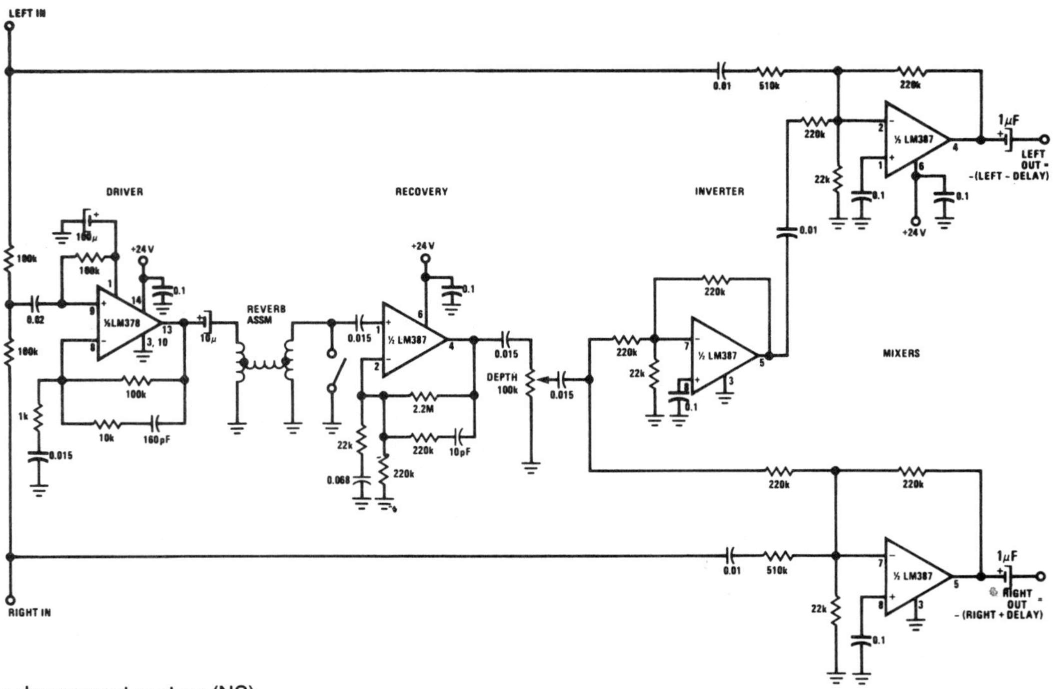

Stereo reverb enhancement system (NS).

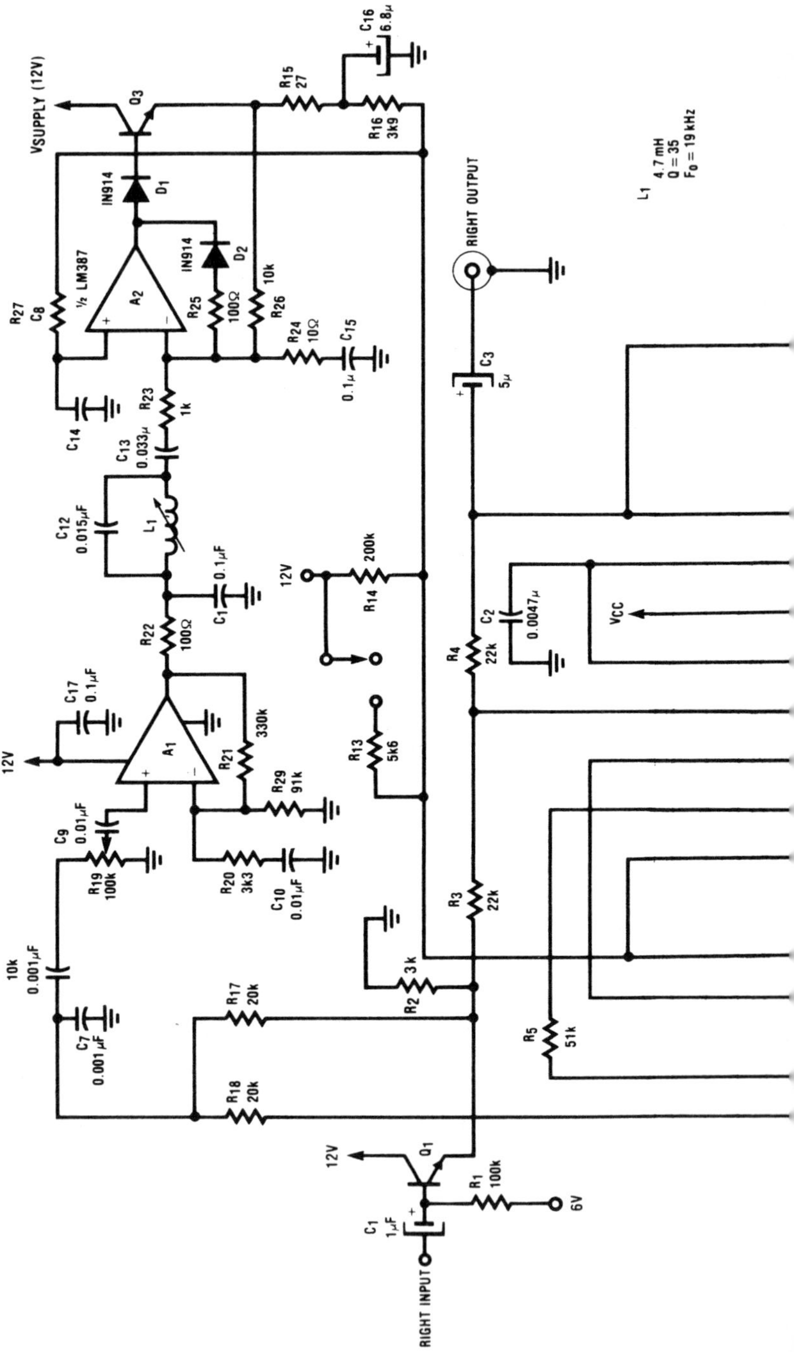

VSUPPLY (12V)
Q3
R15 27
C16 6.8μ
R16 3k9
R27 C8
½ LM387
A2
IN914 D1
IN914 D2
R25 100Ω
R26 10k
R24 10Ω
C15 0.1μ
RIGHT OUTPUT
L1 4.7 mH Q = 35 F0 = 19 kHz
C14
R23 1k
C13 0.033μ
C12 0.015μF
L1
C1 0.1μF
12V
R14 200k
C3 5μ
R22 100Ω
C17 0.1μF
12V
A1
R21 330k
R13 5k6
C2 0.0047μ
VCC
R4 22k
C9 0.01μF
R19 100k
R29 91k
R20 3k3
C10 0.01μF
R3 22k
10k 0.001μF
C7 0.001μF
R17 20k
R2 3k
R5 51k
R18 20k
12V
Q1
C1 1μF
R1 100k
6V
RIGHT INPUT

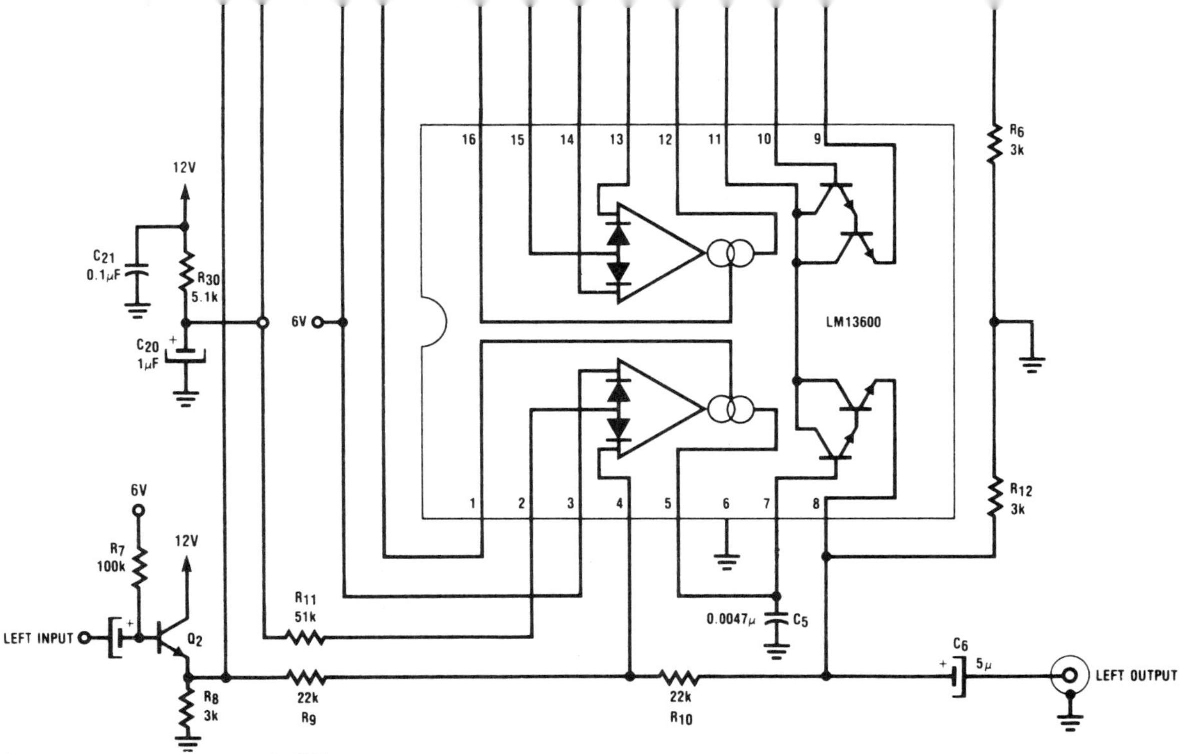

Stereo noise reduction circuit (NS).

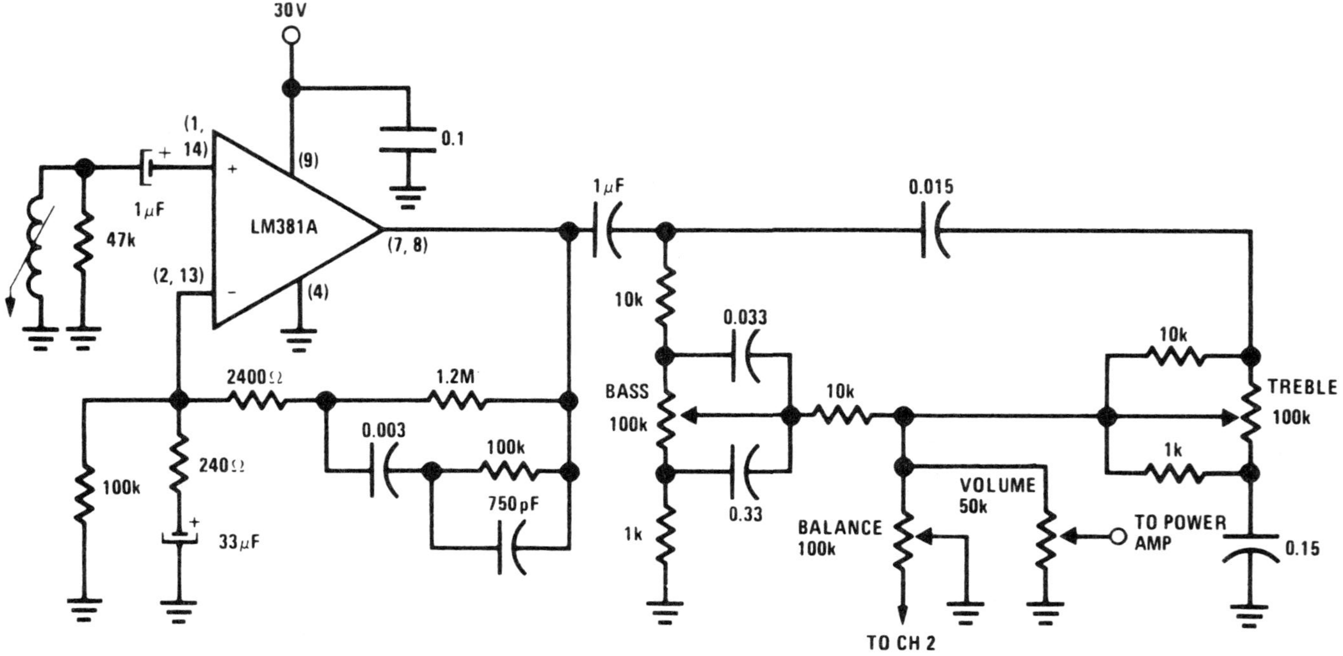

Single channel of complete phono preamp (NS).

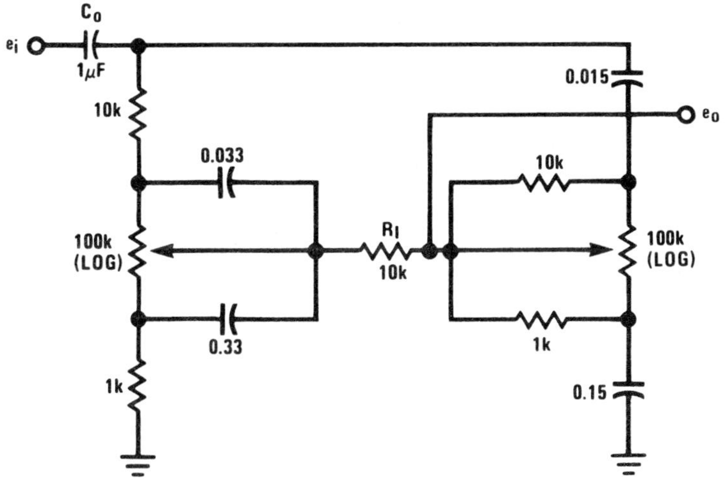

Complete passive bass & treble tone control (NS).

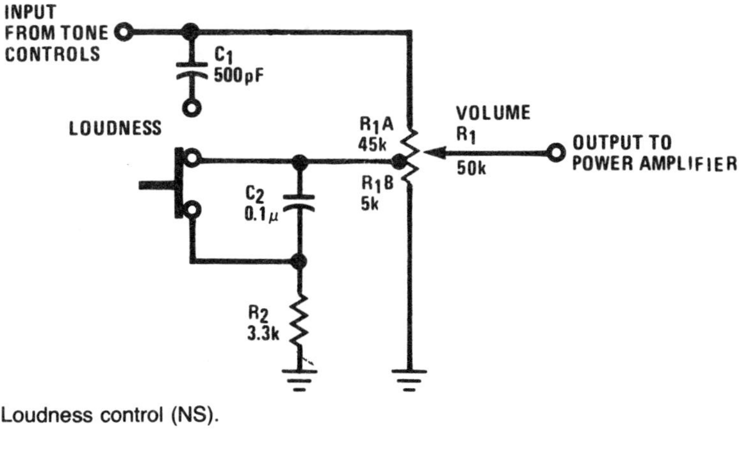

Loudness control (NS).

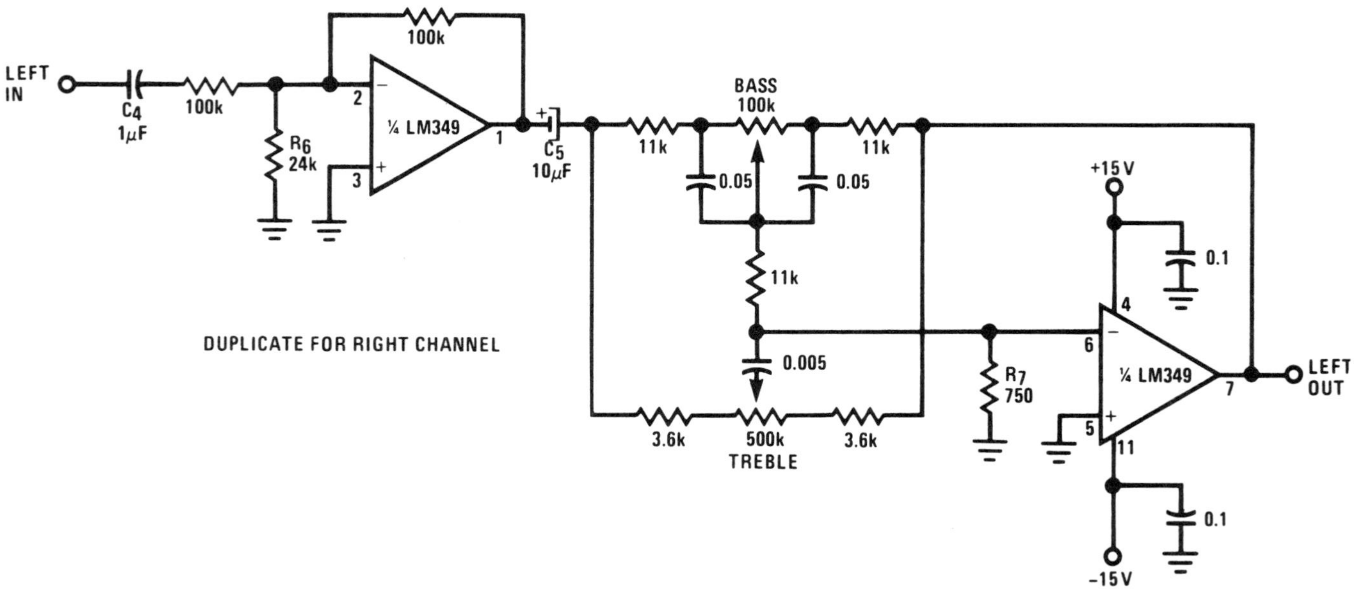

Typical active bass & treble tone control with buffer (NS).

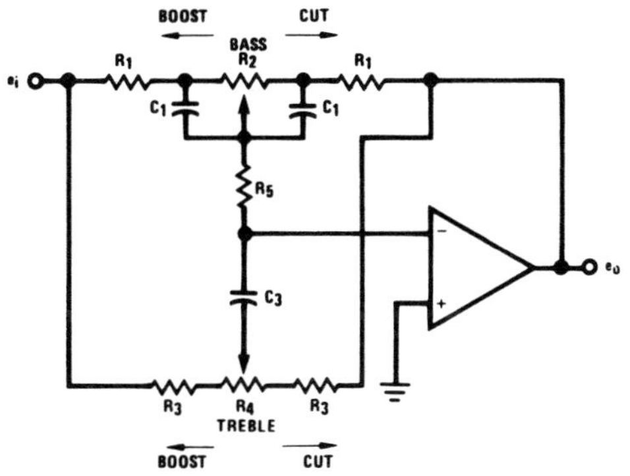

BASS

$$f_L = \frac{1}{2\pi R_2 C_1}$$

$$f_{LB} = \frac{1}{2\pi R_1 C_1}$$

$$A_{VB} = 1 + \frac{R_2}{R_1}$$

ASSUMES $R_2 > R_1$

TREBLE

$$f_H = \frac{1}{2\pi R_3 C_3}$$

$$f_{HB} = \frac{1}{2\pi (R_1 + R_3 + 2R_5) C_3}$$

$$A_{VT} = 1 + \frac{R_1 + 2R_5}{R_3}$$

ASSUMES $R_4 > R_1 + R_3 + 2R_5$

Bass and treble active tone control (NS).

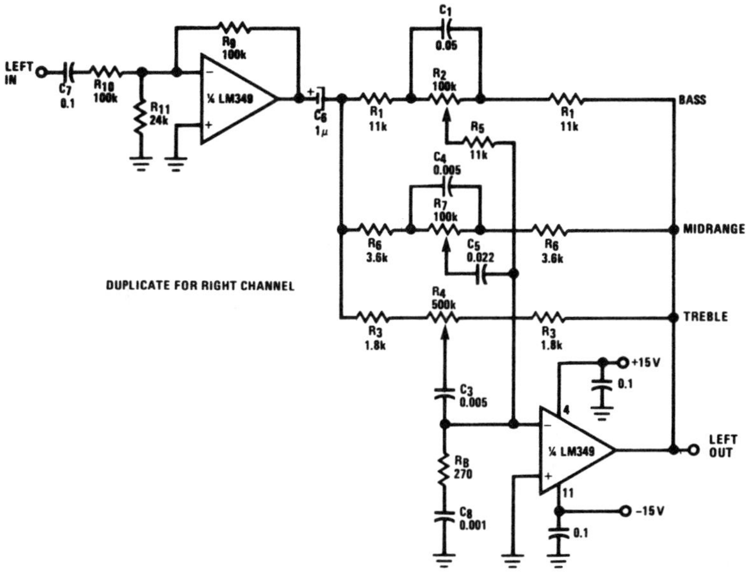

Three band active tone control (bass, midrange & treble) (NS).

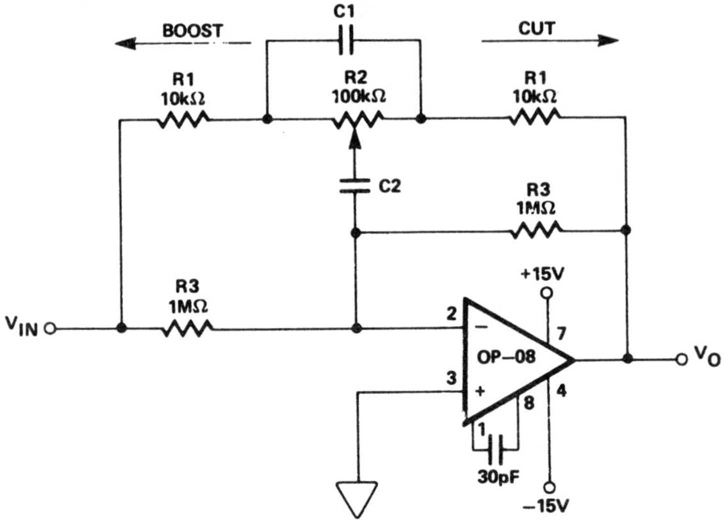

Octave equalizer (PM).

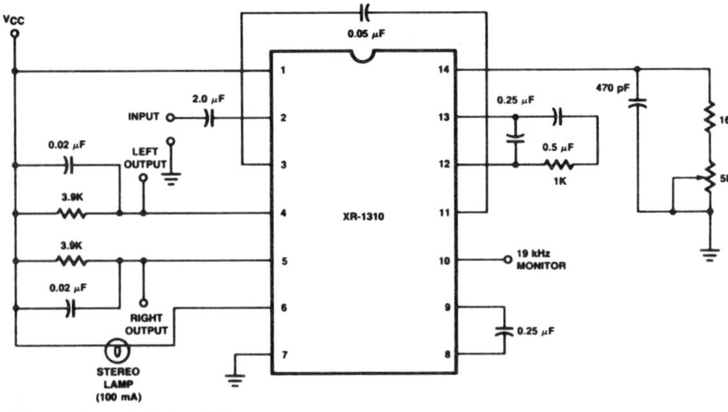

Stereo demodulator (E).

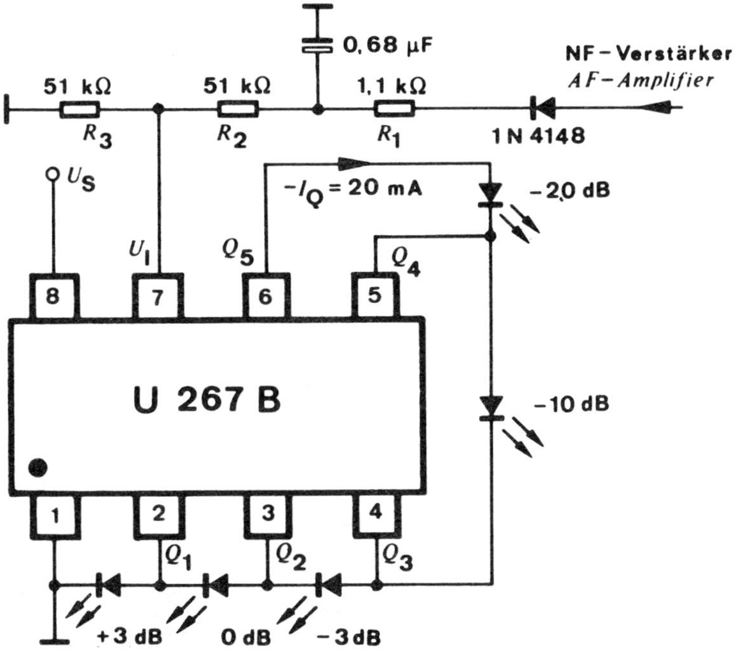

AF recording level display with 5 LED's and logarithmic behavior (AEG).

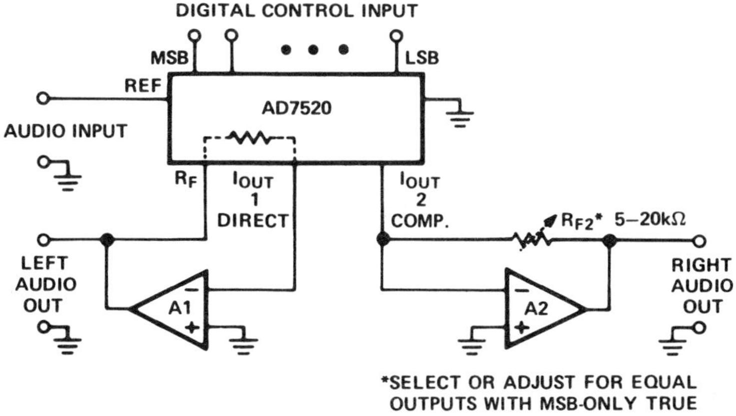

Simplified audio panner (AD).

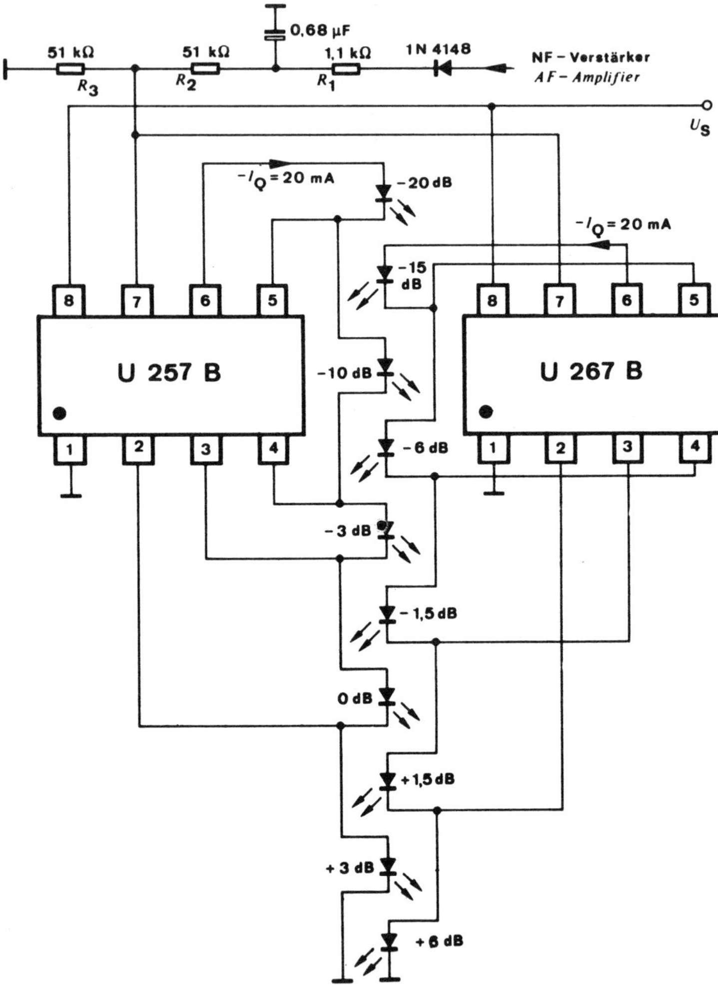

AF recording level control with 10 LED's and logarithmic behavior (AEG).

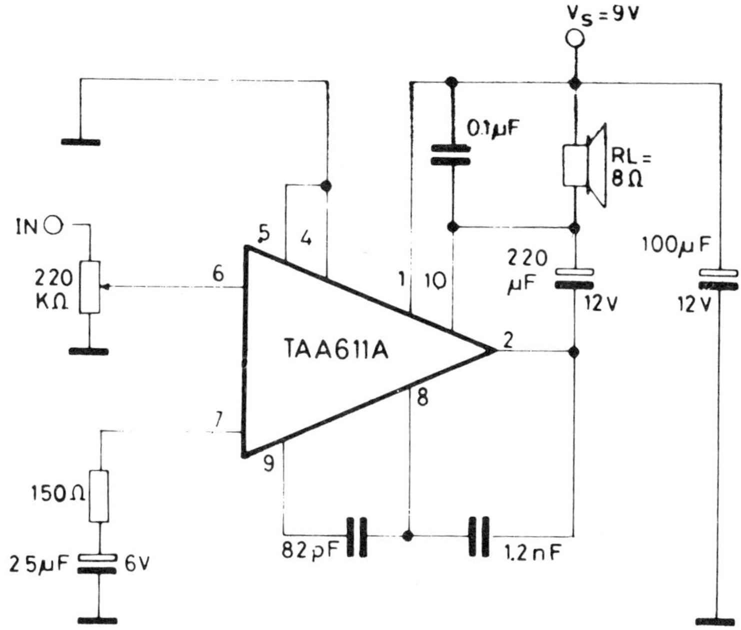

Audio amplifier for record-player (Gv = 34 dB) (SGS).

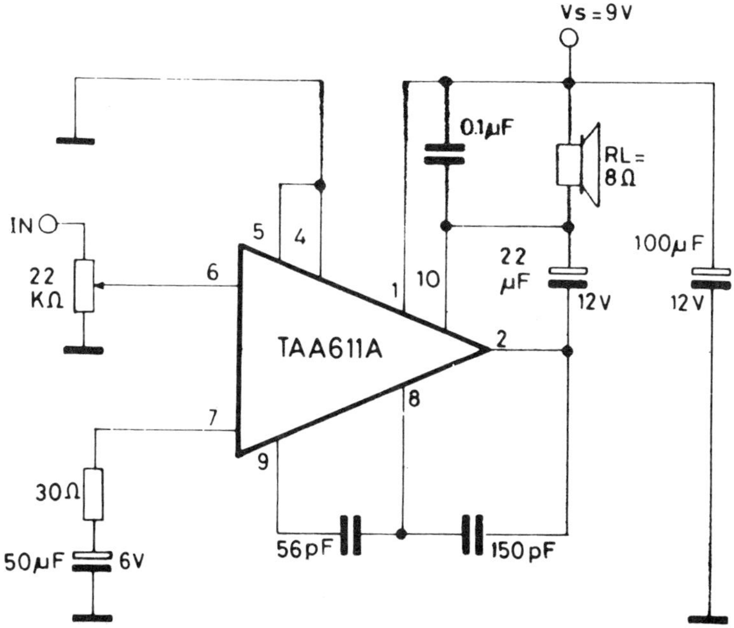

Audio amplifier for radio (Gv = 48 dB) (SGS).

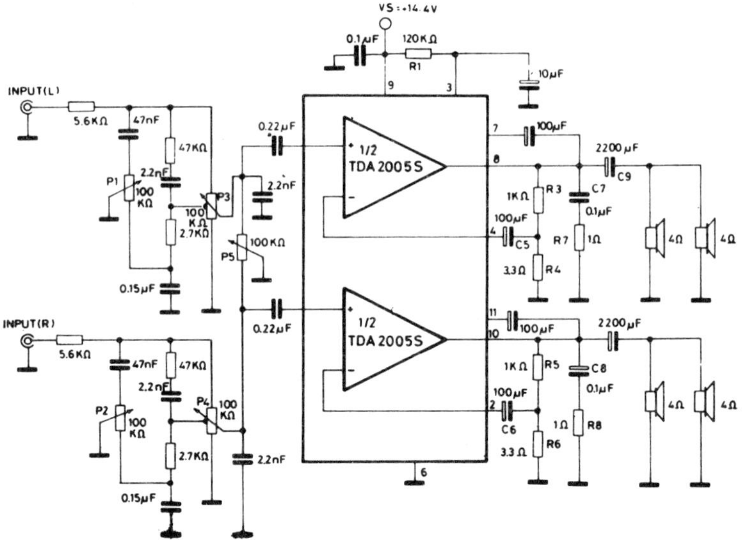

10 + 10W stereo amplifier with tone balance and loudness control (SGS).

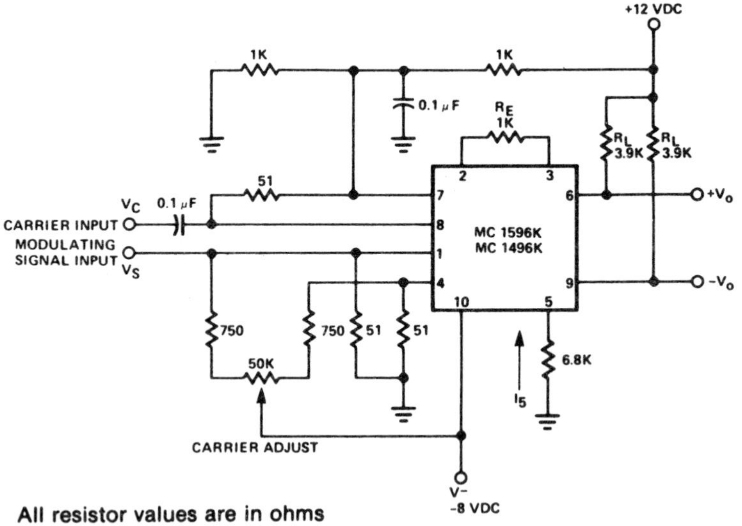

All resistor values are in ohms

AM modulator (S).

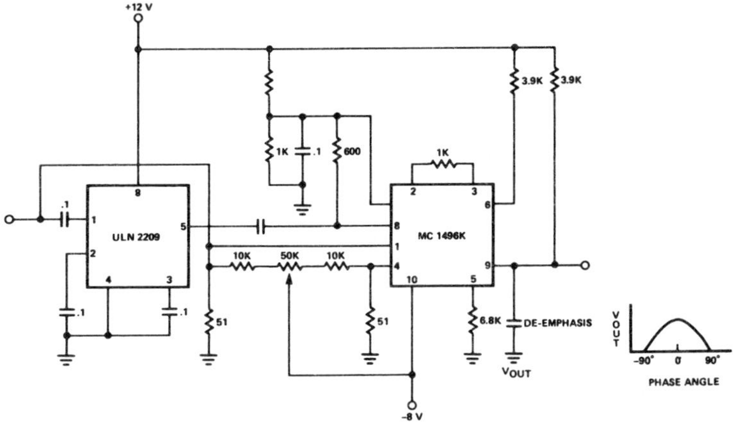

AM demodulator (S).

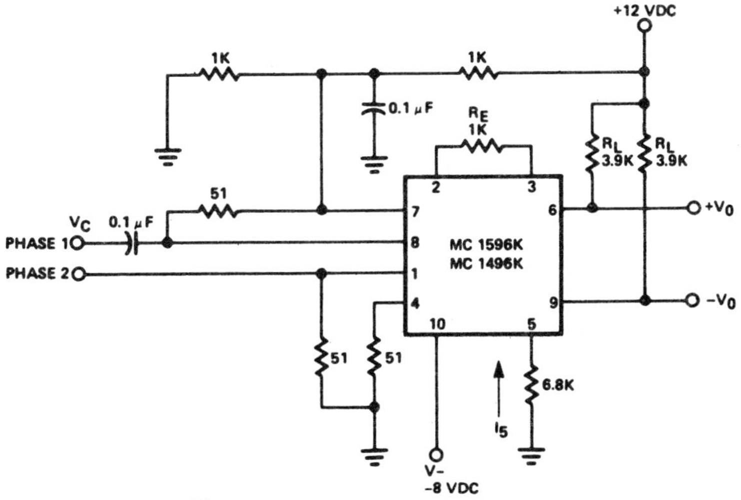

Phase comparator (S).

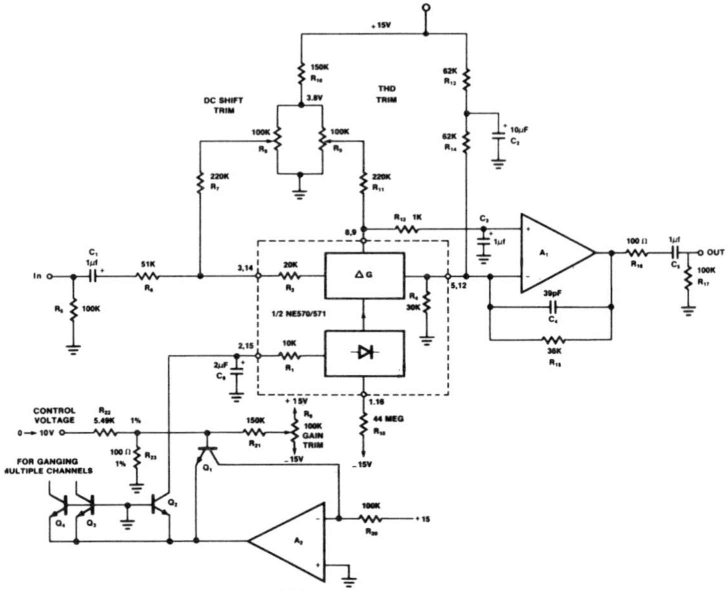

Voltage controlled attenuator (S).

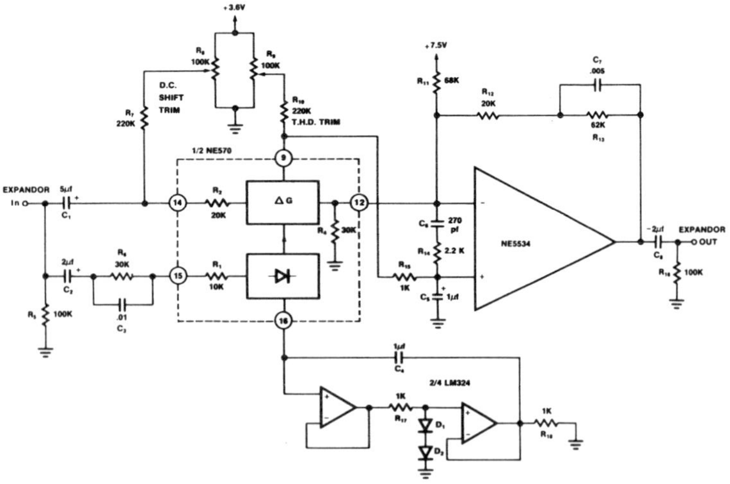

Hi-fi expandor with de-emphasis (S).

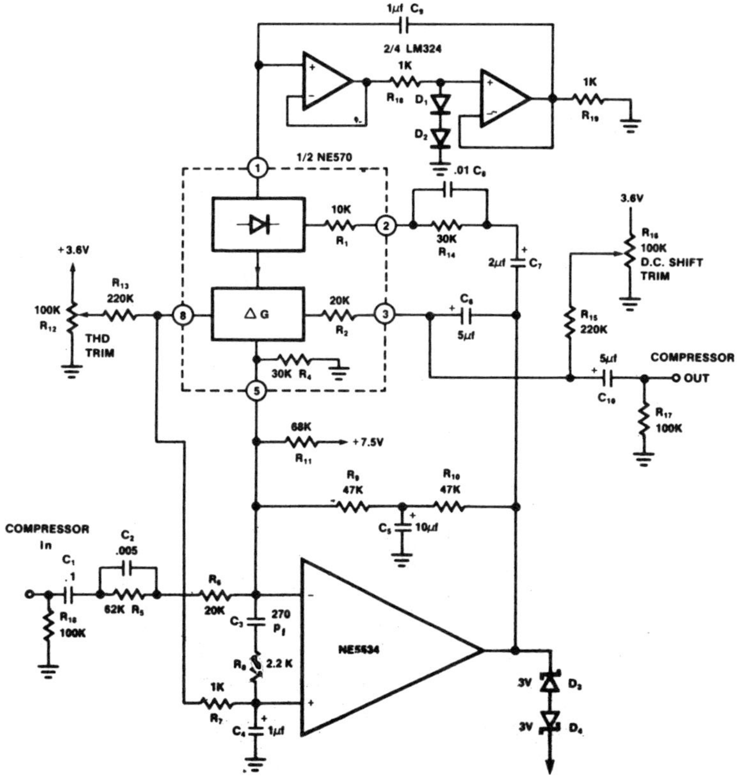

Hi-fi compressor with pre-emphasis (S).

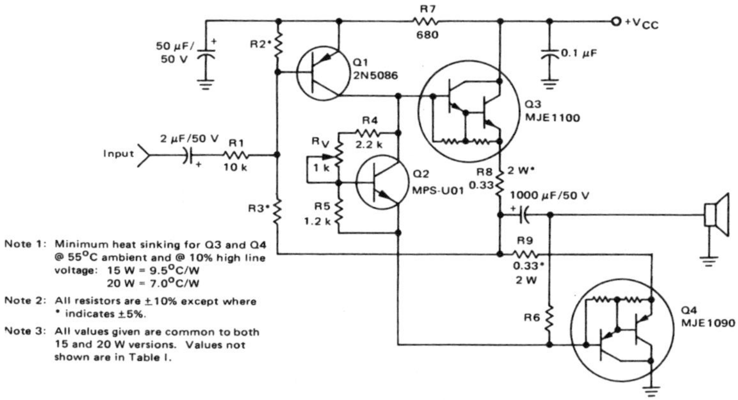

Four transistor audio amplifier for 15 to 20 W output (M).

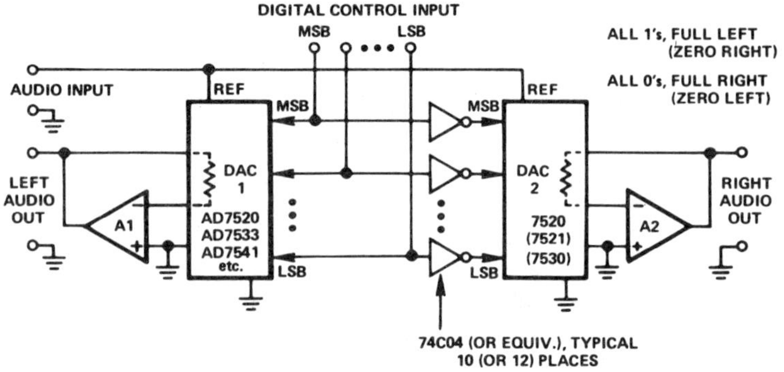

Digitally-programmed audio panner (AD).

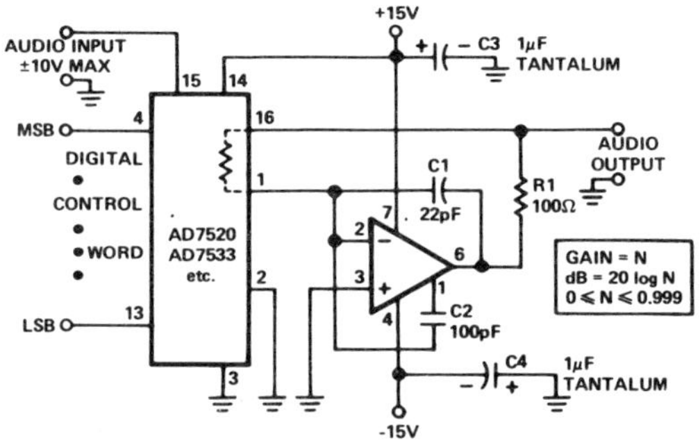

DIGITAL INPUTS FOR TYPICAL ATTENUATIONS

dB ATTEN.	N: INPUT NUMBER – VALUE 1 STAGE	2 STAGES
0	0.999	0.999
-1	0.891	0.944
-2	0.794	0.891
-3	0.708	0.841
-4	0.631	0.794
-5	0.562	0.750
⋮		
-10	0.316	0.562
-15	0.178	0.422
⋮		
-30	0.032	0.178
-40	0.010	0.100
-50	0.003	0.056
-60	0.001	0.032
-70	–	0.018
-80	–	0.010
-90		0.006

Inexpensive, high-performance gain control (AD).

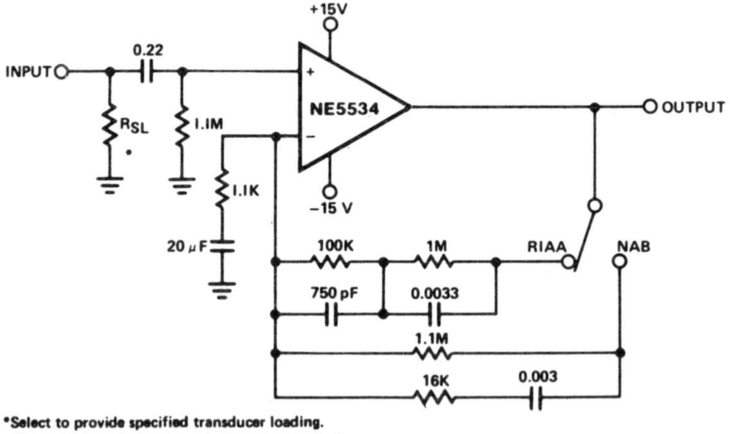

*Select to provide specified transducer loading.
Output Noise ≅ 0.8mV rms (with input shorted)

All resistor values are in ohms.

Preamplifier—RIAA/NAB compensation (S).

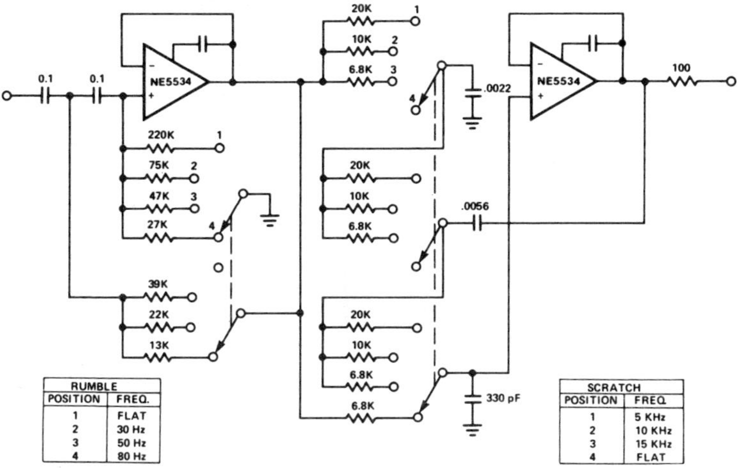

RUMBLE	
POSITION	FREQ.
1	FLAT
2	30 Hz
3	50 Hz
4	80 Hz

SCRATCH	
POSITION	FREQ.
1	5 KHz
2	10 KHz
3	15 KHz
4	FLAT

All resistor values are in ohms.

Rumble/scratch filter (S).

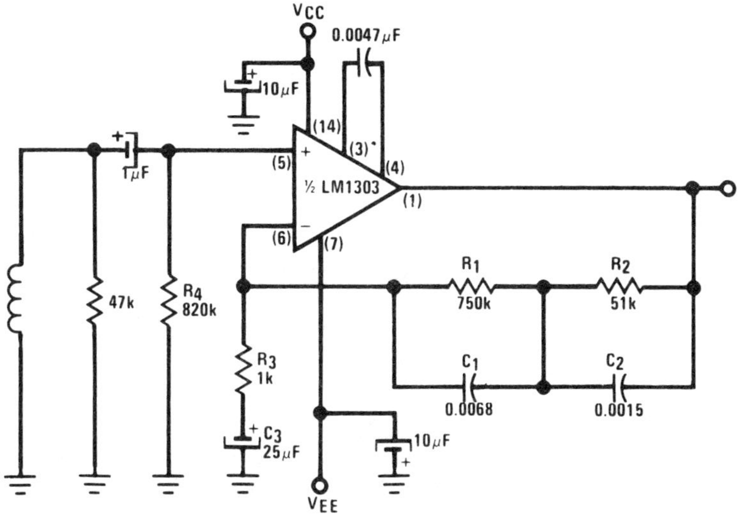

LM1303 Phono Preamp. (RIAA) (NS).

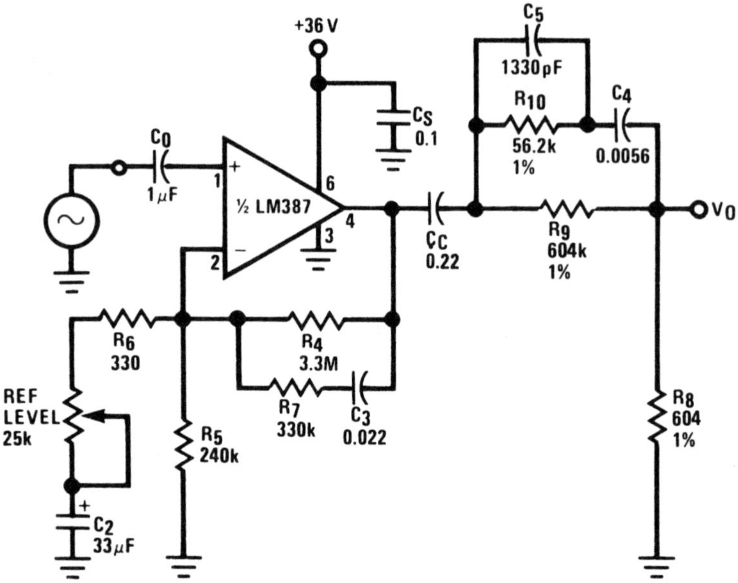

Inverse RIAA response generator (NS).

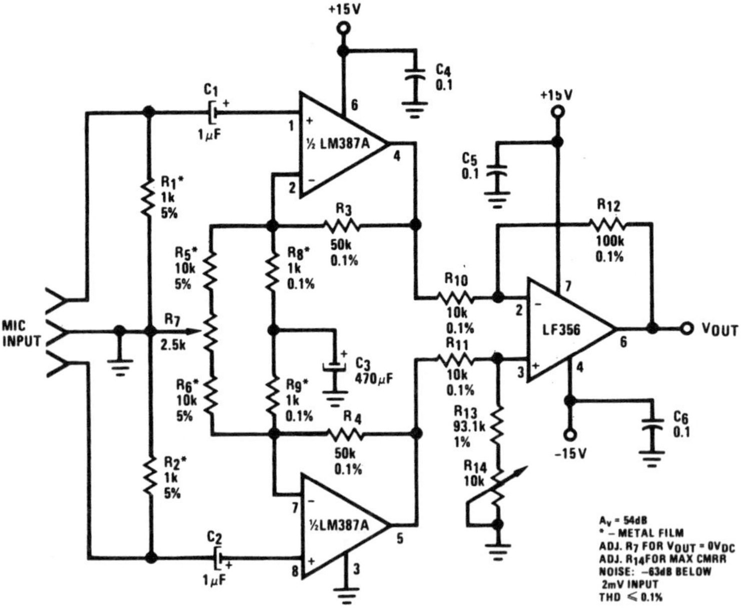

Low noise transformerless balanced mic preamp (NS).

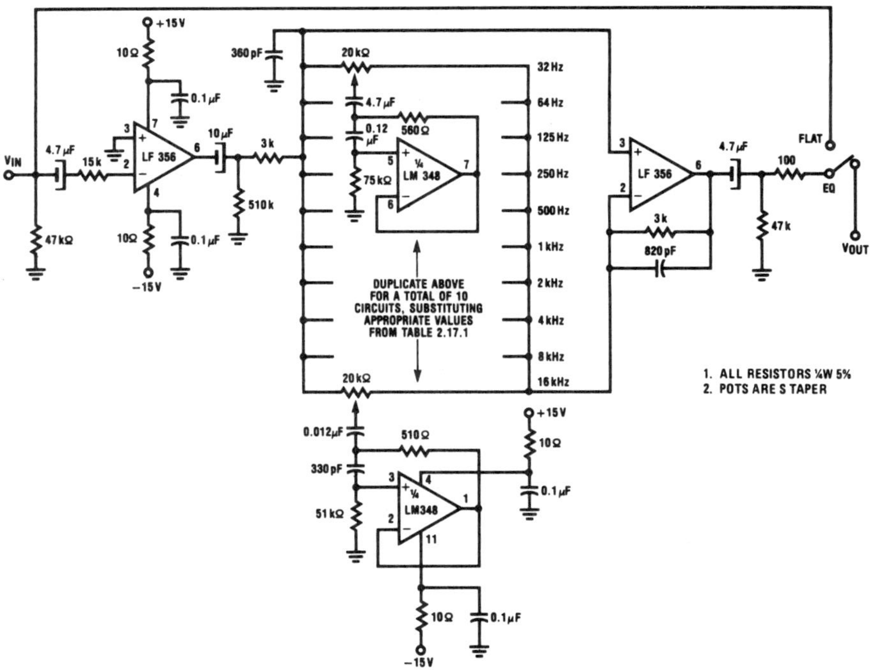

Complete ten band octave equalizer (NS).

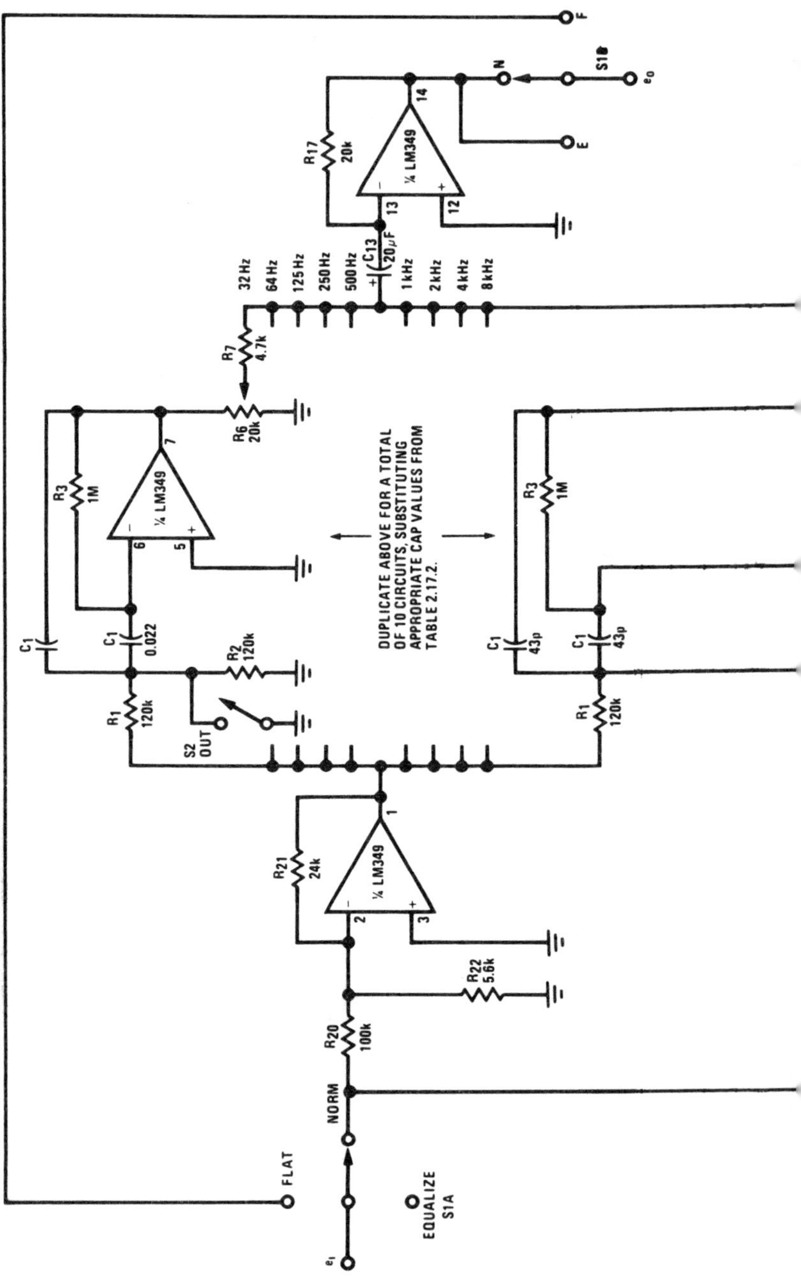
DUPLICATE ABOVE FOR A TOTAL OF 10 CIRCUITS, SUBSTITUTING APPROPRIATE CAP VALUES FROM TABLE 2.17.2.
¼ LM349
R21 24k
R22 5.6k
R20 100k
NORM
FLAT
EQUALIZE S1A
R1 120k
R2 120k
S2 OUT
C1 0.022
C1 43p
R3 1M
R6 20k
R7 4.7k
32 Hz
64 Hz
125 Hz
250 Hz
500 Hz
1 kHz
2 kHz
4 kHz
8 kHz
C13 20μF
R17 20k
S1B

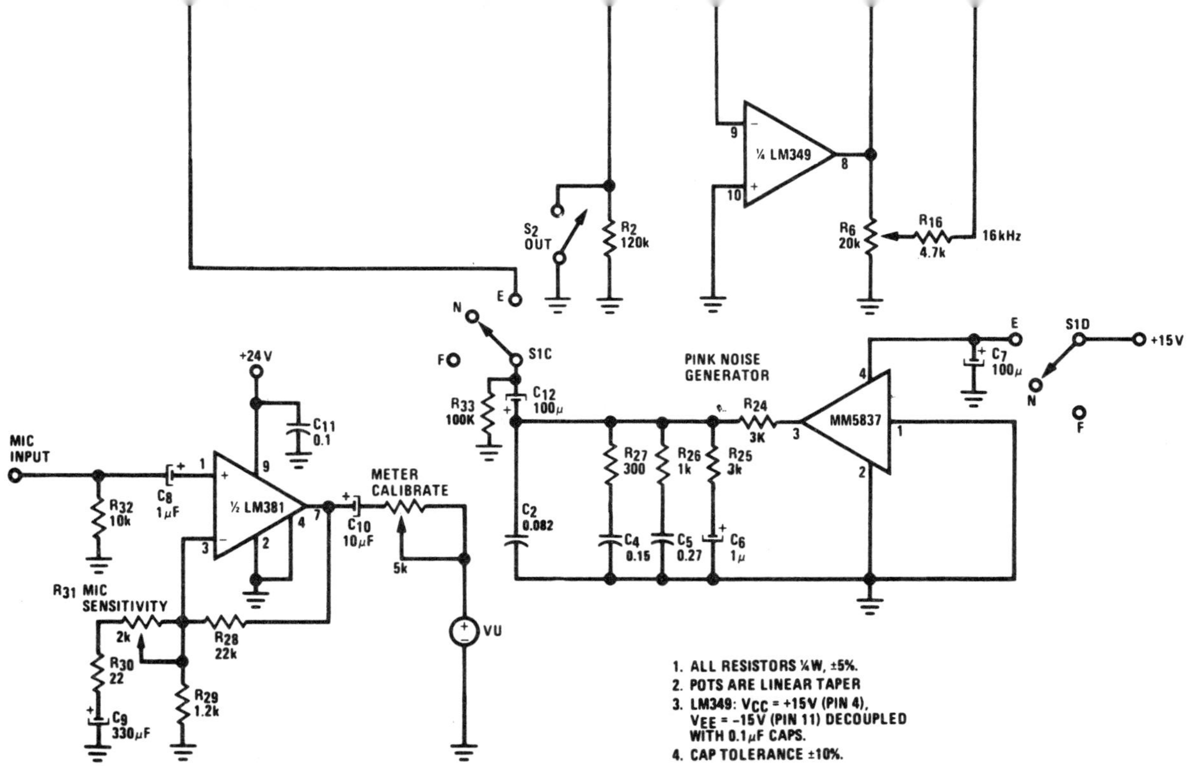

Room equalizing instrument (NS).

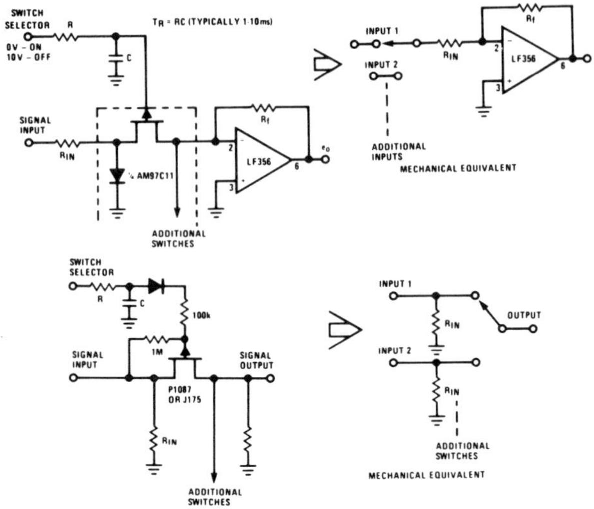

Deglitched voltage mode switch (NS).

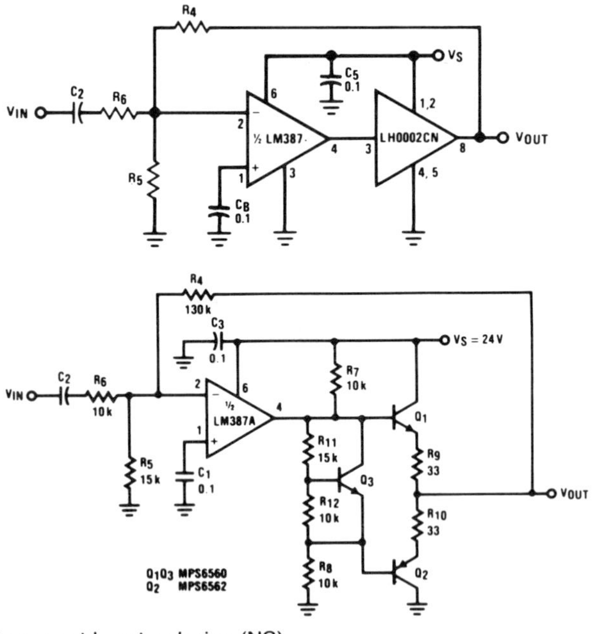

Discrete current booster design (NS).

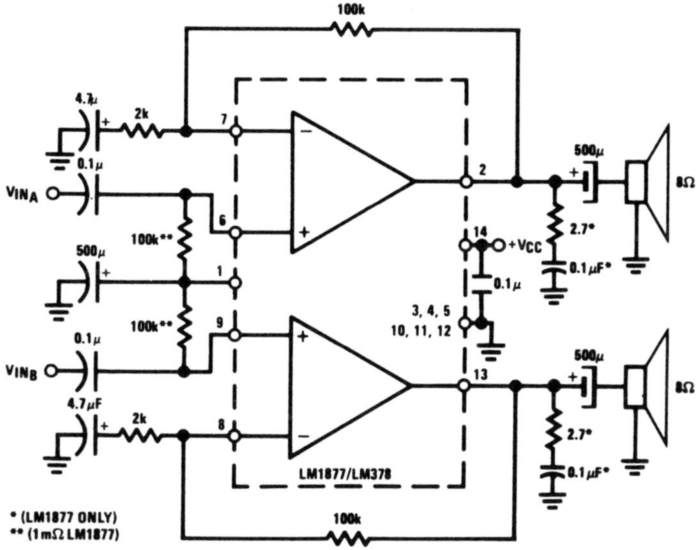

Non-inverting stereo amplifier (NS).

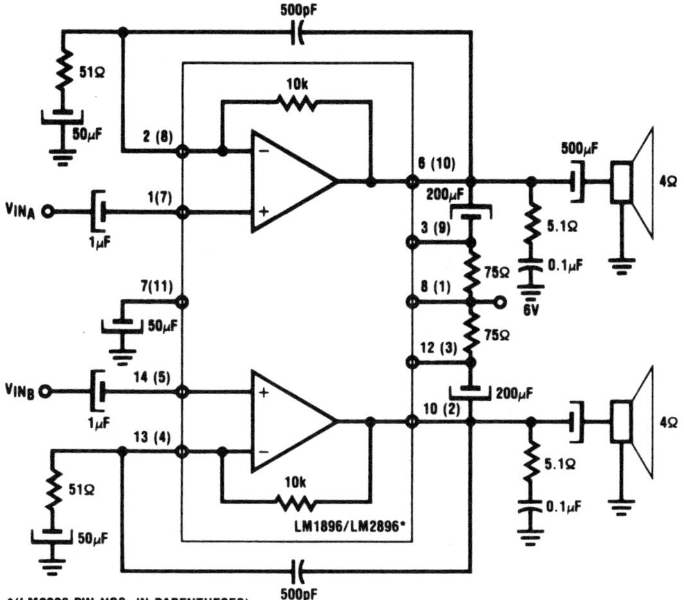

*(LM2896 PIN NOS. IN PARENTHESES)

Low voltage stereo amplifier 1.1W/Ch (NS).

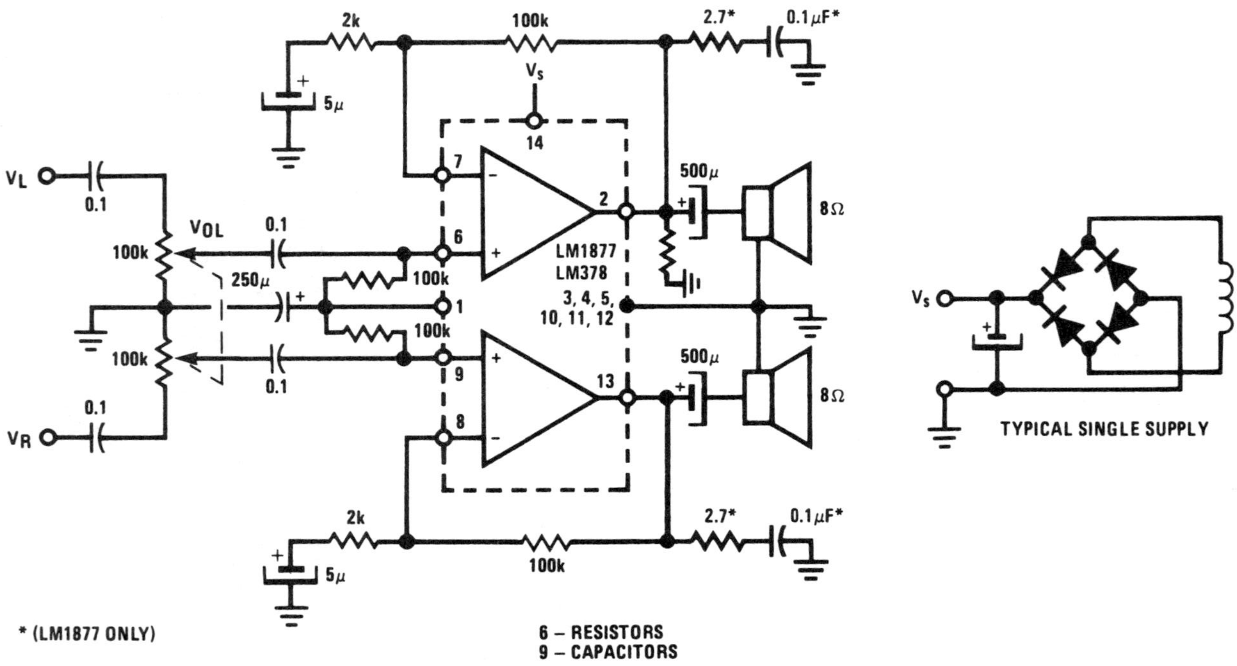

Non-inverting amplifier using single supply (NS).

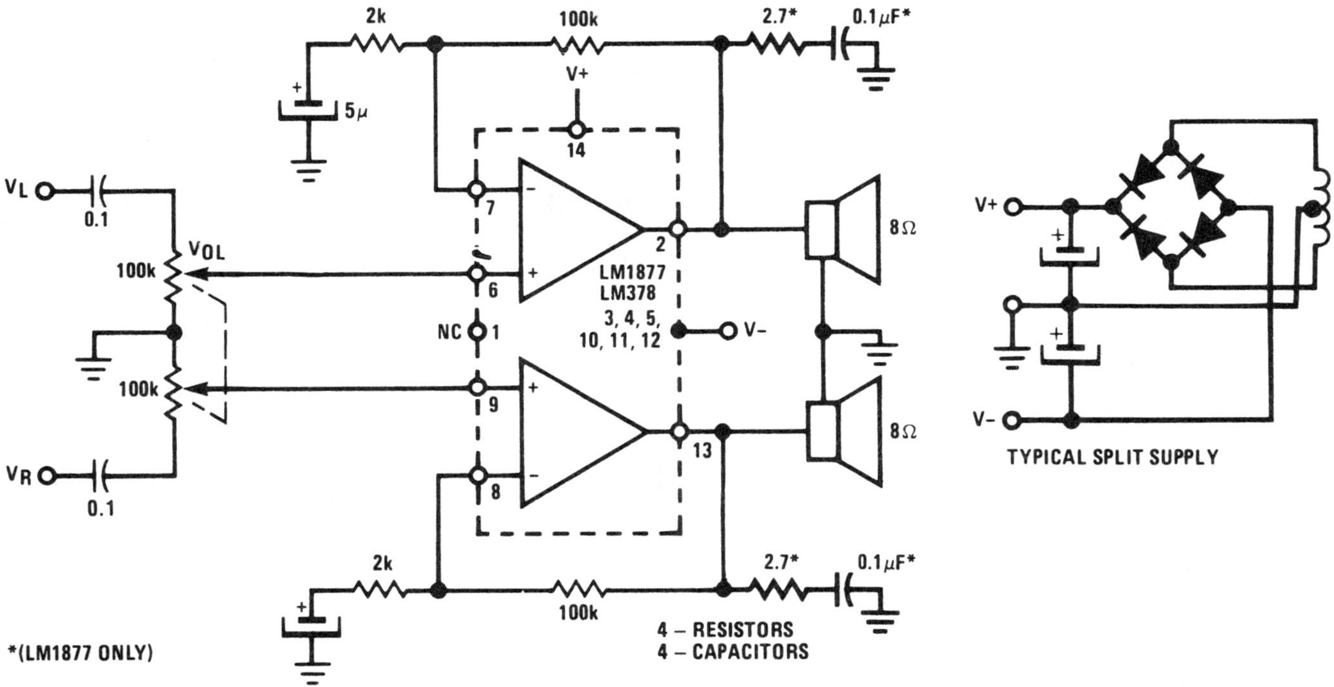

Non-inverting amplifier using split supply (NS).

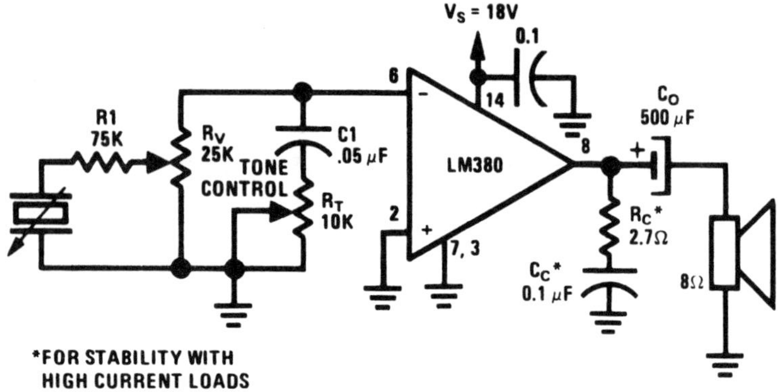

Ceramic phono amp (NS).

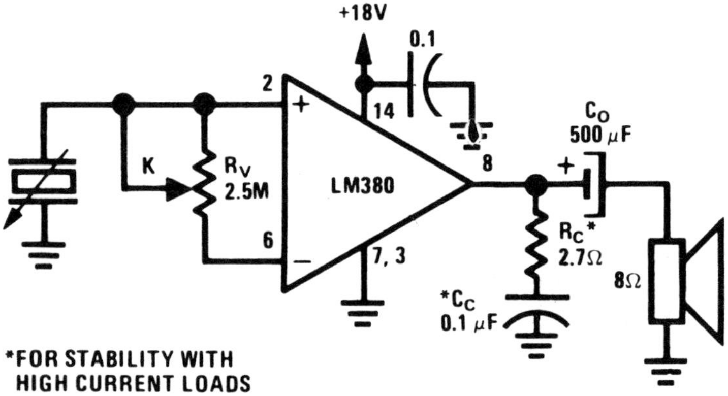

"Common mode" volume control (NS).

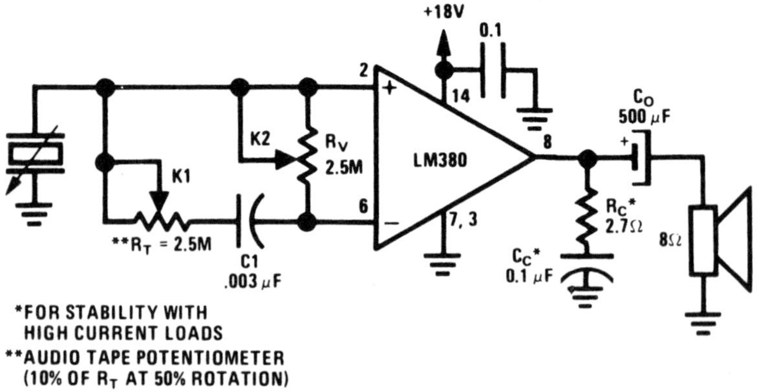

"Common mode" volume and tone control (NS).

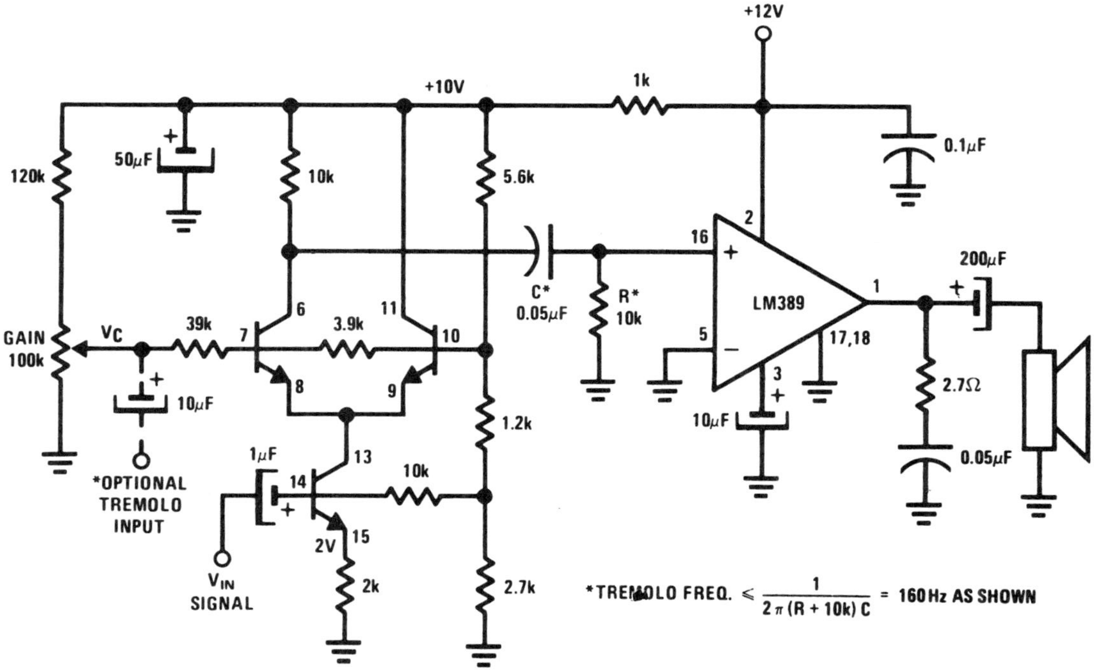

Voltage-controlled amplifier or tremolo circuit (NS).

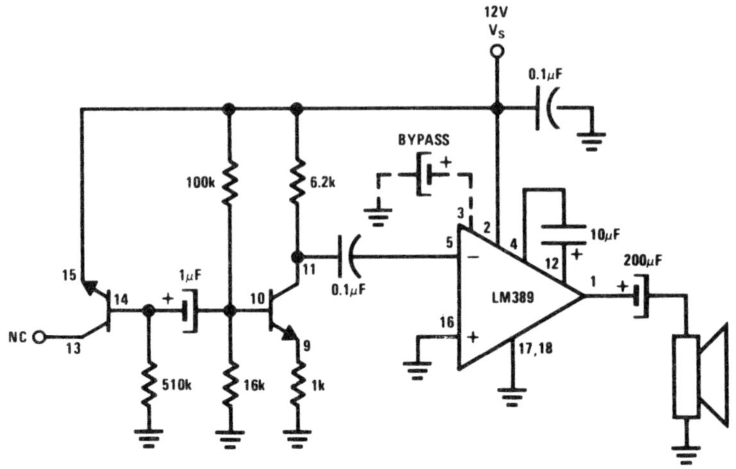

Noise generator using zener diode (NS).

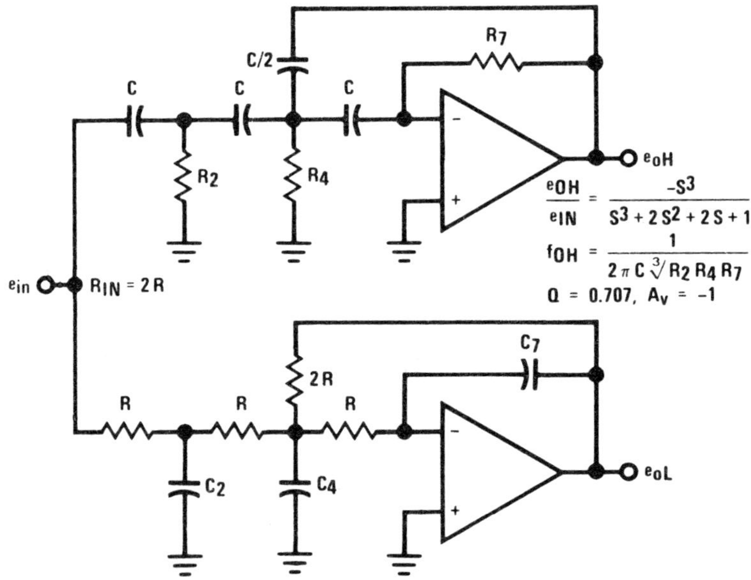

$$\frac{e_{OH}}{e_{IN}} = \frac{-S^3}{S^3 + 2S^2 + 2S + 1}$$

$$f_{OH} = \frac{1}{2\pi C \sqrt[3]{R_2 R_4 R_7}}$$

$Q = 0.707$, $A_V = -1$

$$C_2 = \frac{2.4553}{2\pi f_{OL} R} \qquad R_2 = \frac{0.4074}{2\pi f_{OH} C} \qquad \frac{e_{OL}}{e_{IN}} = \frac{-1}{S^3 + 2S^2 + 2S + 1}$$

$$C_4 = \frac{2.1089}{2\pi f_{OL} R} \qquad R_4 = \frac{0.4742}{2\pi f_{OH} C} \qquad f_{OL} = \frac{1}{2\pi R \sqrt[3]{C_2 C_4 C_7}}$$

$$C_7 = \frac{0.1931}{2\pi f_{OL} R} \qquad R_7 = \frac{5.1766}{2\pi f_{OH} C} \qquad Q = 0.707,\ A_V = -1$$

Complete 3rd order Butterworth crossover network (NS).

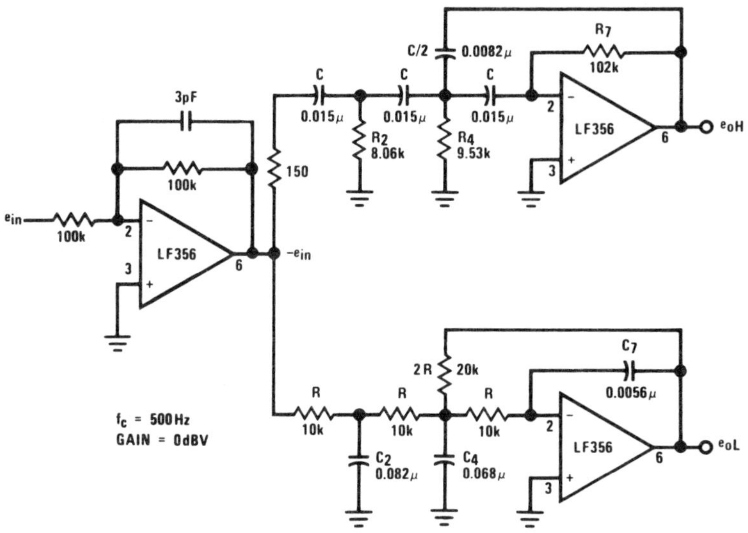

Typical active crossover network example (NS).

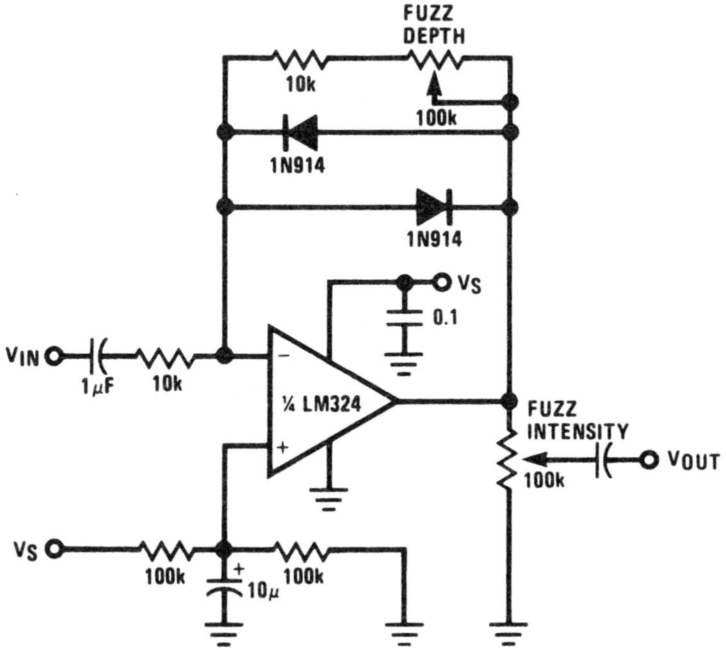

Fuzz circuit (NS).

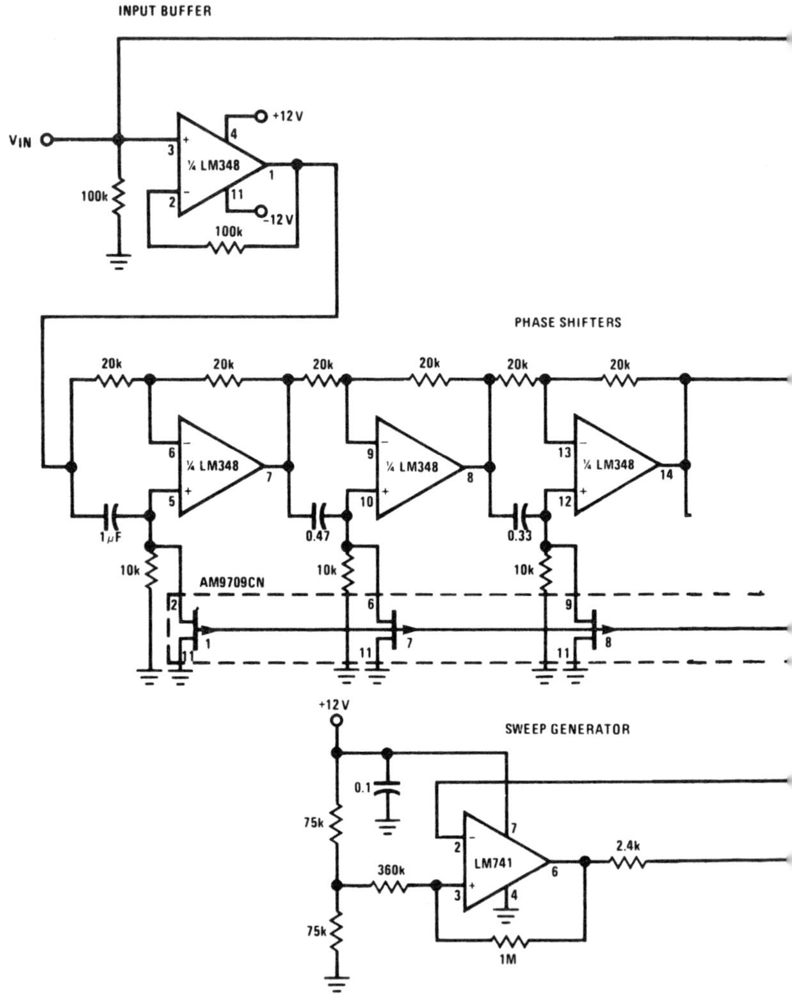

Phase shifter (NS).

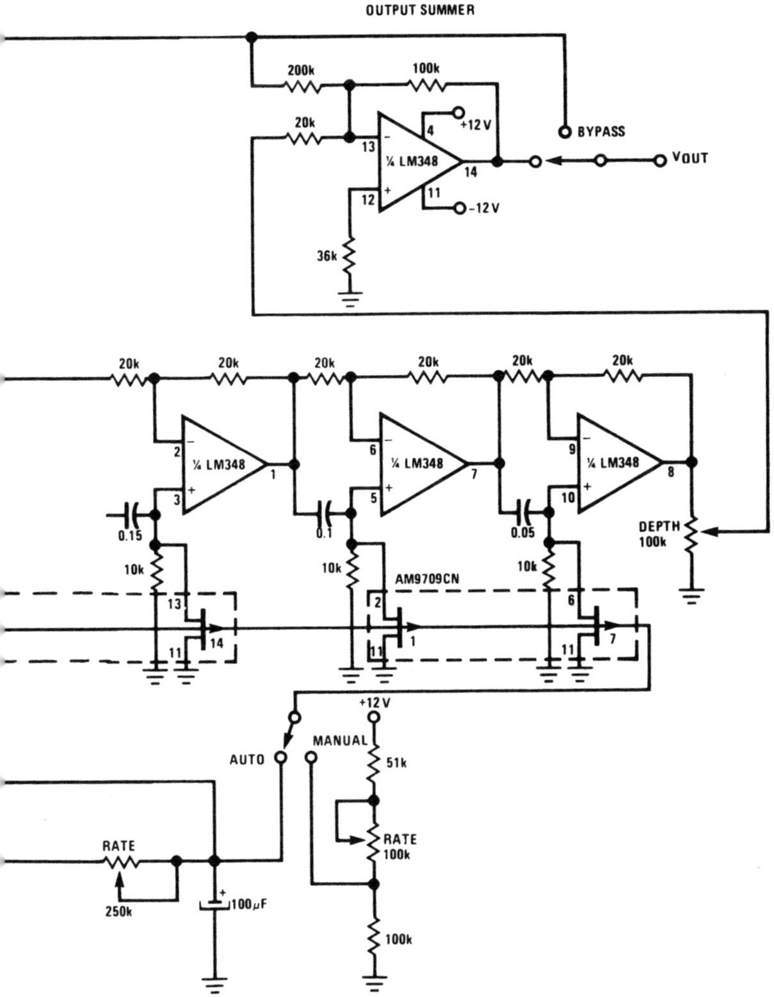
OUTPUT SUMMER
200k
100k
20k
+12 V
4
13
¼ LM348
14
12
11
-12 V
36k
BYPASS
VOUT
20k
20k
20k
20k
20k
20k
2
¼ LM348
1
3
6
¼ LM348
7
5
9
¼ LM348
8
10
0.15
0.1
0.05
DEPTH
100k
10k
10k
10k
AM9709CN
13
14
11
2
1
11
6
7
11
+12 V
AUTO
MANUAL
51k
RATE
100k
RATE
250k
100μF
100k

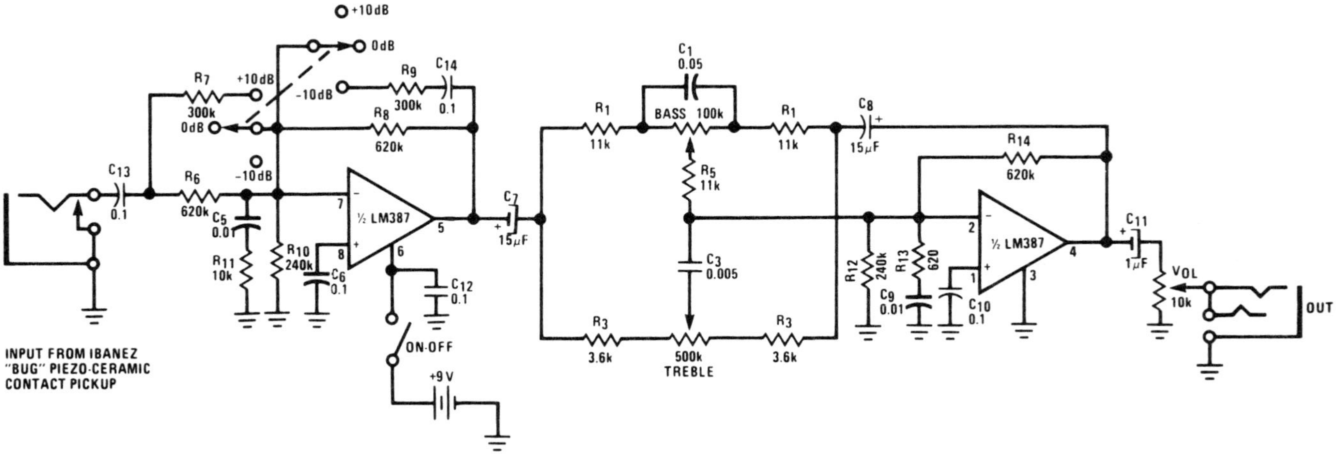

Acoustic Pickup Preamp.

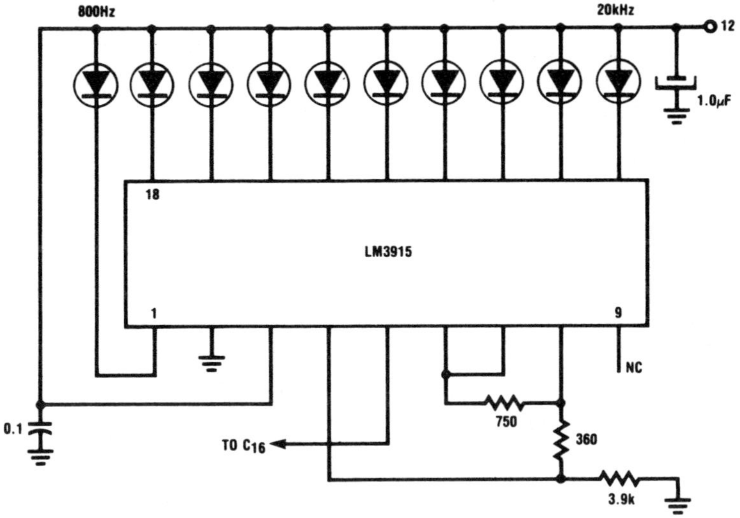

Audio bandwidth display (NS).

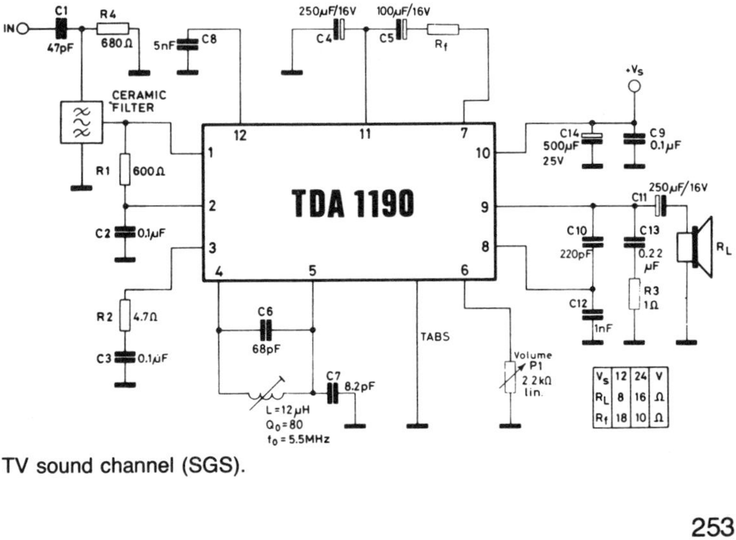

TV sound channel (SGS).

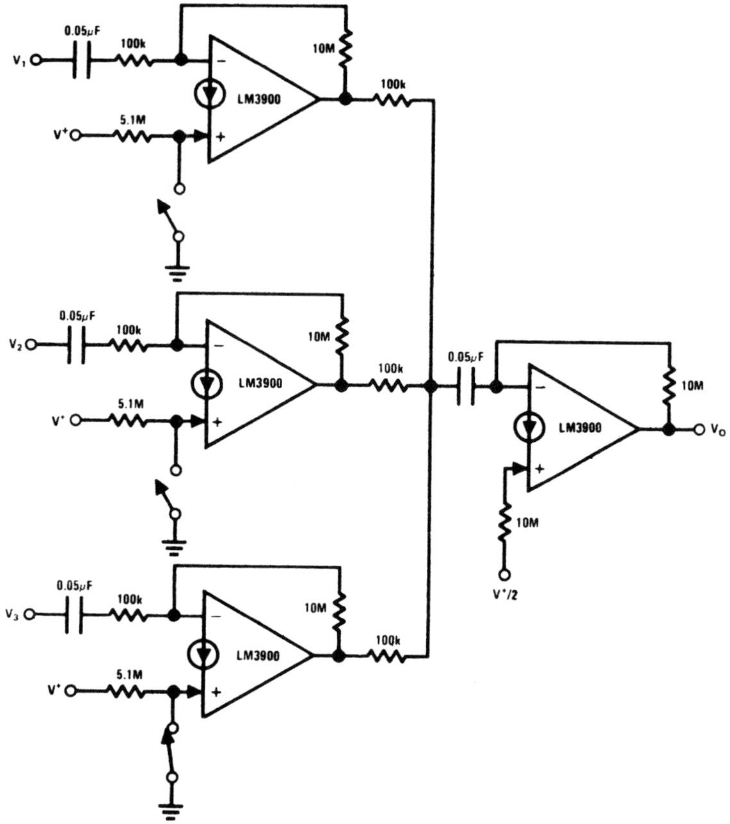

Channel selection by dc control (or audio mixer) (NS).

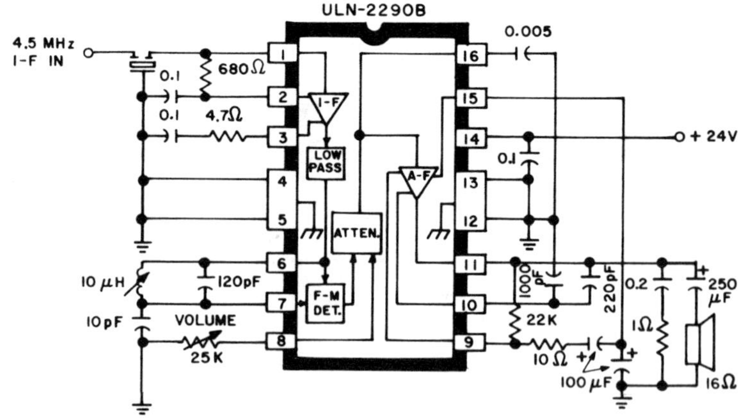

ULN-2290B and ULN-2290Q TV sound channel (SE).

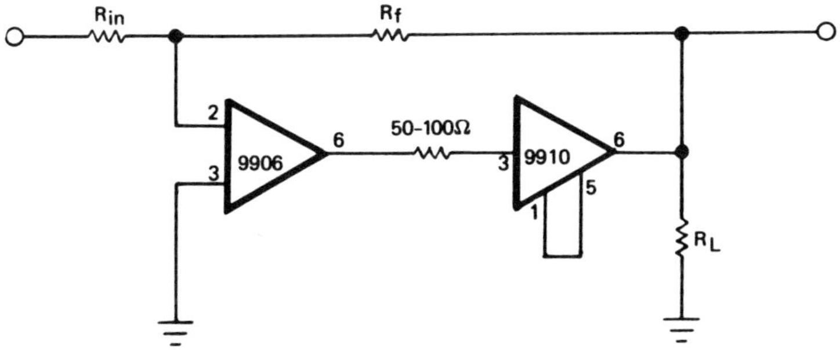

Current booster for video applications (OE).

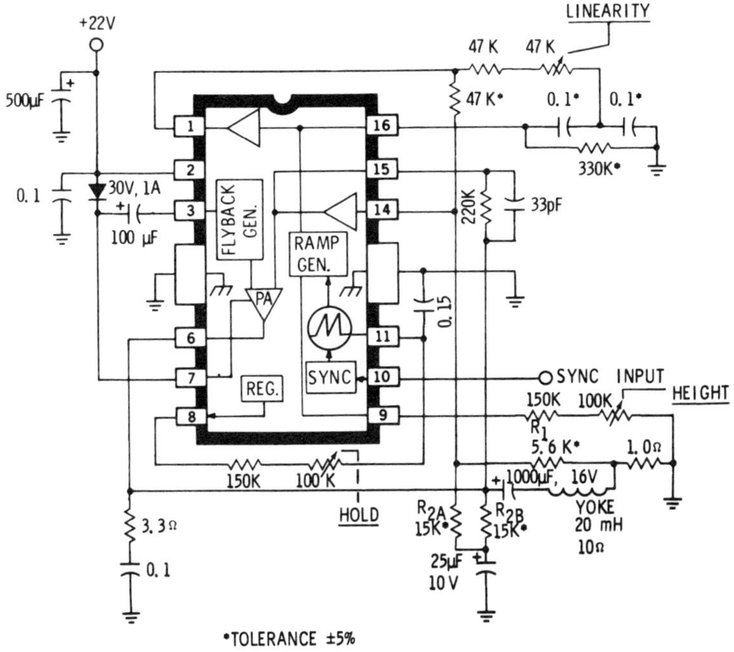

ULN-2270B and ULN-2270Q/TDA1170 vertical deflection system (typical application in large-screen black-and-white TV) (SE).

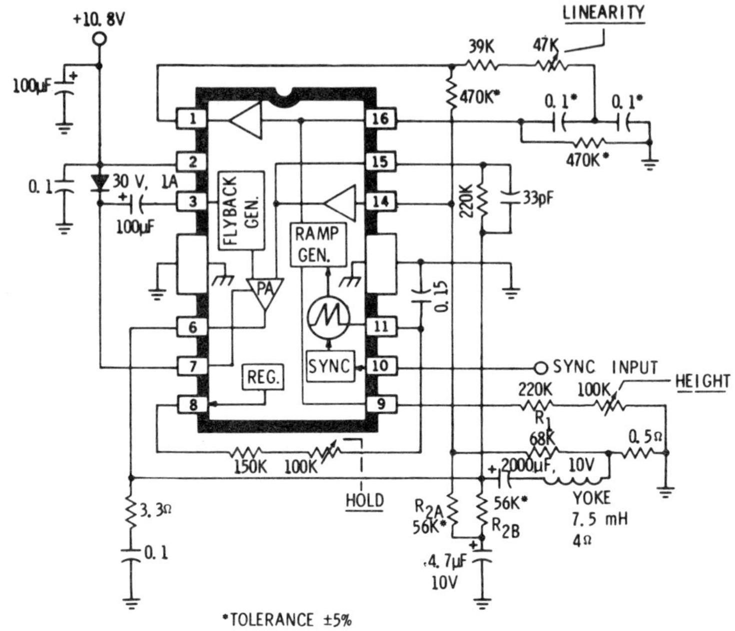

ULN-2270B and ULN-2270Q/TDA1170 vertical deflection system (typical application in small-screen black-and-white TV) (SE).

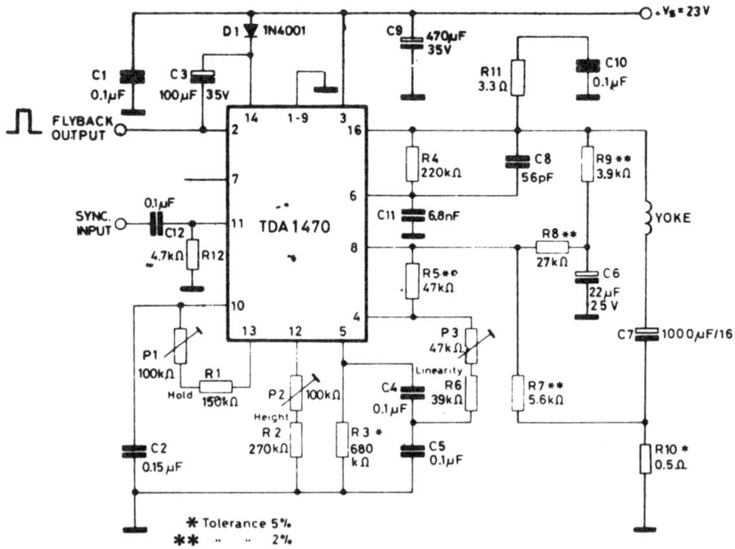

Vertical deflection circuit for PIL 26″ −110° series connected ($R_y = 9.7\Omega$; $L_y = 26.4$ mH; $I_y = 1.18$ App) (SGS).

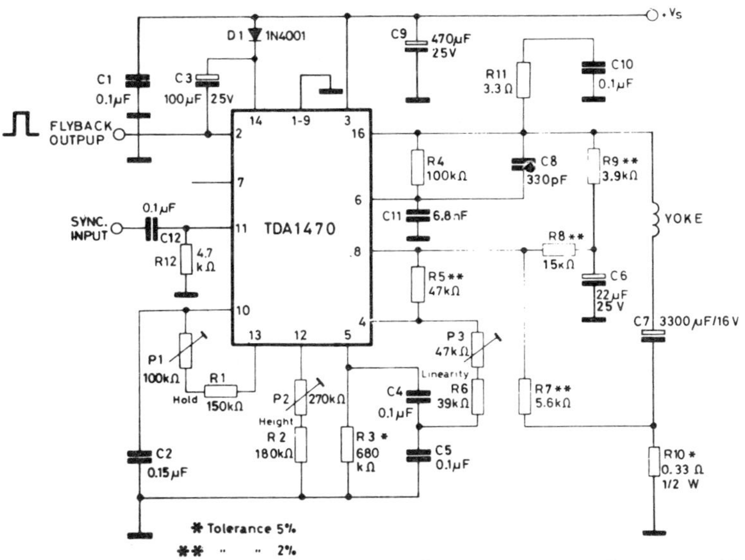

Vertical deflection circuit for large screen 110° TVC set ($R_y = 5.9\Omega$; $L_y = 10$ mH; $I_y = 1.95$ APP) (SGS).

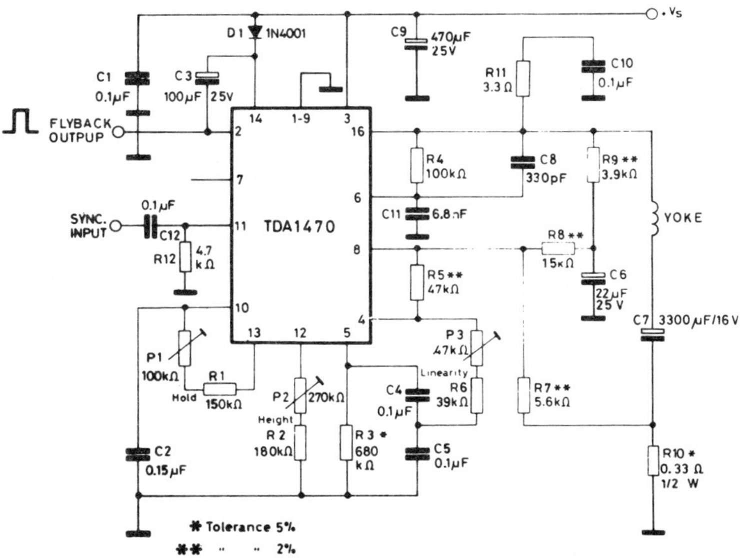

Vertical deflection circuit for PIL26″ – 110° parallel connected ($R_y = 2.5\Omega$; $L_y = 6.6$ mH; $I_y = 2.36$ App) (SGS).

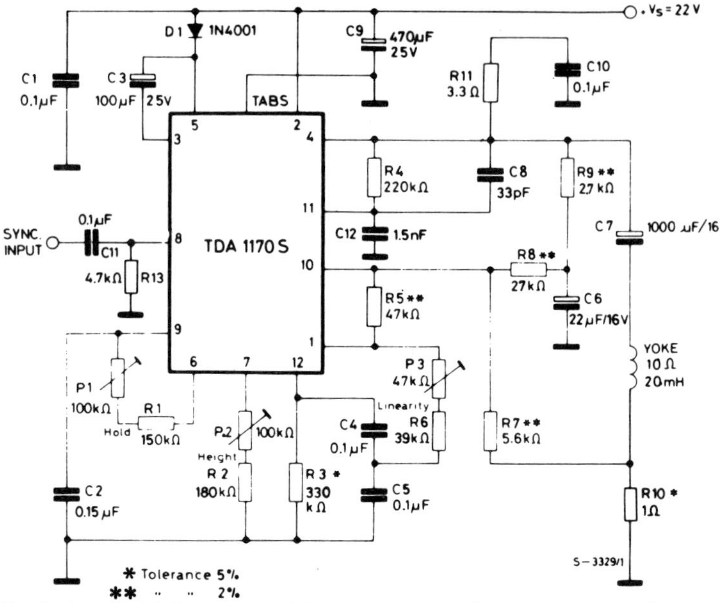

Typical vertical deflection circuit for large screen B/W TV set ($R_y = 10\Omega$; $L_y = 20$ mH; $I_y = 1$ App) (SGS).

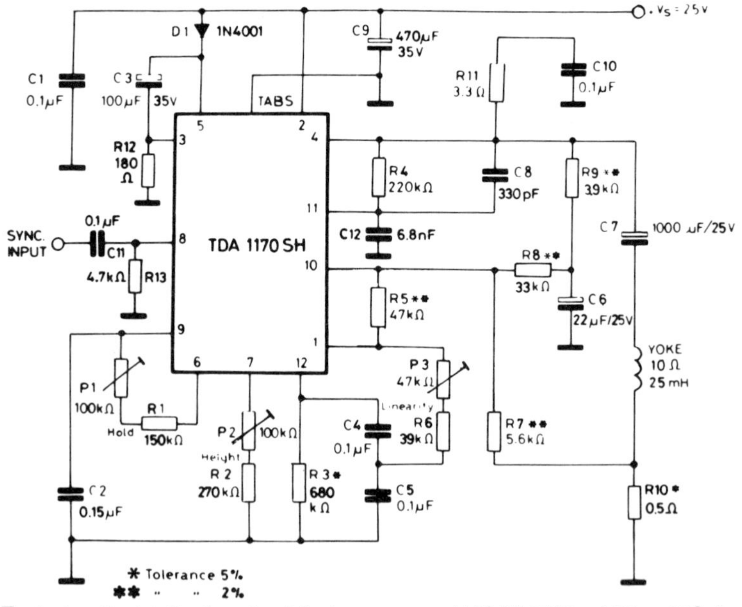

Typical vertical deflection circuit for large screen 110° PIL TVC set ($R_y = 10\Omega$; $L_y = 25$ mH; $I_y = 1.25$ App) (SGS).

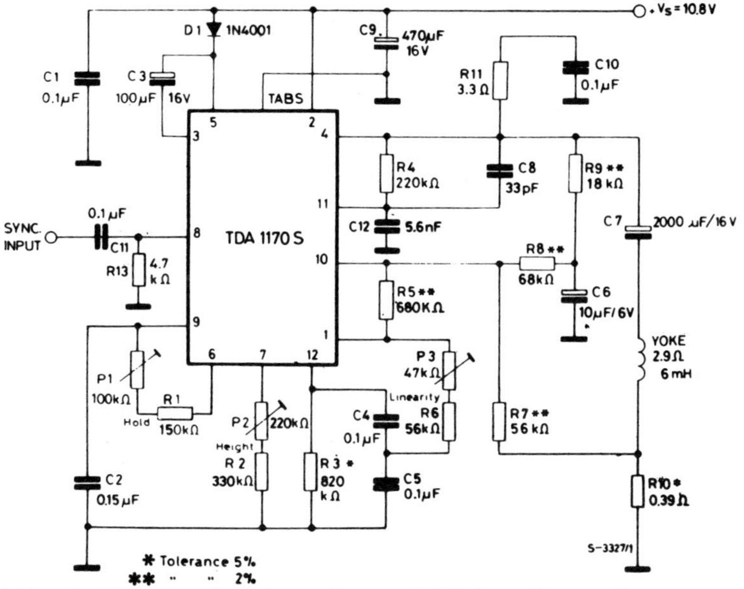

TV vertical deflection circuit for small screen B/W TV set ($R_y = 2.9\Omega$, $L_y = 6$ mH; $I_y = 1.1$ App) (SGS).

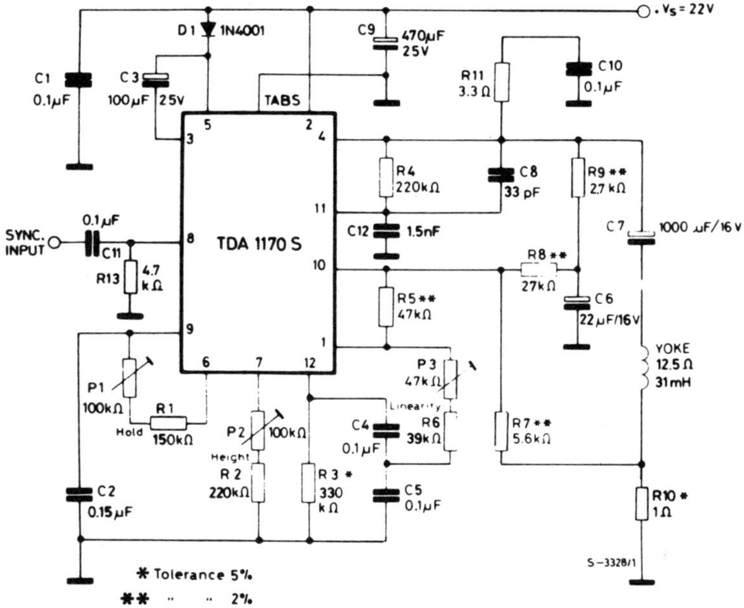

Typical vertical deflection circuit for small screen 90° PIL TVC set ($R_y = 12.5\ \Omega$; $L_y = 31$ mH; $I_y = 0.8$ App) (SGS).

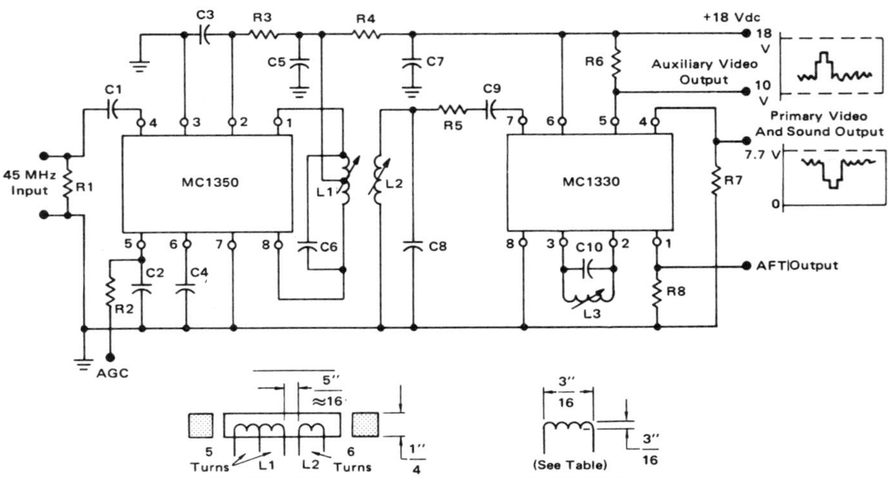

All windings #30 AWG tinned nylon acetate wire tuned with high permiability slugs. Coil Craft #4786 differential transformer.

L3 wound with #26 AWG tinned nylon acetate wire tuned by distorting winding.

C1 = 0.001 μF	C6 = See Table	R1 = 50 Ω	R6 = 3.3 kΩ
C2 = 0.002 μF	C7 = 0.1 μF	R2 = 5 k	R7 = 3.9 kΩ
C3 = 0.002 μF	C8 = See Table	R3 = 470 Ω	R8 = 3.9 kΩ
C4 = 0.002 μF	C9 = 68 pF	R4 = 220 Ω	All Resistors
C5 = 0.002 μF	C10 = See Table	R5 = 22 Ω	1/4-W ±10%

All Caps Marked μF Ceramic HiK
All Caps Marked pF Silver Mica 5%

Table of Component Values

Component	[illegible] MHz	[illegible] MHz	[illegible] MHz
[illegible]	[illegible] pF	[illegible] pF	10 pF
[illegible]	[illegible] pF	12 pF	10 pF
C10	[illegible] pF	[illegible] pF	[illegible] pF
[illegible]	[illegible] Turns	[illegible] Turns	10 Turns

Typical MC1350 video IF amplifier and MC1330 low-level video detector circuit (M).

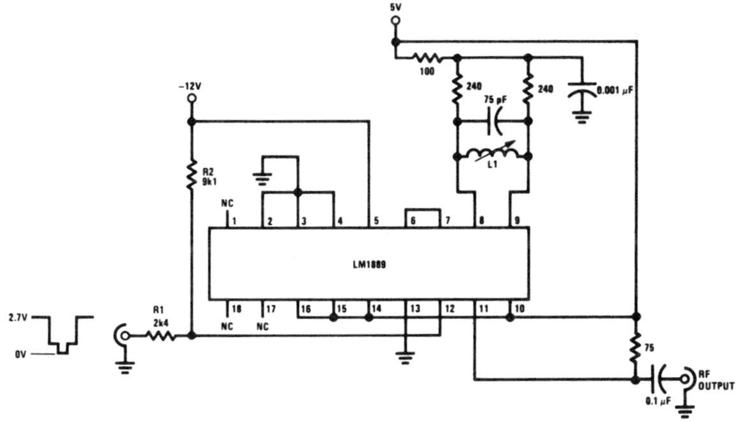

Low-cost monochrome modulator for character generator display (NS).

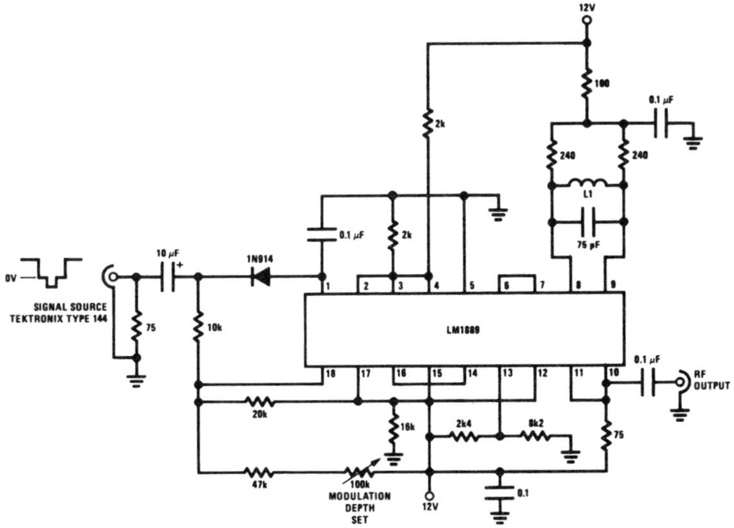

Dc clamped modulator for NTSC pattern generators (NS).

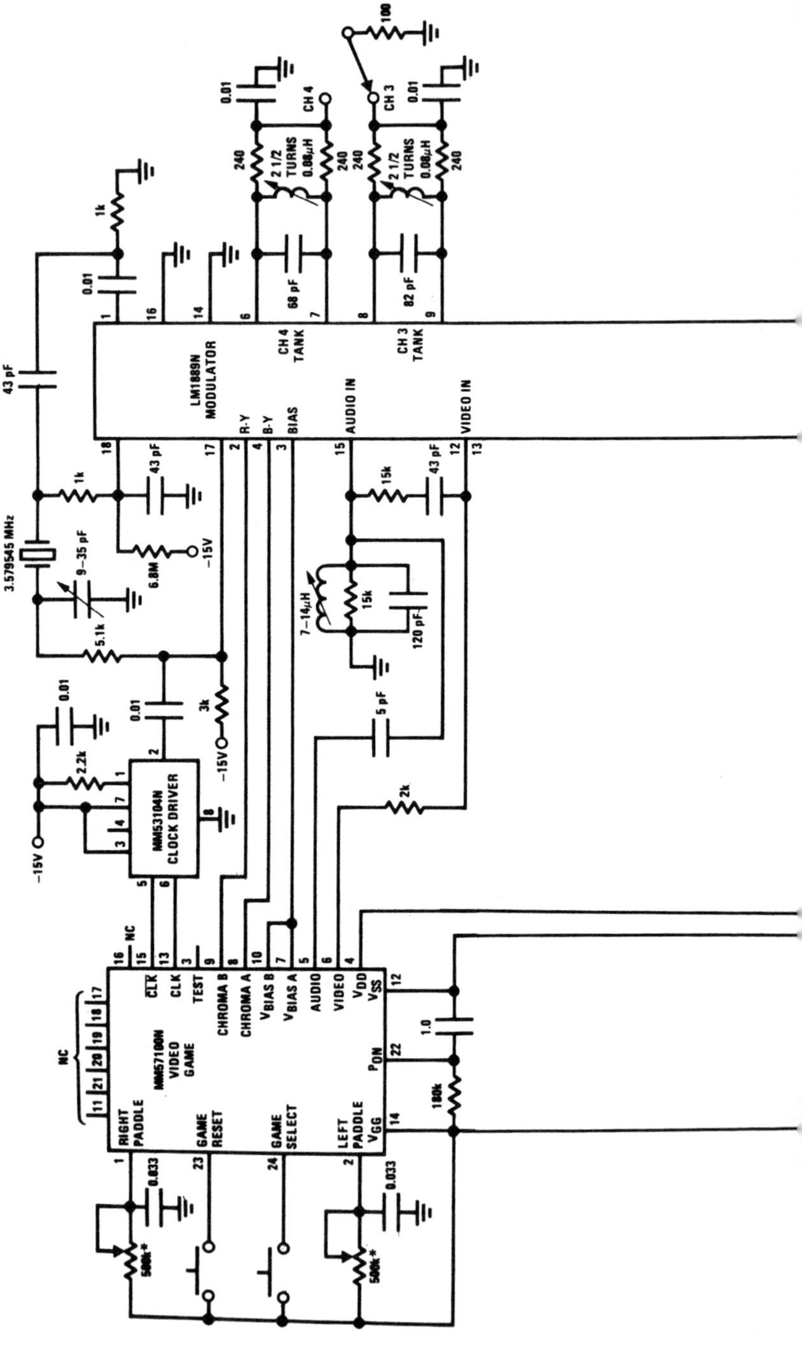
LM1889N
MODULATOR
CH 4
TANK
CH 3
TANK
R-Y
B-Y
BIAS
AUDIO IN
VIDEO IN
MM53104N
CLOCK DRIVER
MM57100N
VIDEO
GAME
RIGHT
PADDLE
GAME
RESET
GAME
SELECT
LEFT
PADDLE
CHROMA B
CHROMA A
AUDIO
VIDEO
TEST
CLK
NC
3.579545 MHz
9–35 pF
43 pF
0.01
1k
6.8M
5.1k
3k
2.2k
−15V
15k
7–14μH
120 pF
5 pF
2k
240
2 1/2
TURNS
0.08μH
68 pF
82 pF
100
1.0
180k
0.033
500k*

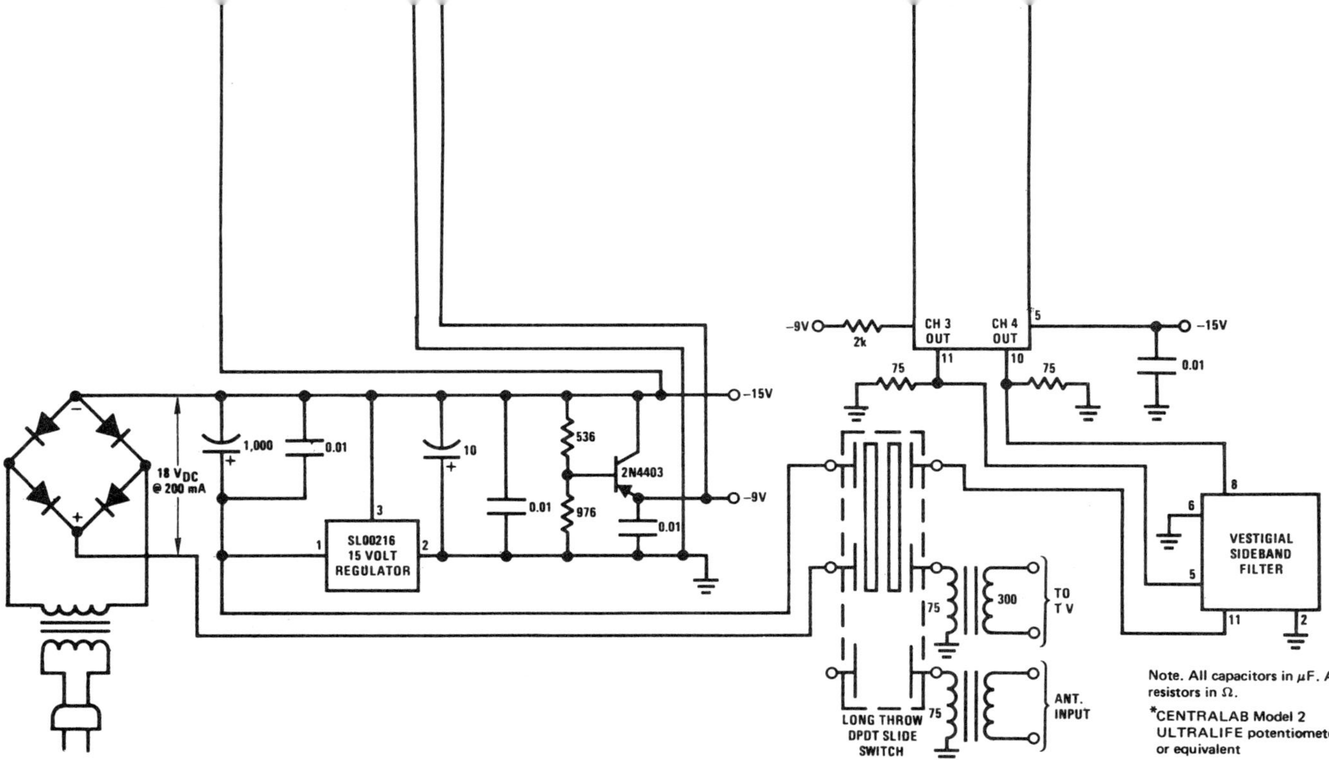

Note. All capacitors in μF. All resistors in Ω.

*CENTRALAB Model 2 ULTRALIFE potentiometer or equivalent

Game schematic (NS).

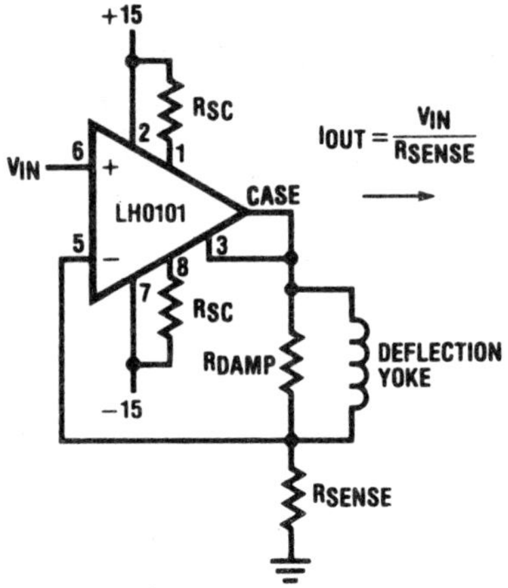

CRT deflection yoke driver (NS).

Chapter 11

Converters and Doublers

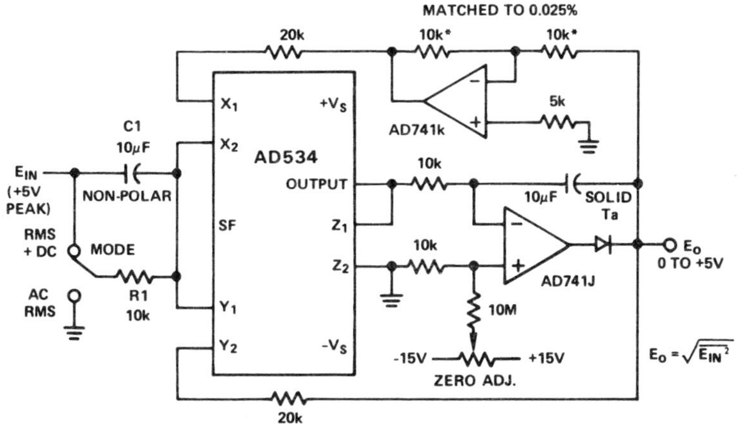

Wideband, high crest, rms-to-dc converter (AD).

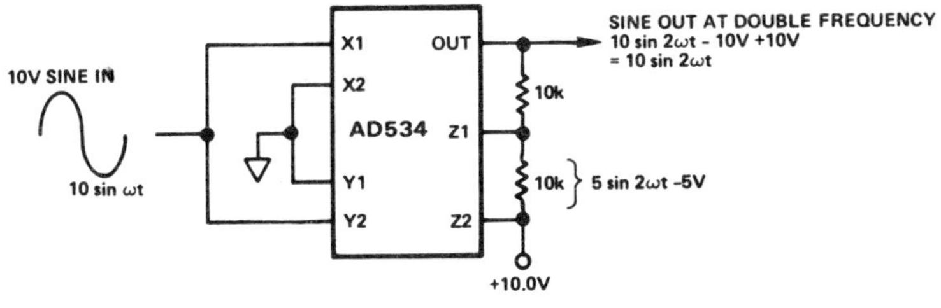

Frequency doubler (AD).

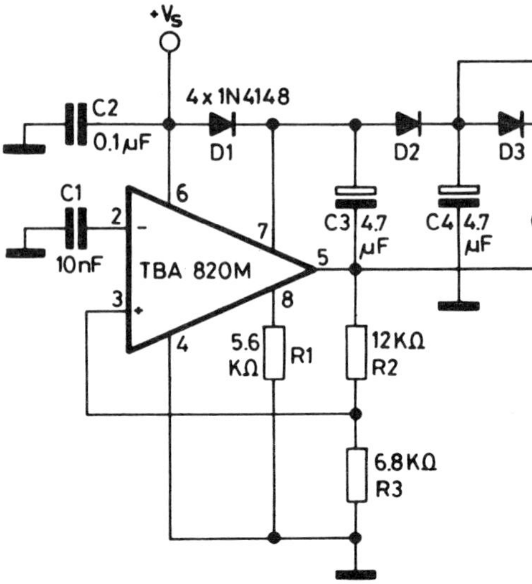

V_S (V)	V1 (V)	I1 (mA)	V2 (V)
8	12	60	19
10	16	60	26
12	20	60	31
16	28	60	43

1.5W dc/dc converter (f = 40 kHz) (SGS).

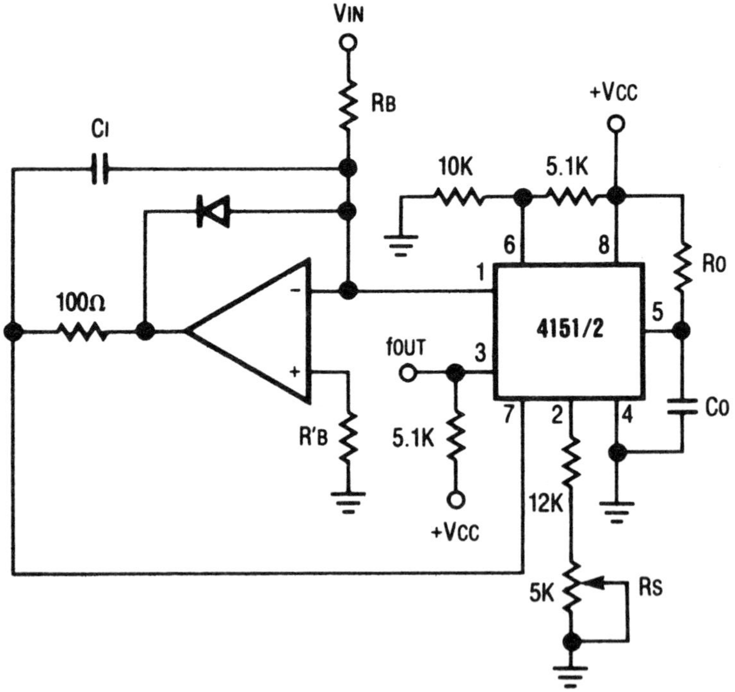

Precision VFC ratio circuit (R).

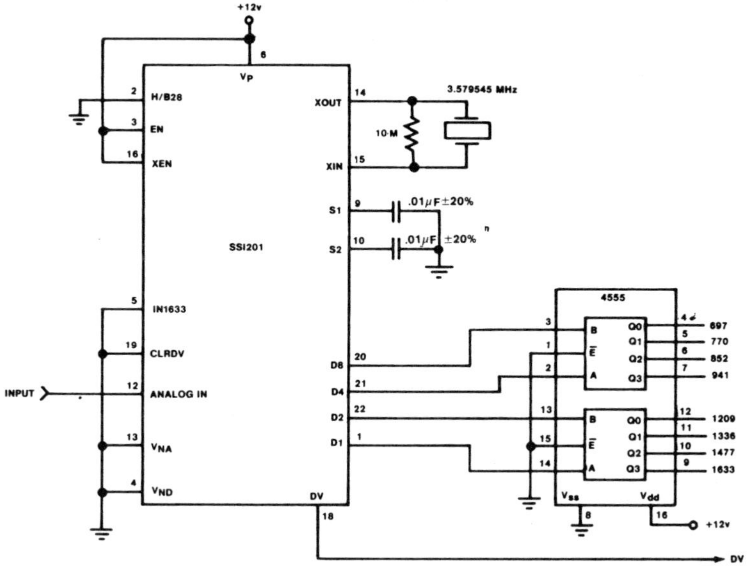

Touch Tone® to 2 of 8 output converter (SIL).

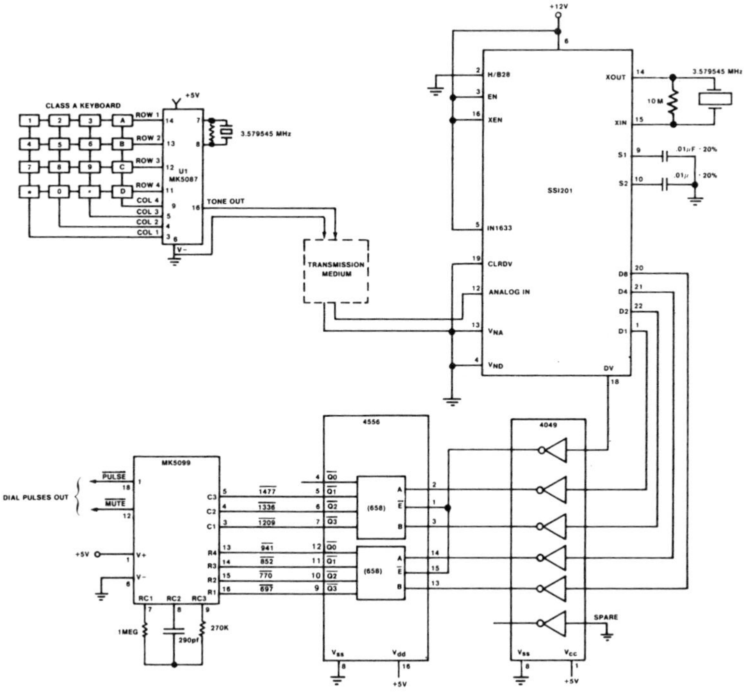

Touch Tone® to rotary converter (SIL).

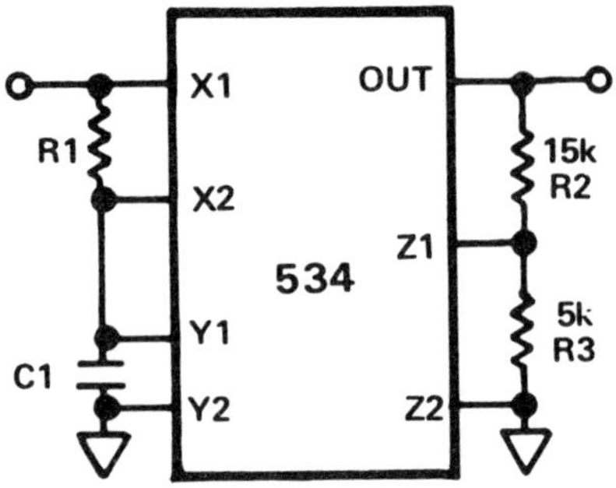

"No-bounce" frequency doubler (AD).

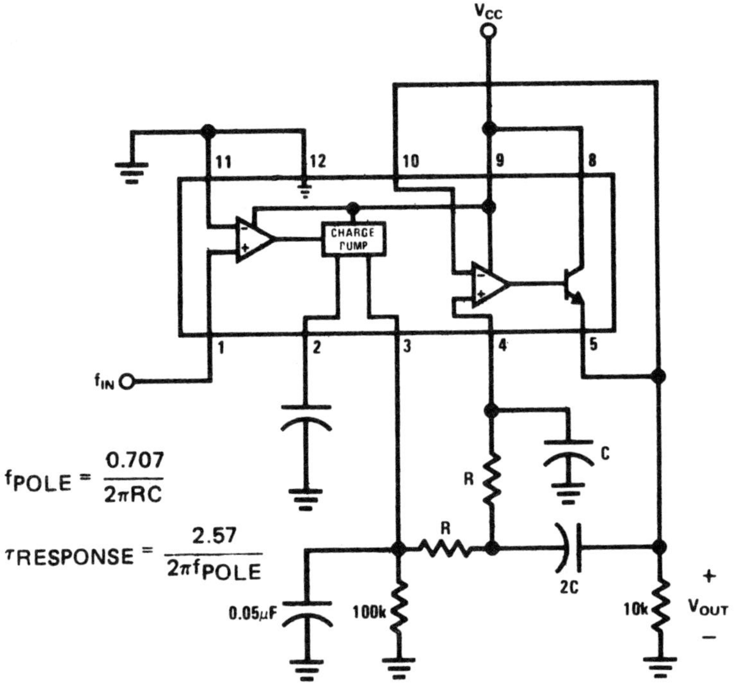

Frequency to voltage converter with 2 pole Butterworth filter to reduce ripple (NS).

Chapter 12

Amplifiers and Preamplifiers for Audio, Video, Amateur Radio, and Other Uses

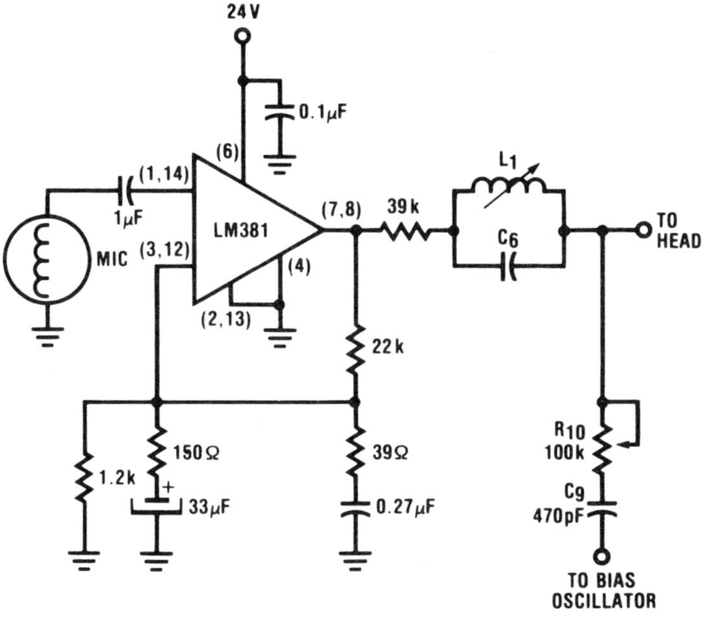

Typical tape recording amplifier (NS).

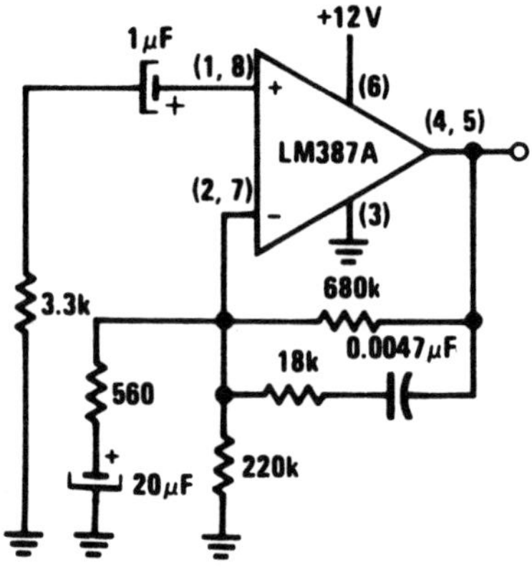

LM387 tape preamp (NS).

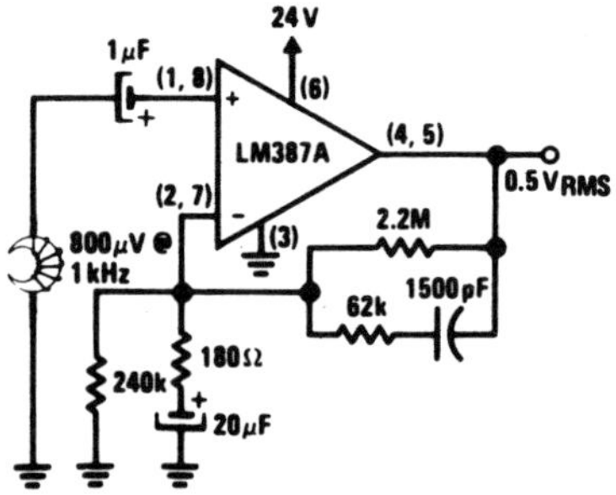

Typical tape playback amplifier (NS).

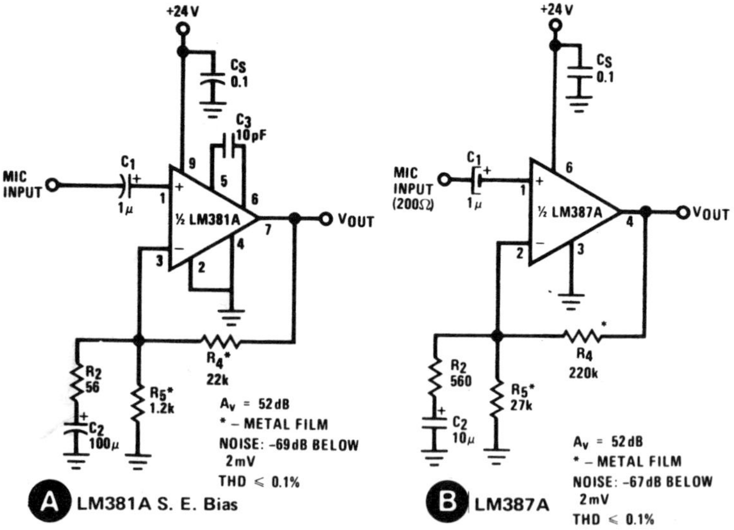

Transformerless mic preamps for unbalanced inputs (NS).

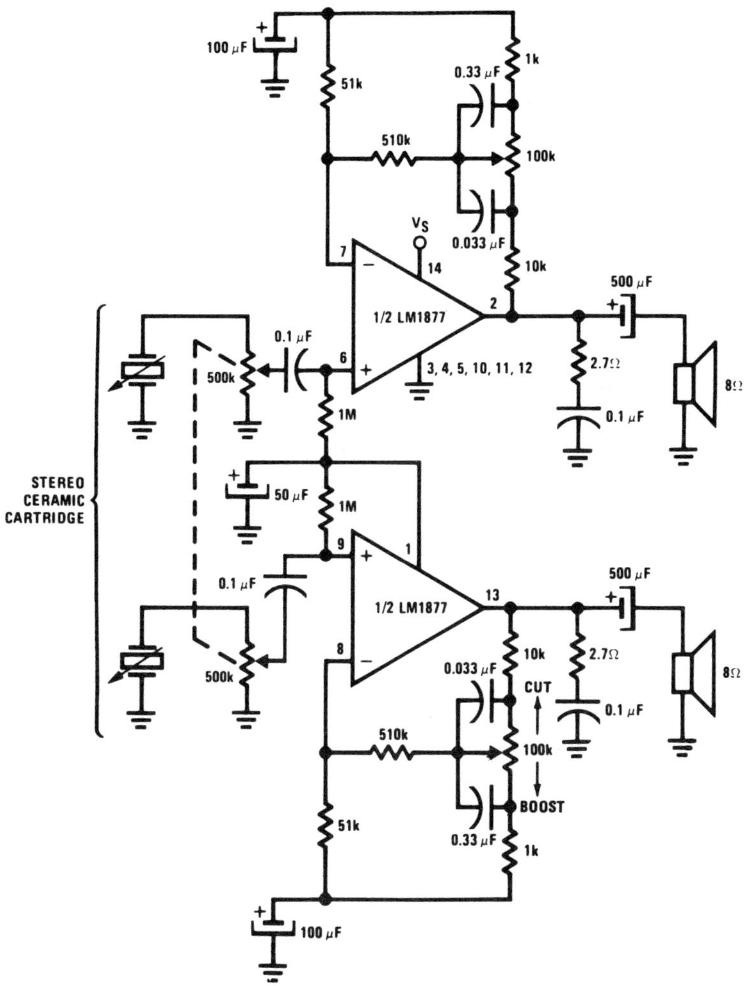

Stereo phonograph amplifier with bass tone control (NS).

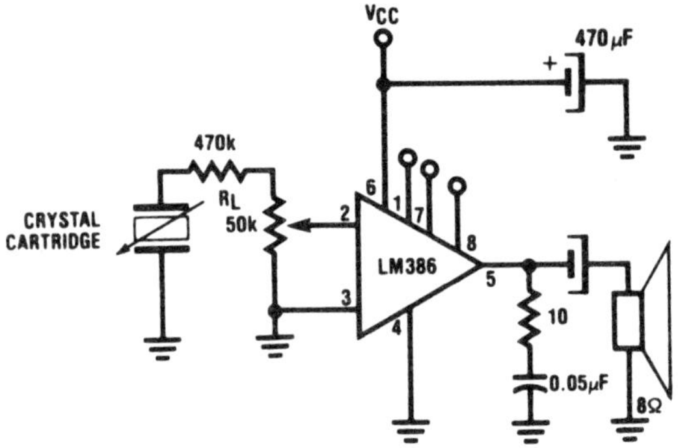

Low cost phono amplifier (NS).

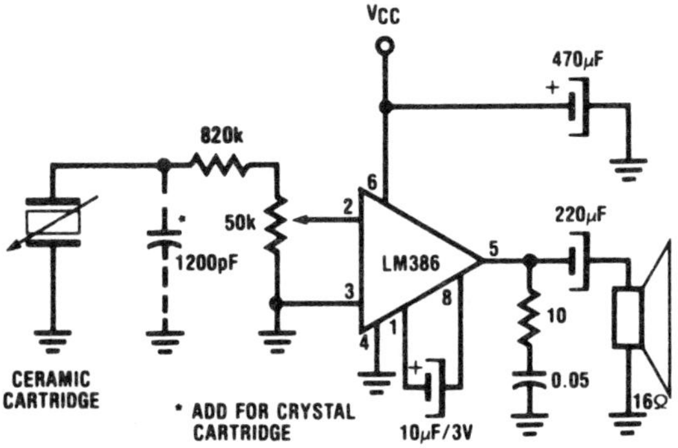

Ceramic cartridge amplifier (NS).

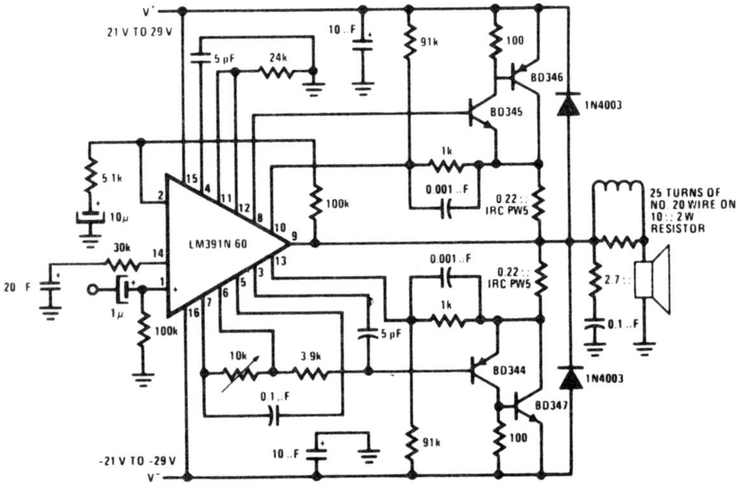

20W-8Ω, 30W-4Ω amplifier (NS).

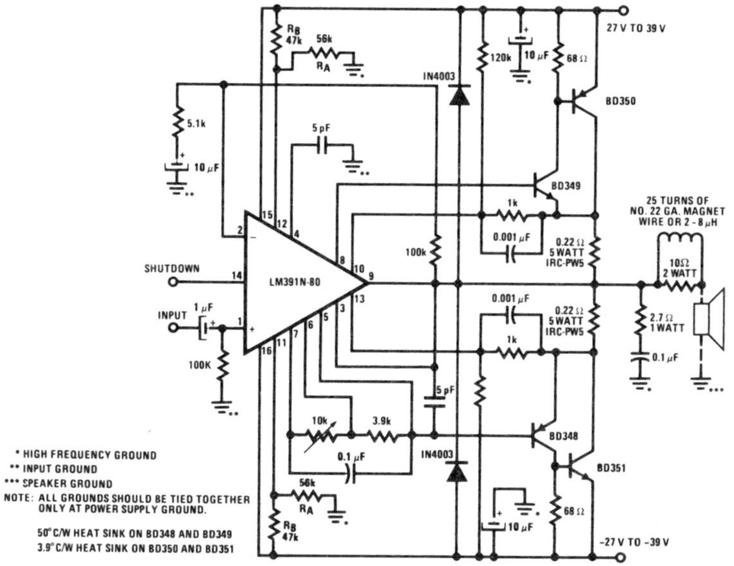

40W-8Ω, 60W-4Ω amplifier (NS).

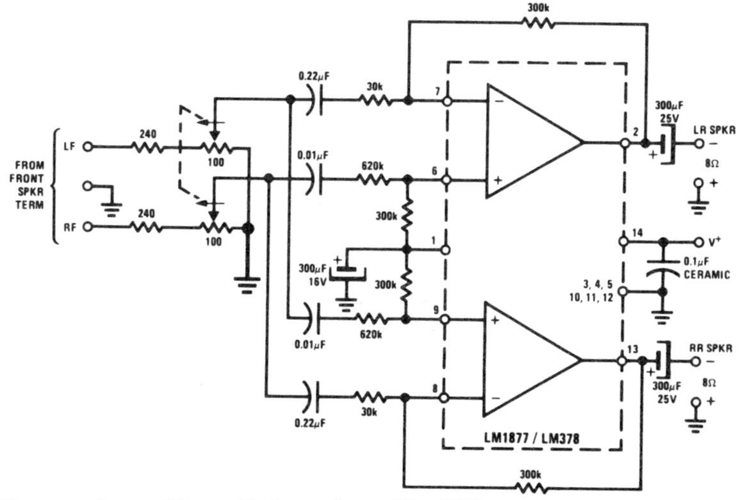

Rear speaker ambience (4-channel) amplifier (NS).

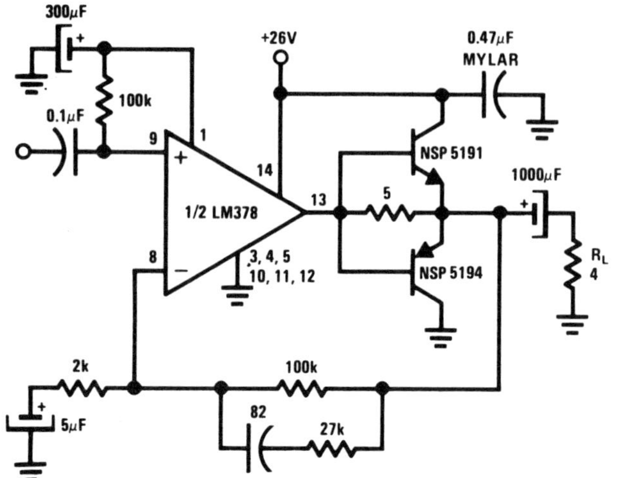

10-watt power amplifier (NS).

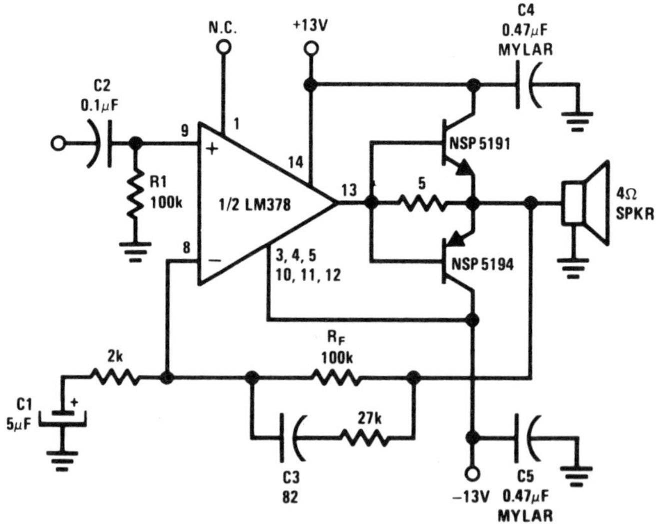

12-watt low-distortion power amplifier (NS).

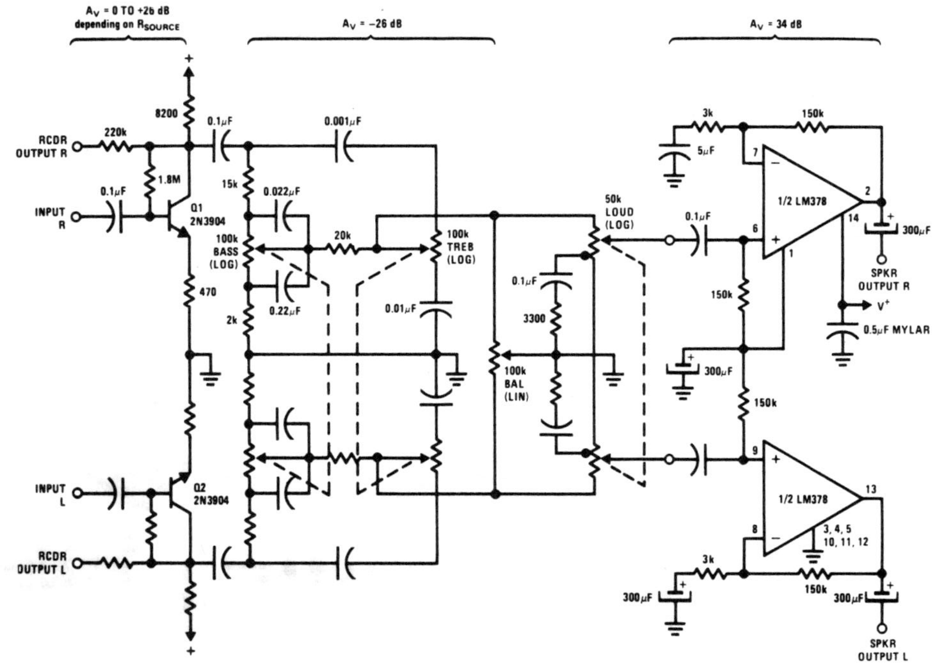

Two-channel power amplifier and control circuits (NS).

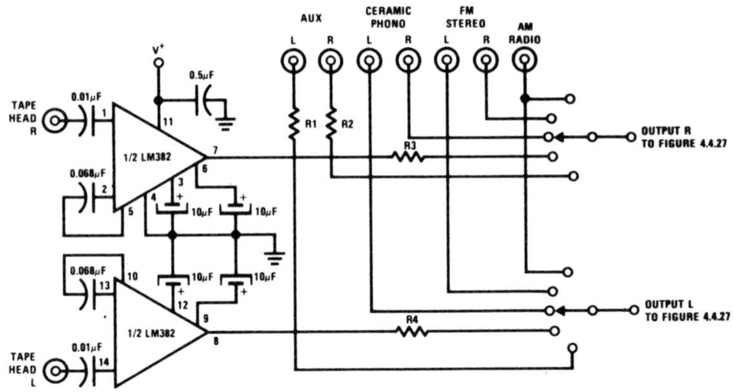

Two-channel tape-playback amplifier and signal switching (NS).

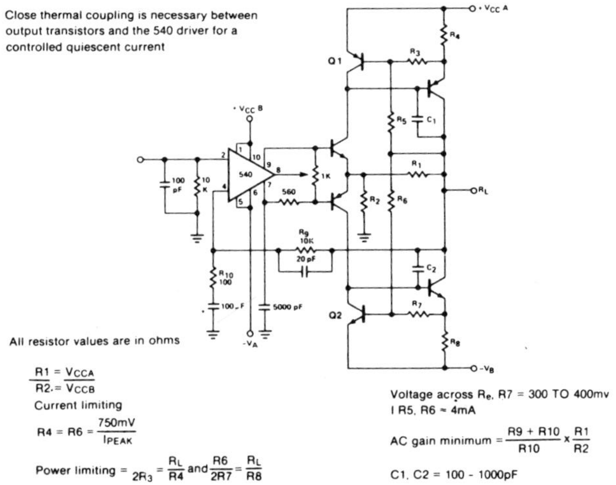

75 watt RMS power amplifier (S).

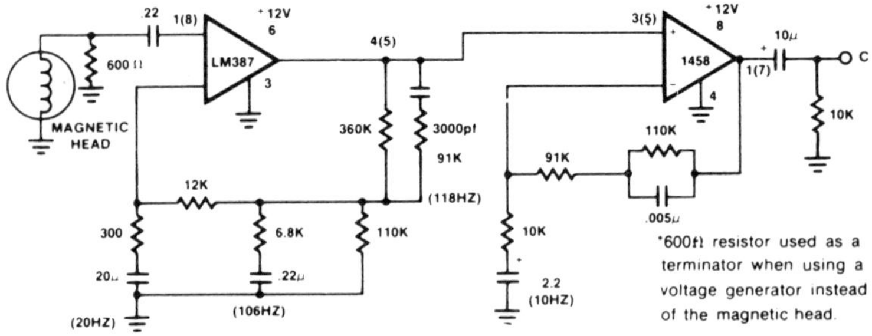

Cassette preamplifier single channel (S).

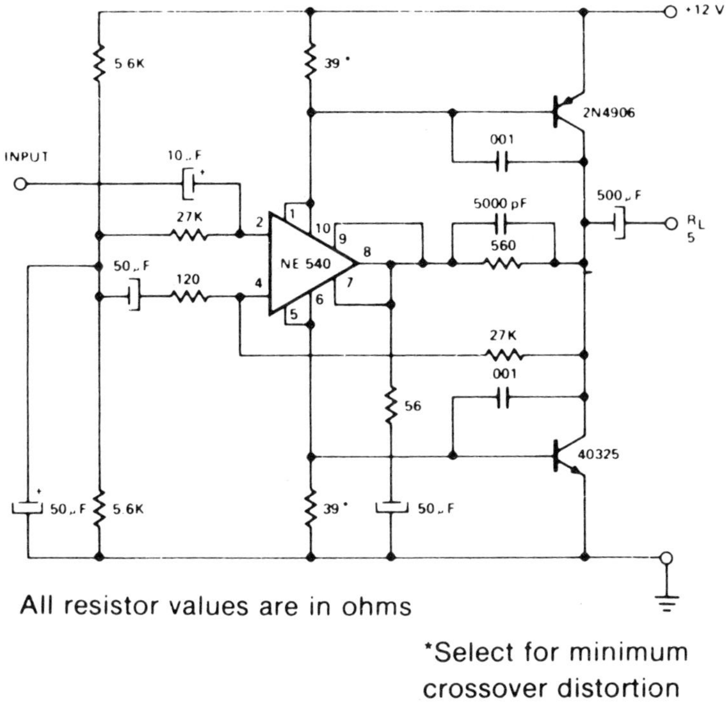

Car radio amplifier (S).

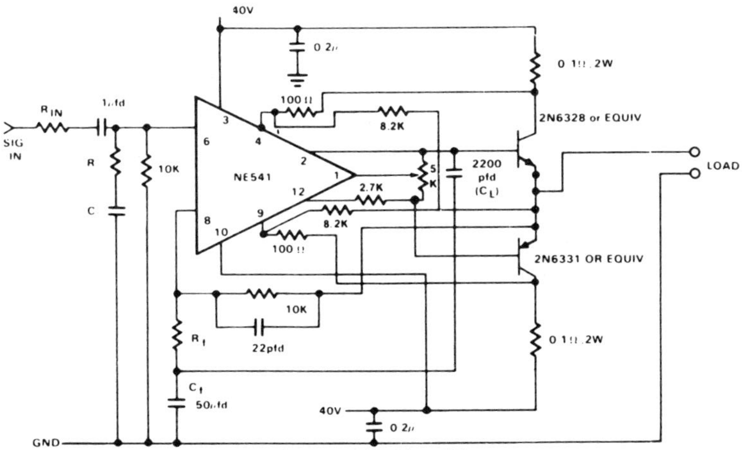

75 watt amplifier with external current limiting (S).

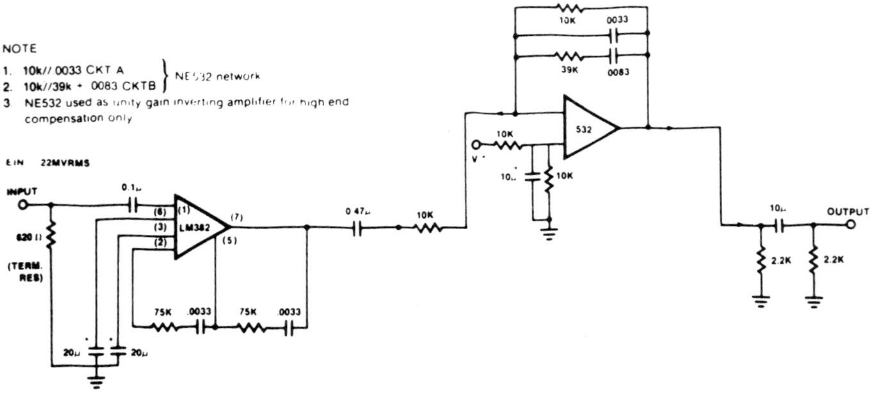

FREQ	GAIN(dB) CKT A	φ (deg) CKT A	GAIN(dB) CKT B	φ (deg) CKT B
20Hz	+68	-15	+68	
40Hz	+76.1	-76	+76	-77
80Hz	+79.4	-135	+79.3	-134
100Hz	+79.8	-147	+79.7	-148
200Hz	+79.7	-187	+79.5	+177
400Hz	+77.6	+157	+76.9	+155
800Hz	+74.6	+144	+73.3	+149
1kHz	+73.9	+144	+72.4	+150
2kHz	+71.8	+134	+70.6	+152
4kHz	+69.2	+110	+69.4	+145
8kHz	+63.9	+76	+67.4	+132
10kHz	+61.5	+65	+66.5	+126
20kHz	+54	-4.0	+62.2	+107

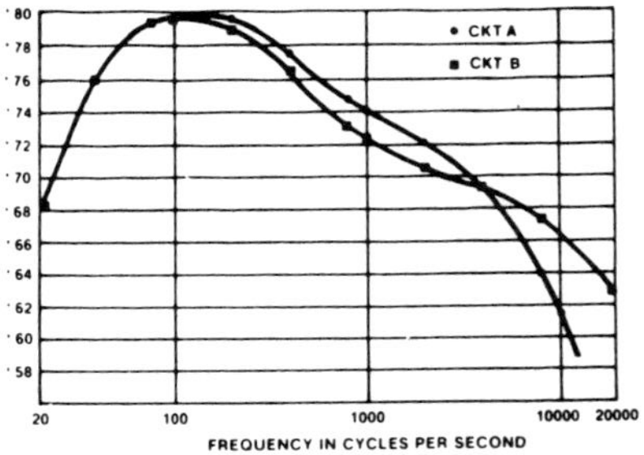

Cassette preamplifier using LM382 and NE532 (S).

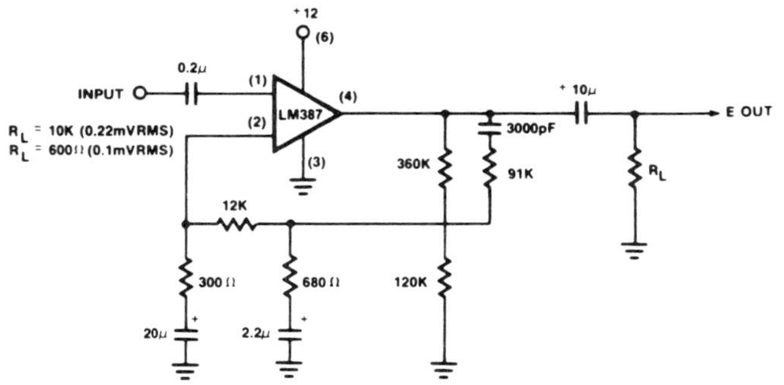

FREQ	GAIN DB RL = 10K	ϕ deg	GAIN dB RL = 600	ϕ deg
20	69		64	
40	76.3		72.4	+115
80	80.4		77.8	+46.3
100	81		78.4	+33
200	80.2		77.8	-0.3
400	77.2		74.9	-18
800	74.8		72.4	-19.8
1kHz	74.2		72.1	-19.3
2kHz	73.3		71.1	-20.4
4kHz	72.3		70	-29.4
8kHz	70.1		68	-44.6
10kHz	69.1		67	-51.1
20kHz	65		62.4	-67.7

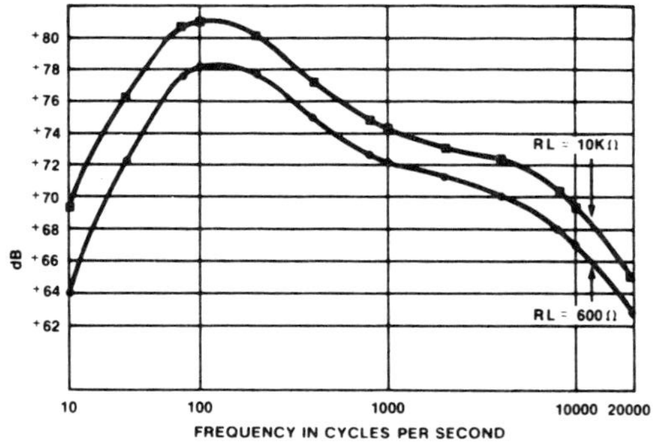

Cassette preamp using LM387 (S).

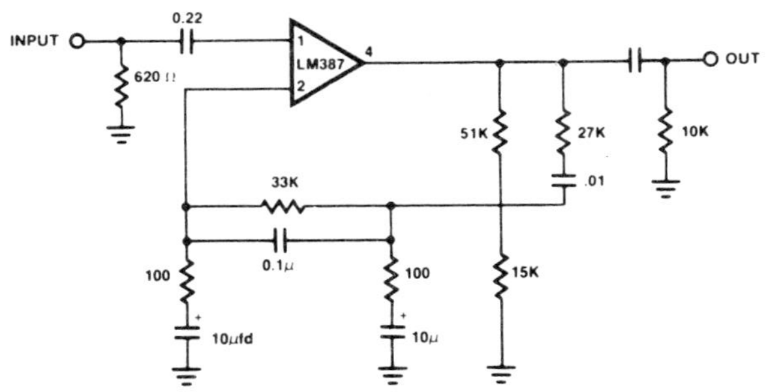

FREQ	GAIN dB
20	79.8db
40	83
80	82.5
100	82
200	76.5
400	68
800	60
1kHz	57.7
2kHz	51.8
4kHz	48.8
8kHz	47.3
10kHz	47
12kHz	47

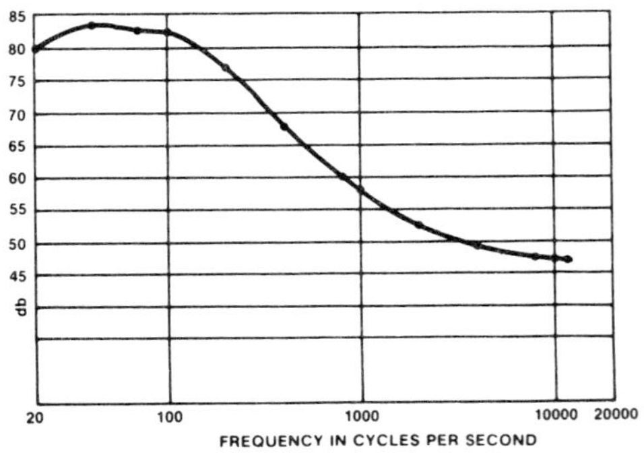

Cassette preamplifier with modified NAB (S).

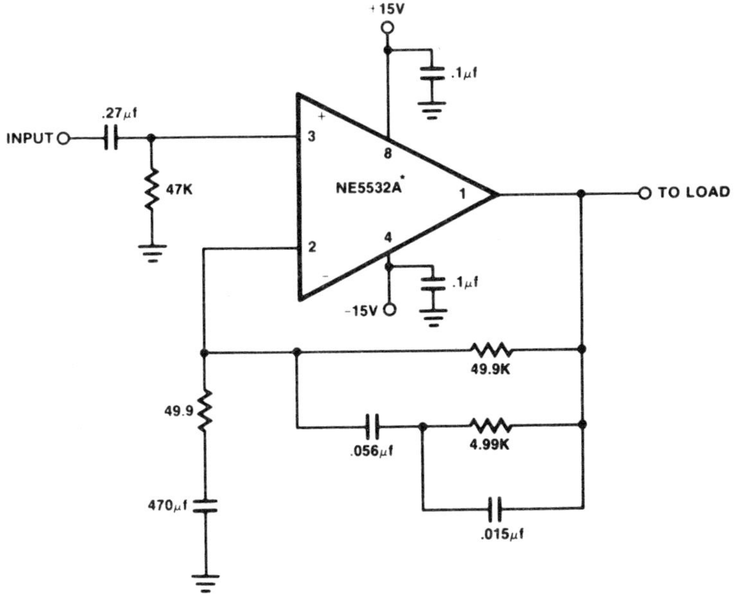

* 1/2 OF DUAL OPERATIONAL AMPLIFIER

NOTE
All resistors are 1% metal film and are valued in ohms.

RIAA phonograph preamplifier using the NE5532A (S).

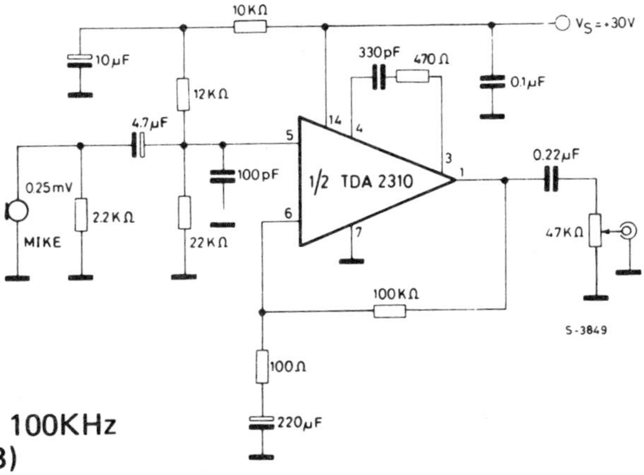

$G_v = 60dB$
B = 15Hz to 100KHz
(± 0.5dB)
SR = 50V/μs
V_o = 5V (d = 0.05% , f = 20KHz)

High slew-rate microphone preamplifier (SGS).

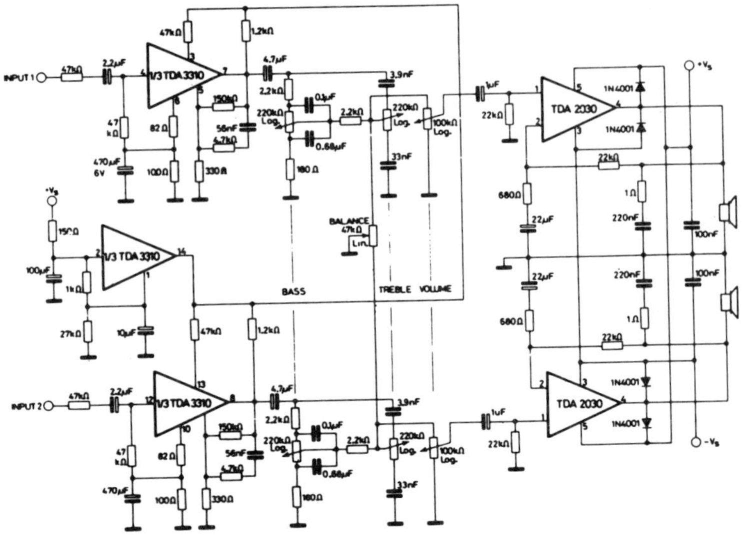

15 + 15 W high quality stereo amplifier with preamplifier equalizer for ceramic pick-ups (SGS).

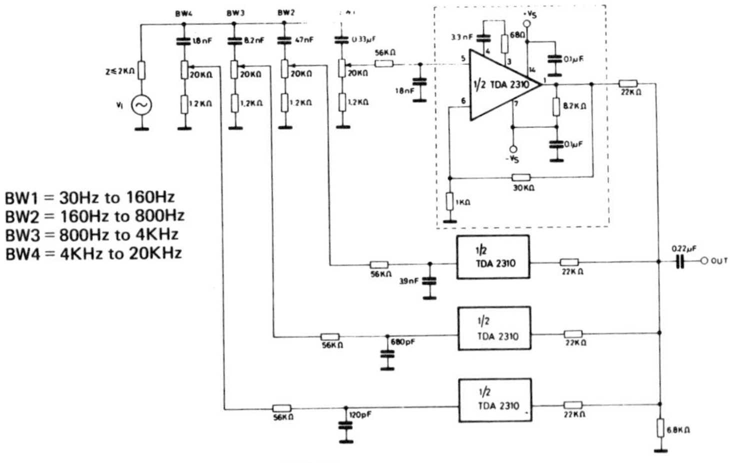

Four band graphic equalizer (SGS).

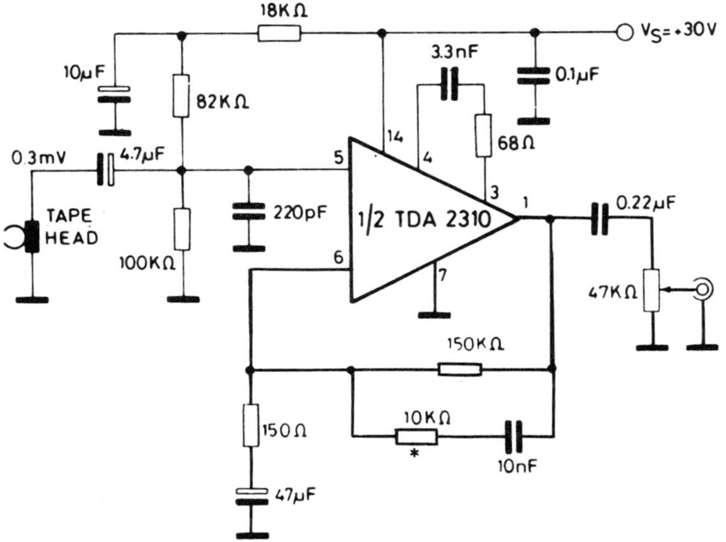

*18KΩ for EQ = 120μs.

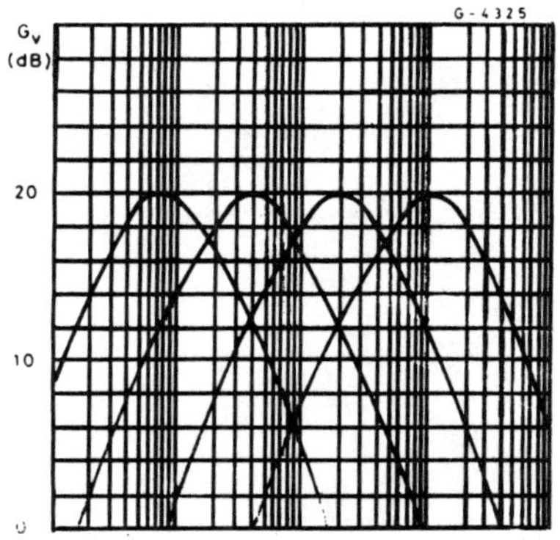

Hi-Fi tape preamplifier (EQ. = 70μs) (SGS).

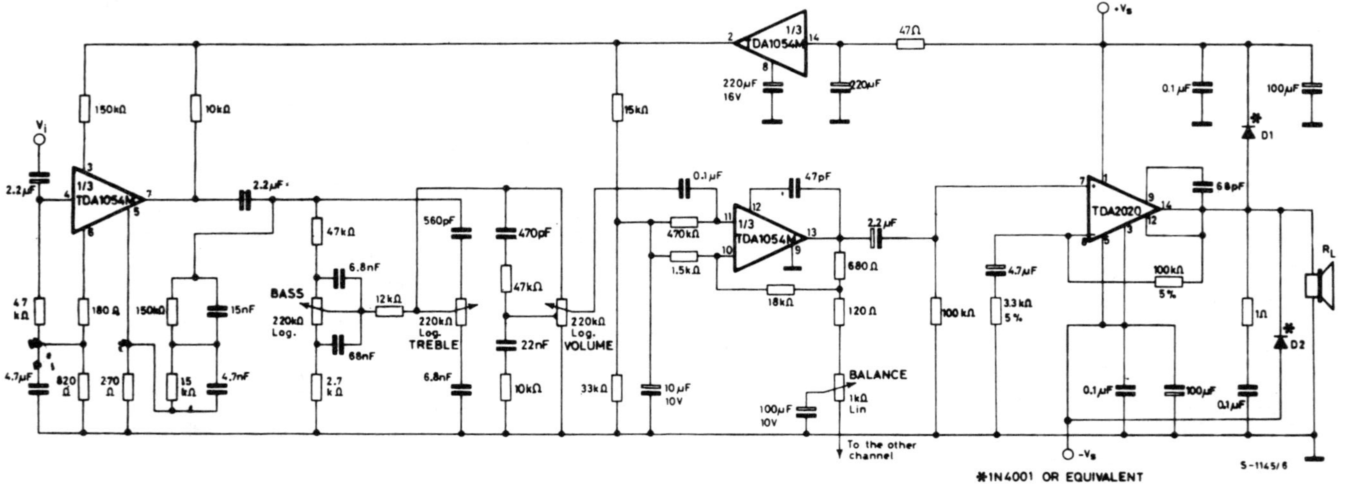

15 W Hi-Fi stereo amplifier with preamplifier-equalizer for magnetic pick-ups (SGS).

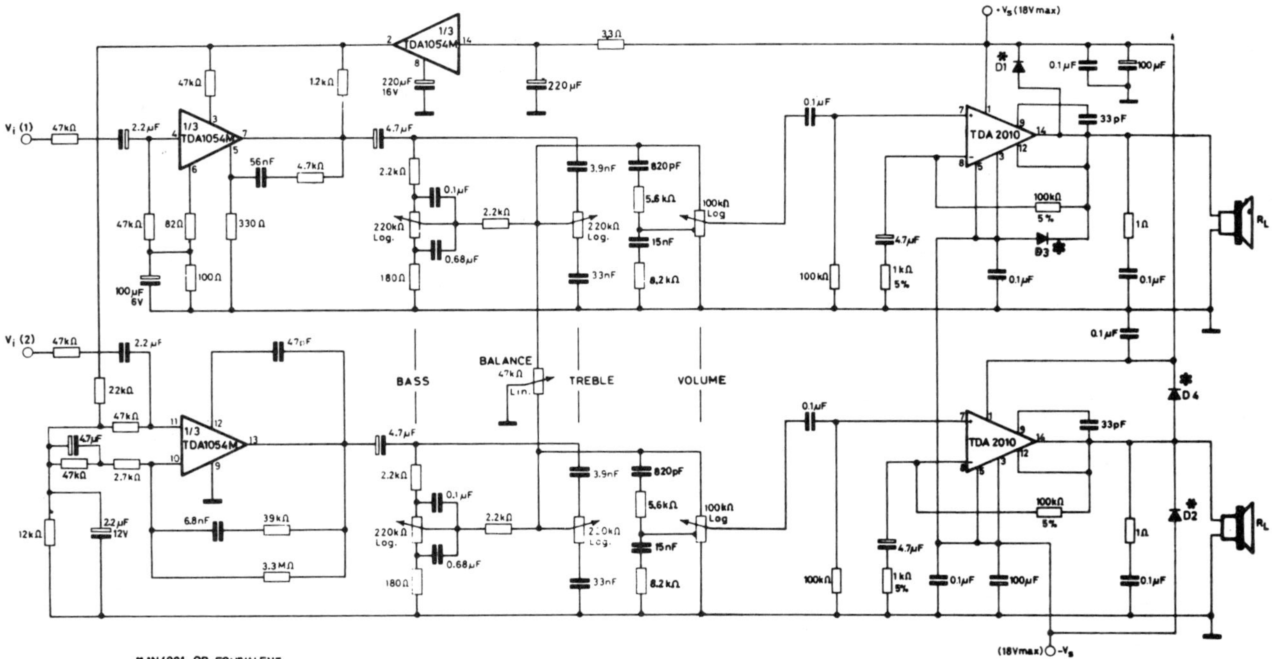

*1N4001 OR EQUIVALENT

10W Hi-Fi stereo amplifier with preamplifier-equalizer for ceramic pick-ups (SGS).

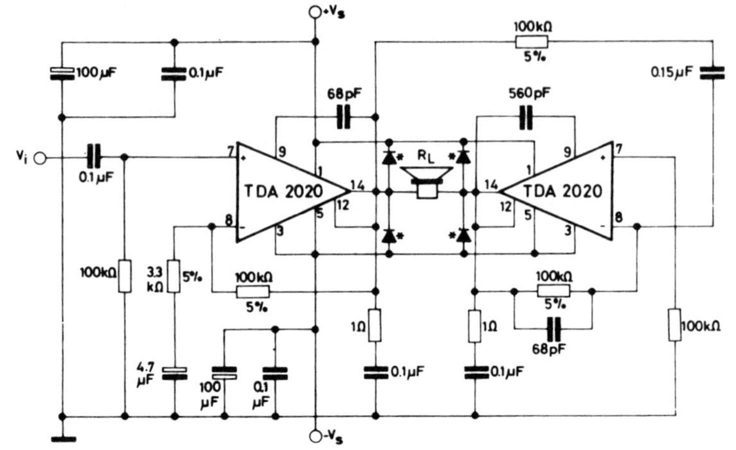

*1N4001 OR EQUIVALENT

Bridge amplifier configuration with split power supply (Po = 24W, Vs = ± 14V, RL = 8 Ω, d ≤ 1%) (SGS).

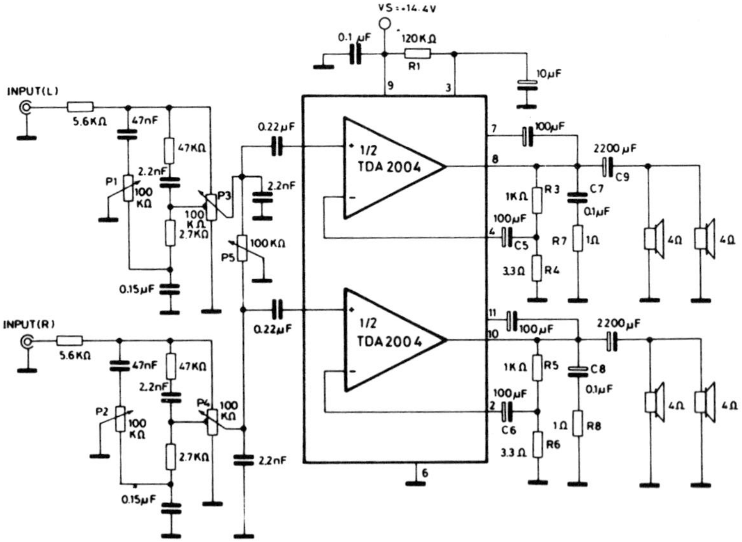

10 + 10W stereo amplifier with tone, balance and loudness control (SGS).

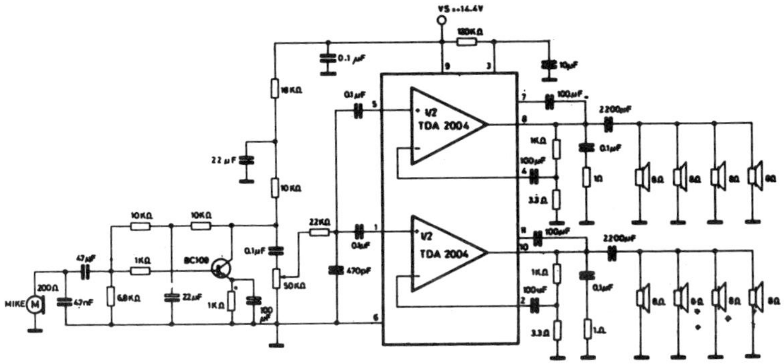

20W bus amplifier (SGS).

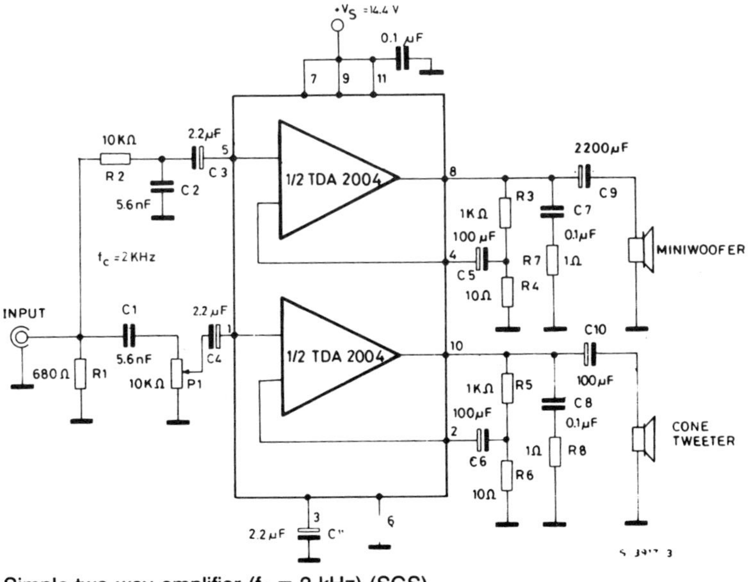

Simple two way amplifier (fc = 2 kHz) (SGS).

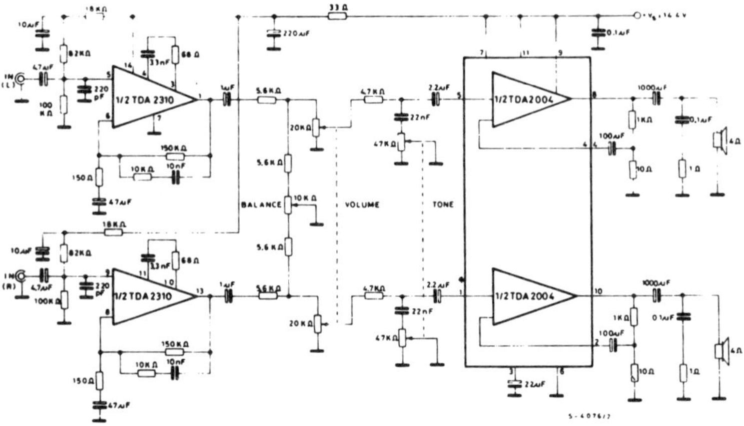

High quality, 10 + 10W cassette player (SGS).

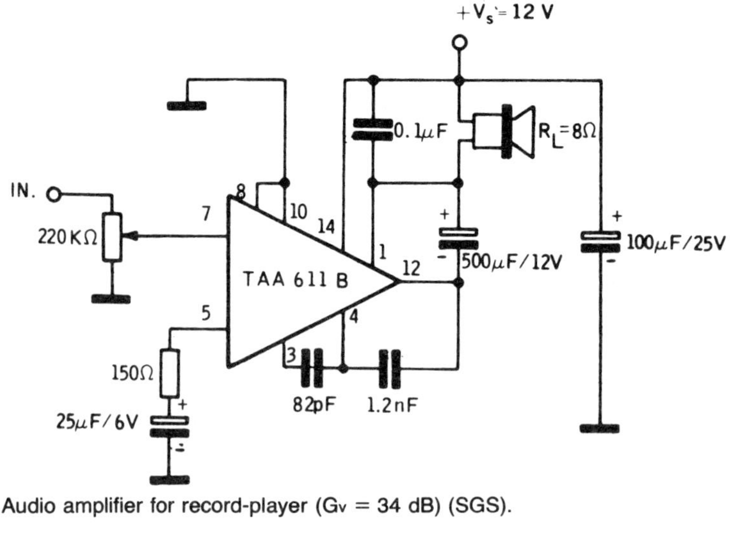

Audio amplifier for record-player (Gv = 34 dB) (SGS).

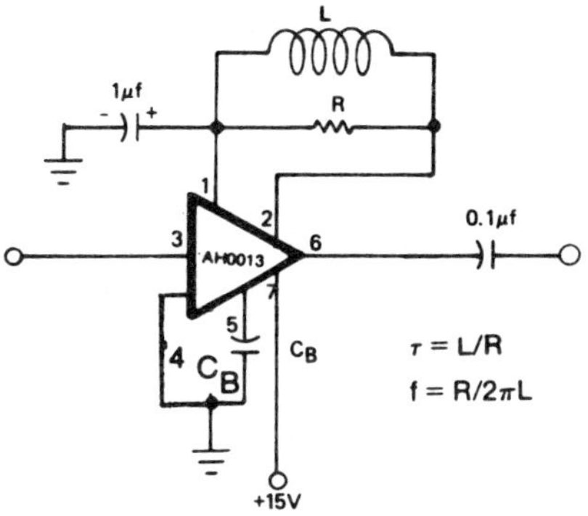

Tuned circuit for ultrasonic applications (OE).

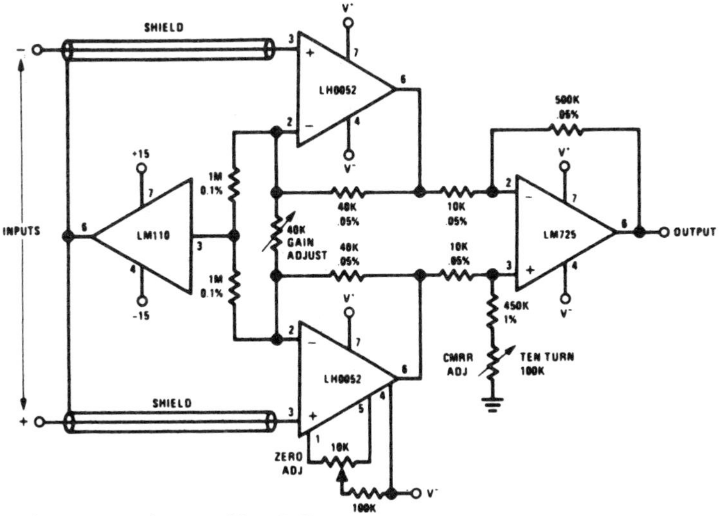

True instrumentation amplifier (NS).

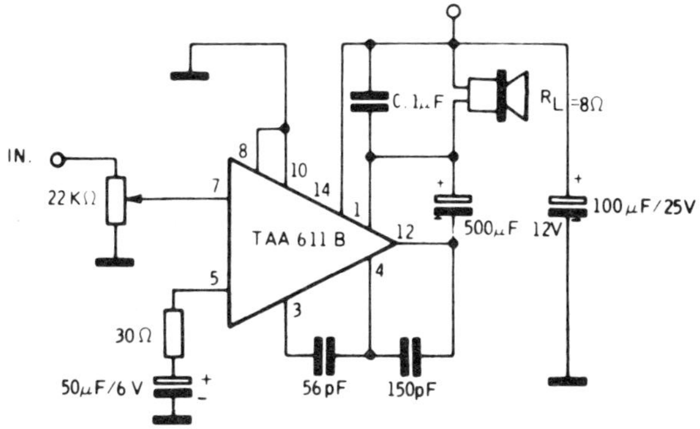

Audio amplifier for radio (G_v = 48 dB) (SGS).

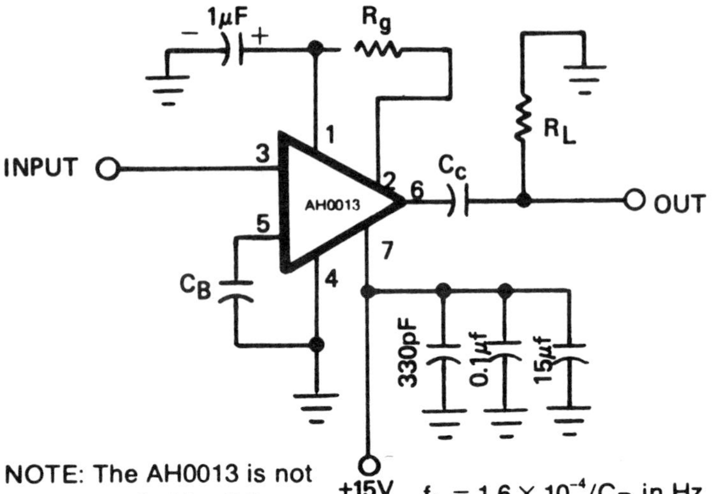

NOTE: The AH0013 is not recommended for DC amplification.

$f_L = 1.6 \times 10^{-4}/C_B$ in Hz
GAIN = 6 Rg
$C_C > 1/2\pi f_L R_L$
Where all C are in Farads,
Rg is in kΩ, R_L is in ohms.

Wideband preamp (OE).

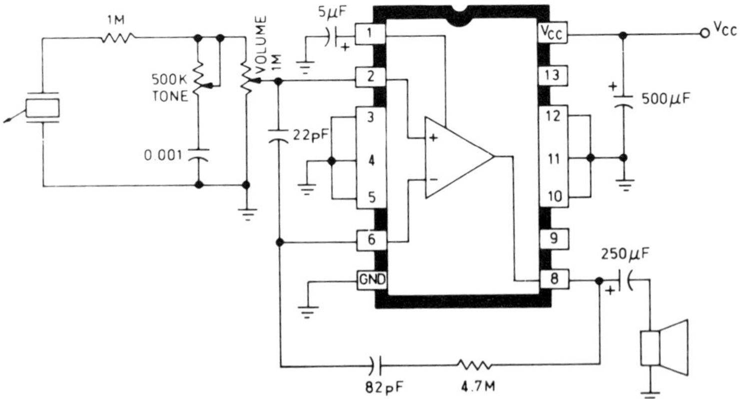

Amplifier with bass boost network (SE).

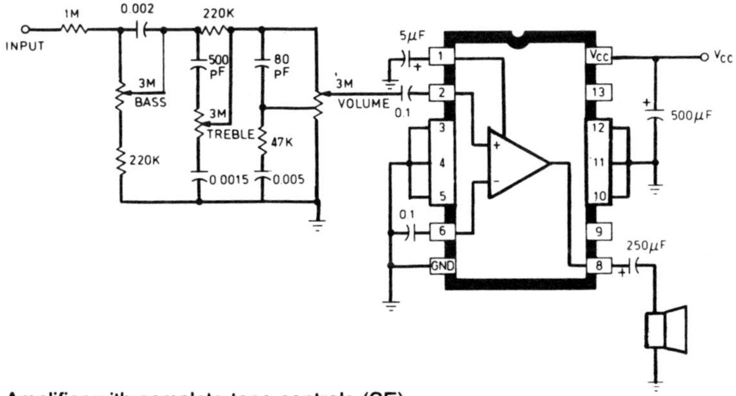

Amplifier with complete tone controls (SE).

D	NOMINAL $\frac{V_{OUT}}{V_{IN}}$
1111111111	$-\frac{1024}{1023}$
1000000000	-2
0000000001	- 1024
0000000000	OPEN LOOP

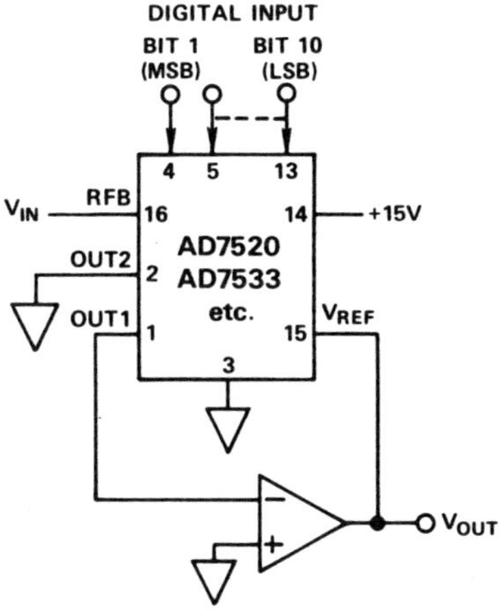

Variable gain amplifier (2-quadrant divider) (AD).

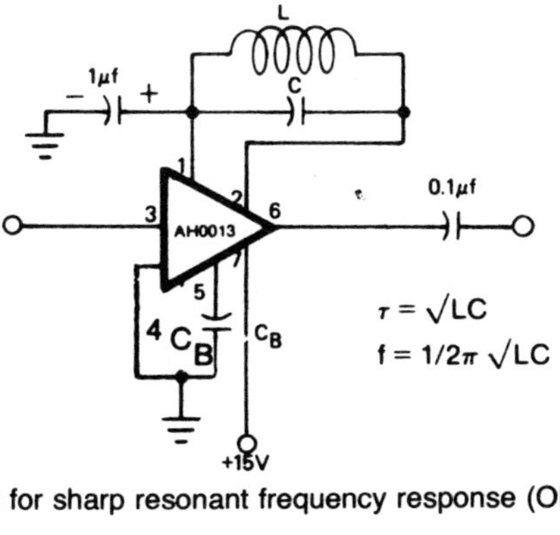

LC gain elements for sharp resonant frequency response (OE).

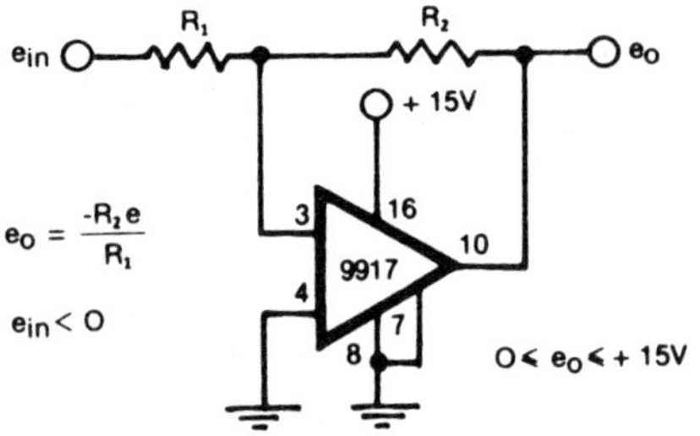

Single supply inverting amplifier (OE).

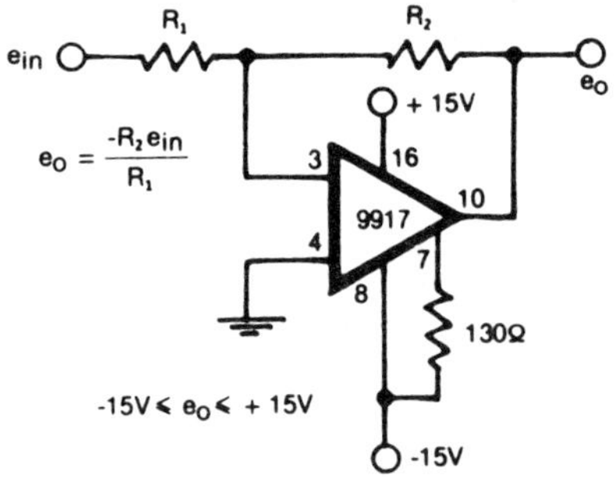

Regular, dual supply operation inverting amplifier (OE).

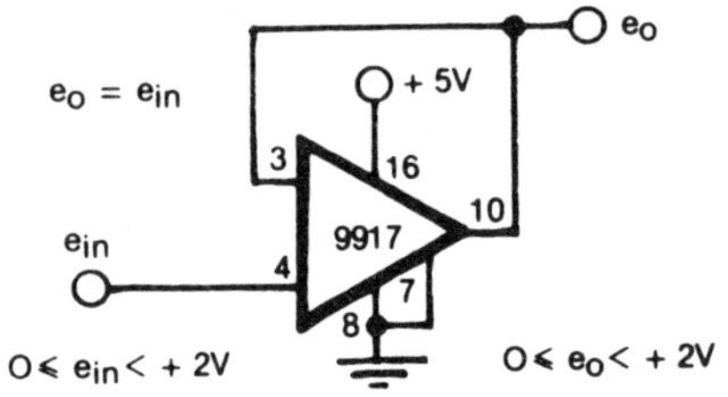

Single supply, non-inverting follower (OE).

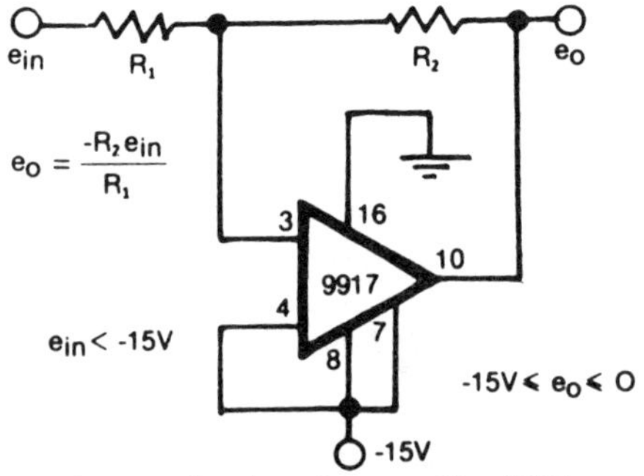

Single negative supply operation inverting amplifier (OE).

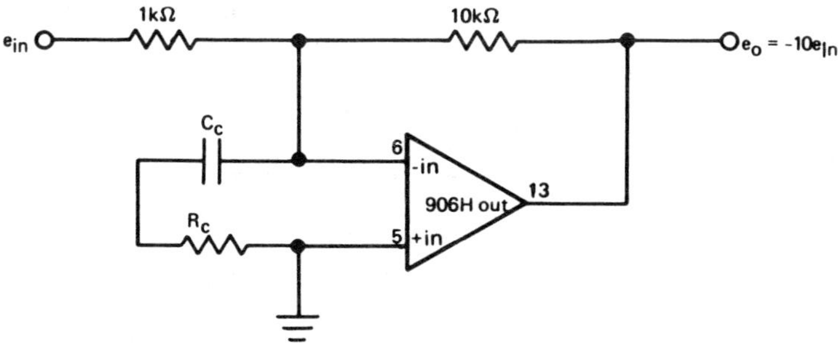

X10 inverting buffer amplifier (OE).

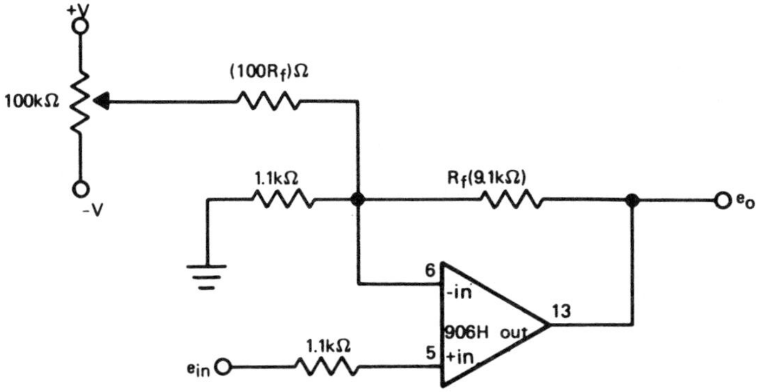

X10 noninverting buffer amplifier with offset correction (OE).

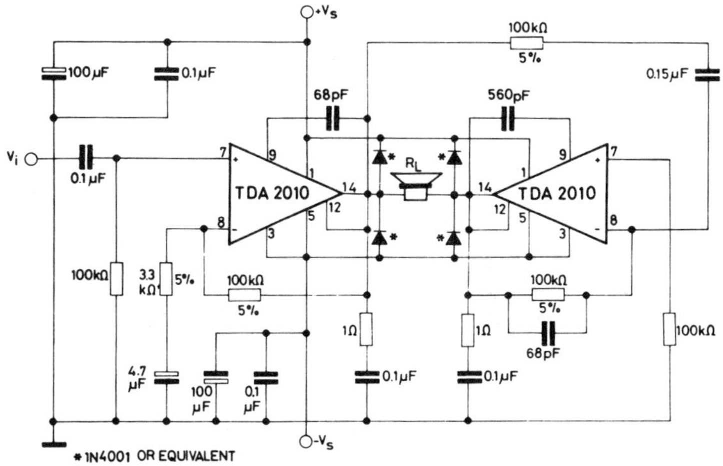

30W bridge amplifier configuration with split power supply ($R_L = 8\,\Omega$ d $\leq 1\%$; $V_s = \pm 17V$) (SGS).

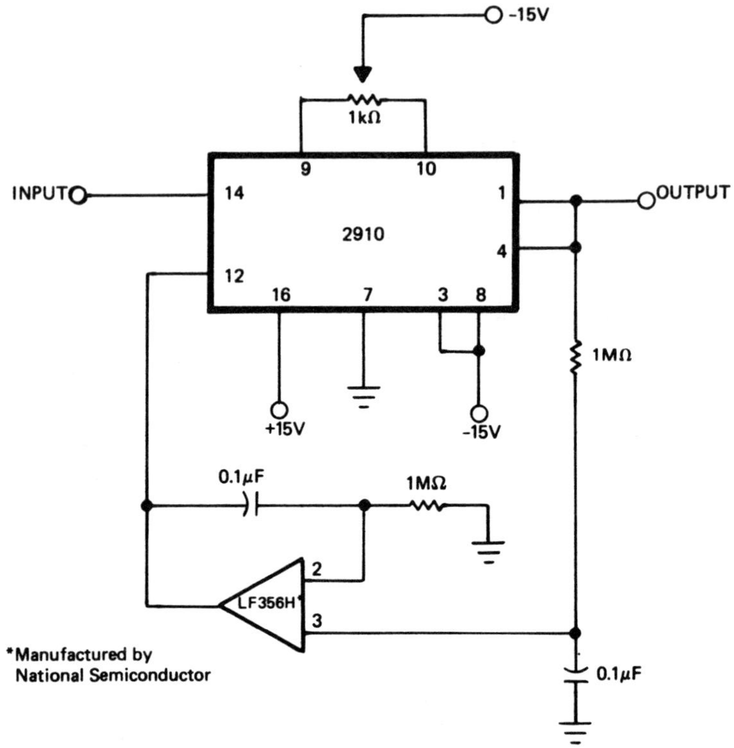

Stabilized bipolar or low duty cycle pulse logarithmic amplifier (OE).

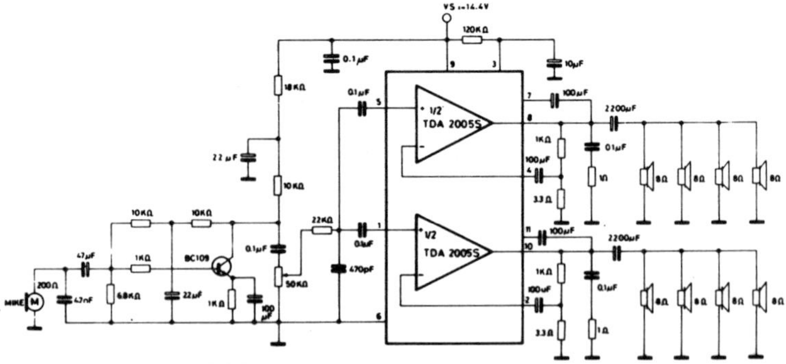

20W bus amplifier (SGS).

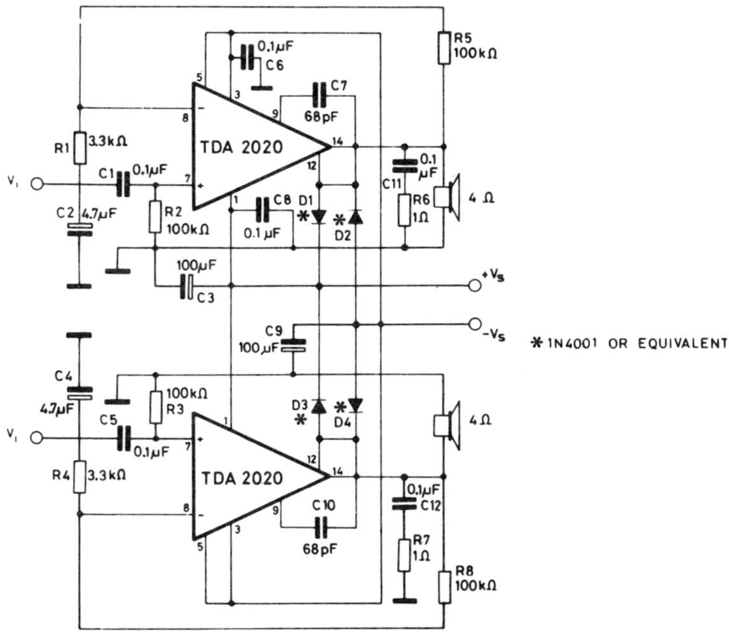

Stereo amplifier with split power supply (SGS).

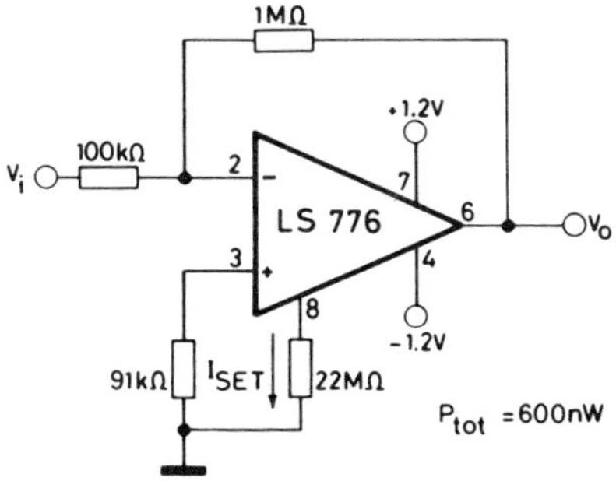

Nanowatt amplifier (SGS).

High input impedance amplifier (SGS).

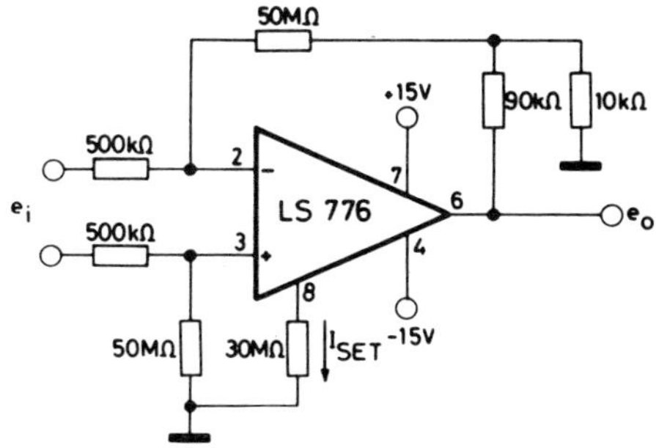

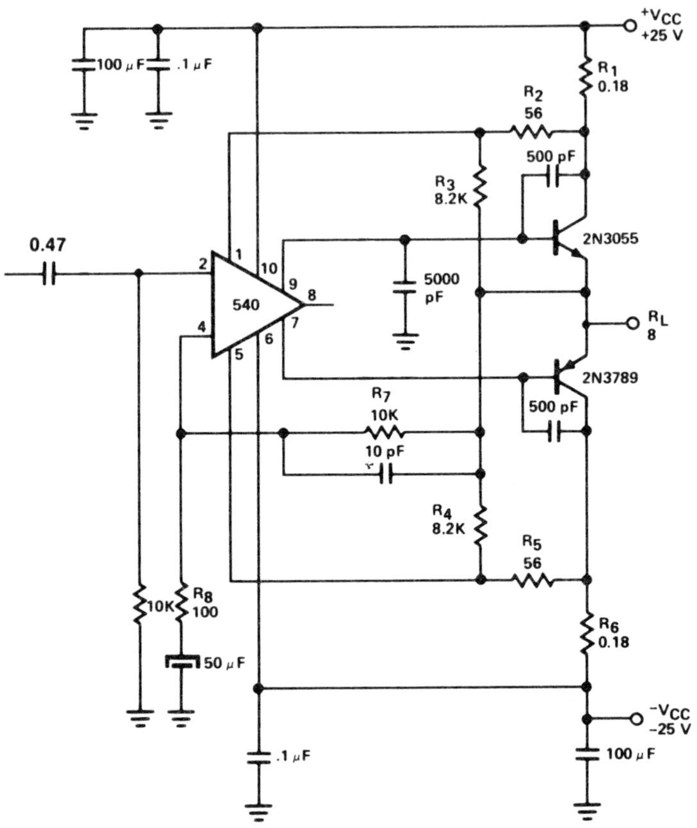

VOLTAGE GAIN

$$A_V = \frac{R_7 + R_8}{R_8}$$

CURRENT LIMITING

$$R_1 = R_6 \cong \frac{650\ \text{mV}}{I_{PEAK}}$$

POWER LIMITING

$$R_2 = R_5 \cong 56\Omega$$

$$R_3 = R_4 \cong \frac{V_{CC}}{3\text{mA}}$$

All resistor values are in ohms.

35 watt power amplifier (S).

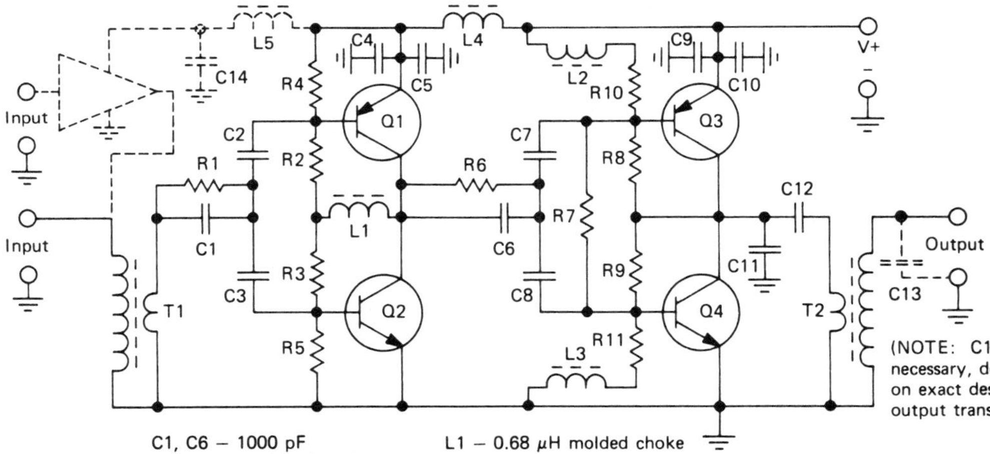

(NOTE: C13 may be necessary, depending on exact design of output transformer.)

C1, C6 – 1000 pF
C2, C3, C7, C8 – 0.1 μF
C4, C9, C14 – 0.15 μF
C5, C10, C12 – 0.68 μF
C11 – 2000 pF
C13 – 100 pF (Typ.)

R1, R4, R5 – 22 ohms/1/4 W
R2, R3 – 330 ohms/1/2 W
R6 – 15 ohms/1/4 W
R8, R9 – 220 ohms/1 W
R10, R11 – 6.8 ohms/1/2 W
R7 – 560–680 ohms/2 W

L1 – 0.68 μH molded choke
L2, L3 – 6.8 μH molded choke
L4 – 4 ferrite beads, Ferroxcube 56 590 65/3B

Q1, Q3 – MRF432
Q2, Q4 – MRF433

T1, T2 – See text

All capacitors are ceramic chips, Union Carbide type 1225 or 1813 or Varadyne size 18 or 14 or equivalent.
All resistors are carbon composite.

Complementary symmetry amplifier schematic (M).

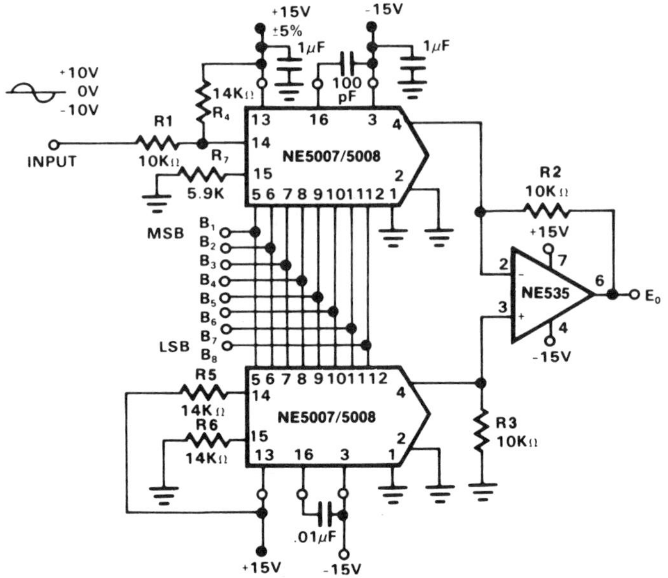

Bipolar input offset binary output } Performs 2 quadrant multiplications—AC input controls output polarity.

NOTES

1. R1 = R2 = R3
2. R4 = R5
3. E_O DC to 20KHz = ±5V
4. E_O DC to 10KHz = ±10V

Dc-coupled digital attenuator/programmable gain amplifier (S).

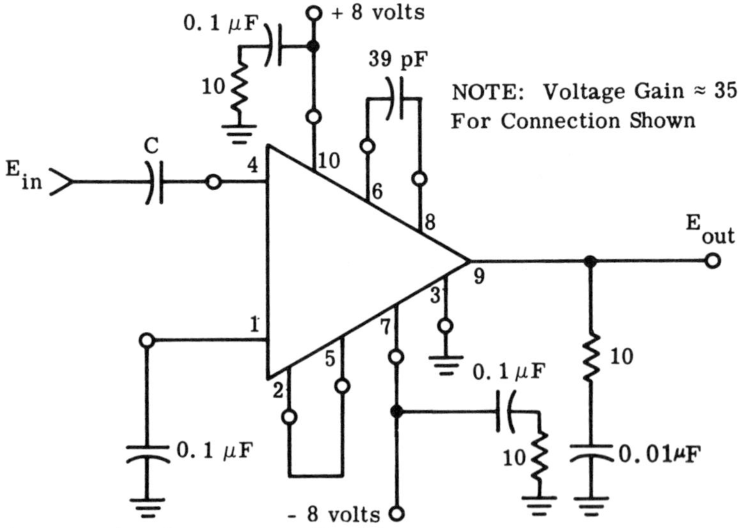

A one-watt inverting power amplifier (M).

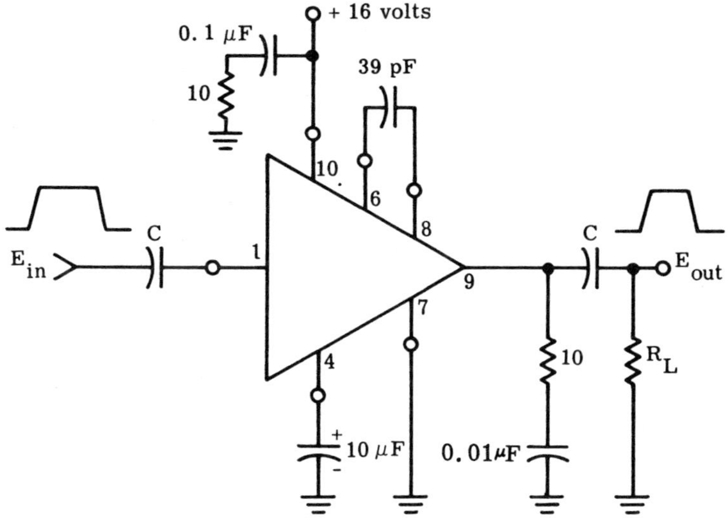

NOTE: Voltage Gain = 18 For Connection Shown

Pulse power amplifier (M).

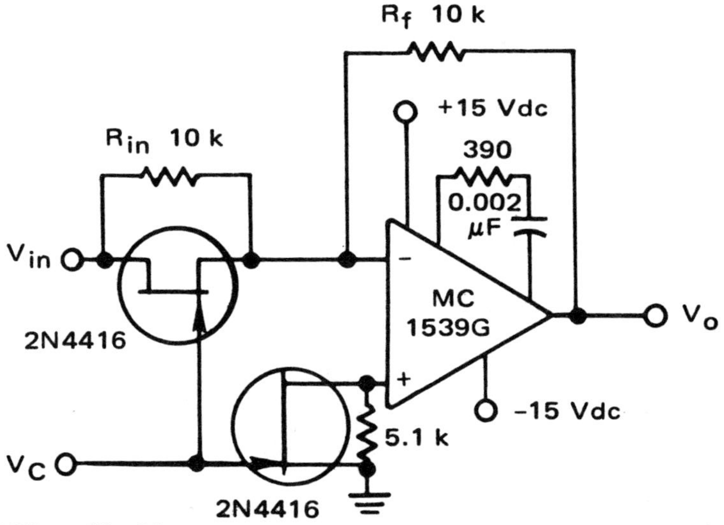

AGC amplifier (M).

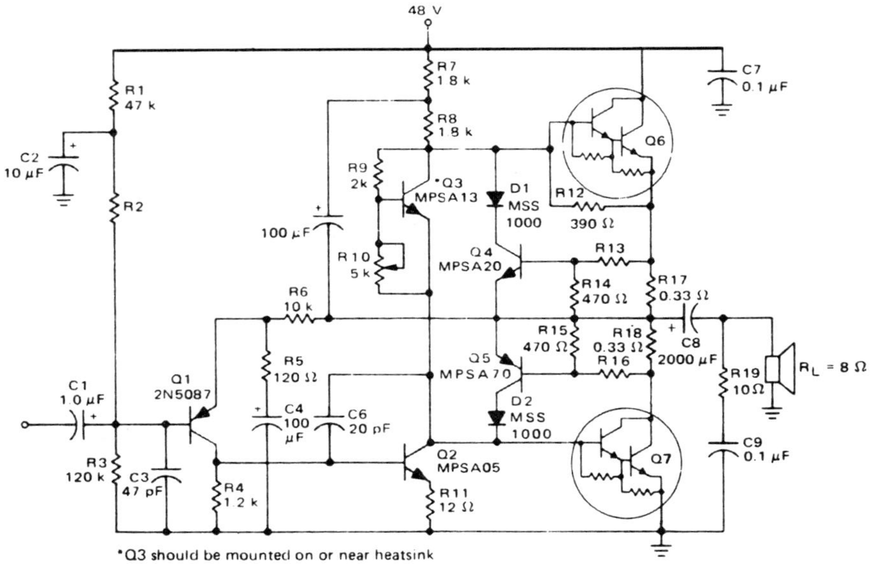

Component	15 W	20 W	25 W
V_{CC}	42 V	46 V	48 V
R2	220 k	160 k	160 k
R5	120 Ω	180 Ω	180 Ω
R13	180 Ω	220 Ω	220 Ω
R16	180 Ω	220 Ω	220 Ω
Q6	MJE800	MJE1100	MJE1100
Q7	MJE700	MJE1090	MJE1090
*Heatsink	6.5°C/W	4.8°C/W	3.6°C/W
R8	Adjusted for 20-30 mA Collector Current for Q6		

*Heatsink size calculated is based on a maximum ambient temperature of 50°C and a load phase angle of 60 degrees (see text for method of calculation). The heatsinks for both devices on one sink.

Circuit diagram for 15 to 25 watt Darlington amplifier (M).

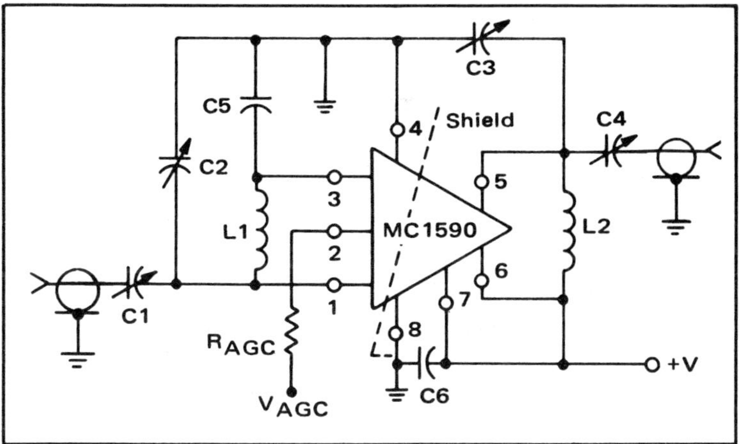

Parameter	30 MHz	60 MHz	100 MHz
Power Gain (dB)	50.8–54	44.2–46.7	31.6–35.7
BW (MHz)	0.7–1.4	1.9–2.4	7.8–9.2
C1 pF	38	1–30	1–30
C2 pF	1–30	1–30	1–10
C3 pF	1–10	1–30	1–15
C4 pF	1–30	1–10	1–10
C5 μF	0.002	0.001	470
C6 μF	0.002	0.001	470
L1 μH	0.6	0.17	0.07
L2 μH	1.35	0.28	0.13
V^+ Vdc	12	12	12

Single-stage tuned amplifier (M).

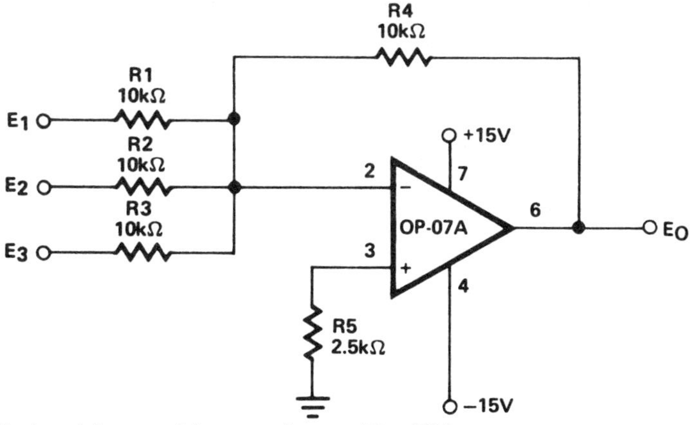

Adjustment-free precision summing amplifier (PM).

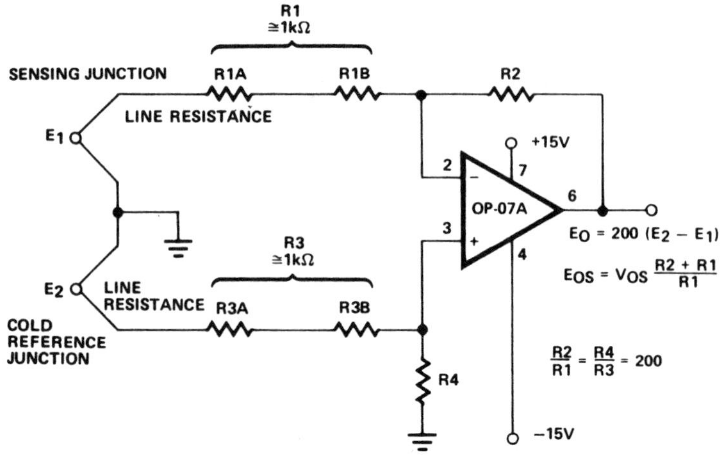

High stability thermocouple amplifier (PM).

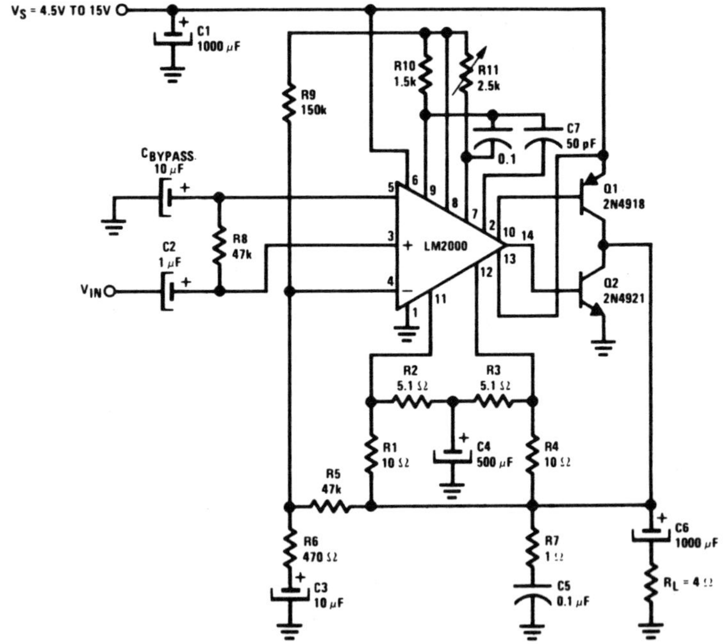

Complementary output amplifier (NS).

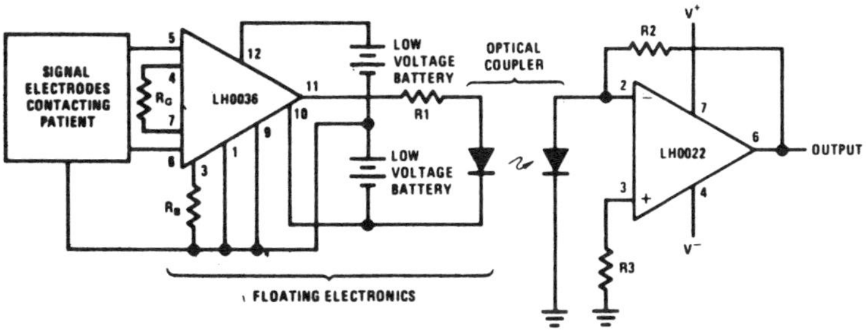

Isolation amplifier for medical telemetry (NS).

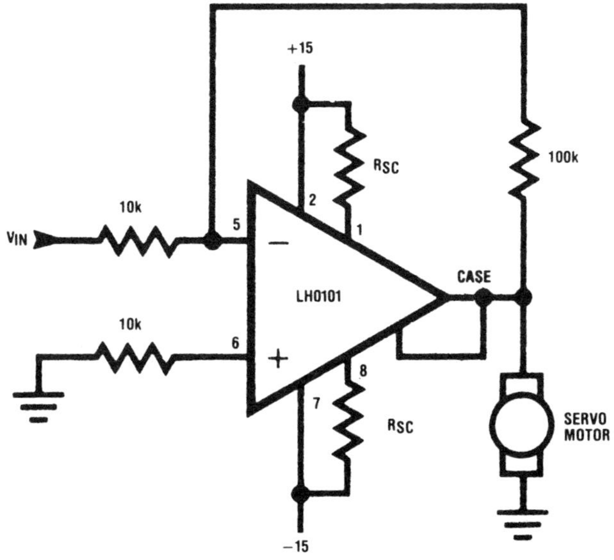

Dc servo amplifier (NS).

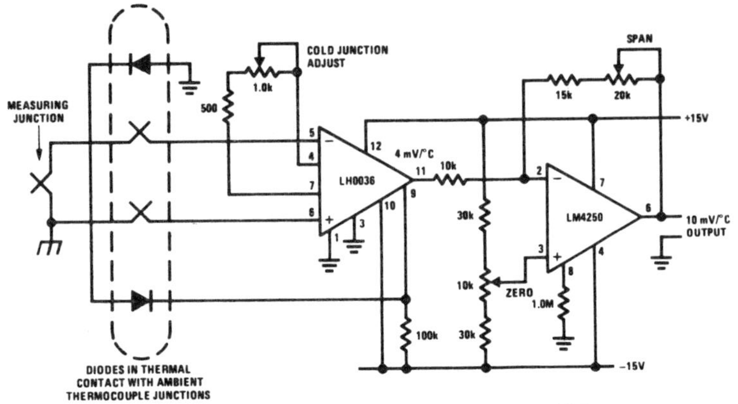

Thermocouple amplifier with cold junction compensation (NS).

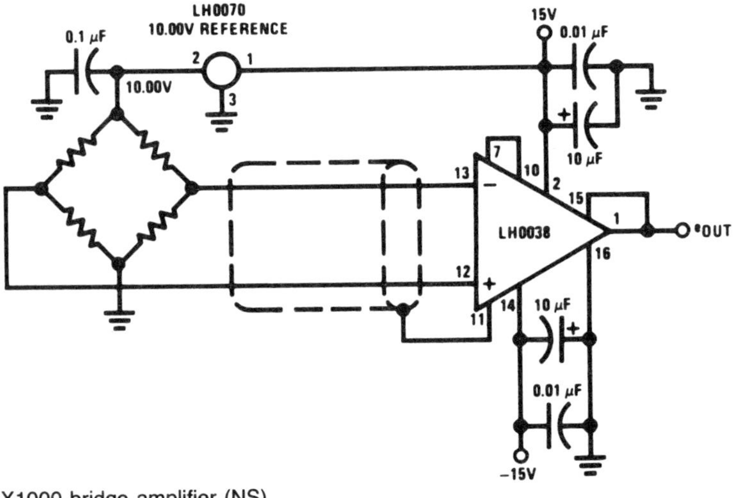

X1000 bridge amplifier (NS).

Picoamp amplifier for pH meters and radiation detectors (NS).

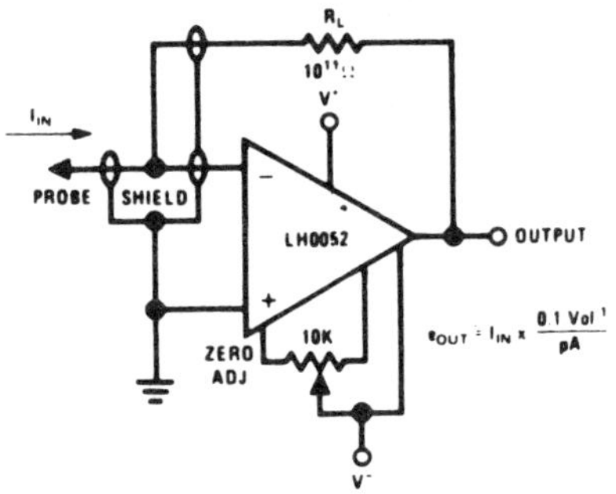

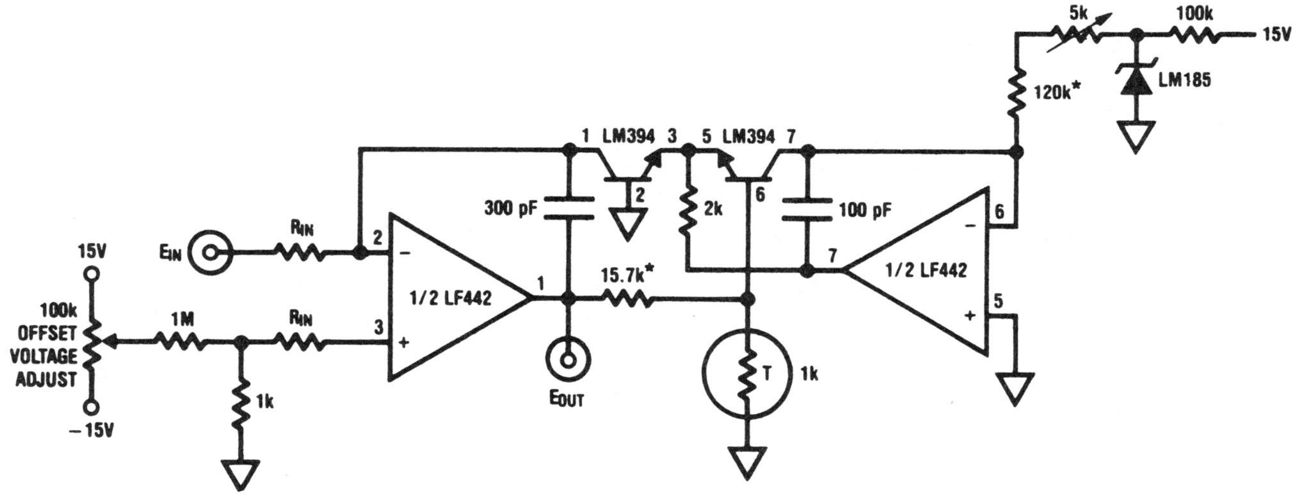

$$E_{OUT} = -\left[\log 10\left(\frac{E_{IN}}{R_{IN}}\right) + 5\right]$$

(R_T) = Tel Labs type Q81

Trim 5k for 10 μA through the 5k-120k combination

*1% film resistor

Conventional log amplifier (NS).

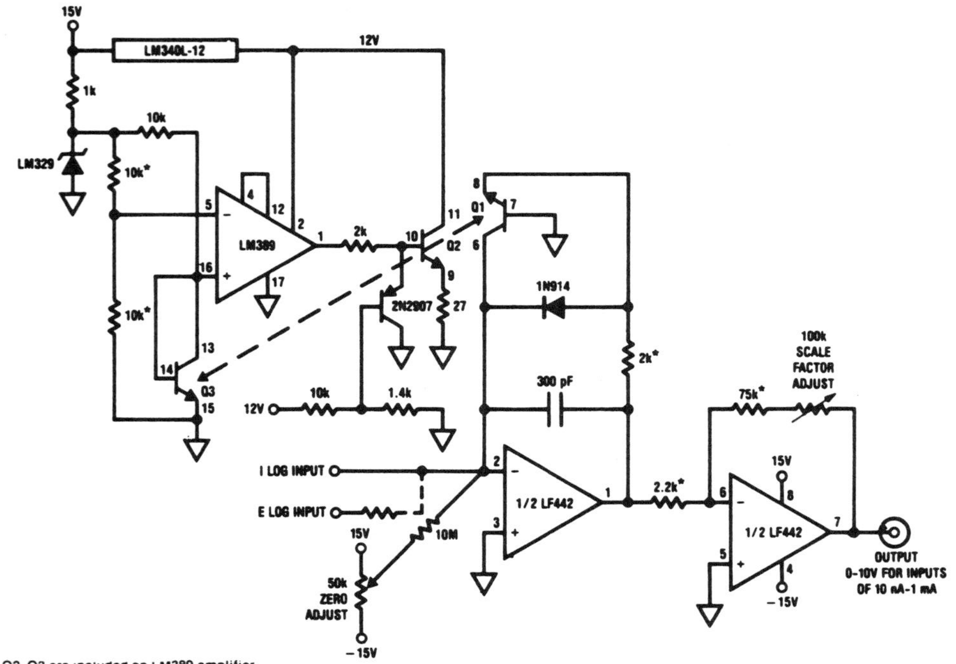

Q1, Q2, Q3 are included on LM389 amplifier chip which is temperature-stabilized by the LM389 and Q2-Q3, which act as a heater-sensor pair. Q1, the logging transistor, is thus immune to ambient temperature variation and requires no temperature compensation at all.

Unconventional log amplifier (NS).

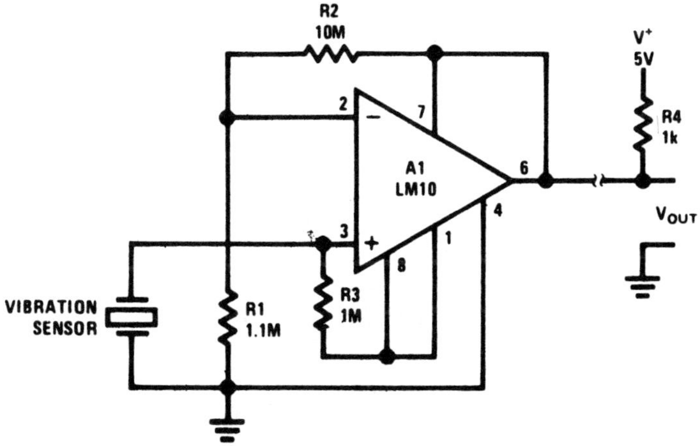

Remote amplifier (NS).

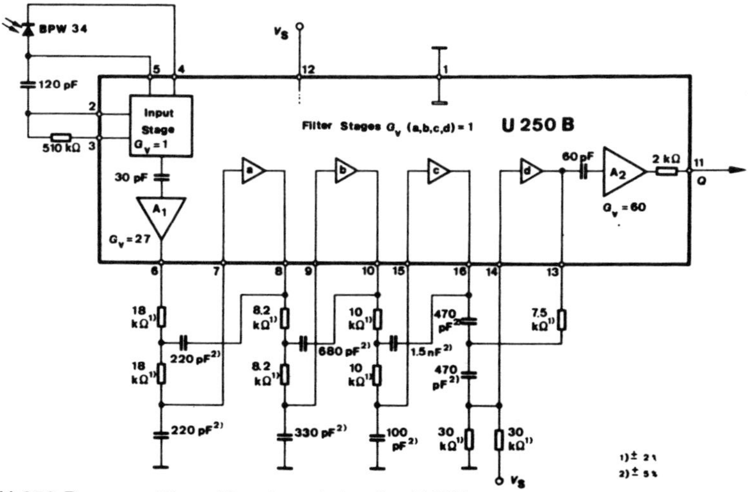

U 250 B preamplifier with external circuitry (AEG).

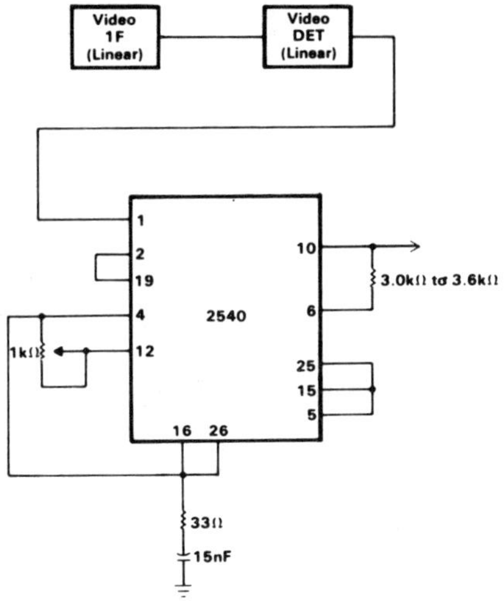

Logarithmic video amplifier (OE).

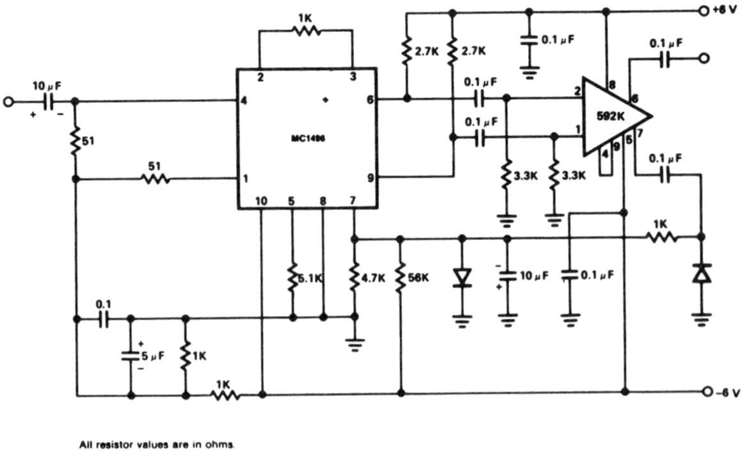

Wide band agc amplifier (S).

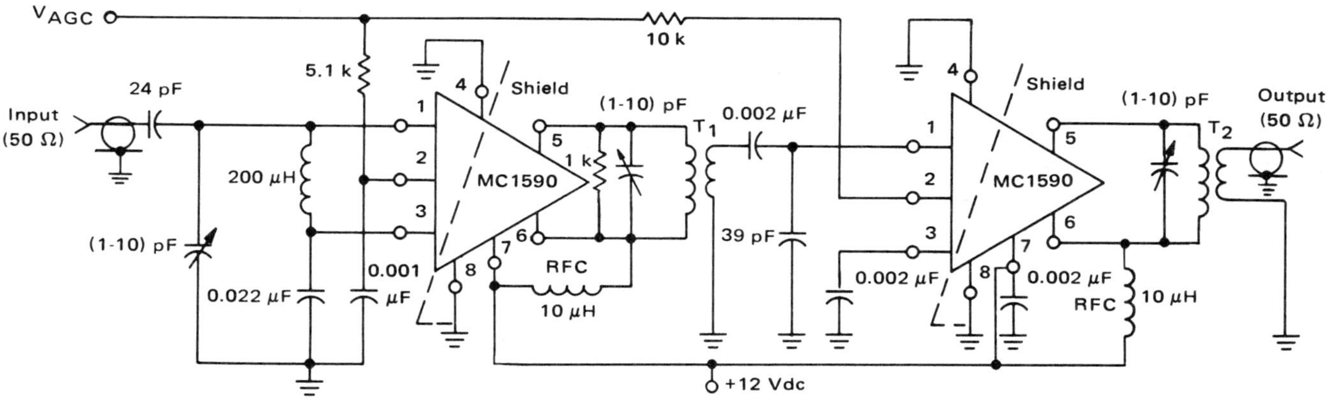

T_1: Primary — 15 turns, No. 22 AWG wire,
¼'' I.D. Air Core Secondary — 4 turns, No. 22 AWG wire, coef. of coupling ≈ 1.0

T_2: Primary — 10 turns, No. 22 AWG wire,
¼'' I.D. Air Core Secondary — 2 turns, No. 22 AWG wire, coef. of coupling ≈ 1.0

Two stage 60 MHz i-f amplifier with power gain ≈ 80 dB, and BW ≈ 1.5 MHz (M).

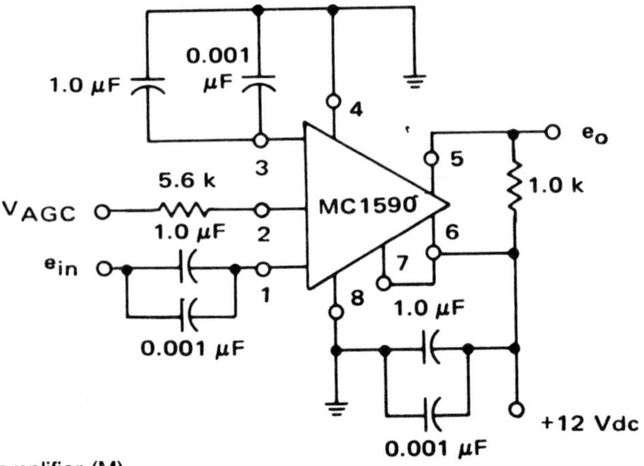

Video amplifier (M).

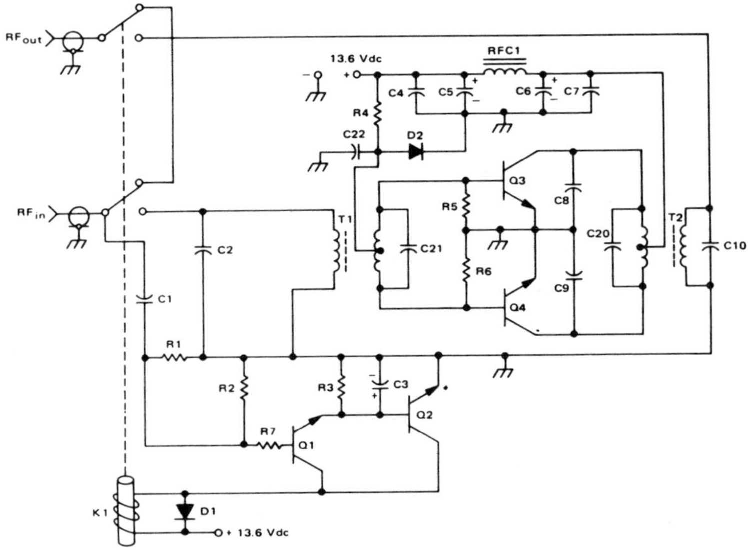

C1	=	33 pF Dipped Mica
C2	=	18 pF Dipped Mica
C3	=	10 μF 35 Vdc for AM operation, 100 μF 35 Vdc for SSB operation.
C4	=	.1 μF Erie
C5	=	10 μF 35 Vdc Electrolytic
C6	=	1 μF Tantalum
C7	=	.001 μF Erie Disc
C8, 9	=	330 pF Dipped Mica
R1	=	100 kΩ 1/4 W Resistor
R2, 3	=	10 kΩ 1/4 W Resistor
R4	=	33 Ω 5 W Wire Wound Resistor
R5, 6	=	10 Ω 1/2 W Resistor
R7	=	100 Ω 1/4 W Resistor
RFC1	=	9 Ferroxcube Beads on #18 AWG Wire
D1	=	1N4001
D2	=	1N4997
Q1, Q2	=	2N4401
Q3, 4	=	MRF454
T1, T2	=	16:1 Transformers
C20	=	910 pF Dipped Mica
C21	=	1100 pF Dipped Mica
C10	=	24 pF Dipped Mica
C22	=	500 μF 3 Vdc Electrolytic
K1	=	Potter & Brumfield KT11A 12 Vdc Relay or Equivalent

140 W (PEP) amateur radio linear amplifier 2-30 MHz (M).

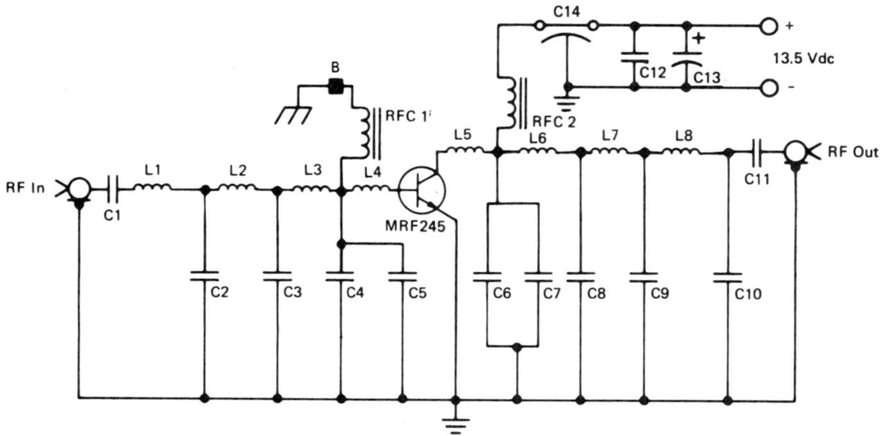

C1, 11 – 500 pF Dipped mica
C2, 9 – 10 pF UNELCO
C3 – 60 pF UNELCO
C4, 5 – 250 pF UNELCO
C6, 7 – 250 pF UNELCO
C8 – 80 pF UNELCO
C10 – 40 pF UNELCO
C12 – 0.1 μF Erie Redcap
C13 – 1 μF Tantalum
C14 – 680 pF Allen Bradley Feed-Thru

RFC 1 – 0.15 μH Molded choke
RFC 2 – 10 T NO. 18 AWG Enameled Wire, 1/4" I.D.
B – Ferroxcube Bead 56-590-65, 3 Beads

L1 – 1.2 X 0.3 cm Airline Inductor
L2 – 3.5 X 0.3 cm Airline Inductor
L3 – 4.0 X 0.3 cm Airline Inductor
L4, L5 – 0.3 X 0.3 cm Airline Inductor
L6 – 2.7 X 0.3 cm Airline Inductor
L7 – 0.8 X 0.3 cm Airline Inductor
L8 – 3.0 X 0.3 cm Airline Inductor

Board: G10, $\epsilon_r \approx 5$, t = 0.16 cm, 57 gm, Copper-Clad connectors = BNC

Circuit schematic of 80-watt, 143-156-MHz power amplifier (M).

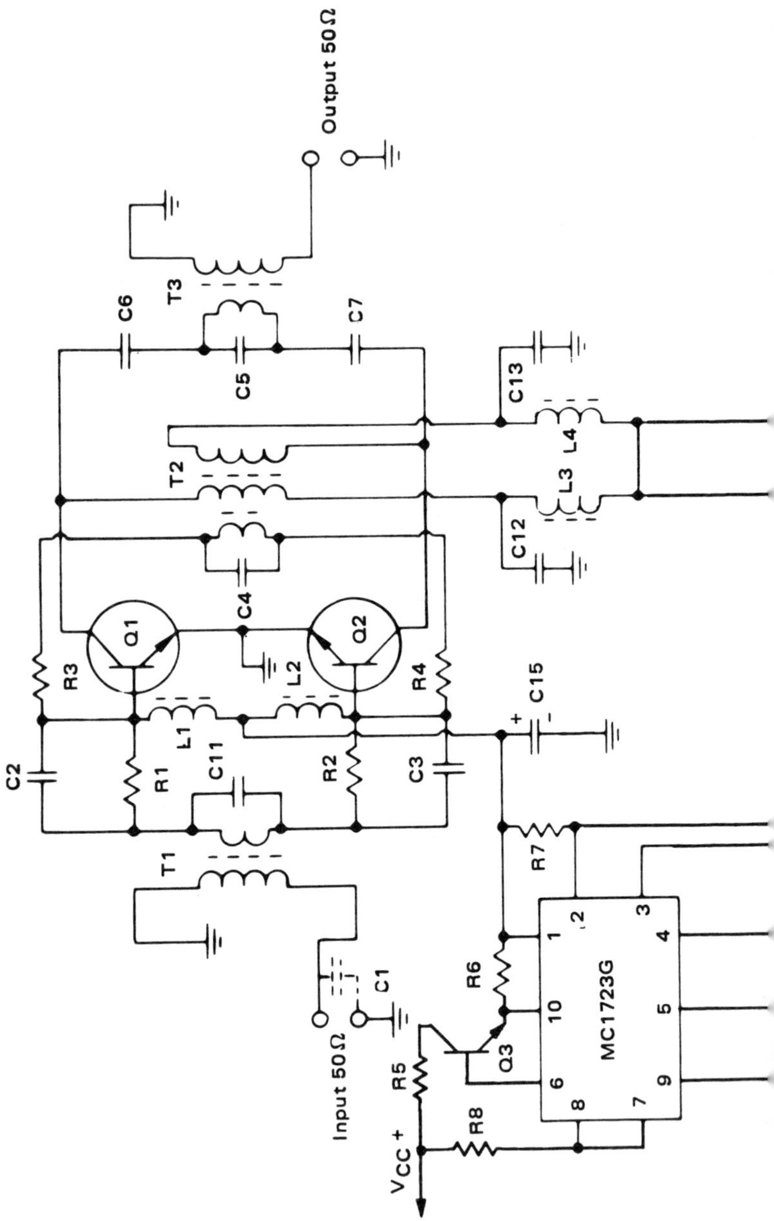

Output 50Ω
Input 50Ω
T1
T2
T3
C1
C2
C3
C4
C5
C6
C7
C11
C12
C13
C15
L1
L2
L3
L4
R1
R2
R3
R4
R5
R6
R7
R8
Q1
Q2
Q3
MC1723G
V_{CC} +

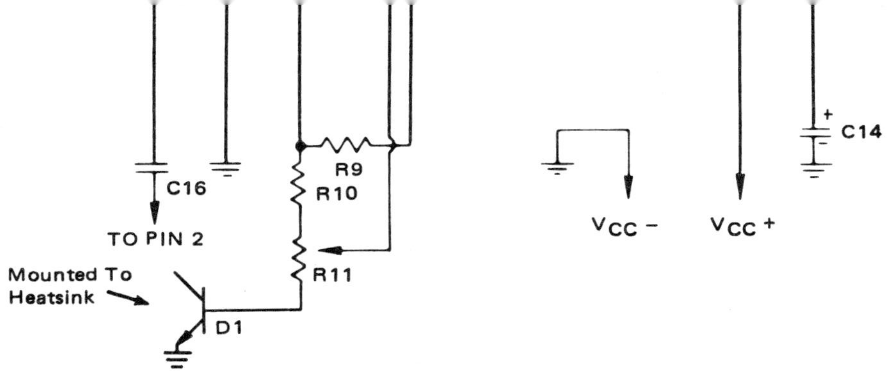

C1 – 100 pF
C2, C3 – 5600 pF
C4, C5 – 680 pF
C6, C7 – 0.10 μF
C11 – 470 pF
C12, C13 – 0.33 μF
C14 – 10 μF – 50 V electrolytic
C15 – 500 μF – 3 V electrolytic
C16 – 1000 pF

R1, R2 – 2 X 3.3 Ω, 1/2 W in parallel
R3, R4 – 2 X 3.9 Ω, 1/2 W in parallel
R5 – 47 Ω, 5 W
R6 – 1.0 Ω, 1/2 W
R7, R8 – 1.0 k, 1/2 W
R9 – 18 k, 1/2 W
R10 – 8.2 k, 1/2 W
R11 – 1.0 k Trimpot
D1 – 2N5190
L1, L2 – Ferroxcube
VK200 20/4B
L3, L4 – 6 ferrite beads
each, Ferroxcube
56590 65/3B

Q1, Q2 – MRF422, Q3 – 2N5990
T1, T2, T3 – See text

All capacitors except electrolytics and C16
are chips –

Union Carbide type 1813 and 1225,
or Varadyne size 18 or 14, or equivalent

300-watt linear amplifier (M).

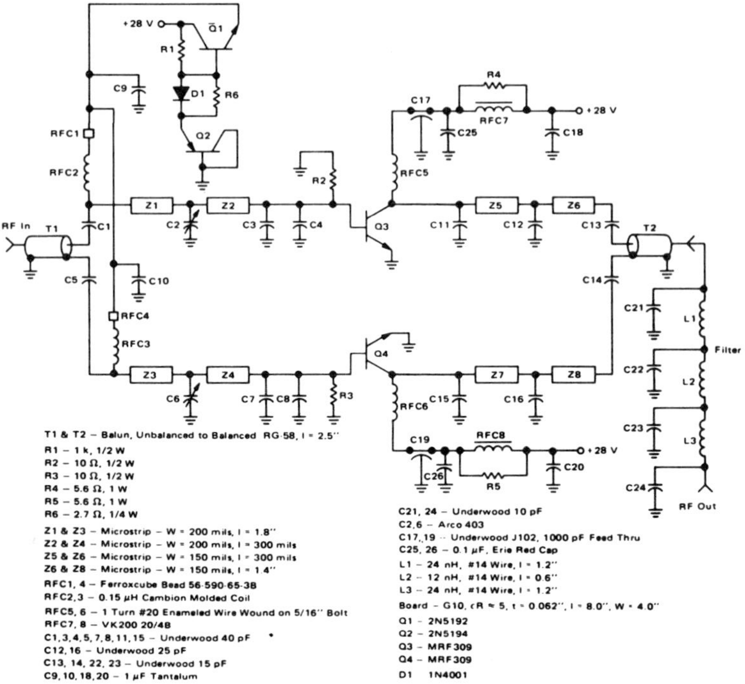

100 W (PEP) 420-450 MHz push-pull linear amplifier (M).

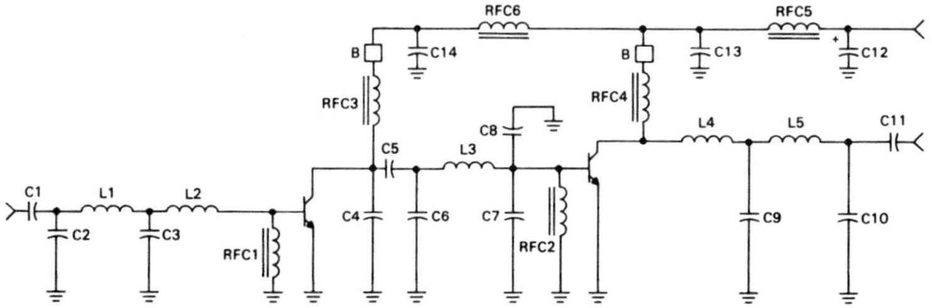

Schematic diagram of dipped silvered mica capacitor vhf amplifier (136-160 MHz) (M).

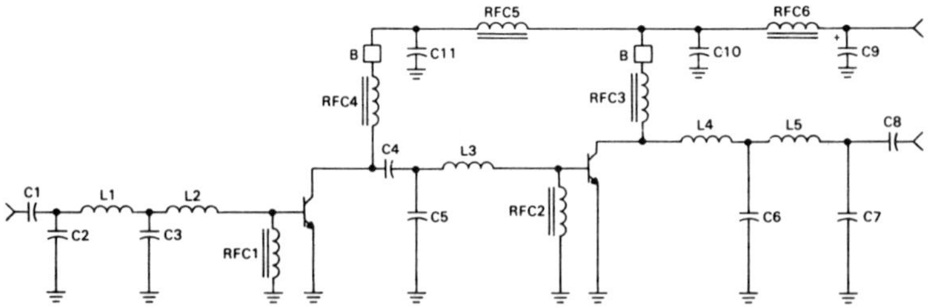

Chip Capacitor vhf amplifier (160-174 MHz) (M).

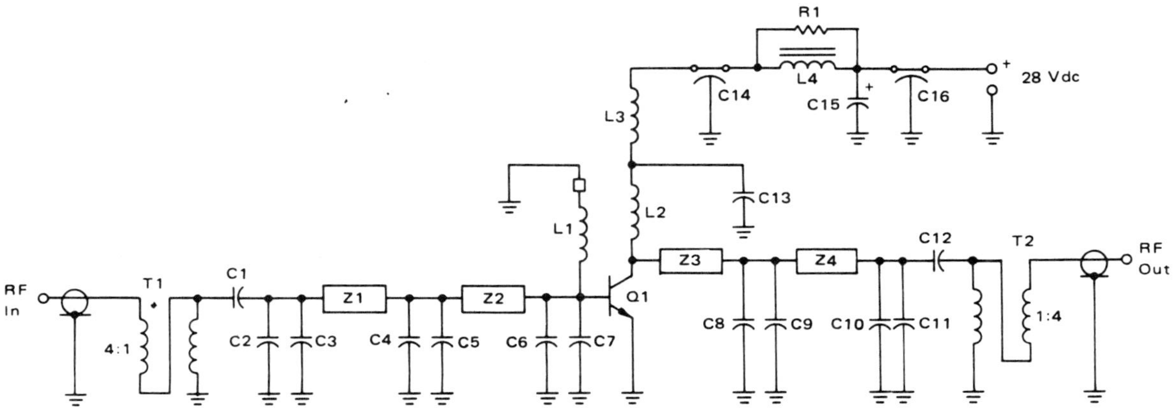

C1 – 63 pF Chip
C2, C8 – 27 pF Chip
C3 – 24 pF Chip
C4 – 15 pF Chip
C5, C9 – 30 pF Chip
C6, C7 – 50 pF Chip
C10 – 10 pF Chip
C11 – 5.1 pF Chip
C12 – 150 pF Chip
C13 – 270 pF Chip
C14, C16 – 680 pF Feedthru
C15 – 1.0 μF 50 V Tantalum

All Chip Capacitors are 100 mil TDK-ACI Co, Style FC282 BAG
L1 – 0.15 μH Molded Choke with Ferroxcube Bead #56-590-65/4B on ground end of coil
L2 – 1 Turn #22 AWG, 1/8″ ID
L3 – 0.15 μH Molded Choke
L4 – Ferroxcube VK-200-19/4B

Q1 – 2N6439

R1 – 10 Ω 2 Watt

T1, T2 – 25 Ω Subminiature Coax (Type UT25) 2.25 inches (57.15 mm) long

Z1 – Microstrip Line
800 mils L X 225 mils W
20.32 mm L X 5.715 mm W
Z2 – Microstrip Line
200 mils L X 225 mils W
5.08 mm L X 5.715 mm W
Z3, Z4 – Microstrip Line
550 mils L X 125 mils W
13.97 mm L X 3.175 mm W

Board – 0.031″ (0.787 mm) Glass Teflon
$\epsilon_r = 2.56$

2N6439 60 watt building block 225-400 MHz (M).

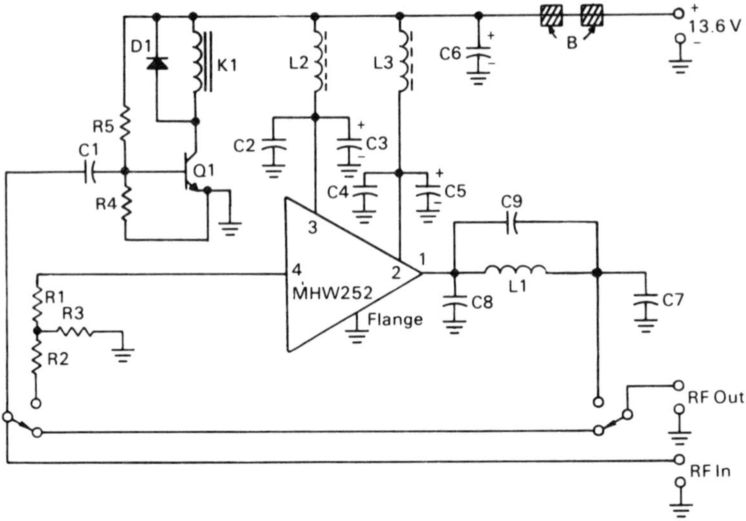

R1, R2, R3 — See Table 1
R4 — 1.0 kΩ/1/4 W
R5 — 27 kΩ/1/4 W
C1 — 5.0 pF Dipped Mica
C2, C4 — 0.01 μF Ceramic Disc
C3, C5 — 1.0 μF Dipped Tantalum
C6 — 10 μF/15 V Electrolytic
C7 — 24 pF Dipped Mica
C8, C9 — 10 pF Dipped Mica

L1 — 1-1/4 Turns AWG #18 Enameled Wire 5/16" ID
L2, L3 Ferroxcube VK200 — 20/4B or Equivalent
K1 — Arrow M NFZ - 12, Omron LZN2-UA-DC12 or Equivalent
Q1 — 2N4401
D1 — 1N4001
B — Ferrite Beads, Ferroxcube 5659065/3B or Fair-Rite Products Corp. 2673000101 or Equivalent.

25-watt add-on amplifier for hand held vhf radios (M).

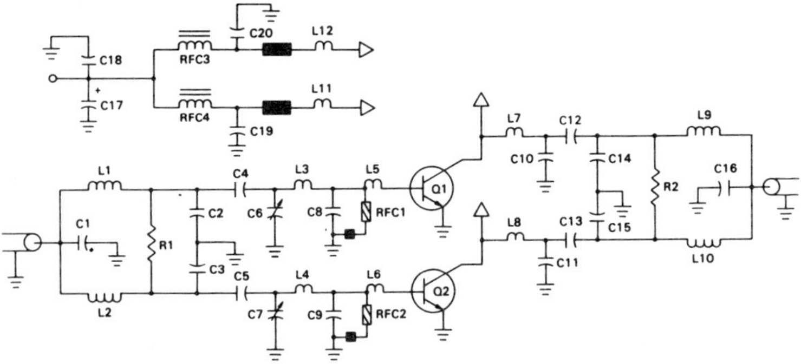

C1, C16 — 25 pF Unelco (J101)
C2, C3 — 15 pF CM04 Mica
C4, C5 — 68 pF Standex
C6, C7 — Arco 404 Variable
C8, C9 — 150 pF Standex
C10, C11 — 56 pF Standex
C12, C13 — 39 pF Standex
C14, C15 — 15 pF Standex
C17 — 100 μF @ 16 V Electrolytic
C18, C19, C20 — 680 pF Allen Bradley Feedthru

L1, L2 — 7 Turns #18, 0.125" ID
L3, L4, L5, L6 — Printed Inductors
L7, L8 — Printed Inductors
L9, L10 — 7 Turns #18 AWG, 0.125 ID
L11, L12 — 4 Turns #18 AWG, 0.250 ID w/Bead
Q1, Q2 — MRF264
RFC1, RFC2 — 0.15 μH Molded Choke w/Bead, Ferroxcube 56-590 65/3B
RFC3, RFC4 — 4 Ferrite Beads each on #18 AWG
R1 — 100 Ω 1/2 W Carbon
R2 — 100 Ω 2.0 W Carbon

Schematic - 60 W vhf amplifier (M).

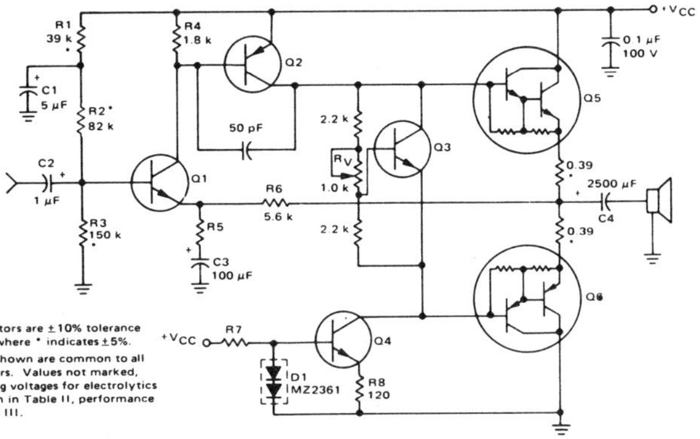

Note 1: All resistors are ±10% tolerance except where * indicates ±5%.

Note 2: Values shown are common to all amplifiers. Values not marked, including voltages for electrolytics are given in Table II, performance in Table III.

15 to 60 watt power amplifier with ac coupled output (M).

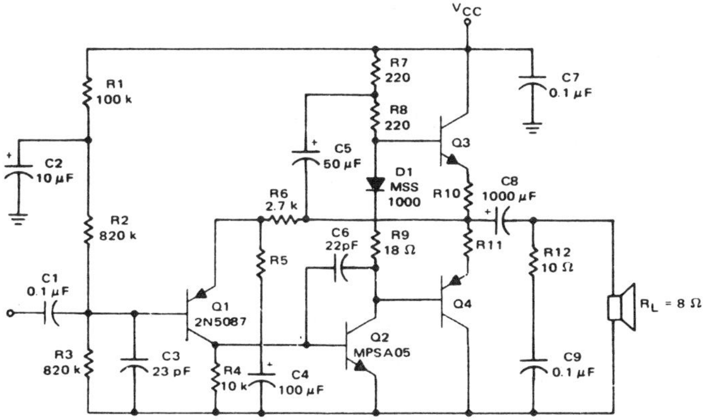

Component	3 W	5 W
V_{CC}	17	22
R_5	120 Ω	100 Ω
R10 & R11	0.82	0.56
Q3*	MPSU01 MJE200	MPSU01 MJE200
Q4*	MPSU51 MJE210	MPSU51 MJE210
**Heatsink	MPSU01/51 27.5°C/W	MPSU01/51 16.8°C/W
**Heatsink	MJE200/210 36°C/W	MJE200/210 19.7°C/W

*Parts in same block are interchangeable. P.C. board is for MJE200/210, but can be easily changed to MPSU01/51.

**Heatsink size calculation is based on a maximum ambient temperature of 50°C and a load phase angle of 60 degrees (see test for method of calculation) Heat sink is for both devices on one sink.

3 to 5 watt amplifier (NPN driver) (M).

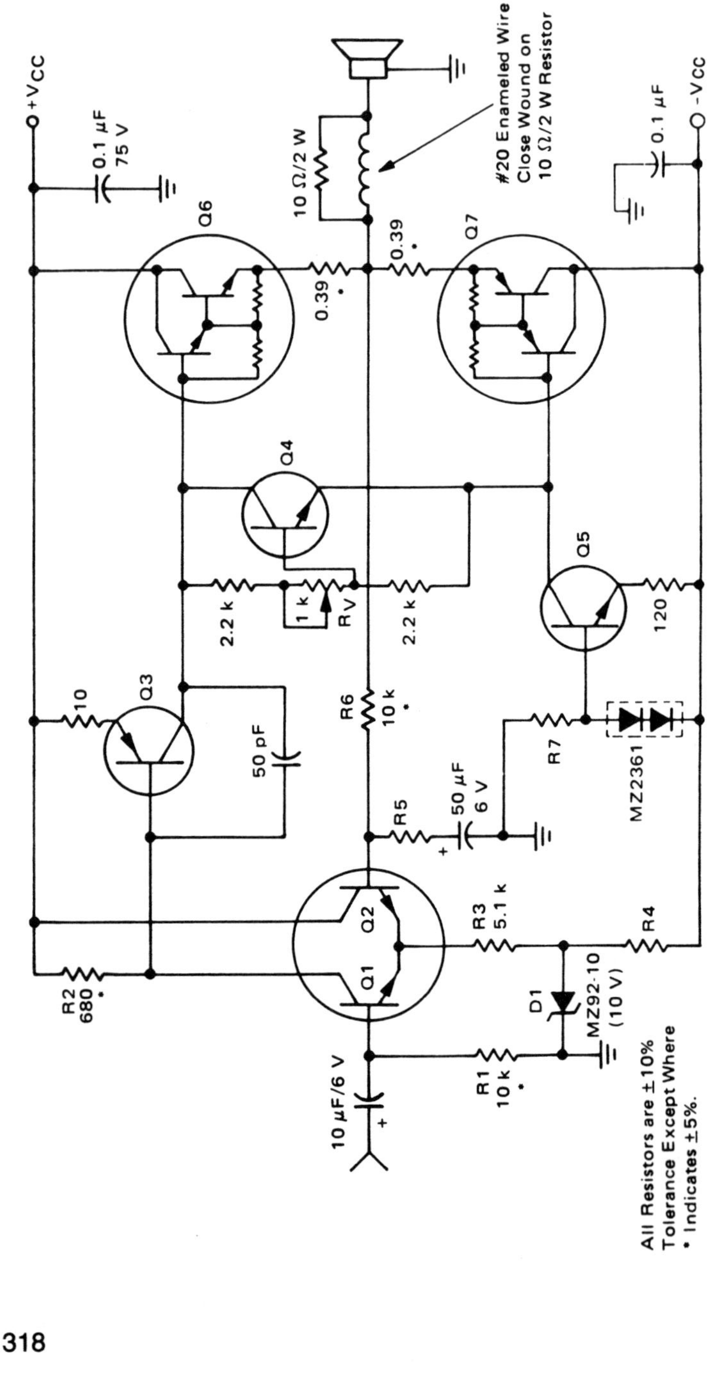
+VCC
0.1 µF
75 V
Q6
10 Ω/2 W
#20 Enameled Wire
Close Wound on
10 Ω/2 W Resistor
0.1 µF
-VCC
0.39
0.39
Q7
Q4
Q5
2.2 k
1 k
RV
2.2 k
120
10
Q3
50 pF
R6
10 k
R7
MZ2361
R5
50 µF
6 V
Q2
Q1
R3
5.1 k
R4
R2
680
D1
MZ92-10
(10 V)
R1
10 k
10 µF/6 V
All Resistors are ±10%
Tolerance Except Where
• Indicates ±5%.

Power Watts (RMS)	15		20		25		35		50		60	
Load Impedance	4	8	4	8	4	8	4	8	4	8	4	8
V_{CC}	±16 V	±19 V	±18 V	±23 V	±19 V	±24 V	±22 V	±28 V	±25 V	±33 V	±28 V	±36 V
R4 (ohms)	1.5 k	2.2 k	2.0 k	3.3 k	2.2 k	3.3 k	3.0 k	3.9 k	3.6 k	5.6 k	3.9 k	6.2 k
R5 (ohms)	1.2 k	820	1.0 k	750	1.0 k	680	820	560	680	470	620	430
R7 (ohms)	15 k	18 k	18 k	22 k	18 k	22 k	22 k	27 k	22 k	33 k	27 k	33 k
Q1,Q2 Dual Transistors	MD 8001	MD 8001	MD 8001	MD 8001	MD 8001	MD 8001	MD 8001	MD 8001	MD 8001	MD 8002	MD 8001	MD 8002
Q3	MPS A55	MPS A55	MPS A55	MPS A55	MPS A55	MPS A55	MPS A55	MPS A56	MPS A55	MPS A56	MPS A56	MPS A56
Q4	MPS A13	MPS A13	MPS A13	MPS A13	MPS A13	MPS A13	MPS A13	MPS A13	MPS A13	MPS A13	MPS A13	MPS A13
Q5	MPS A05	MPS A05	MPS A05	MPS A05	MPS A05	MPS A05	MPS A05	MPS A06	MPS A05	MPS A06	MPS A06	MPS A06
Q6	MJE 1100	MJE 1100	MJE 1100	MJE 1100	MJE 1102	MJE 1100	MJE 6043	MJE 6044	MJE 6043	MJE 6044	MJE 6044	MJE 6044
Q7	MJE 1090	MJE 1090	MJE 1090	MJE 1090	MJE 1092	MJE 1090	MJE 6040	MJE 6041	MJE 6040	MJE 6041	MJE 6041	MJE 6041
Min. heat sink for outputs @ 55°C ambient temperature and 10% high line voltage	9.5°C/W		7.0°C/W		5.0°C/W		6.0°C/W	5.5°C/W	4.0°C/W		3.0°C/W	

15 to 60 Watt power amplifier with dc coupled output (M).

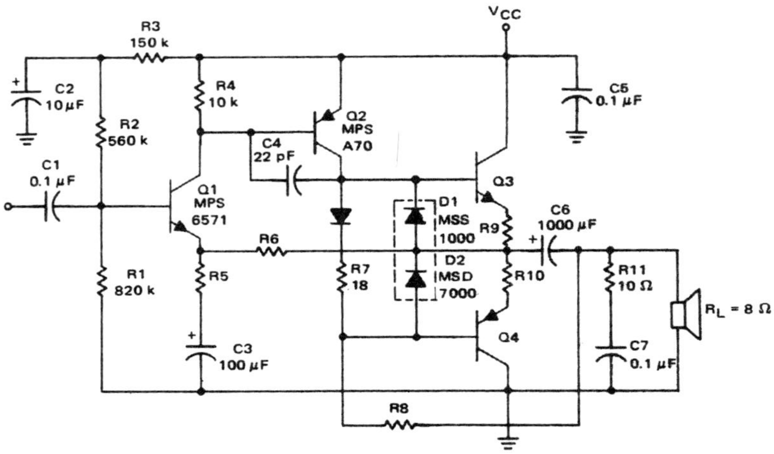

3 W and 5 W amplifiers (PNP driver) (M).

	3 W	5 W
V_{CC}	18 V	22 V
R_5	180	150
R_6	470	390
R9 & R10	0.82	0.56
**Q3	MPSU01 or MJE200	MPSU01 or MJE200
**Q4	MPSU51 or MJE210	MPSU51 or MJE210
Heatsink	with MPSU01/51 27.5°C/W	with MPSU01/51 16.8°C/W
	with MJE200/210 36°C/W	with MJE200/210 19.7°C/W

*Heatsink size calculation is based on a maximum ambient temperature of 50°C and a load phase angle of 60 degrees (see test for method of calculation). Heatsink is for both devices on one sink.

**Parts in same block are interchangeable.
P.C. board is for MPSU01/51, but can be changed to MJE200/210

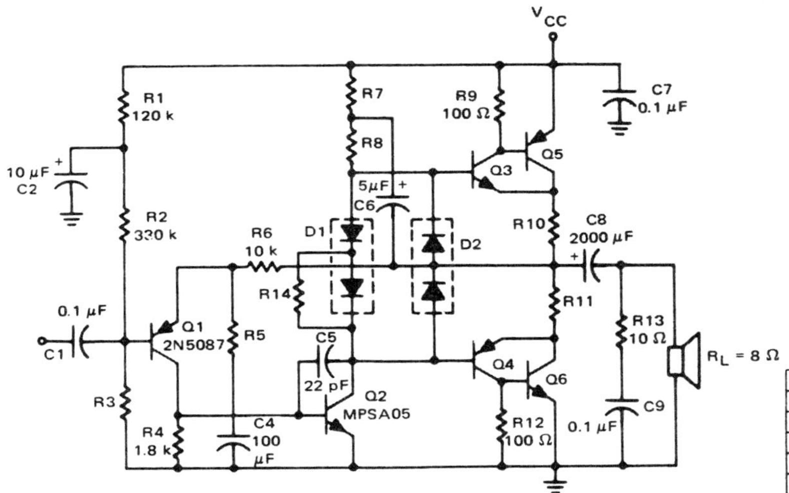

Components	7 W	10 W	15 W	20 W	25 W	35 W
V_{CC}	25 V	30 V	36 V	43 V	48 V	54 V
R3	560 k	560 k	620 k	620 k	680 k	680 k
R5	180 Ω	180 Ω	120 Ω	120 Ω	120 Ω	120 Ω
R7 & R8	3.9 k	3.9 k	3.9 k	3.9 k	3.9 k	5.6 k
R10 & R11	0.47	0.47	0.47	0.47	0.47	0.33
R14	—	—	—	—	—	470 Ω
Q3	MPSA05	MPSA05	MPSU05 MJE180	MPSU05 MJE181	MPSU05 MJE181	MPSU05 MJE181
Q4	MPSA55	MPSA55	MPSU55 MJE170	MPSU55 MJE171	MPSU55 MJE171	MPSU55 MJE171
Q5	MJE170	MJE170	MJE105	MJE105	MJE2901	MJE2901
Q6	MJE180	MJE180	MJE205	MJE205	MJE2801	MJE2801
D1 & D2	MSD7000	MSD7000	MSD7000	MSD7000	MSD7000	MSD7000
*Heatsink	15.8°C/W	9.8°C/W	6.2°C/W	4.7°C/W	3.4°C/W	2°C/W

*Heatsink size calculation is based on a maximum ambient temperature of 50°C and a load phase angle of 80 degrees (see text for methods of calculation). The heat sink is for both devices on one sink.

7 to 35 watt amplifiers (M).

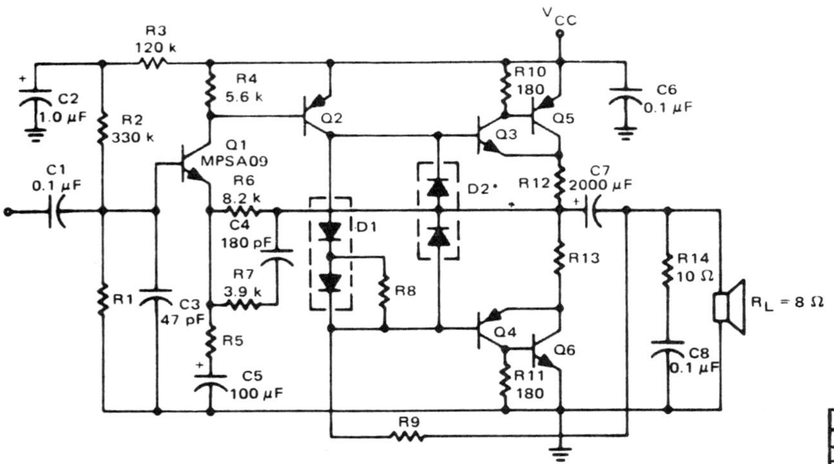

**D2 provides Overload Current Limiting. Provision for D2 was not made on PC Board Layout.

Component	7 W	10 W	15 W	20 W	25 W	35 W
V_{CC}	25 V	30 V	38 V	46 V	48 V	54 V
R1	560 k	560 k	560 k	620 k	620 k	620 k
R5	100 Ω	82 Ω	100 Ω	100 Ω	150 Ω	180 Ω
R8	Value Selected to Provide 30 mA Collector Current in Q5					
R9	4.7 k	4.7 k	8.2 k	8.2 k	8.2 k	8.2 k
R12 & R13	0.47	0.47	0.47	0.47	0.33	0.33
Q2	2N5087	2N5087	2N5087	2N5087	MPSA55	MPSA56
Q3	MPSA05	MPSA05	MPSU05 MJE220	MPSU05 MJE224	MPSU05 MJE224	MPSU05 MJE224
Q4	MPSA55	MPSA55	MPSU55 MJE230	MPSU55 MJE234	MPSU55 MJE234	MPSU55 MJE234
Q5	MJE230	MJE230	MJE105	MJE105	MJE2901	MJE2901
Q6	MJE220	MJE220	MJE205	MJE205	MJE2801	MJE2801
D1 & D2	MSD7000	MSD7000	MSD7000	MSD7000	MSD7000	MSD7000
*Heatsink	15.3°C/W	9.8°C/W	6.2°C/W	4.7°C/W	3.4°C/W	2°C/W

*Heatsink size calculation is based on a maximum ambient temperature of 50°C and a load phase angle of 60 degrees (see text for method of calculation). The heatsink is for both devices on one sink.

7 to 35 watt amplifier (PNP driver) (M).

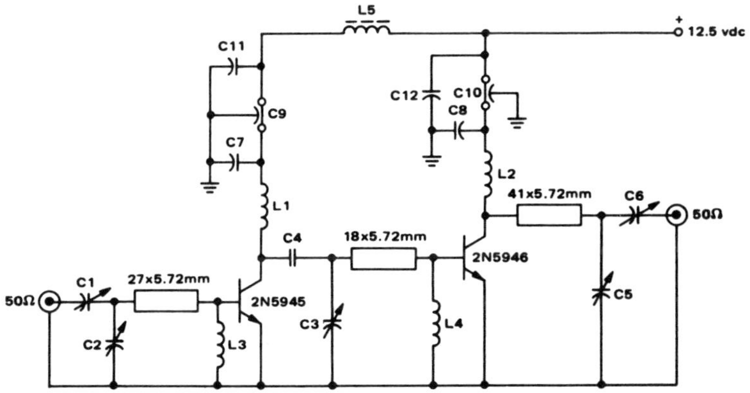

C1,2,3,5,6 – 1.5-20 pF, Arco 402 or equiv.
C4 – 10 pf dipped mica
C7, 8 – 0.1 μF ceramic
C9, 10 – 470 pf Feed thru
C11, 12 – 1 μf Tantalum
L1, 2 – 5 turns #20 AWG Closewound 3/16" I.D.
L3, 4 – 3.9 μhy molded choke w/ferrite core
L5 – Ferroxcube VK200 20/4B or equiv.
Board is 1/16" thick epoxy-glass
"G-10" Dielectric with 1oz copper on both sides

Schematic diagram of 10 watt amplifier 450-470 MHz (M).

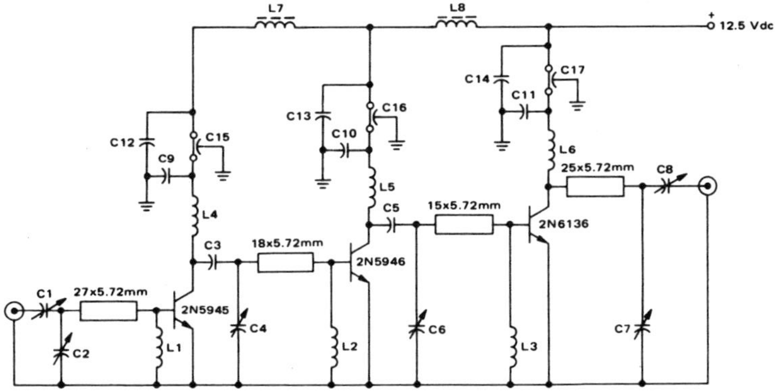

C1,2,4,6,7,8 – 1.5-20pf Arco 402 or equiv.
C3, 5 – 10pf dipped mica
C9,10,11 – 0.1 μf ceramic
C12,13,14 1μf Tantalum
C15,16,17 470pf feed thru
L1,2,3 – 3.9μhy molded choke w/ferrite core
L4,5,6 – 5 turns #20 awg closewound 3/16" I.D.
L7, 8 – Ferroxcube VK200 20/4B or equiv.
Board is 3/16" "G-10" epoxy fiberglass Dielectric with 1 oz copper on both sides.

25 watt amplifier 450-470 MHz (M).

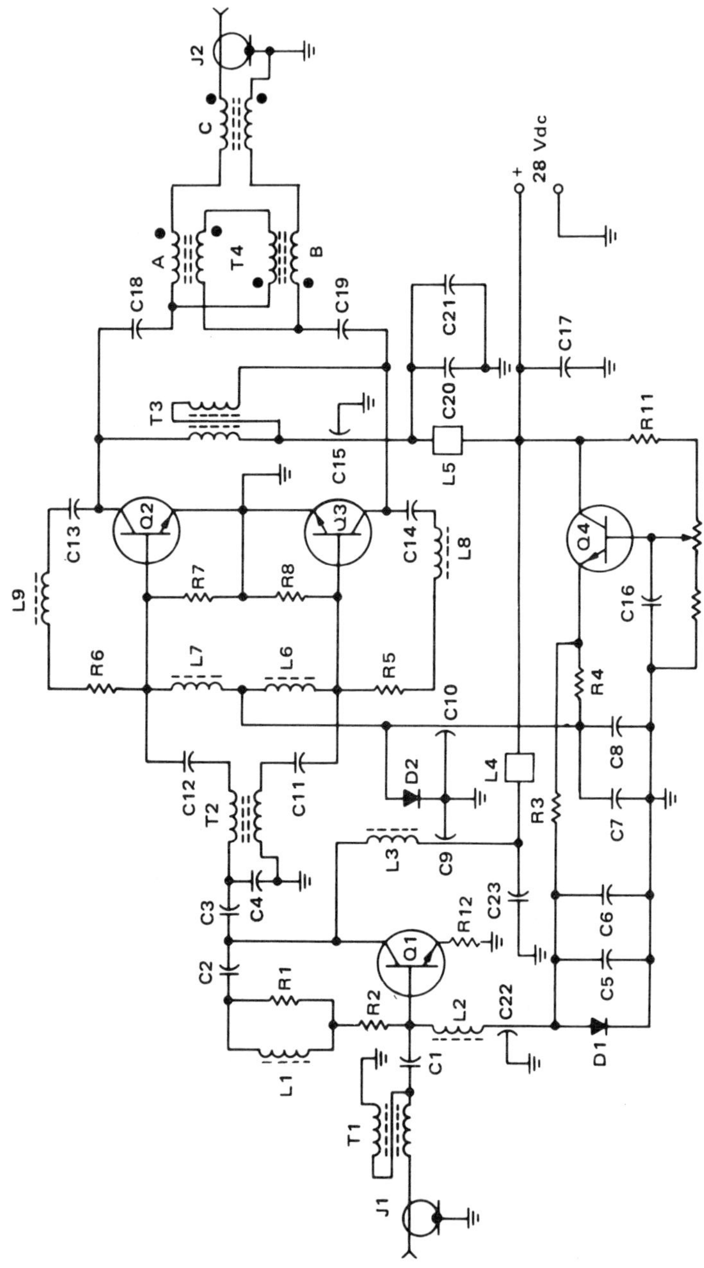
J1
T1
L1
C1
R1
R2
C2
C3
C4
Q1
R12
L2
C22
D1
C5
C6
C23
L3
C9
T2
C12
C11
D2
C10
L4
R3
C7
C8
R4
C16
R9
R10
Q4
R11
L5
C15
C20
C21
C17
+
28 Vdc
T3
Q2
Q3
C13
C14
L9
L8
R6
R7
R8
L7
L6
R5
C18
C19
A
B
T4
C
J2

PARTS LIST

C1 – 0.033 μF mylar
C2, C3 – 0.01 μF mylar
C4 – 620 pF dipped mica
C5, C7, C16 – 0.1 μF ceramic
C6 – 100 μF/15 V electrolytic
C8 – 500 μF/6 V electrolytic
C9, C10, C15, C22 – 1000 pF feed through
C11, C12 – 0.01 μF
C13, C14 – 0.015 μF mylar
C17 – 10 μF/35 V electrolytic
C18, C19, C21 – Two 0.068 μF mylars in parallel
C20 – 0.1 μF disc ceramic
C23 – 0.1 μF disc ceramic
R1 – 220 Ω, 1/4 W carbon
R2 – 47 Ω, 1/2 W carbon
R3 – 820 Ω, 1 W wire W
R4 – 35 Ω, 5 W wire W
R5, R6 – Two 150 Ω, 1/2 W carbon in parallel
R7, R8 – 10 Ω, 1/2 W carbon
R9, R11 – 1 k, 1/2 W carbon
R10 – 1 k, 1/2 W potentiometer
R12 – 0.85 Ω (6 5.1 Ω or 4 3.3 Ω 1/4 W resistors in parallel, divided equally between both emitter leads)

T1 – 4:1 Transformer, 6 turns, 2 twisted pairs of #26 AWG enameled wire (8 twists per inch)

T2 – 1:1 Balun, 6 turns, 2 twisted pairs of #24 AWG enameled wire (6 twists per inch)

T3 – Collector choke, 4 turns, 2 twisted pairs of #22 AWG enameled wire (6 twists per inch)

T4 – 1:4 Transformer Balun, A&B – 5 turns, 2 twisted pairs of #24, C – 8 turns, 1 twisted pair of #24 AWG enameled wire (All windings 6 twists per inch). (T4 – Indiana General F624-19Q1, – All others are Indiana General F627-8Q1 ferrite toroids or equivalent.)

L1 – .33 μH, molded choke
L2, L6, L7 – 10 μH, molded choke
L3 – 1.8 μH (Ohmite 2-144)
L4, L5 – 3 ferrite beads each
L8, L9 – .22 μH, molded choke

Q1 – 2N6370
Q2, Q3 – 2N5942
Q4 – 2N5190

D1 – 1N4001
D2 – 1N4997

J1, J2 – BNC connectors

160 watt (PEP) broadband linear amplifier (M).

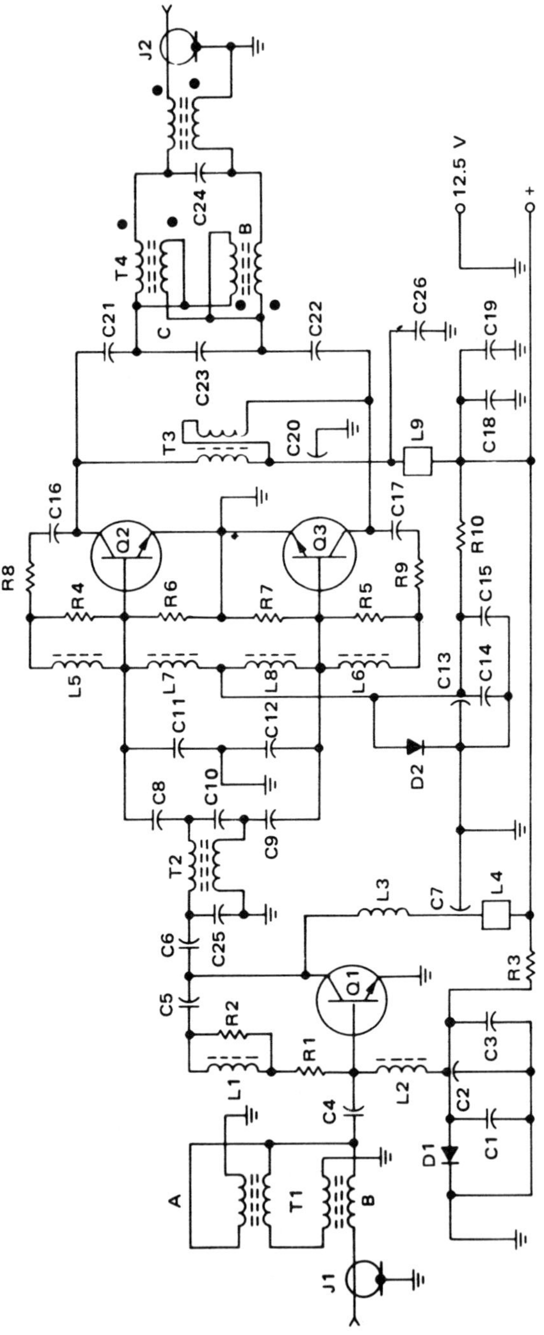
J1
J2
A
B
C
T1
T2
T3
T4
Q1
Q2
Q3
D1
D2
R1
R2
R3
R4
R5
R6
R7
R8
R9
R10
L1
L2
L3
L4
L5
L6
L7
L8
L9
C1
C2
C3
C4
C5
C6
C7
C8
C9
C10
C11
C12
C13
C14
C15
C16
C17
C18
C19
C20
C21
C22
C23
C24
C25
C26
12.5 V
+

C1, C14, C18 – 0.1 μF ceramic.
C2, C7, C13, C20 – 0.001 μF feed through.
C3 – 100 μF/3V.
C4, C6 – 0.033 μF mylar
C5 – 0.0047 μF mylar.
C8, C9 – 0.015 and 0.033 μF mylars in parallel.
C10 – 470 pF mica.
C11, C12 – 560 pF mica.
C15 – 1000 μF/3 V
C16, C17 – 0.015 μF mylar
C19 – 10 pF 15 V
C21, C22 – two 0.068 μF mylars in parallel.
C23 – 330 pF mica
C24 – 39 pF mica
C25 – 680 pF mica
C26 – .01 μF ceramic

R1, R6, R7 – 10 Ω, 1/2 W carbon.
R2 – 51 Ω, 1/2 W carbon
R3 – 240 Ω, 1 wire W
R4, R5 – 18 Ω, 1 W carbon
R8, R9 – 27 Ω, 2 W carbon
R10 – 33 Ω, 6 W wire W

L1 – 0.22 μh molded choke
L2, L7, L8 – 10 μh molded choke
L5, L6 – 0.15 μh
L3 – 25 t, #26 wire, wound on a 100 Ω, 2 W resistor. (1.0 μh)
L4, L9 – 3 ferrite beads each.

T1 – 2 twisted pairs of #26 wire, 8 twists per inch. A = 4 turns, B = 8 turns. Core- -Stackpole 57-9322-11, Indiana General F627-8Q1 or equivalent

T2 – 2 twisted pairs of #24 wire, 8 twists per inch, 6 turns. (Core as above.)

T3 – 2 twisted pairs of #20 wire, 6 twists per inch, 4 turns. (Core as above.)

T4 – A and B = 2 twisted pairs of #24 wire, 8 twists per inch. 5 turns each. C = 1 twisted pair of #24 wire, 8 turns. Core - -Stackpole 57-9074-11, Indiana General F624-19Q1 or equivalent.

Q1 – 2N6367
Q2, Q3 – 2N6368

D1 – 1N4001
D2 – 1N4997

J1, J2 – BNC connectors

Schematic diagram of 12.5 V amplifier (M).

A 25 Watt Amplifier Schematic

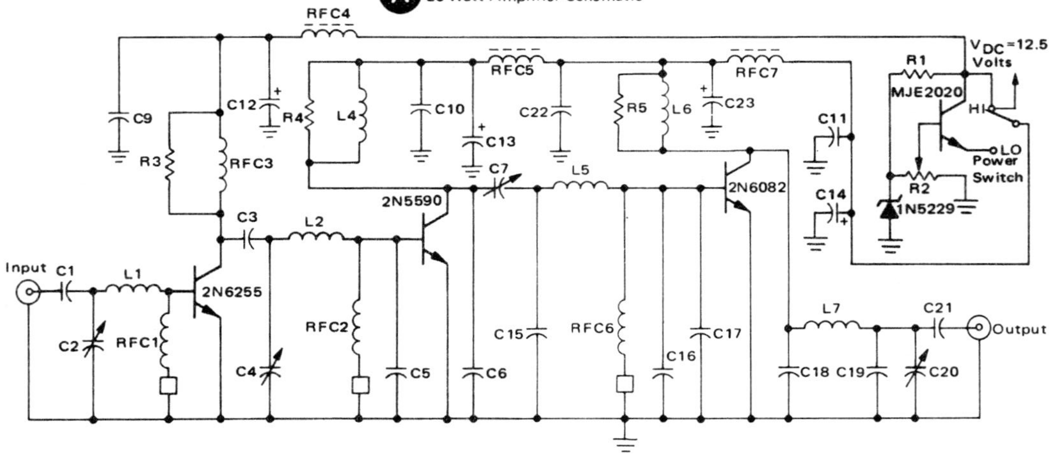

NOTES:

1. Shaded components are common to both 25 Watt and 10 Watt systems.
2. Component designations C8 and L3 are not used in 25 Watt system.
3. Amplifiers constructed on 0.62'', single sided, G10, printed circuit board.

C1, C3	– 10 pF	Dipped Silvered Mica
C5, C6, C16, C17, C18	– 56 pF	
C8	– 36 pF	
C15, C21	– 22 pF	
C19	– 33 pF	
C9, C10, C22	– 0.001 μF Ceramic Disc	

L1, L2	– 1-1/2 T, #16 AWG Wire, 0.25'' I.D. (30 nH)
L3	– 1-1/2 T, #16 AWG Wire, 0.30'' I.D. (35 nH)
L4, L6	– 3 T, #16 AWG Wire, Wound on 100 Ohm Resistor (45 nH)
L5	– #16 AWG Wire, 0.8'' Long, ''U'' Shaped (12 nH)
L7	– #16 AWG Wire, 1.1'' Long, Formed Around 0.6'' Dia. Cyl. (15 nH)

C11	– 0.01 µF Ceramic Disc	RFC1, 2, 6	–	0.15 µH Molded Choke with Ferroxcube 5659065/3B Ferrite Bead on Ground Lead
C12, C13, C23	– 1.0 µF, 35 V, Tantalum	RFC3	–	7 T, #20 AWG Wire, Wound on R3 (100 nH)
C14	– 10 µF, 25 V, Aluminum Electrolytic	RFC4, 5, 7	–	Ferroxcube VK200 19/4B Ferrite Choke
C2, C4, C7, C20	– 8-60 pF Compression Mica Trimmer ARCO #404 or Equivalent	R1	–	91 Ohm, 2 W, ±5% Carbon Resistor
		R2	–	100 Ohm, 0.25 W, Potentiometer, CTS Type R101B or Equivalent
		R3	–	560 Ohm, 1 W, ±10% Carbon Resistor
		R4, R5	–	100 Ohm, 1 W, ±10% Carbon Resistor

B 10 Watt Amplifier Schematic

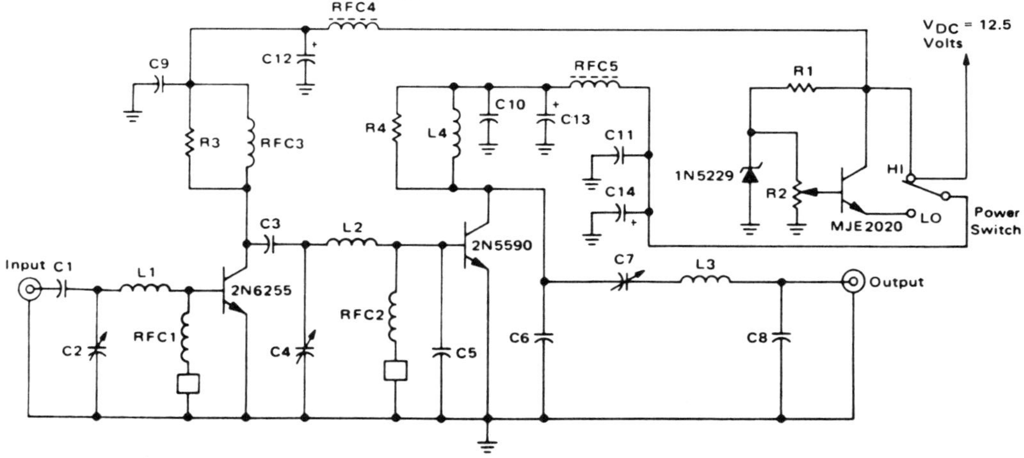

Schematic diagrams for 25 watt and 10 watt amplifiers (M).

Chapter 13

Oscillators

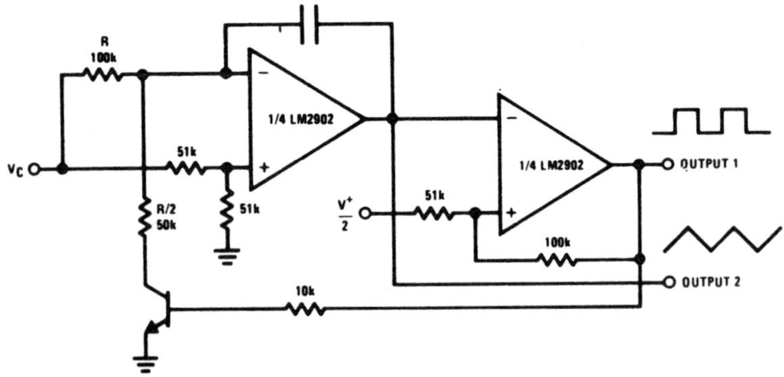

Voltage controlled oscillator circuit (NS).

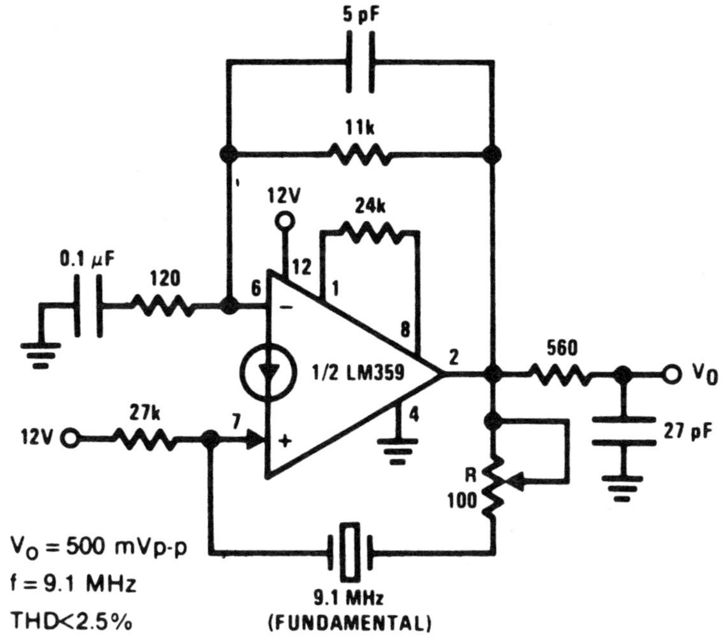

Crystal controlled sinewave oscillator (NS).

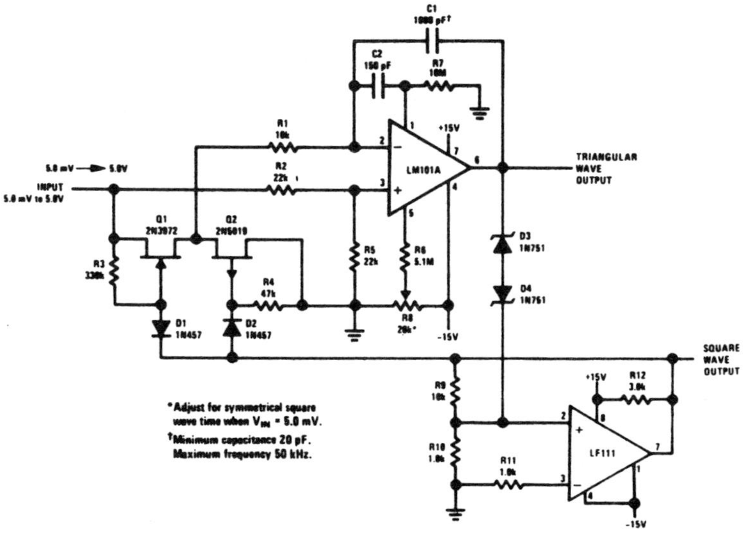

10 Hz to 10 kHz Voltage controlled oscillator (NS).

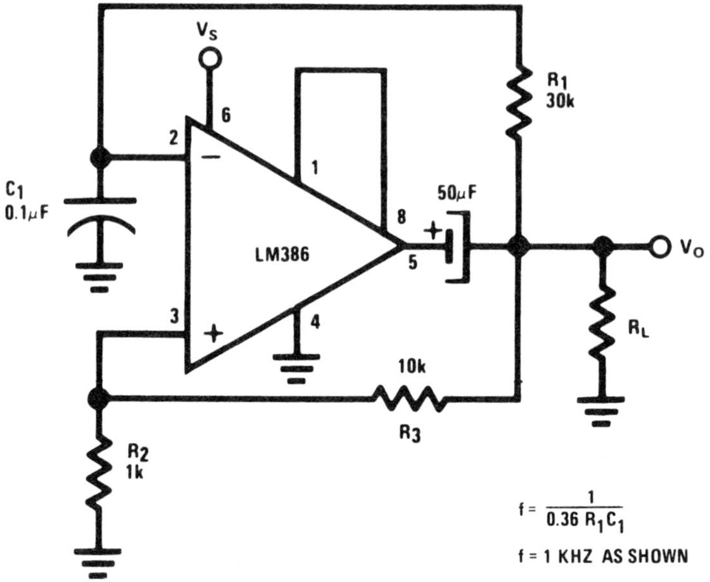

Squarewave oscillator (NS).

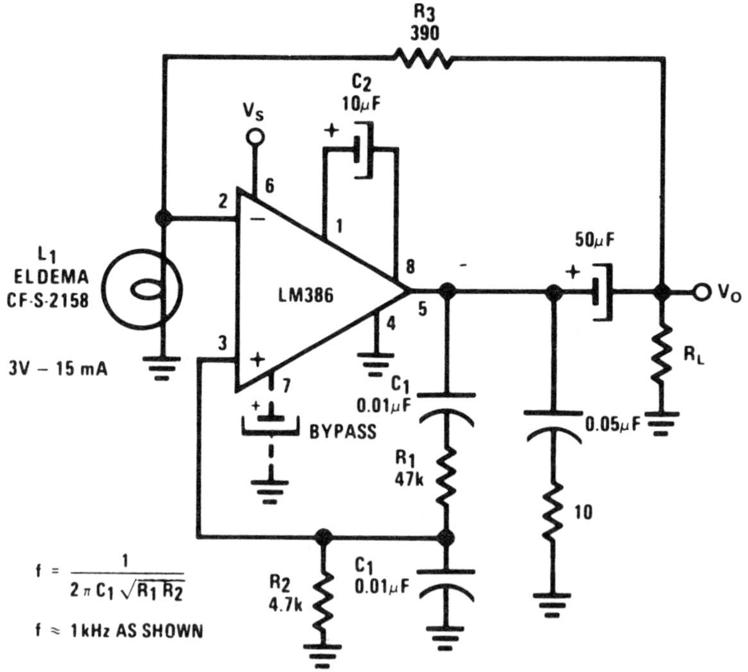

Low distortion power Wien bridge oscillator (NS).

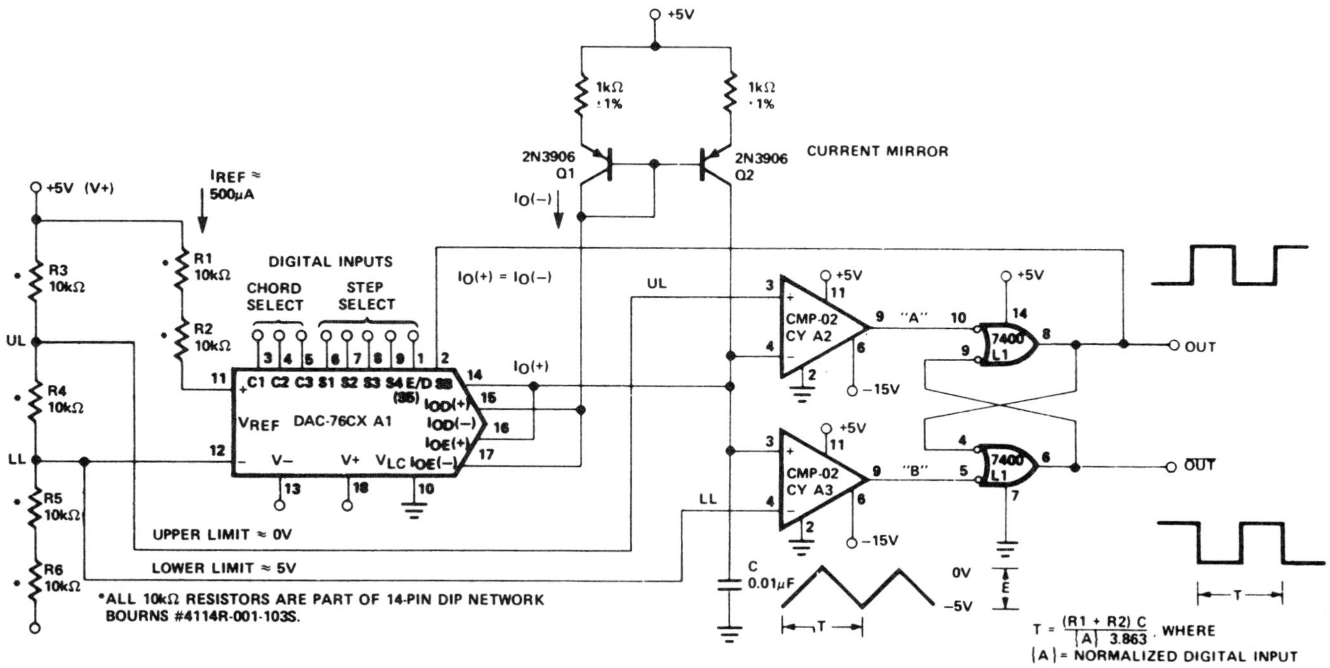

Circuit diagram exponential digitally-controlled oscillator (PM).

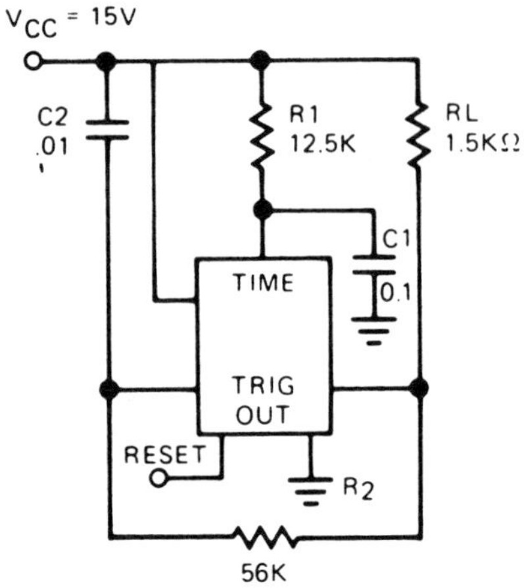

NE558 400 Hz Squarewave oscillator (S).

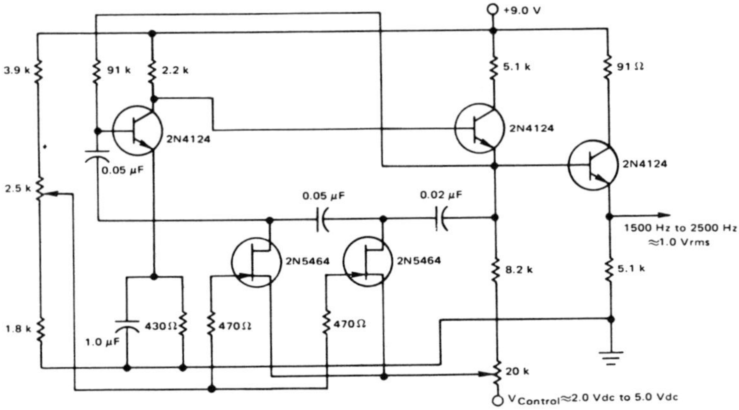

Voltage-controlled oscillator (M).

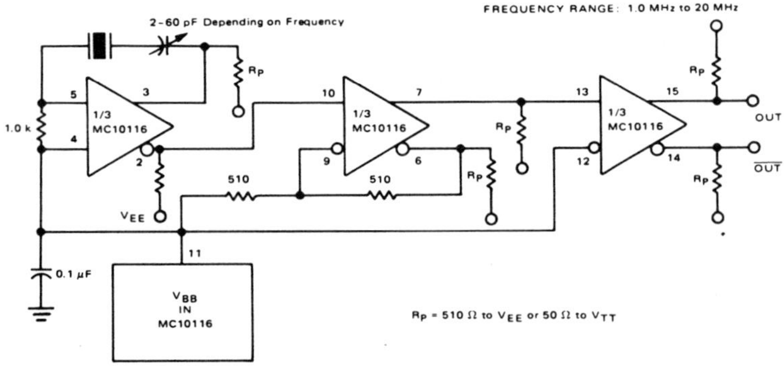

MECL fundamental crystal oscillator (M).

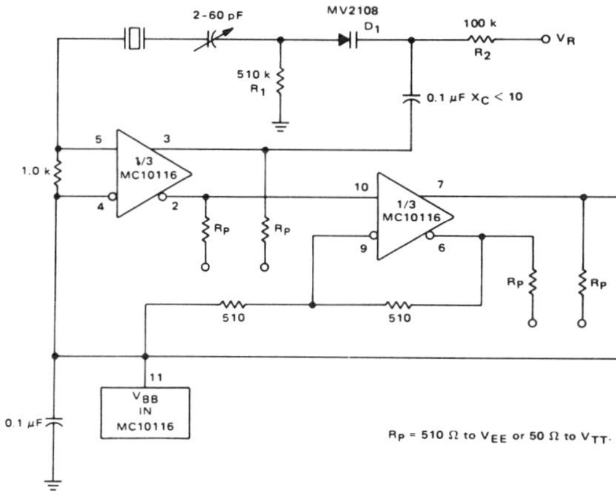

MECL voltage controlled crystal oscillator (M).

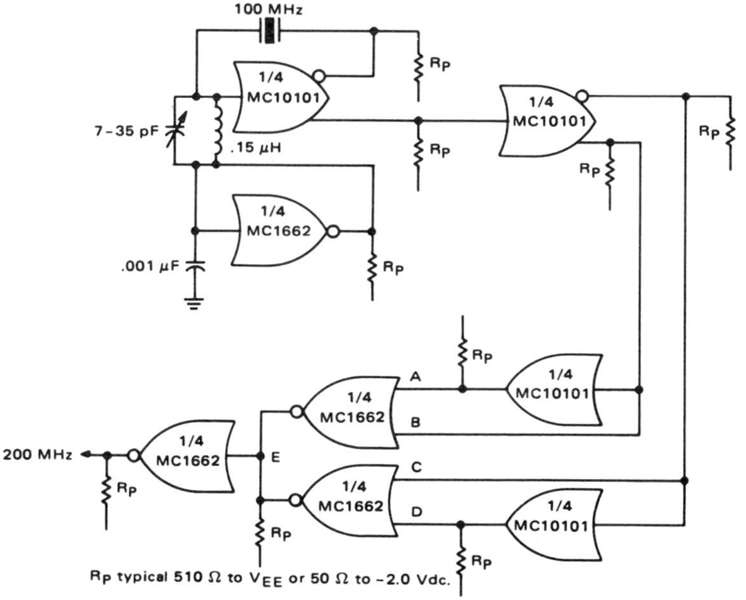

200 MHz crystal oscillator (M).

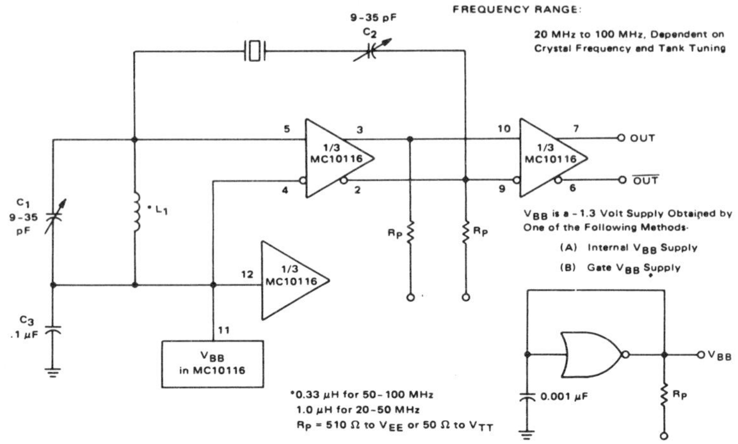

MECL overtone crystal oscillator (M).

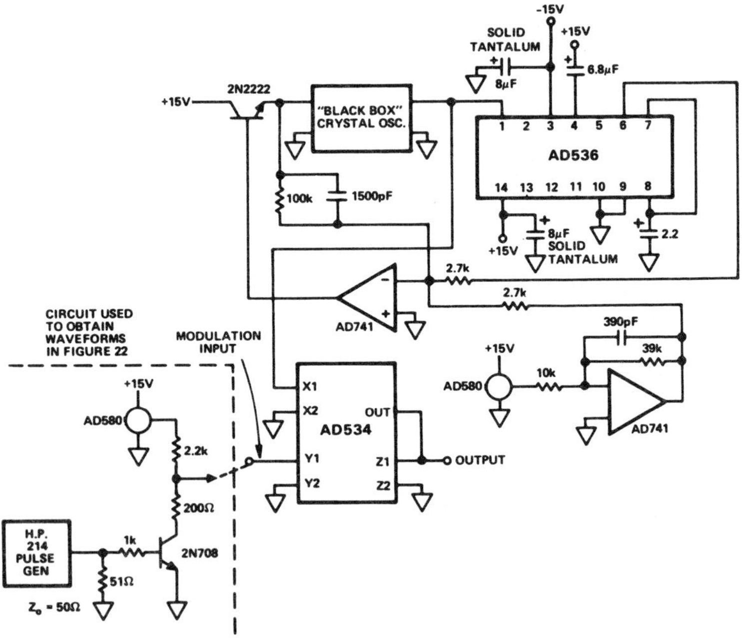

Crystal oscillator circuit, showing high-speed gain adjustment (AD).

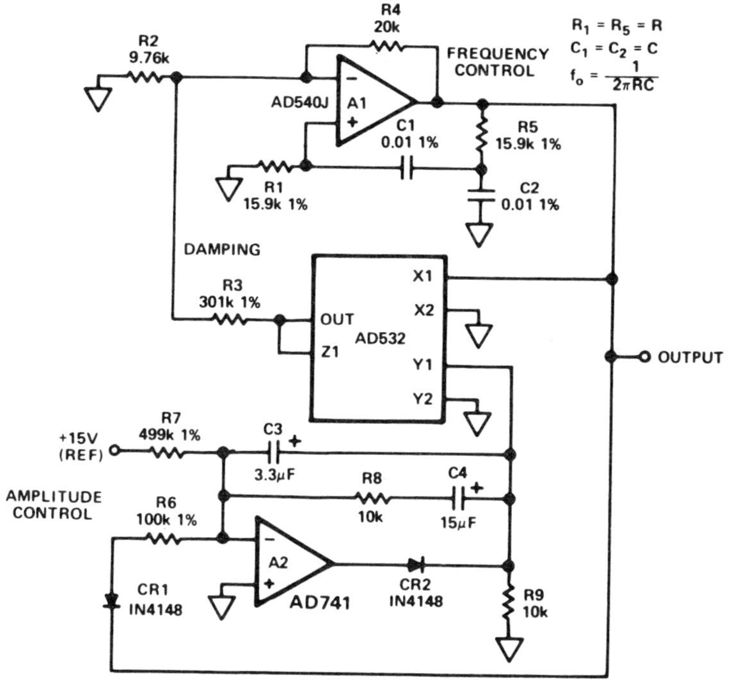

Low-distortion sine wave oscillator (AD).

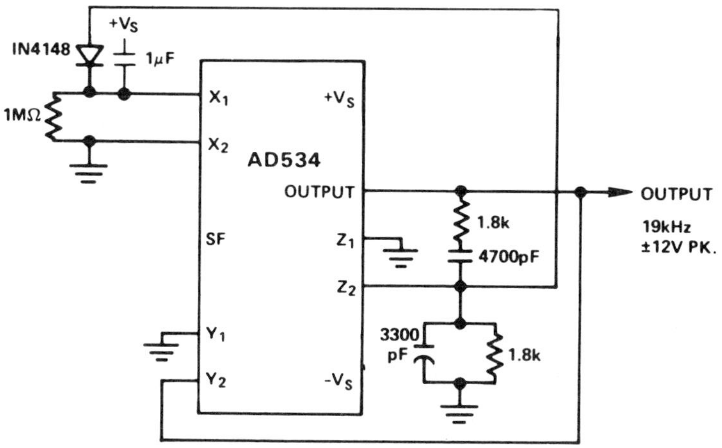

Stabilized Wien-bridge oscillator (AD).

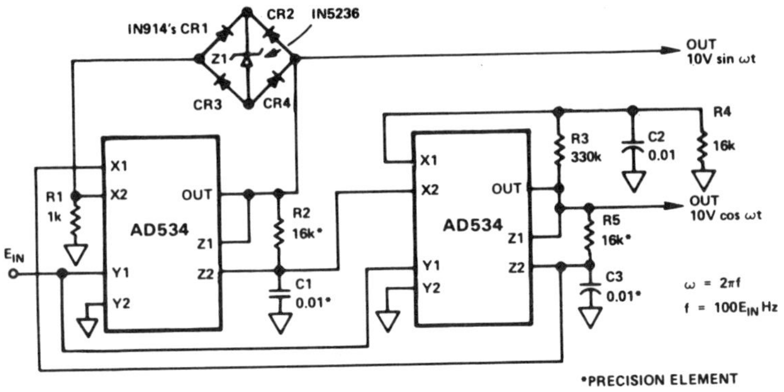

Voltage-controlled sine-wave oscillator (AD).

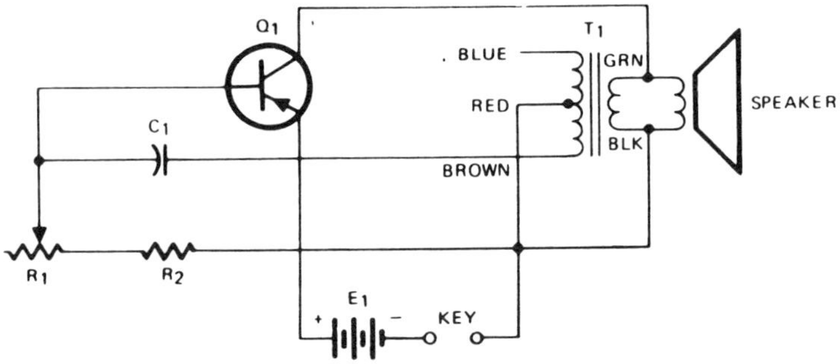

Q_1	PNP Power Transistor (WEP230)
C_1	0.1 μF Paper or Ceramic Capacitor
T_1	500 Ohm Center-tapped Primary and 8 Ohm Secondary (Argonne AR-164 or Stancor TA-42)
R_1	10K Ohm Small Potentiometer WEP F70-334
R_2	1 K Ohm, ½ Watt Resistor WEP22-2096
Speaker	4 to 8-Ohm PM Loudspeaker WEP N70-293
E_1	6 Volt Battery (or four 1.5 Volt Batteries in Series)
Key	WEP D70-045 telegraph-type key
	Printed Circuit Board
	Transistor Socket (WEP8051)

Code practice oscillator (W).

Chapter 14

Audio and Video Receivers and Related Circuits

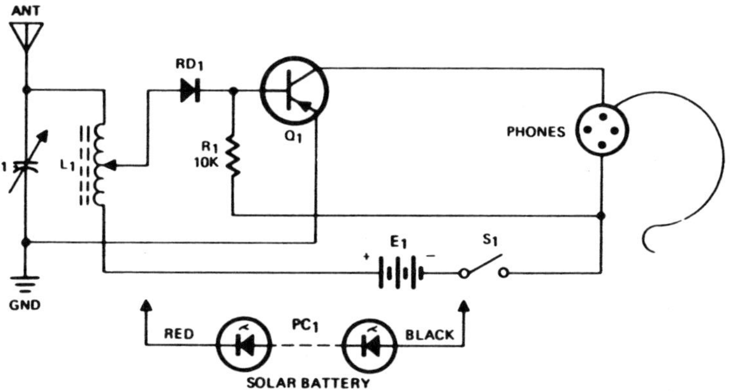

Q_1	PNP Transistor WEP254
L_1	Antenna Coil (J.W. Miller No. 2004, or equivalent)
C_1	365pf Tuning Condenser WEP F70-452
RD_1	Germanium Diode (WEP134)
R_1	10K Ohm, ½ Watt Resistor WEP22-2120
S_1	SPST Switch WEP TS101
E_1	3 Volt Battery (Two 1.5 Volt Batteries in Series)
PC_1	8 series-connected Solar Cells (WEP8501)
	Printed Circuit Board
	Transistor Socket (WEP8050)

Single stage transistor radio (W).

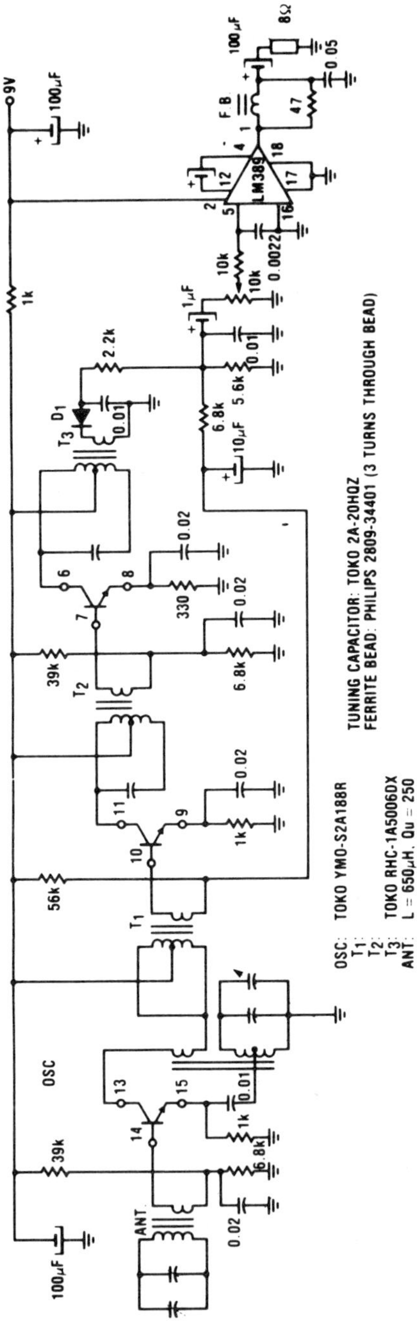

9V
100μF
1k
LM389
10k
0.0022
1μF
2.2k
D1
T3
0.01
6.8k
5.6k
10μF
330
0.02
39k
T2
56k
T1
OSC
ANT
0.02
1k
8Ω
0.05
F.B.
47
OSC: TOKO YMO-S2A188R
T1:
T2:
T3: TOKO RHC-1A5006DX
ANT: L = 650μH, Qu = 250
TUNING CAPACITOR: TOKO 2A-20HQZ
FERRITE BEAD: PHILIPS 2809-34401 (3 TURNS THROUGH BEAD)

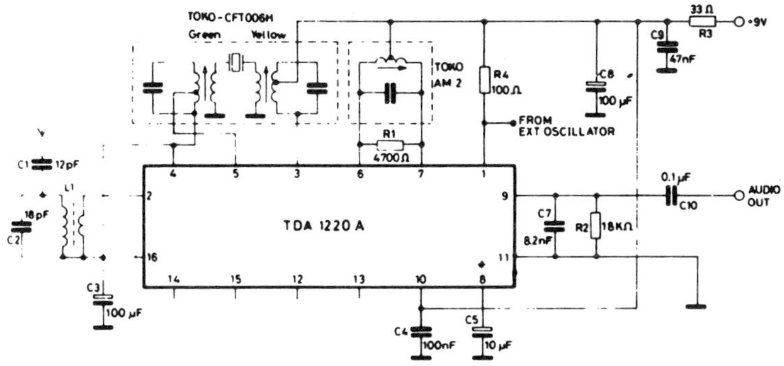

Low cost 27 MHz receiver, with external XTAL oscillator (SGS).

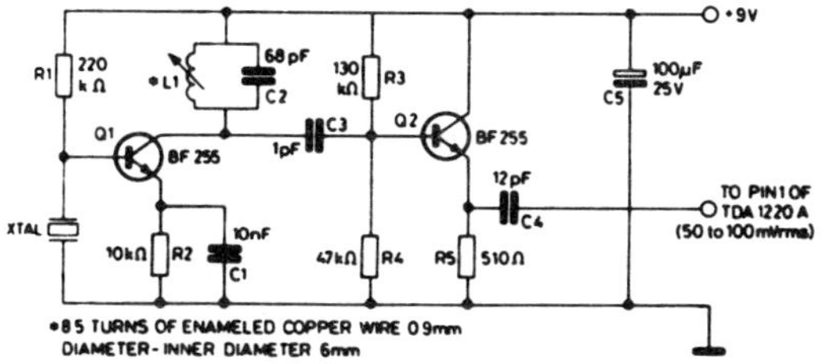

External XTAL oscillator for low cost 27 MHz receiver (SGS).

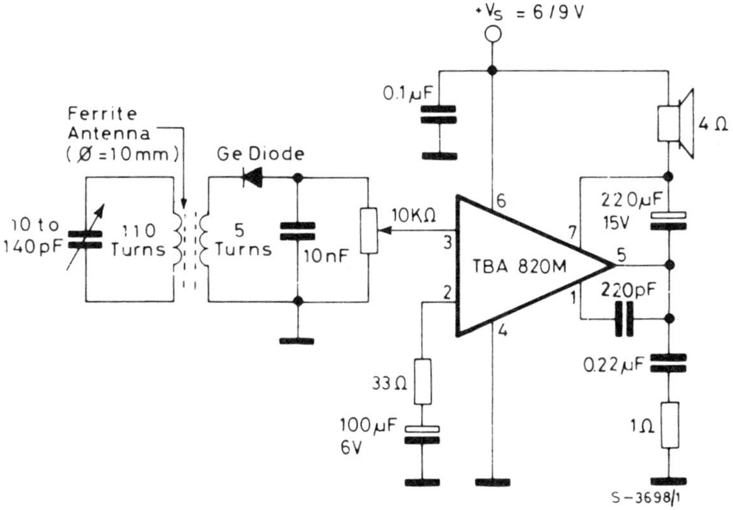

Low cost toy AM radio (0.5 to 1.5 MHz) (SGS).

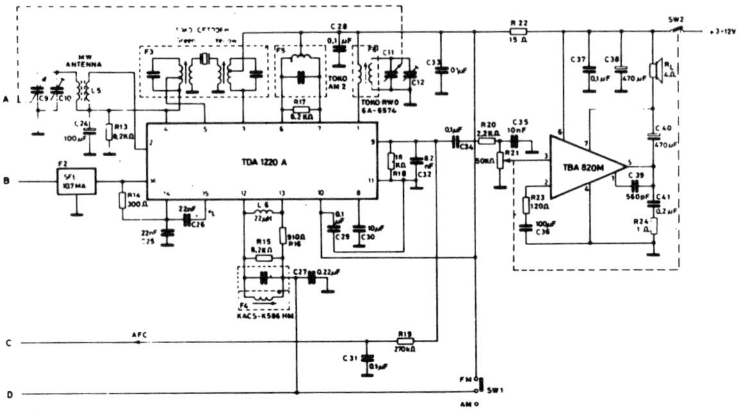

Low cost AM-FM radio (SGS).

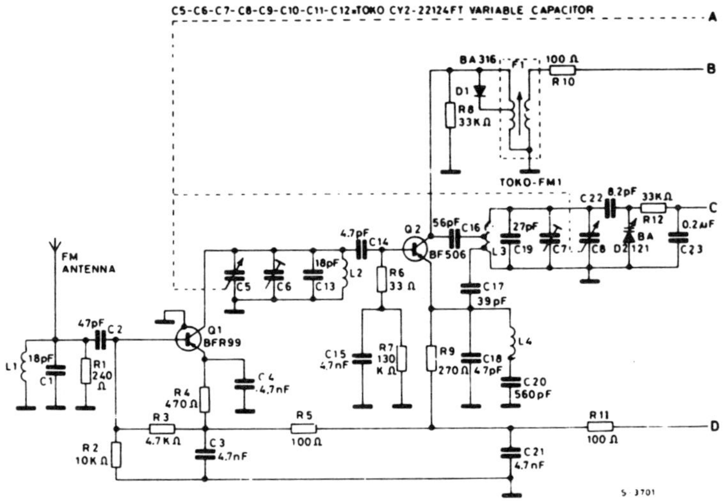

COILS

L1 FM Antenna coil - 6 Turns copper wire 0.9 mm diameter. Inner diameter 4 mm. Winding pitch 1 mm.
L2 FM Tuning coil - 5 Turns copper wire 0.9 mm diameter. Inner diameter 4 mm Winding pitch 0.5 mm.
L4 - 18 Turns copper wire 0.6 mm diameter. Inner diameter 2.5 mm. Closely wound.
L5 MW Antenna coil - Televox.
L3 FM osc. coil - 4 Turns copper wire 0.9 mm diameter. Inner diameter 4 mm. Winding pitch 2 mm.

FM front end

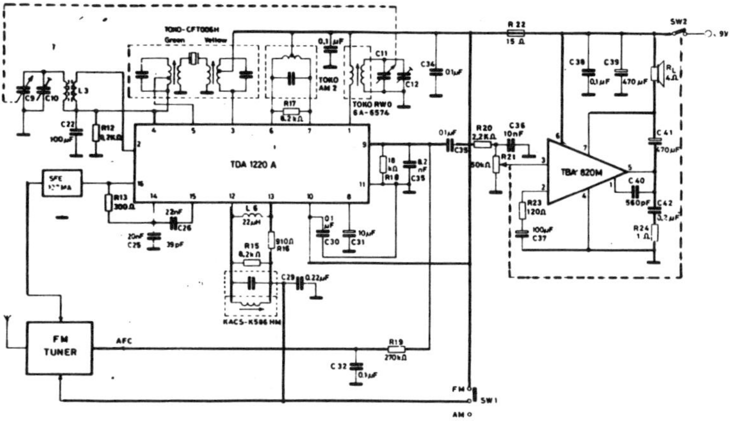

Low cost AM-FM radio with 1.6W output power (V_s = 9V) (SGS).

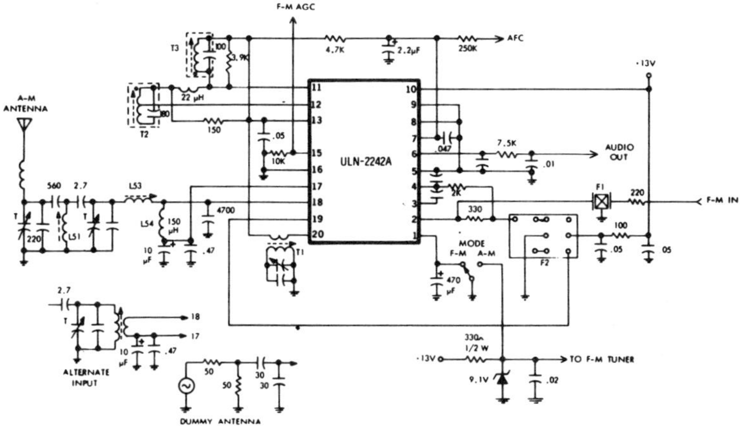

F1 SFE10.7MA, F-M I-F filter
F2 CFZ455C, A-M I-F filter
T1 RWO-6A7640BM, A-M Osc., Qu = 50, 11:1 turns ratio
T2 RLCS-4A7893GO, A-M Det., 256 μH, Qu = 75, 5.4:1 turns ratio
T3 BKAC-K3651HM, F-M Det., Qu = 65

Low-cost automobile radio (SE).

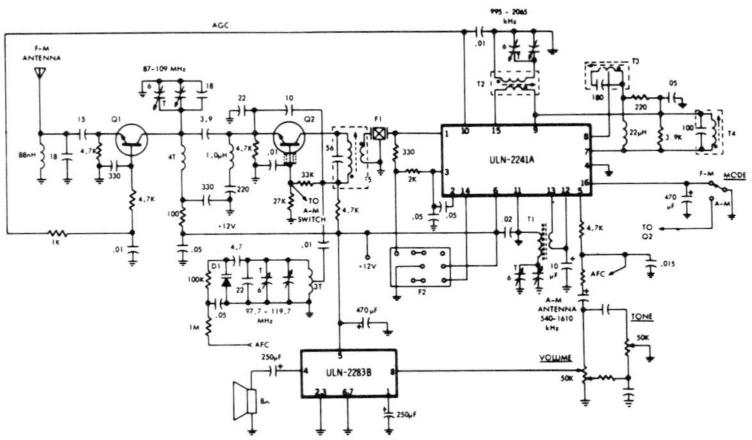

D1 Siemens TA314, F-M varactor diode
F1 SFE10.7MA, F-M I-F filter
F2 CFZ455C, A-M I-F filter
Q1 MPS3563, F-M R-F amplifier
Q2 MPS3563, F-M converter
T1 Ferrite loop, 571 μH, Qu = 150, 17:1 turns ratio
T2 RWO-6A7640BM, A-M osc., Qu = 50, 11:1 turns ratio
T3 RLCS-4A7893GO, A-M det., 256 μH, Qu = 75, 5.4:1 turns ratio
T4 BKAC-K3651HM, F-M det., Qu = 65
T5 BKACS-K4551AO, F-M input, Qu = 90, 7:1 turns ratio

Low cost AM/FM tuner (SE).

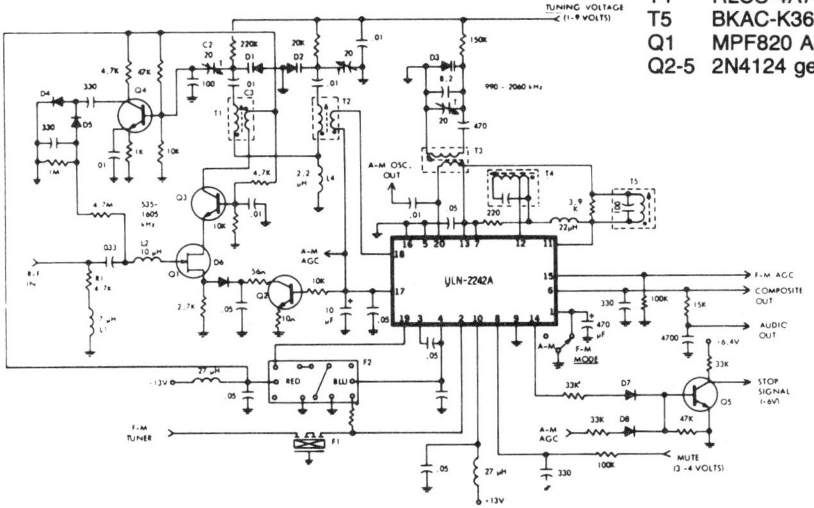

D1-3	Toko KV1215 or Sanyo SVC311 A-M varactor diode (3)
D4-7	1N914 or 1N4148 general purpose diode (4)
F1	SFJ10.7MA, F-M I-F filter
F2	CFU455C, A-M I-F filter
T1	RWOS-6A7894AO, A-M R-F, 178 μH, Qu = 120, 10:1 turns ratio
T2	RWOS-6A7894AO, A-M R-F, 178 μH, Qu = 120, 10:1 turns ratio
T3	RWOS-6A7892AO, A-M Osc., Qu = 120, Q1 = 80, 5:1 turns ratio
T4	RLCS-4A7893GO, A-M Det., 256 μH, Qu = 75, 5.4:1 turns ratio
T5	BKAC-K3651HM, F-M Det., Qu = 65
Q1	MPF820 A-M R-F amplifier
Q2-5	2N4124 general purpose amplifier

Varactor-tuned automobile radio (SE).

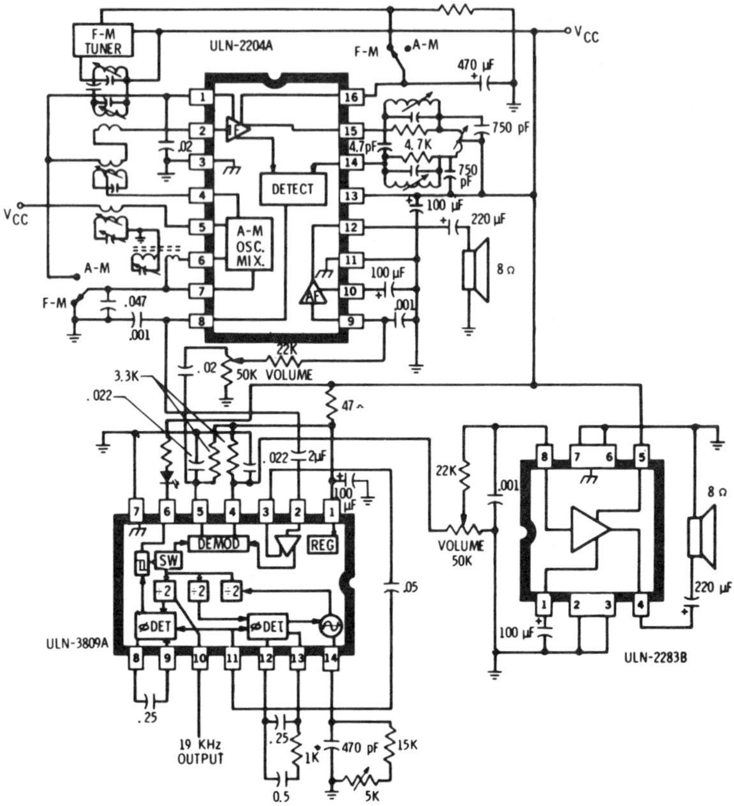

Minimum-cost application in A-M/F-M stereo radio (SE).

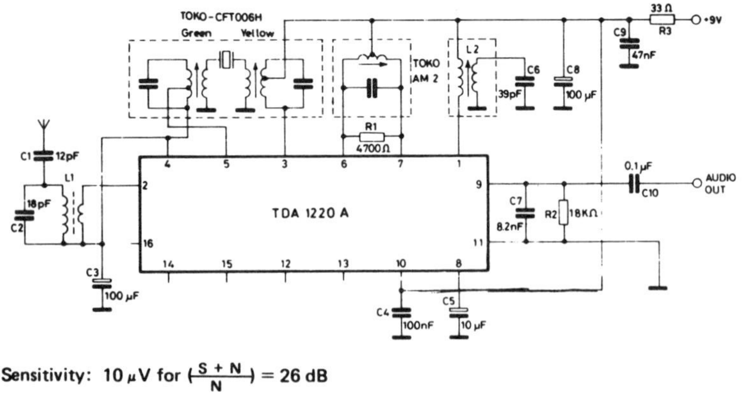

Sensitivity: 10 µV for $\left(\frac{S+N}{N}\right) = 26$ dB

Low cost 27 MHz receiver (SGS).

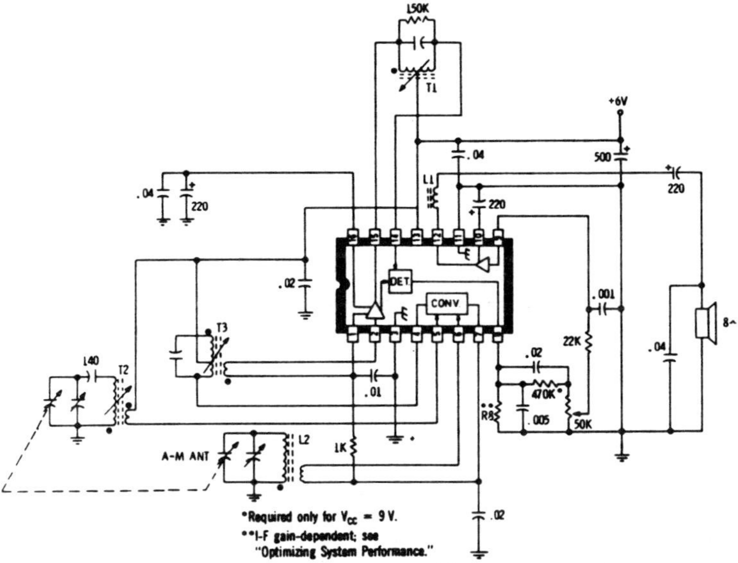

NOTE: Dress speaker wires as far as possible from antenna. L1 and 0.04 μF speaker bypass capacitor may be deleted, depending on speaker location.

L1 Audio choke	10 μH, Qu = 2@2.52 MHz, 3 turns through ferrite bead
L2 A-M antenna coil	Qu = 250, 110:10 turns ratio, Q2B core, 3.5″ (90 mm) × 0.394″ (10 mm) ϕ
T1 A-M detector coil	390 pF, Qu = 130@455 kHz, 100 turns, center-tapped
T2 A-M oscillator	460 μH, Qu = 120@796 kHz, 110:11 turns ratio
T3 A-M I-F transformer	180 pF, Qu = 145@455 kHz, 155:10 turns ratio, primary tapped at 127 turns

ULN-3838A A-M radio system (SE).

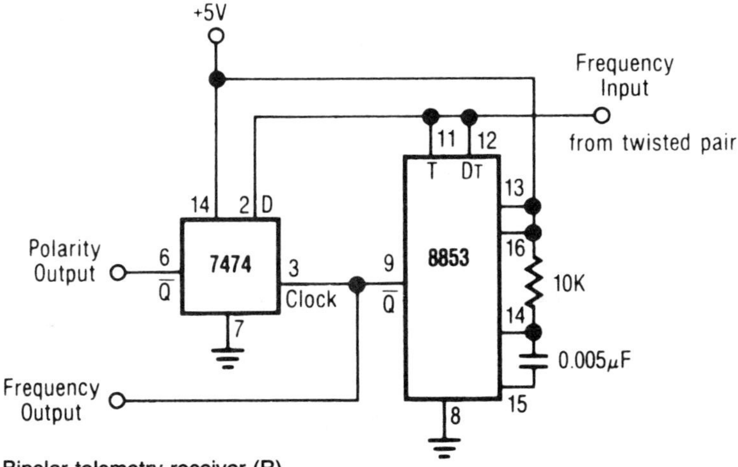

Bipolar telemetry receiver (R).

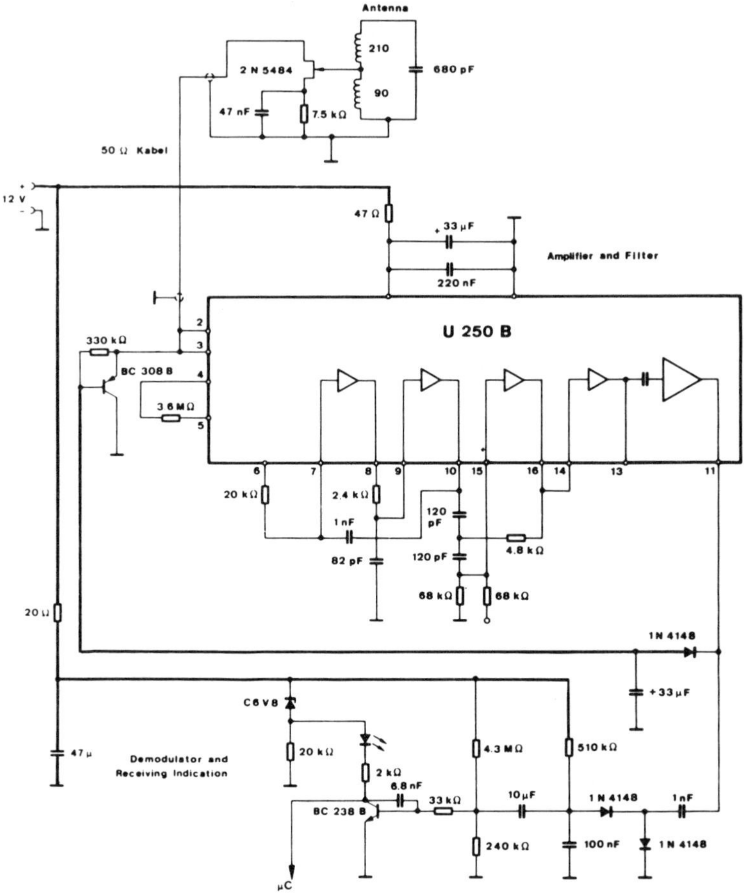

U 250 B as DCF 77 receiver (AEG).

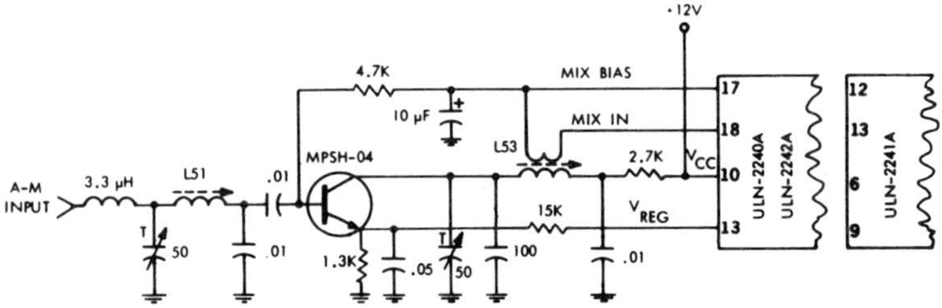

Automobile tuner with AM/rf stage (SE).

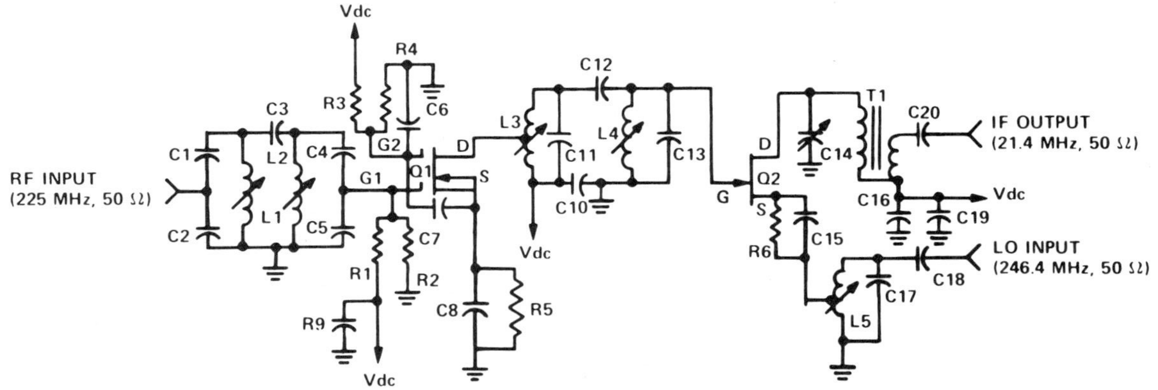

C1 – 8.2 pF	Ceramic disc capacitor
C2 – 43 pF	Dipped, silvered mica capacitor
C3 – 0.2 pF	Ceramic tubular capacitor
C4 – 12 pF	Ceramic disc capacitor
C5 – 15 pF	Dipped, silvered mica capacitor
C6, C7, C8, C9, C10 C16 – 220 pF	Ceramic disc capacitor
C11 – 10 pF	Dipped, silvered mica capacitor
C12 – 0.47 pF	Ceramic tubular capacitor
C13 – 3.6 pF	Ceramic tubular capacitor
C14 – 8–60 pF	Mica compression trimmer, Arco 404
C15, C19 – 0.01 μF	Ceramic disc capacitor
C17, C18 – 3 pF	Dipped, silvered mica capacitor
C20 – 24 pF	Dipped, silvered mica capacitor

R1 – 300 kilohm, 1/4 W carbon resistor
R2 – 27 kilohm, 1/4 W carbon resistor
R3 – 62 kilohm, 1/4 W carbon resistor
R4 – 56 kilohm, 1/4 W carbon resistor
R5 – 220 ohm, 1/4 W carbon resistor
R6 – 910 ohm, 1/4 W carbon resistor
L1 – 2-1/2 turns no. 20 enameled wire on 9/32" plastic form with brass slug (70 nH)
L2, L4 – 2-1/6 turns no 20 enameled wire on 9/32" plastic form with brass slug (60 nH)
L3 – 1-5/6 turns no. 20 enameled wire on 9/32" plastic form with brass slug, tapped 1/2 turn from ground (55 nH)
L5 – 2-1/2 turns no. 20 enameled wire on 9/32" plastic form with brass slug, tapped 2/3 turn from ground (70 nH)
Q1 – 3N201 MOSFET
Q2 – 2N5486 JFET
T1 – Primary = 18 turns no. 24 enameled wire on T44-6 Micrometals toroid core
Secondary = 4 turns no. 24 enameled wire twisted around last 4 turns of ground end of primary

225-MHz receiver front-end schematic diagram (M).

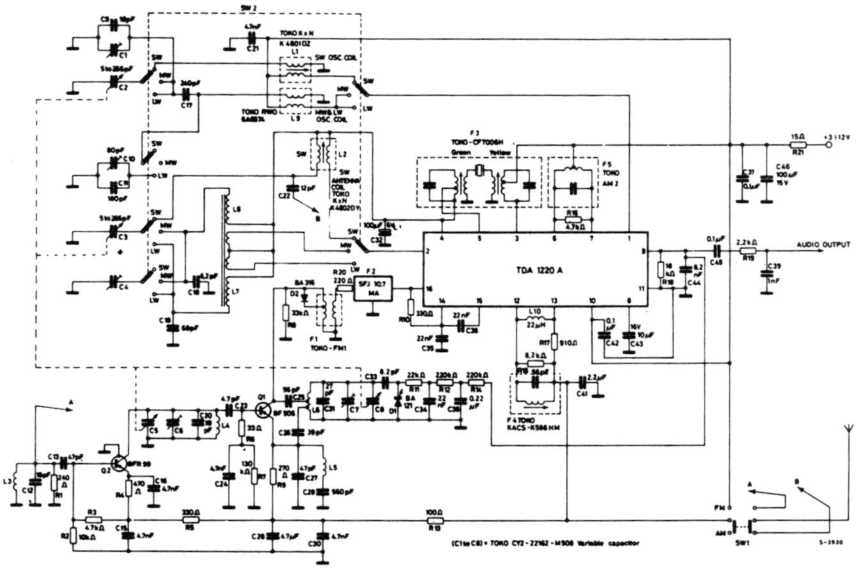

L3 FM Antenna Coil - 6 Turns copper wire 0.9 mm diameter. Inner diameter 4 mm - winding pitch 1 m.
L4 FM Tuning Coil - 5 Turns copper wire 0.9 mm diameter. Inner diameter 4 mm - winding pitch 0.5 mm.
L5 - 18 Turns copper wire 0.6 mm. diameter. Inner diameter 2.5 mm. Closely wound.*
L6 FM oscillator Coil - 4 Turns silvered copper wire 1 mm. diameter. Inner diameter 4 mm. - winding pitch 2.5 mm.
L7 - LW antenna coil - Televox.
L8 - MW antenna coil - Televox.
SW3 - ALPS type SSA 0.63.

LW - MW - SW - FM - tuner (SGS).

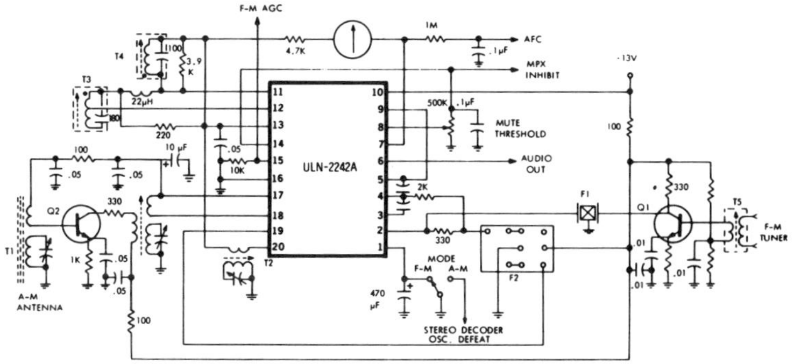

F1 SFE10.7MA, F-M I-F filter
F2 CFZ455C, A-M I-F filter
Q1 MPS3563, F-M converter
Q2 MPSH-04, A-M R-F amplifier
T1 Ferrite loop, 571 μH, Qu = 150, 17:1 turns ratio
T2 RWO-6A7640BM, A-M Osc., Qu = 50, 11:1 turns ratio
T3 RLCS-4A7893GO, A-M Det., 256 μH, Qu = 75, 5.4:1 turns ratio
T4 BKAC-K3651HM, F-M Det., Qu = 65
T5 BKACS-K4551AO, F-M Input, Qu = 90, 7:1 turns ratio

High-performance A-M/F-M tuner (SE).

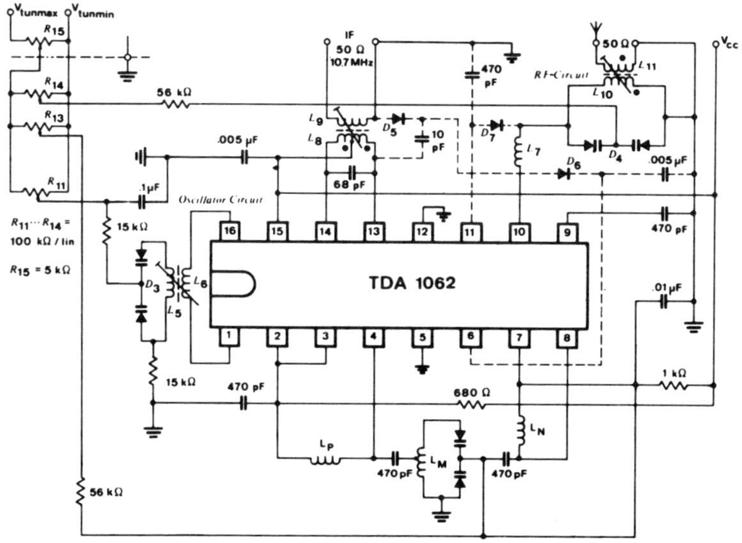

Basic FM Front End design incorporating TDA 1062 (AEG).

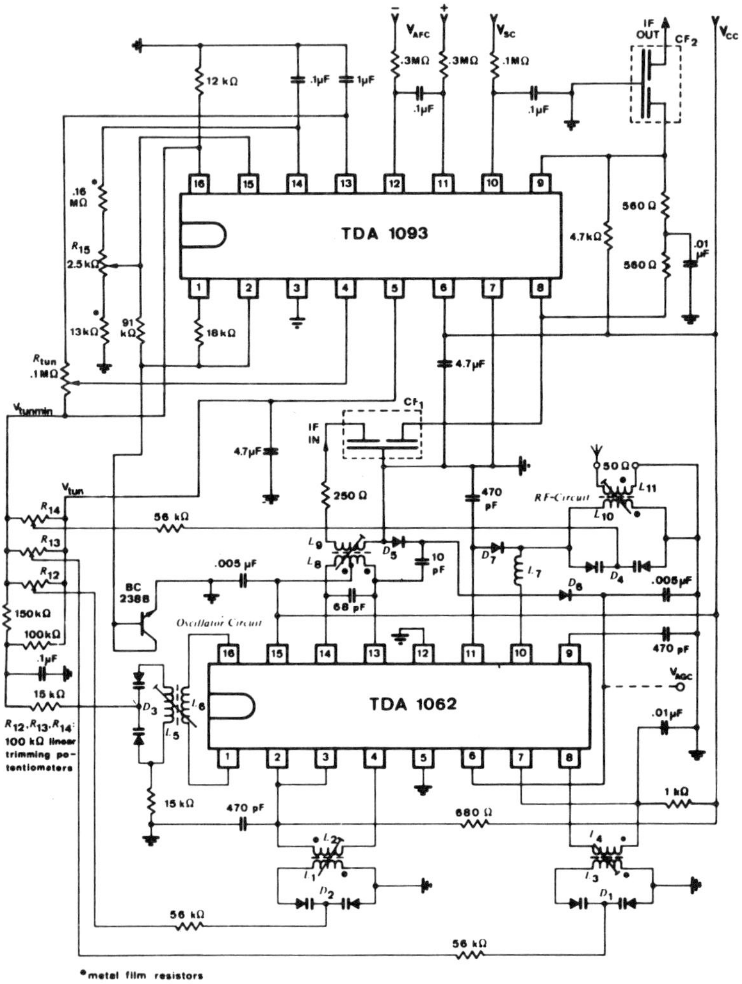

High performance FM front end system (AEG).

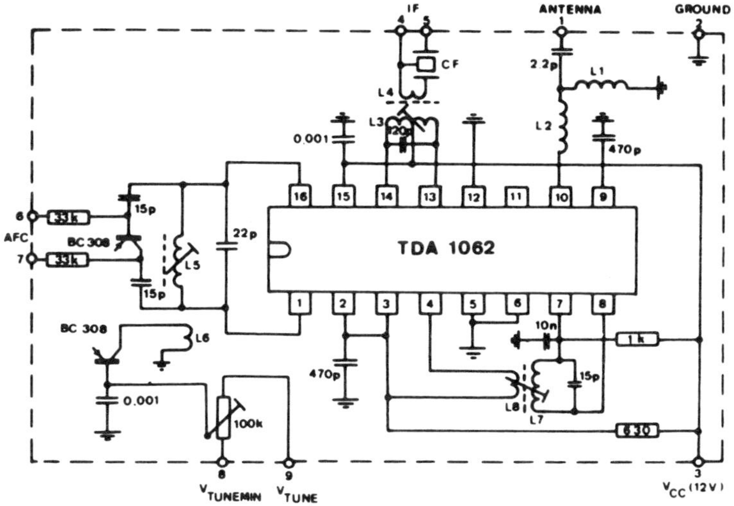

Designing weather band front ends around TDA 1062 (AEG).

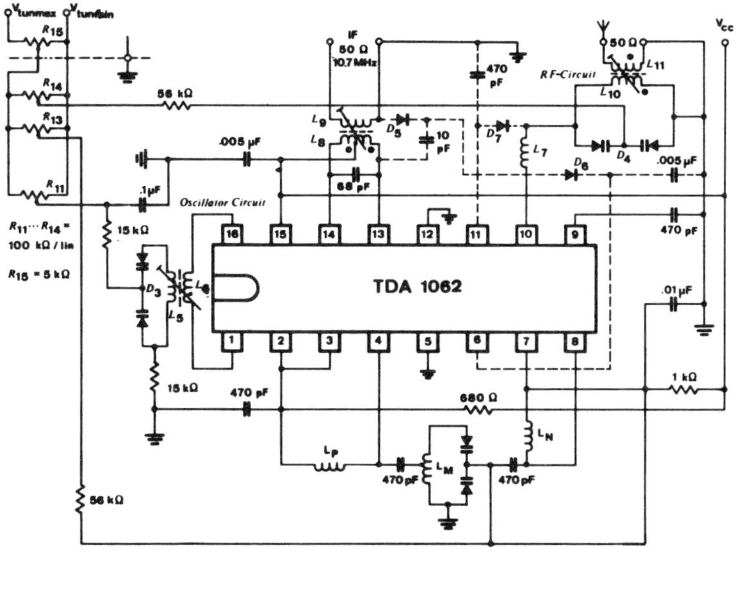

A low cost FM tuner for 88-108 MHz U.S. FM band (AEG).

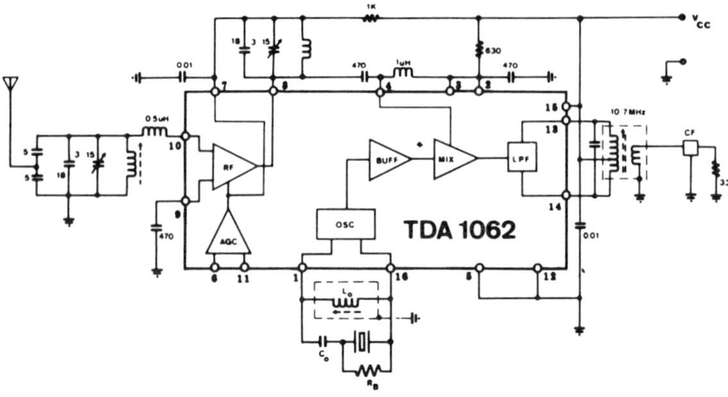

Circuit diagram of a crystal-controlled VHF front end, built around the TDA 1062 IC (AEG).

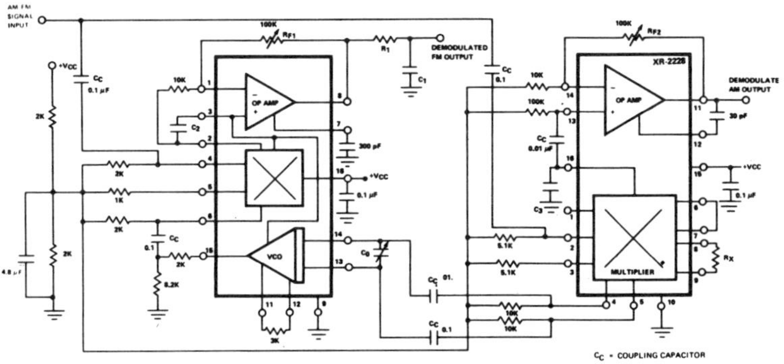

High-frequency AM and FM detector using the XR-215 and the XR-2228 (E).

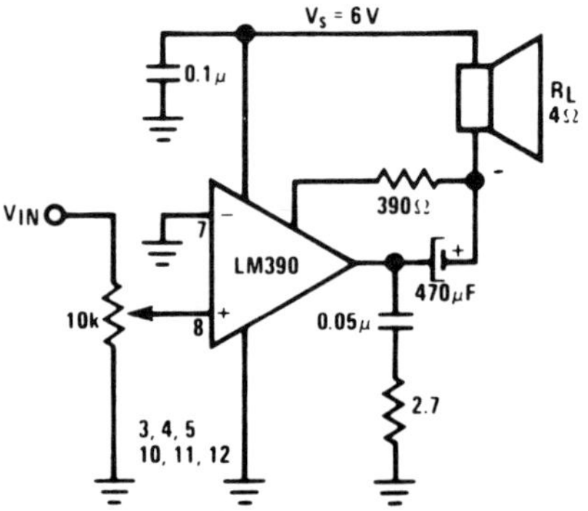

1 Watt power amplifier for 6 Volt systems (NS).

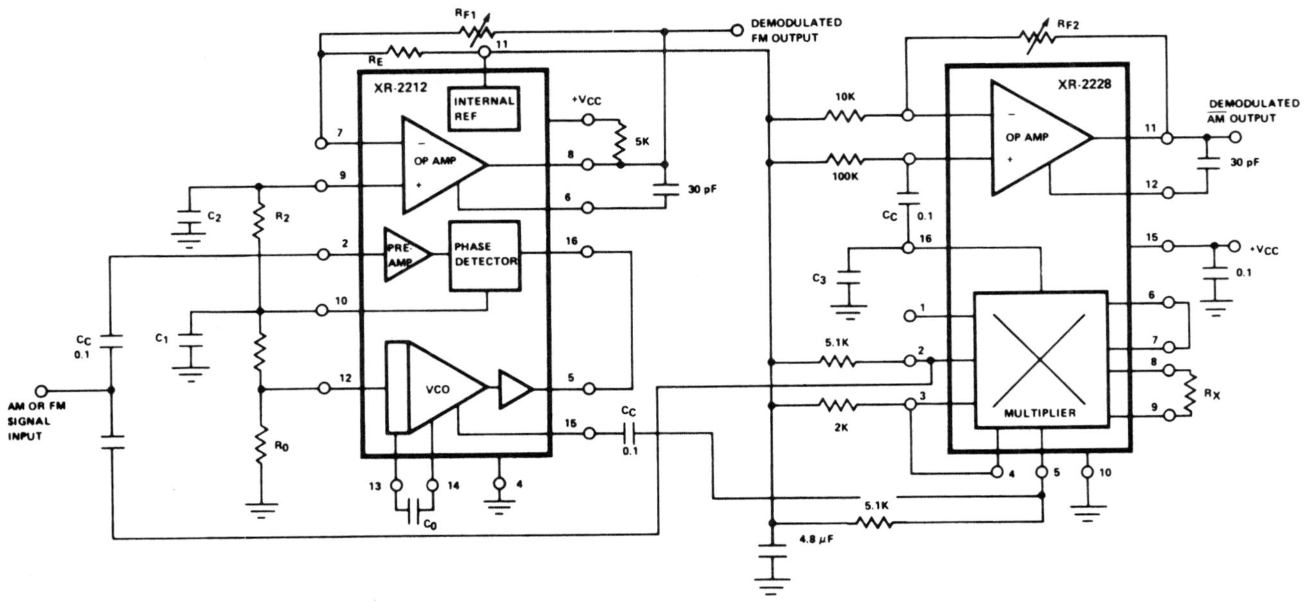

A two-chip AM/FM detector system using the XR-2212 phase-locked loop and the XR-2228 multiplier/detector (E).

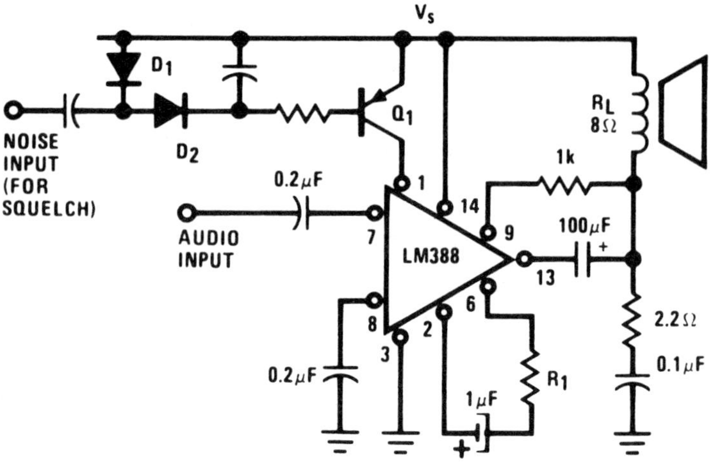

LM388 squelch circuit for FM scanners and walkie talkies (NS).

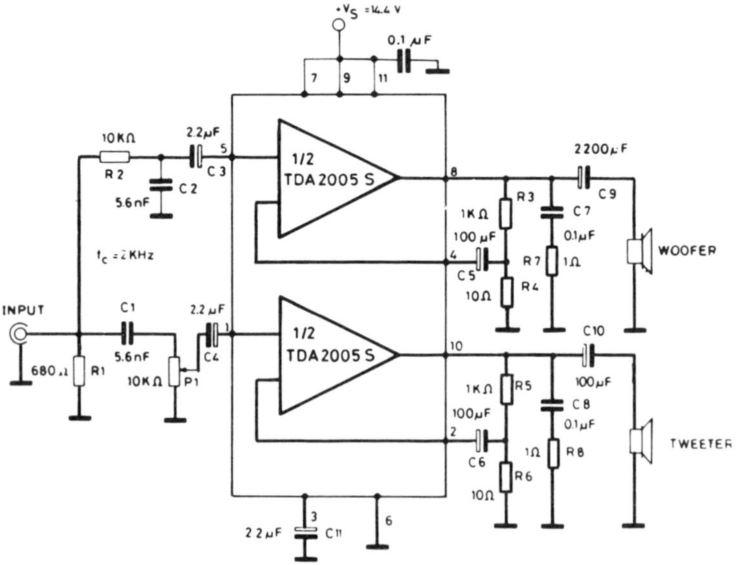

Simple 20W two way amplifier (fc = 2 kHz) (SGS).

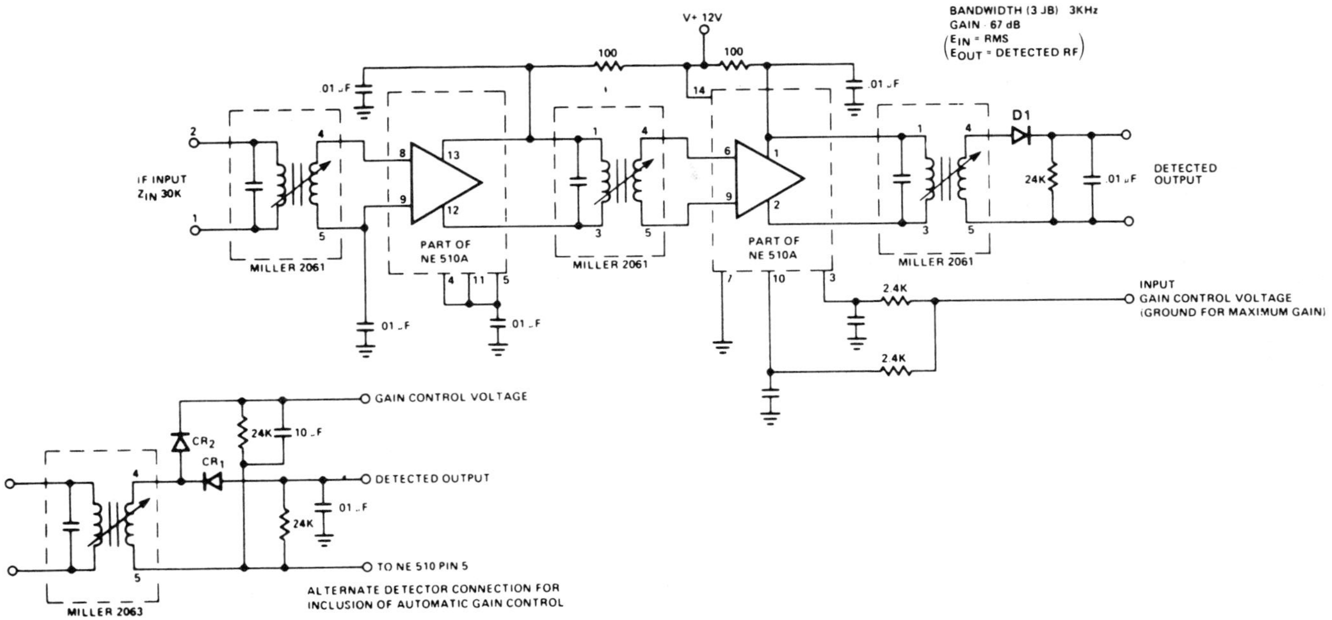

All resistor values are in ohms

455kHz IF amplifier (cascode connection) (S)

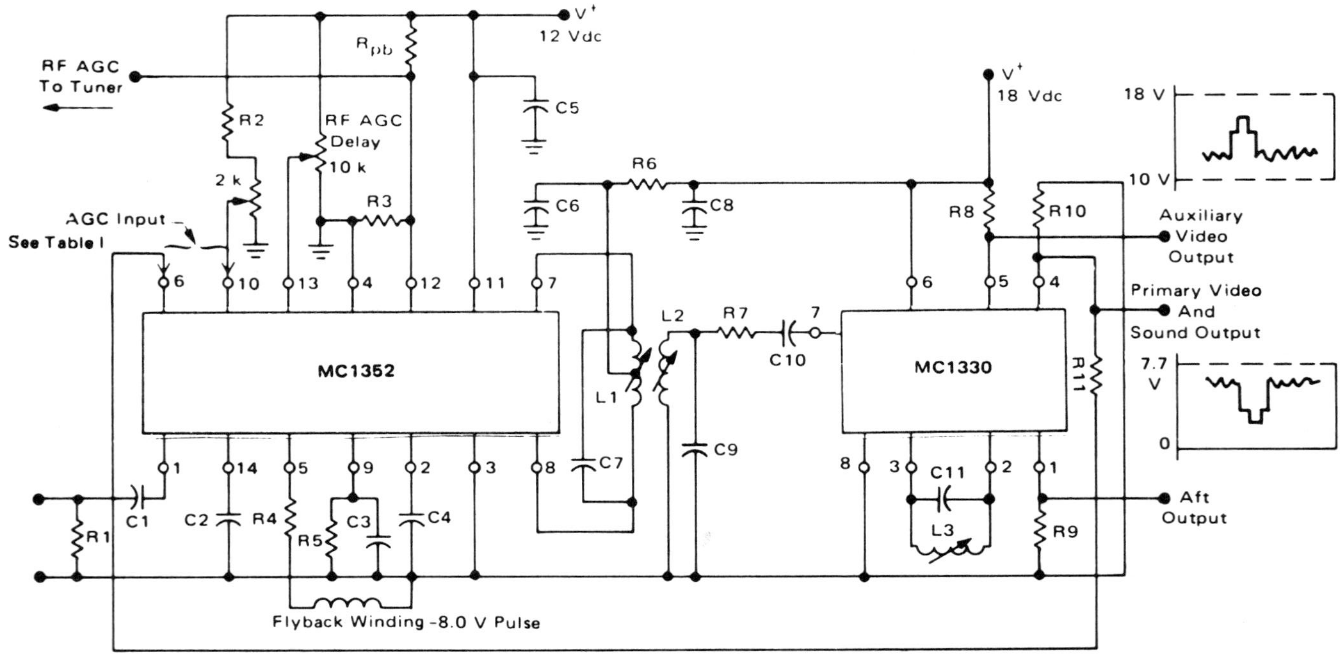
V+
12 Vdc
Rpb
RF AGC
To Tuner
R2
RF AGC
Delay
10 k
2 k
R3
AGC Input
See Table I
C5
R6
C6
C8
MC1352
L2
R7
7
C10
L1
C7
C9
MC1330
R11
V+
18 Vdc
18 V
10 V
R8
R10
Auxiliary
Video
Output
Primary Video
And
Sound Output
7.7
V
0
C11
L3
R9
Aft
Output
C1
R1
C2
R4
R5
C3
C4
Flyback Winding -8.0 V Pulse

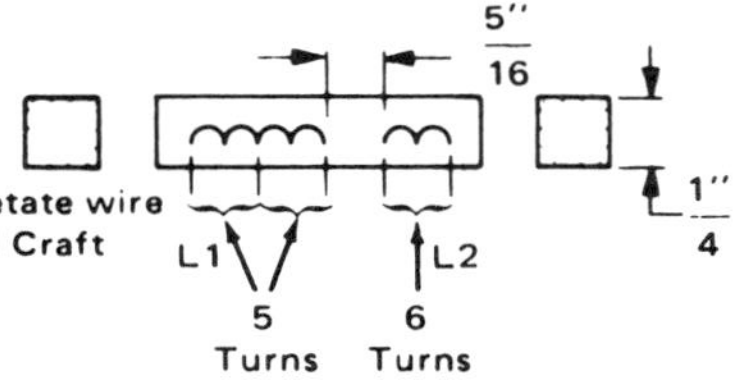

All windings #30 AWG tinned nylon acetate wire tuned with high permiability slugs. Coil Craft #4786 differential transformer.

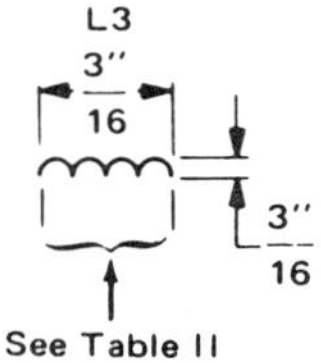

Wound with #26 AWG tinned nylon acetate wire tuned by distorting winding.

C10 = 62 pF
C11 = (See Table II)
All Resistors 1/4-Watt ±5%

TABLE I

Video Polarity	Pin 6 Voltage	Pin 10 Voltage	R4
Negative-Going Sync.	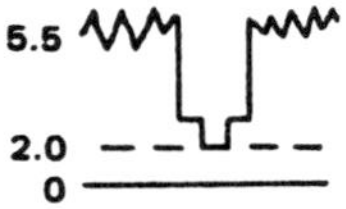	Adj. 1.0–4.0 Vdc Nom 2.0 V	0
Positive-Going Sync.	Adj. 1.0–8.0 Vdc Nom 4.5 V	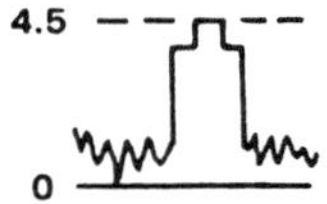	3.9 k

TABLE II

Component	36 MHz	45 MHz	58 MHz
C7	24 pF	15 pF	10 pF
C9	18 pF	12 pF	10 pF
C11	33 pF	33 pF	18 pF
L3	12 Turns	10 Turns	10 Turns

R_{pb} (See Text)
R1 = 50 Ω
R2 = 3.9 kΩ
R3 = (See Text)
R4 = (See Table I)
R5 = 220 kΩ
R6 = 220 Ω
R7 = 22 Ω
R8 = 3.3 kΩ
R9 = 3.9 kΩ
R10 = 3.9 kΩ
R11 = 4.7 kΩ
C1 = 0.001 μF
C2 = 0.1 μF
C3 = 0.25 μF
C5 = 0.1 μF
C6 = 0.1 μF
C7 = (See Table II)
C8 = 0.1 μF
C9 = (See Table II)

Television IF amplifier and detector using an MC1330 and an MC1352 (M).

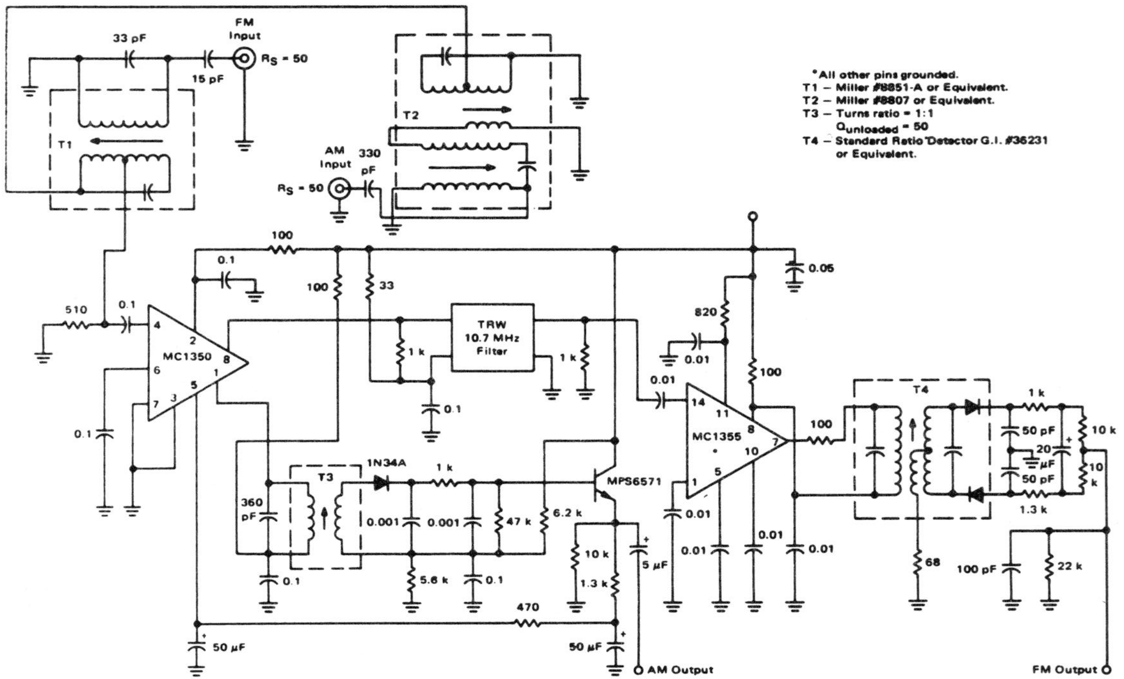

MC1350/MC1355 AM/FM IF (M).

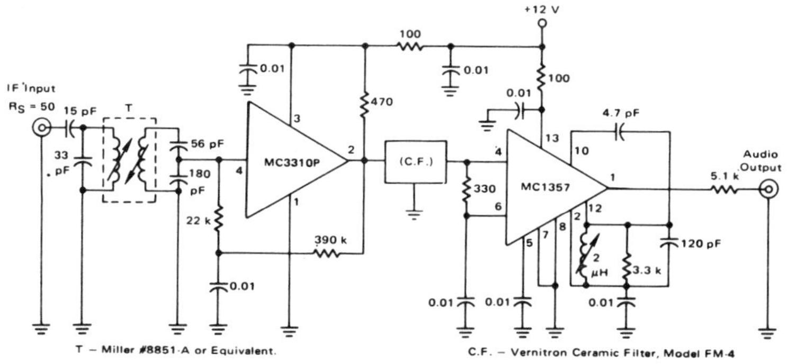

MFC4010/MC1357 IF(M).

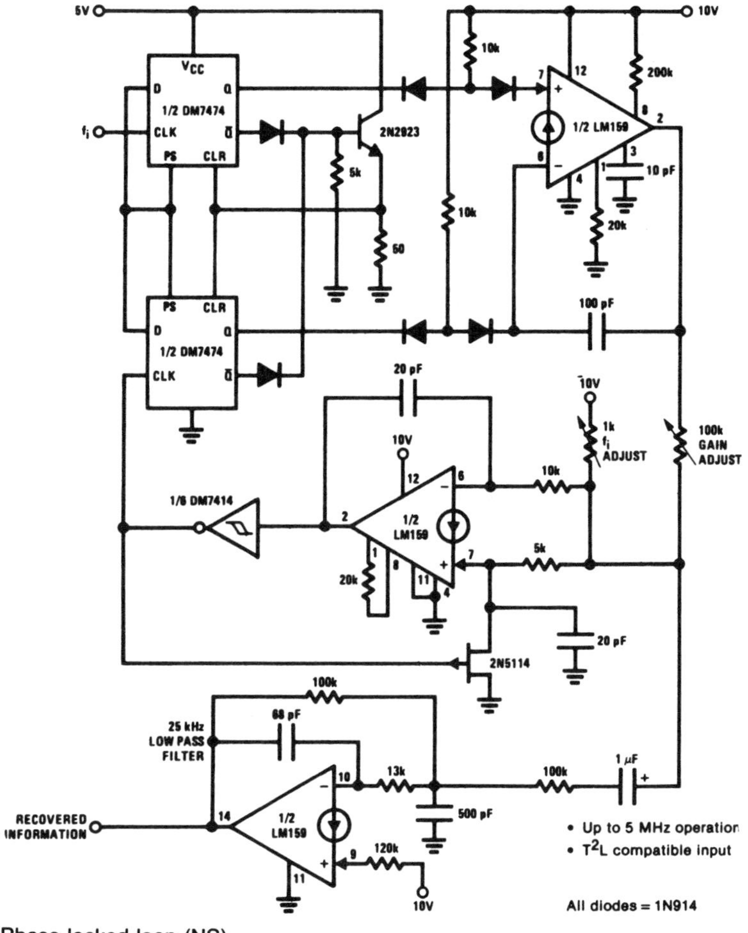

Phase locked loop (NS).

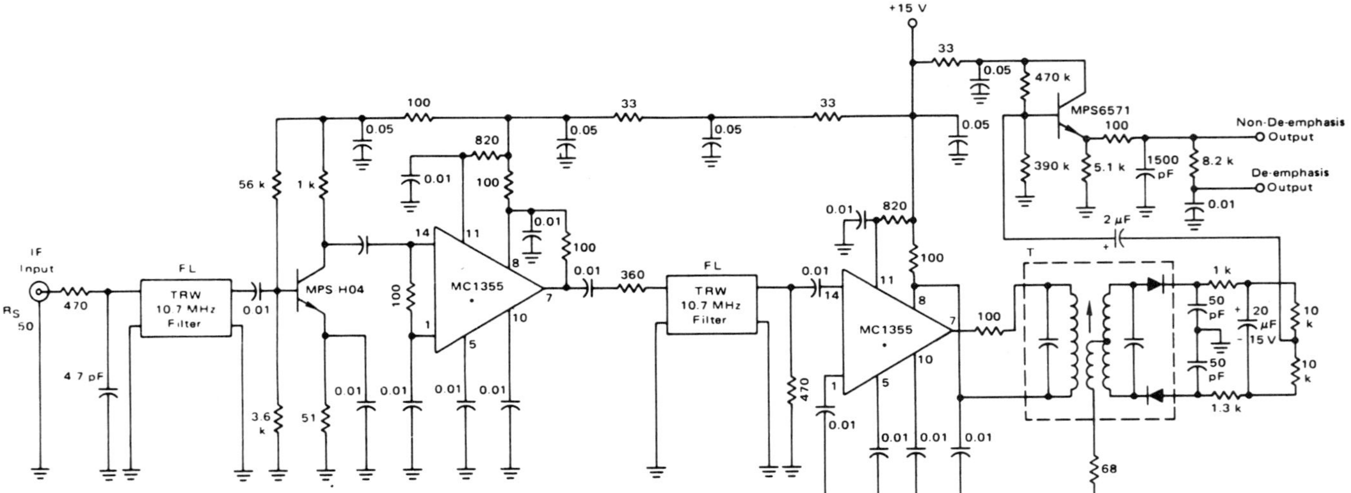

*All other pins grounded.

T Ratio Detector (Input impedance ≃ 1.5 k) G.I. #36231 or equivalent.

FL See Figure 2, Text.

Dual MC1355 FM IF (M).

Chapter 15

Transmitters, Transceivers and Related Circuits

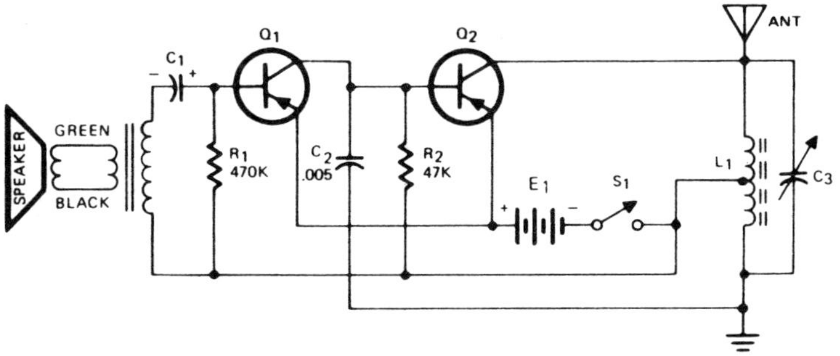

Q_1, Q_2	PNP Transistors (WEP254 or WEP620)
C_1	10 μF, 16 Volt Electrolytic Capacitor (WEP AX10-16)
C_2	0.005 μF Ceramic or Paper Capacitor
C_3	365pf Tuning Capacitor WEP F70-452
L_1	Tapped Antenna Coil (J.W. Miller No. 2004, or equivalent)
T_1	500 Ohm Primary and 8 Ohm Secondary (See text) (Argonne AR-164 or Stancor TA-42)
R_1	470K Ohm, ½ Watt resistor WEP22-2160
R_2	47K Ohm, ½ Watt resistor WEP22-2136
E_1	6 Volt battery (or four 1.5 Volt Batteries in Series)
S_1	SPST Pushbutton Switch (WEP PB532)
Speaker	8 Ohm, PM speaker WEP N70-293
	Transistor Sockets (WEP8051)
	Printed Circuit Board

Wireless microphone (W).

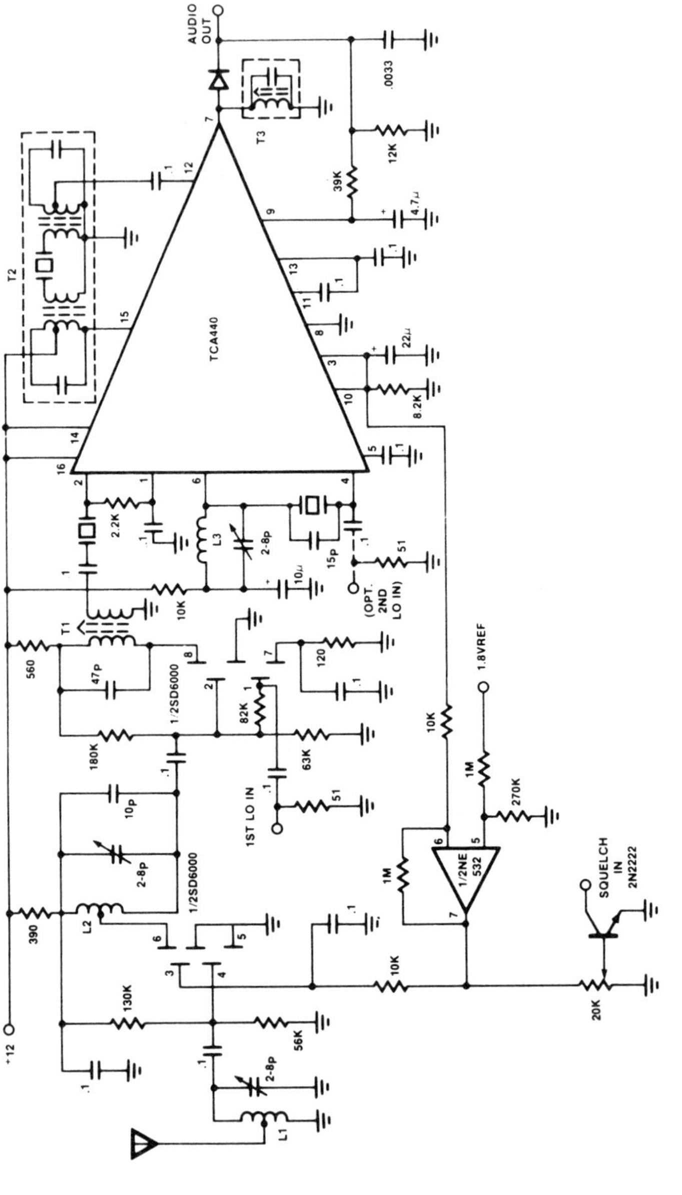
AUDIO OUT
T2
T3
TCA440
.0033
12K
39K
4.7μ
22μ
8.2K
2.2K
L3
2-8p
15p
10μ
51
(OPT. 2ND LO IN)
10K
T1
560
47p
1/2SD6000
82K
120
63K
180K
10p
1ST LO IN
1.8VREF
1M
270K
1/2NE 532
SQUELCH IN 2N2222
L2
390
130K
56K
20K
+12
L1

CB radio (S).

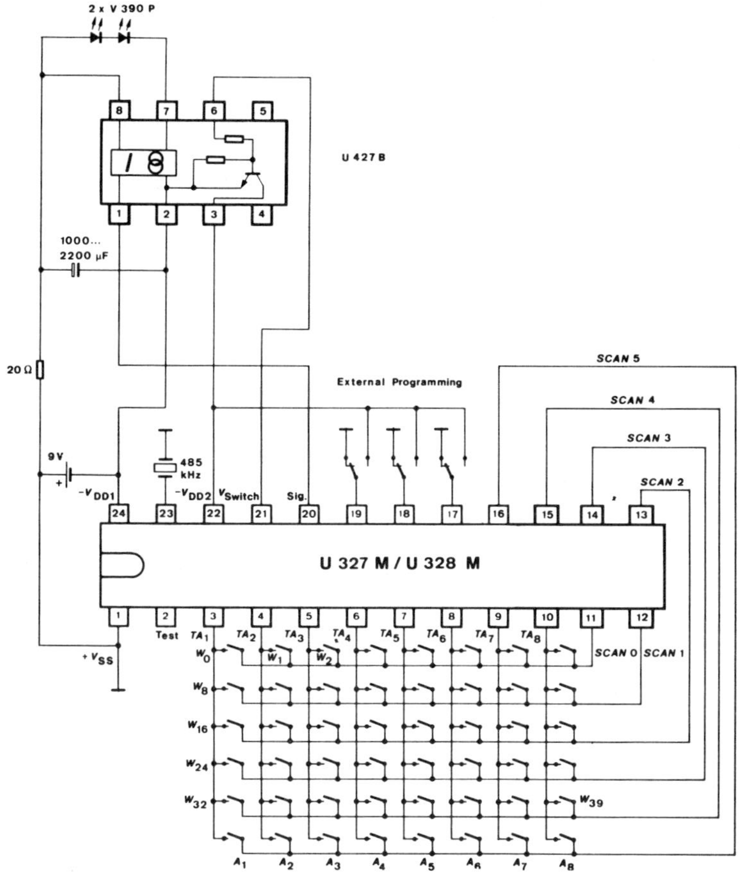

Infrared transmitter with external circuitry (AEG).

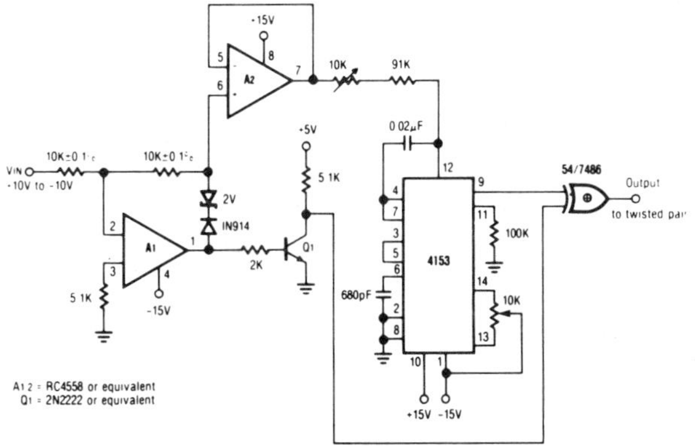

Bipolar telemetry transmitter (R).

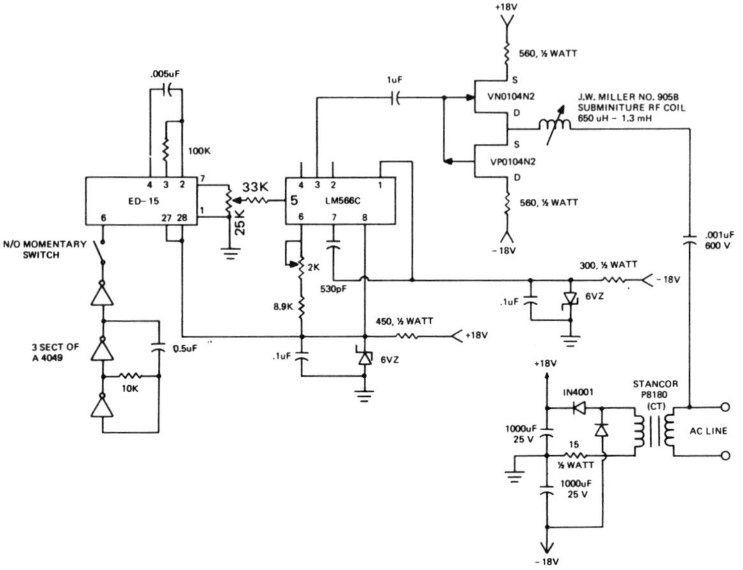

Carrier current transmitter (SU).

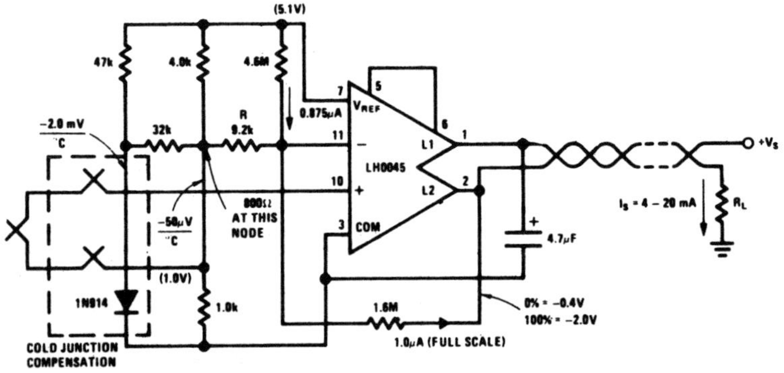

FOR 1μA FULL SCALE, R_{IN} = V_{IN}/1μA = SOURCE IMPEDANCE @ PIN 11
e.g., V_{IN} (FULL SCALE) = 10 mV, R_{IN} = 10k
BRIDGE IMPEDANCE = 0.8k, ∴ R = 10k – 0.8k = 9.2k

Thermocouple input transmitter (NS).

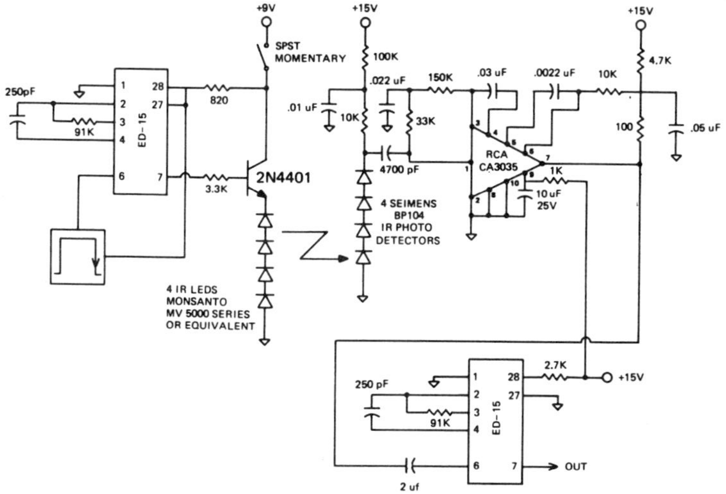

IR remote control transmitter/receiver (SU).

Chapter 16

Assorted Communications Circuits

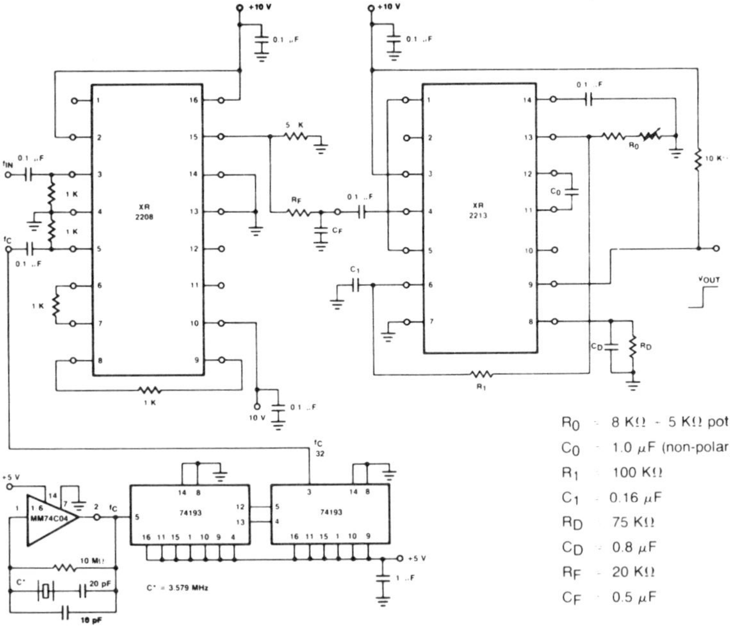

R_0 = 8 KΩ - 5 KΩ pot
C_0 = 1.0 µF (non-polar)
R_1 = 100 KΩ
C_1 = 0.16 µF
R_D = 75 KΩ
C_D = 0.8 µF
R_F = 20 KΩ
C_F = 0.5 µF

Narrow band tone decoder (E).

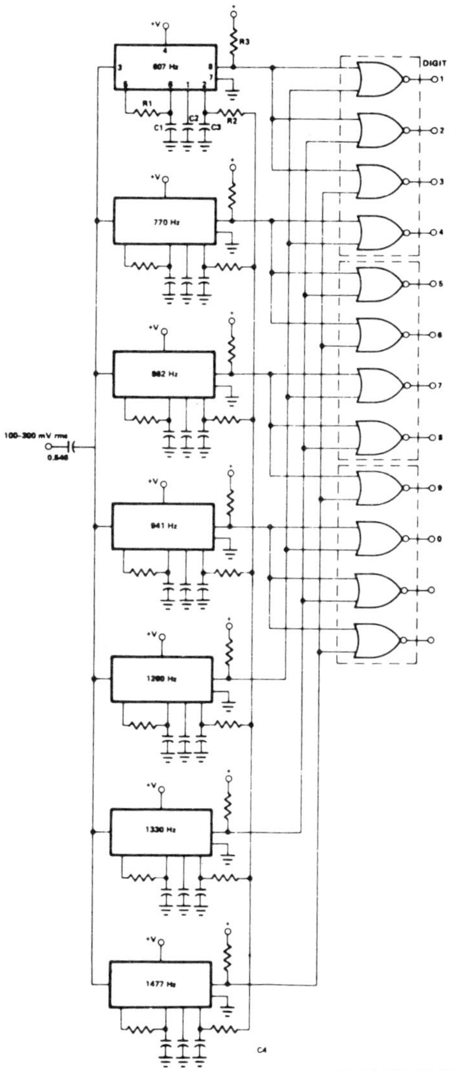

COMPONENT VALUES (TYPICAL)

R_1	6.8 to 15K ohm	
R_2	4.7K ohm	
R_3	2.0K ohm	
C_1	0.1 mfd	
C_2	1.0 mfd	6V
C_3	2.2 mfd	6V
C_4	250	6V

Low-cost touch tone® decoder (E).

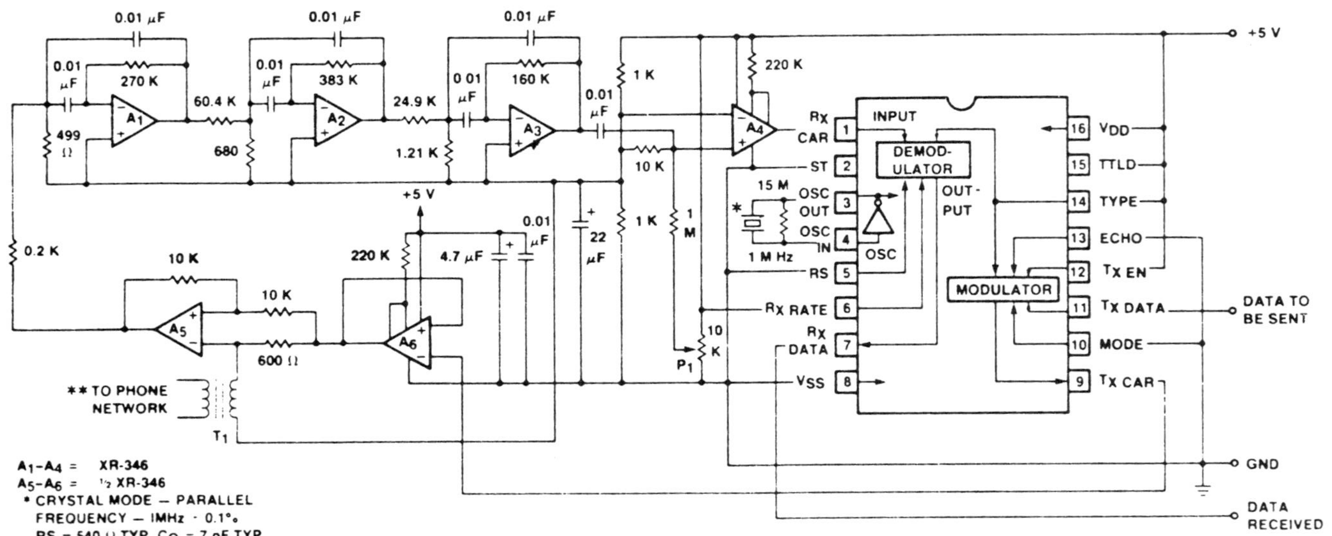

A_1–A_4 = XR-346
A_5–A_6 = ½ XR-346
* CRYSTAL MODE — PARALLEL
FREQUENCY — IMHz · 0.1%
RS = 540 Ω TYP. C_O = 7 pF TYP
SUGGESTED SUPPLIERS
TYCO, CTS KNIGHT, MOTOROLA CRYSTAL PRODUCTS
** FOR DIRECT CONNECT TO PHONE LINE, SYSTEM MUST BE APPROVED BY FCC
T_1 = MICROTRAN T1104 OR EQUIVALENT
P_1 = ADJUSTED FOR 50% SQUARE WAVE AT Rx CAR INPUT AT MINIMUM RECEIVED CARRIER

Complete 300 Baud, Answer mode, FSK Modem (E).

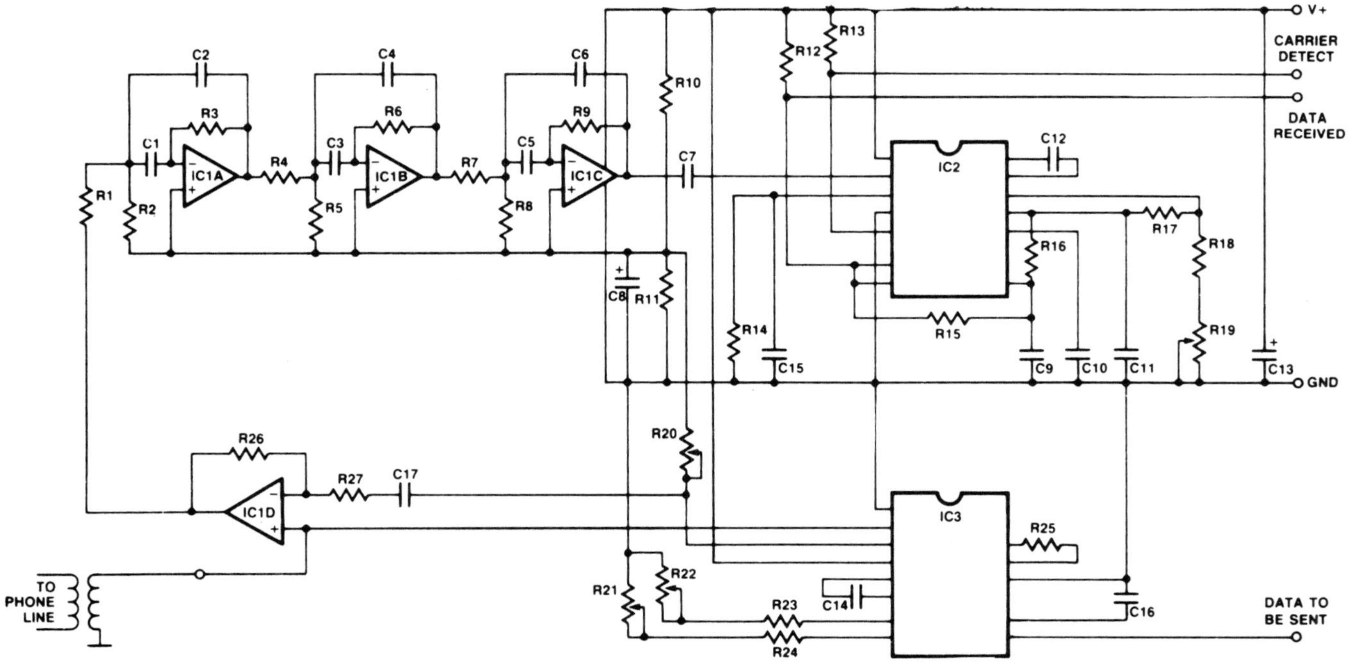

Complete FSK Modem Using XR-2211 and XR-2206 (E).

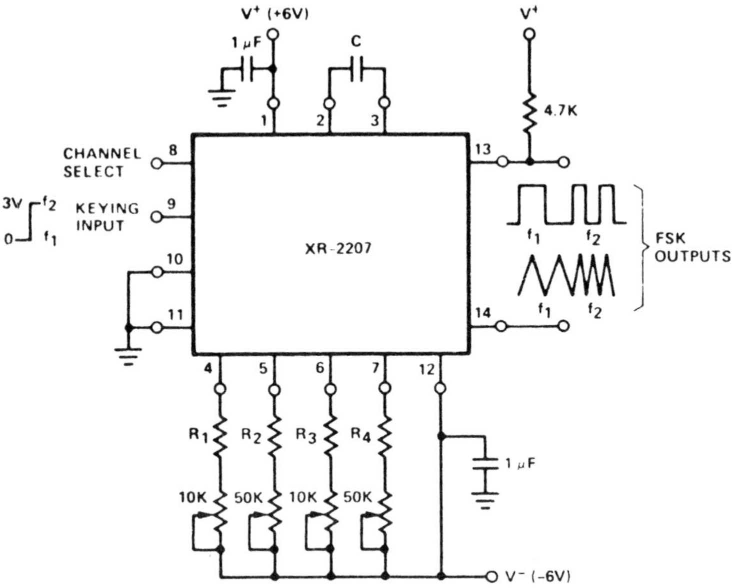

Multi-channel FSK generation (E).

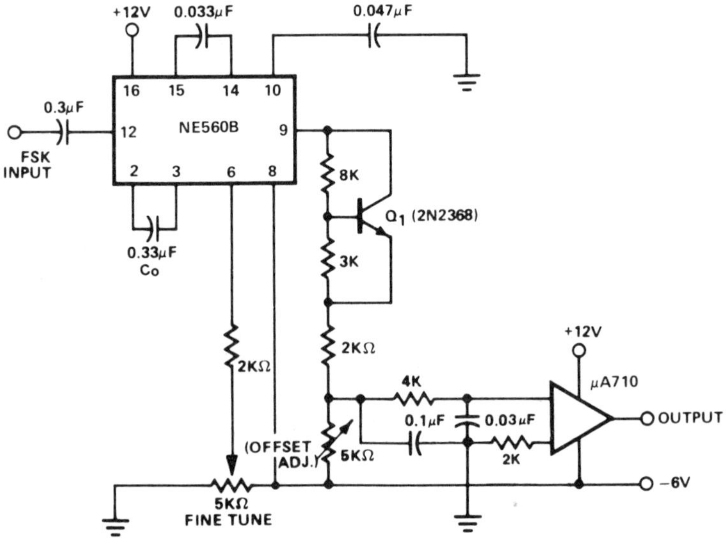

FSK demodulator using the 560 (S).

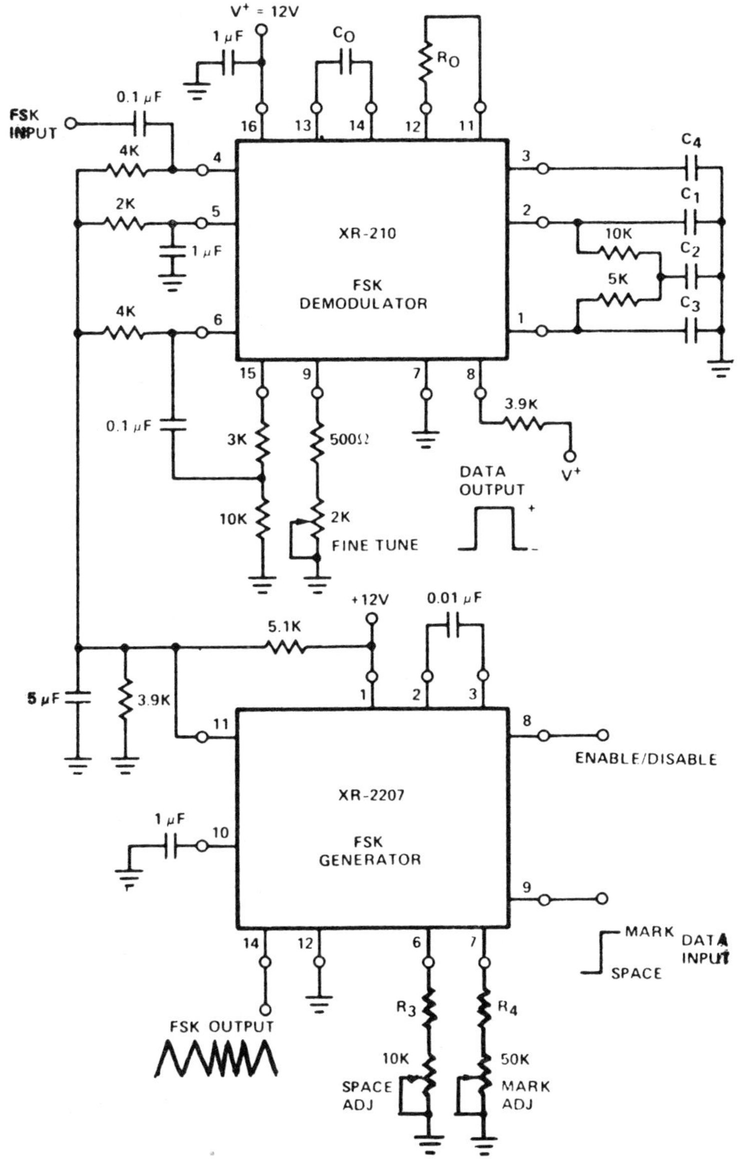

Full duplex FSK modem using XR-210 and XR-2207 (See Table component values on page 374).

OPERATING CONDITIONS	TYPICAL COMPONENT	VALUES
300 Baud	XR-210	XR-2207
Low Band: f_1 = 1070 Hz, f_2 = 1270 Hz	R_0 = 5.1 kΩ, C_0 = 0.22 μF; $C_1 = C_2$ = 0.047 μF, C_3 = 0.033 μF	R_3 = 91k; R_4 = 470k
High Band: f_1 = 2025 Hz, f_2 = 2225 Hz	R_0 = 8.2 kΩ, C_0 = 0.1 μF; $C_1 = C_2 = C_3$ = 0.033 μF	R_3 = 47k; R_4 = 470k
1200 Baud: f_1 = 1200 Hz, f_2 = 2200 Hz	R_0 = 2 kΩ, C_0 = 0.14 μF; C_1 = 0.033 μF, C_3 = 0.02 μF; C_2 = 0.01 μF	R_3 = 75k; R_4 = 91k

Component values for full duplex FSK modem on page 373.

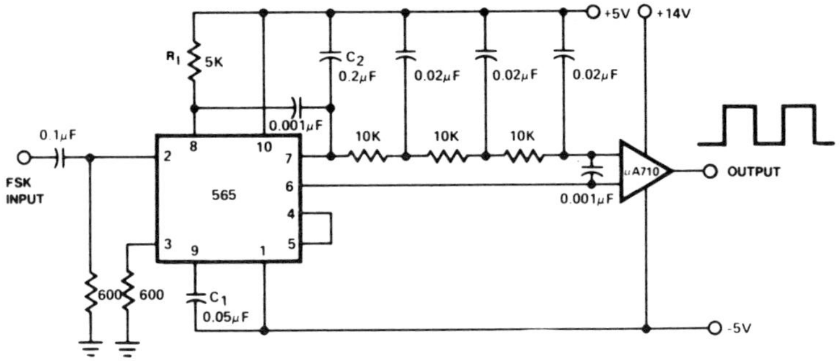

FSK decoder using the 565 (S).

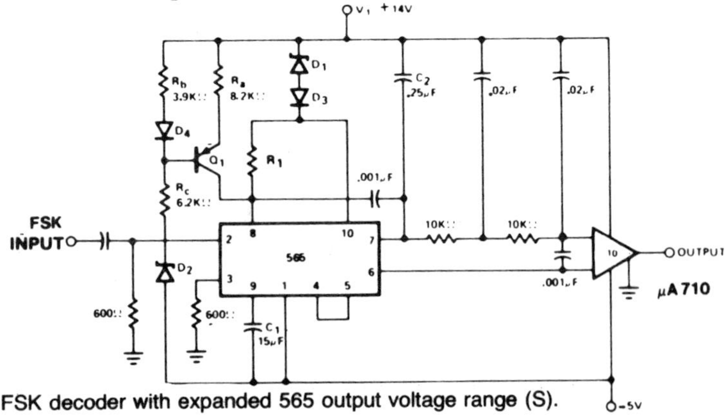

FSK decoder with expanded 565 output voltage range (S).

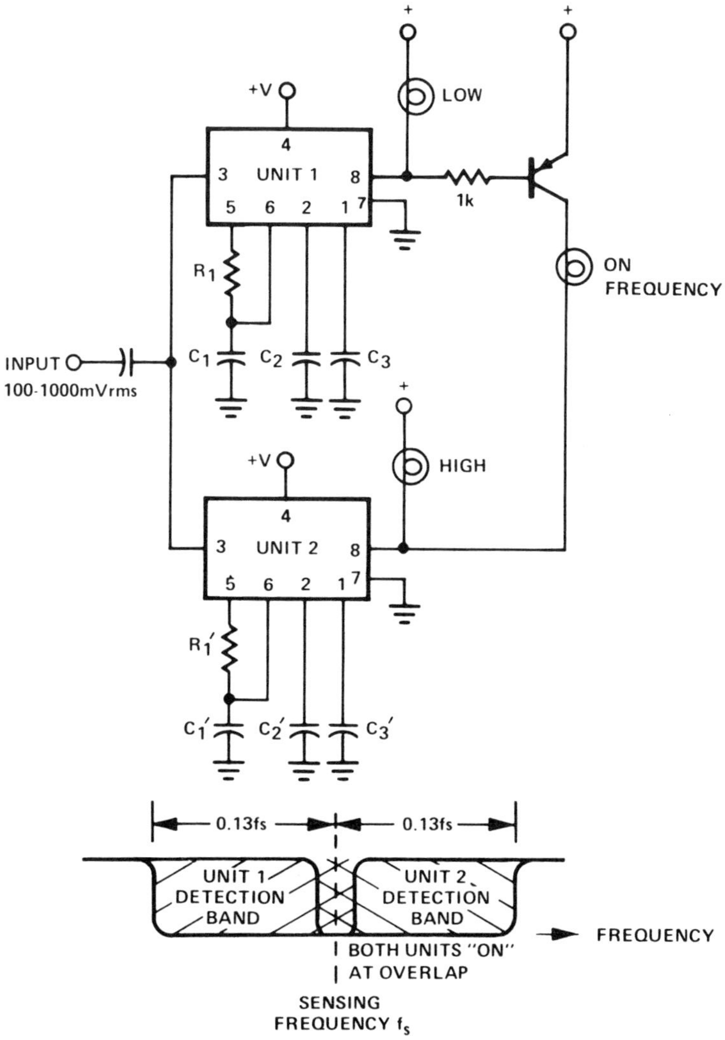

Frequency detector with low-cost, lamp readout (S).

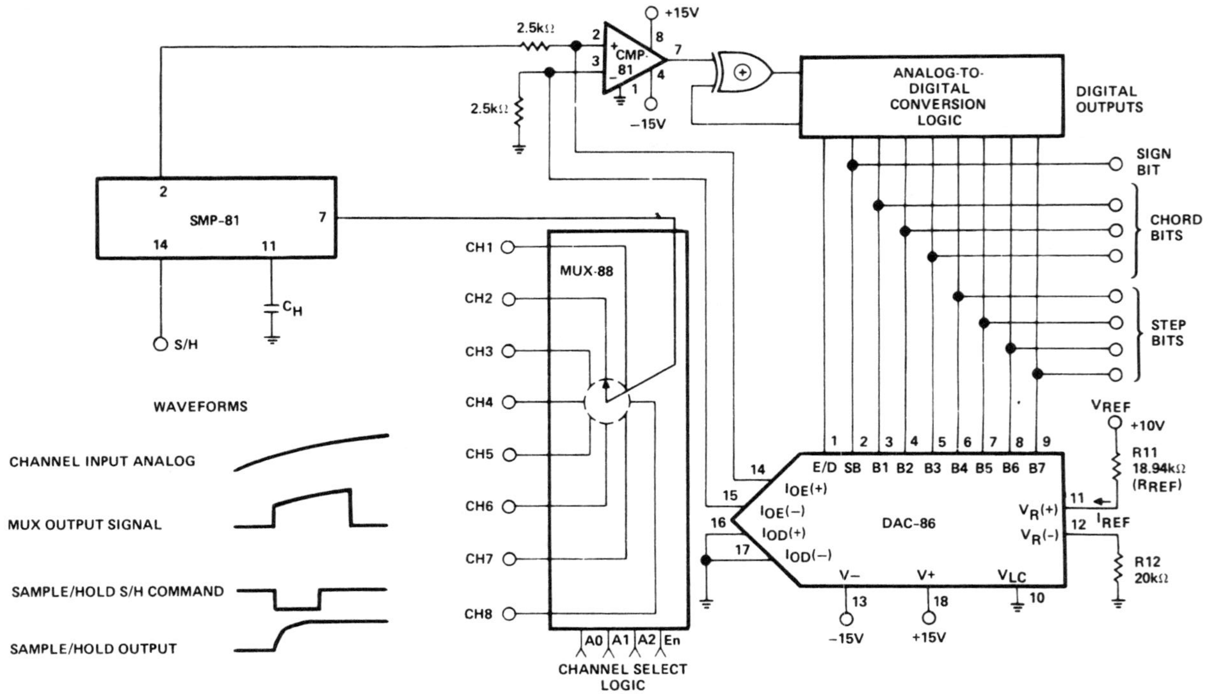

Eight-channel shared codec PCM encoder (PM).

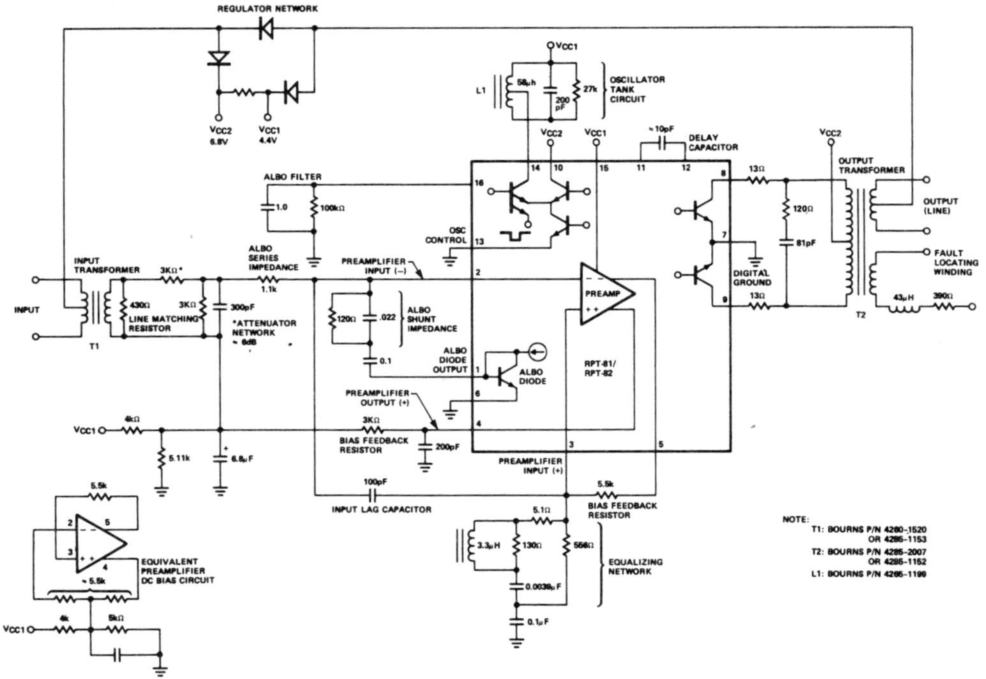

RPT-81/RPT-82 in typical 1.544MHz T1 repeater system (PM).

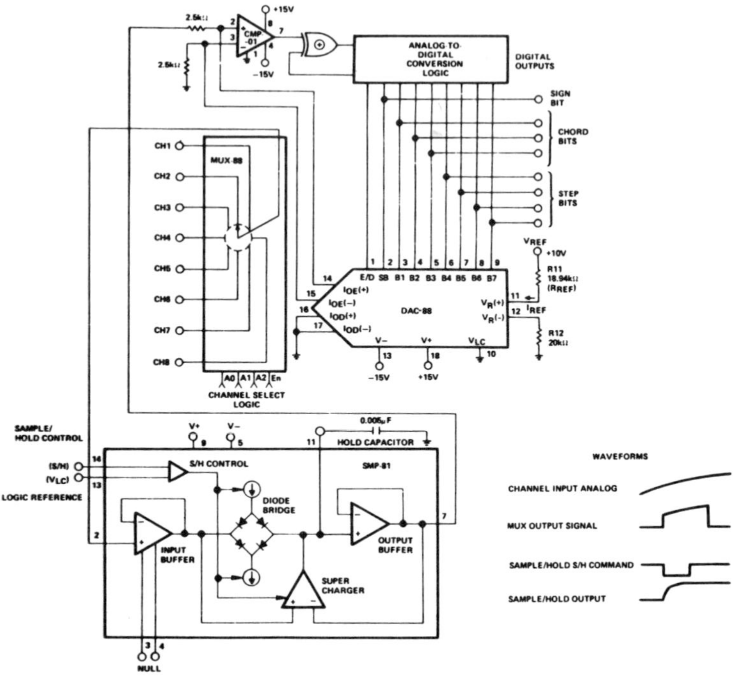

Eight channel shared codec PCM encoder (PM).

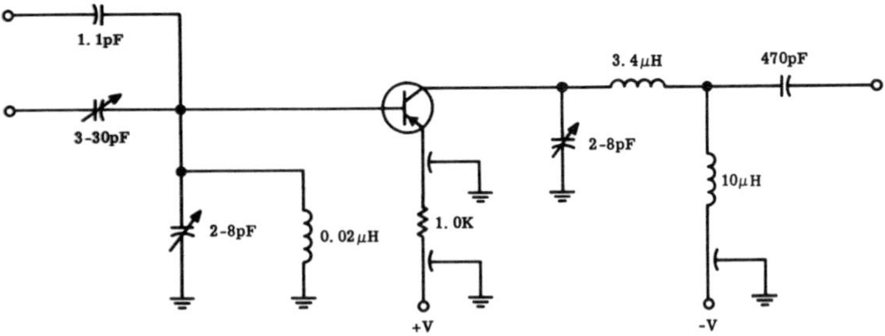

2N3308 mixer 300, 250 MHz→50 MHz (M).

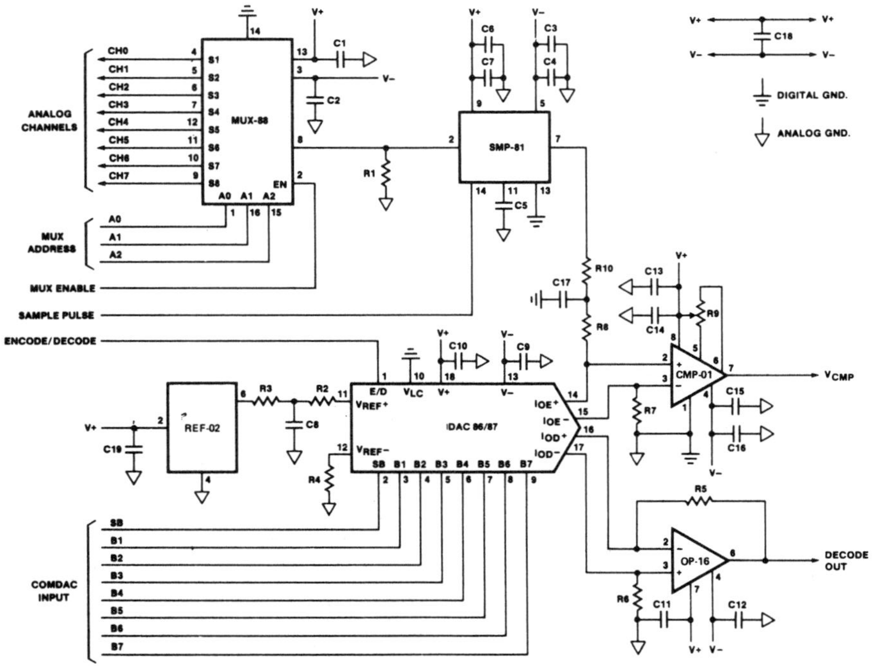

*COMDAC is a registered trademark.

COMDAC* encoder/decoder analog circuit board (PM).

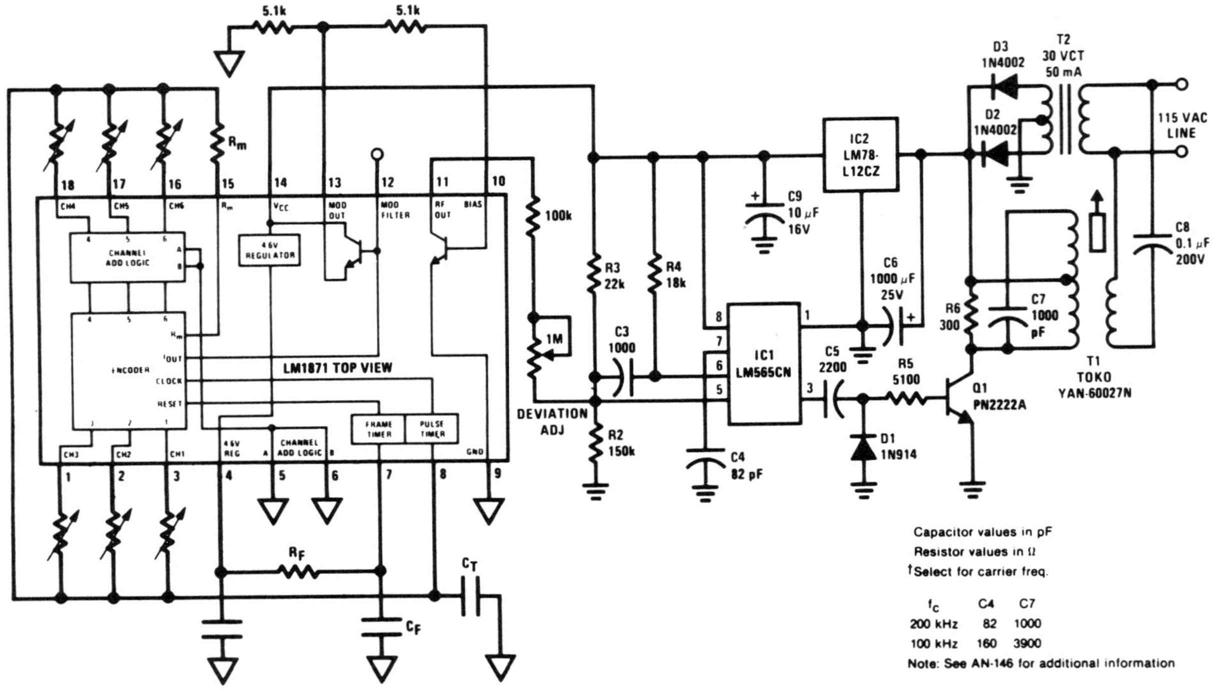

Capacitor values in pF
Resistor values in Ω
†Select for carrier freq.

f_c	C4	C7
200 kHz	82	1000
100 kHz	160	3900

Note: See AN-146 for additional information

LM1871, LM566 200 kHz line carrier transmitter with FSK carrier modulation (NS).

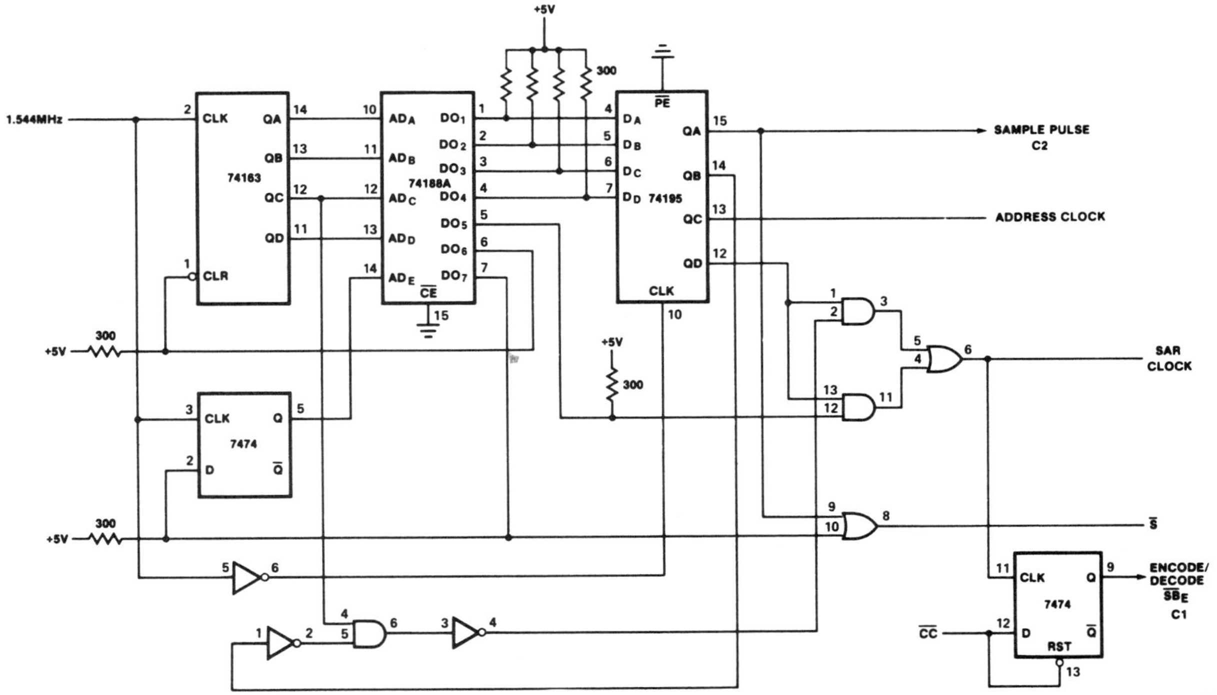

Encode clock (PM).

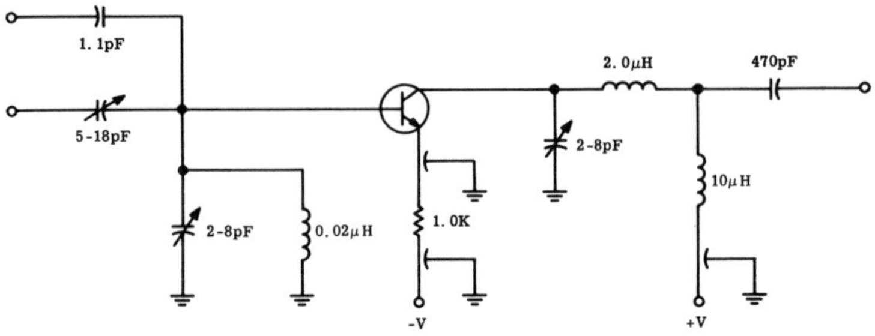

MM1941 mixer 300, 250 MHz→50 MHz (M).

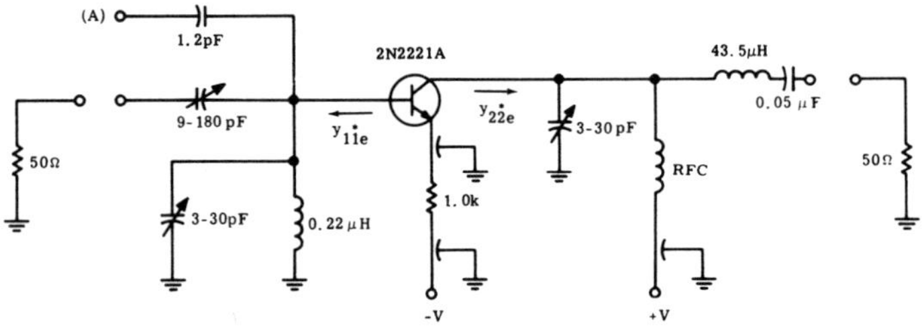

A 30 MHz rf to 5 MHz IF mixer (M).

Chapter 17

Gadgets and Useful Devices

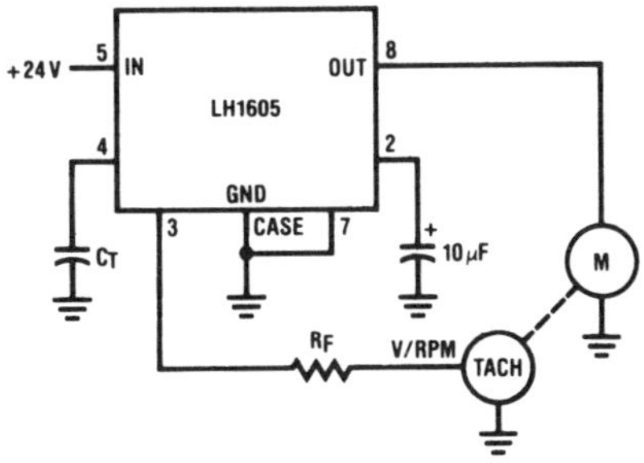

Motor speed regulation system (NS).

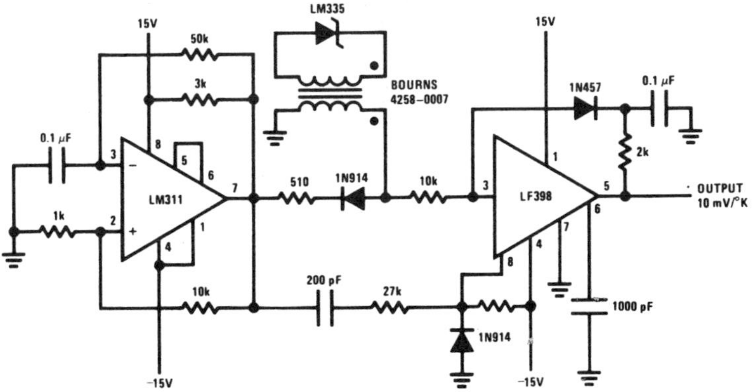

Isolated temperature sensor (NS).

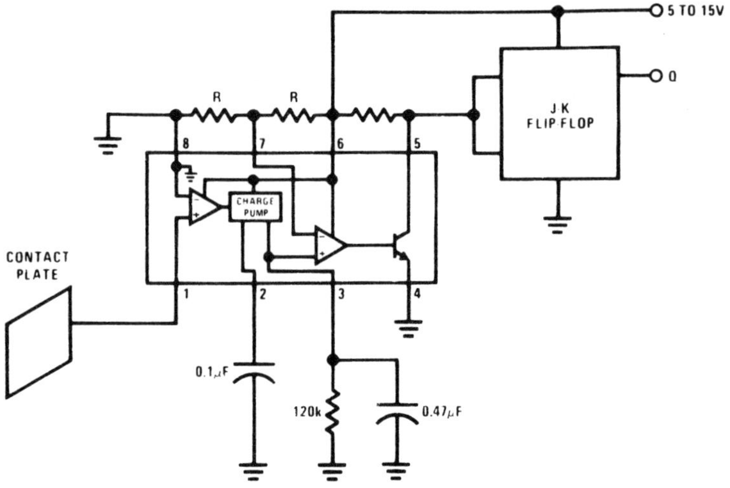

Finger touch or contact switch (NS).

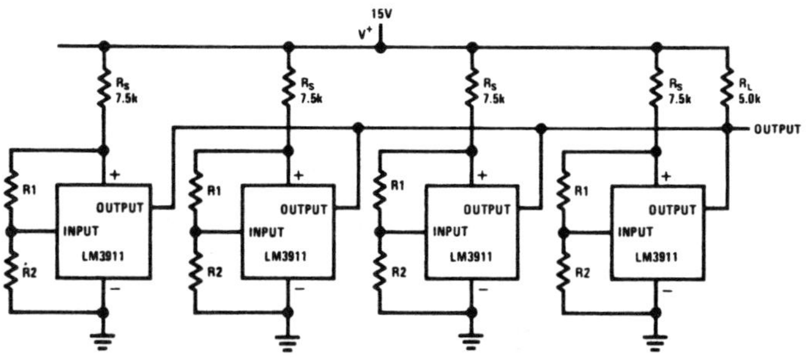

$$\text{Trip Point} = V_Z \frac{R1}{R1 + R2}$$

$$R_S = \frac{(V^+ - 6.8V)}{0.001A + \frac{6.8V}{R1 + R2}}$$

Over temperature detectors with common output (NS).

$$f_{SWITCH} = \frac{1}{1.4\ R1C1} = 1.9\ Hz$$

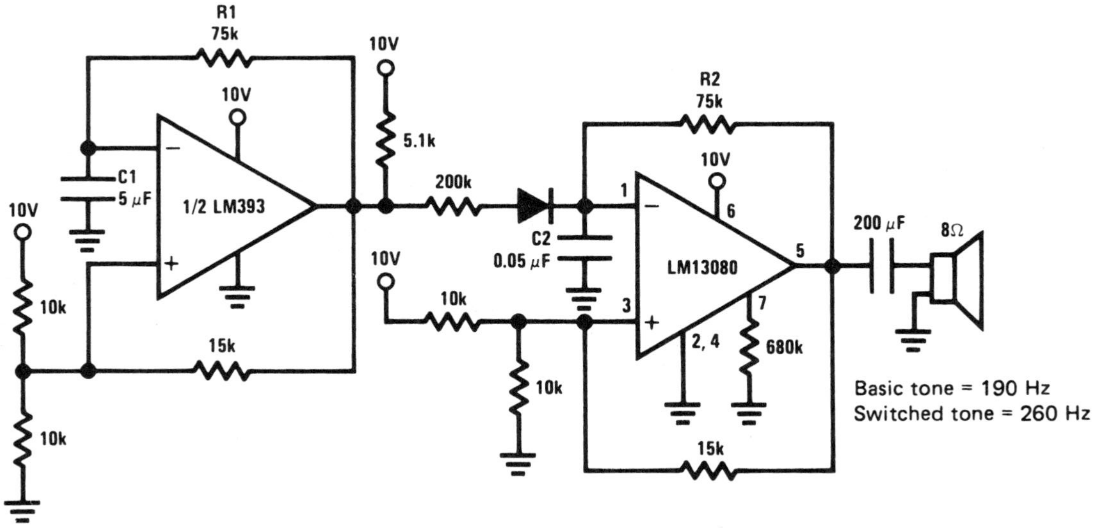

2-tone siren (NS).

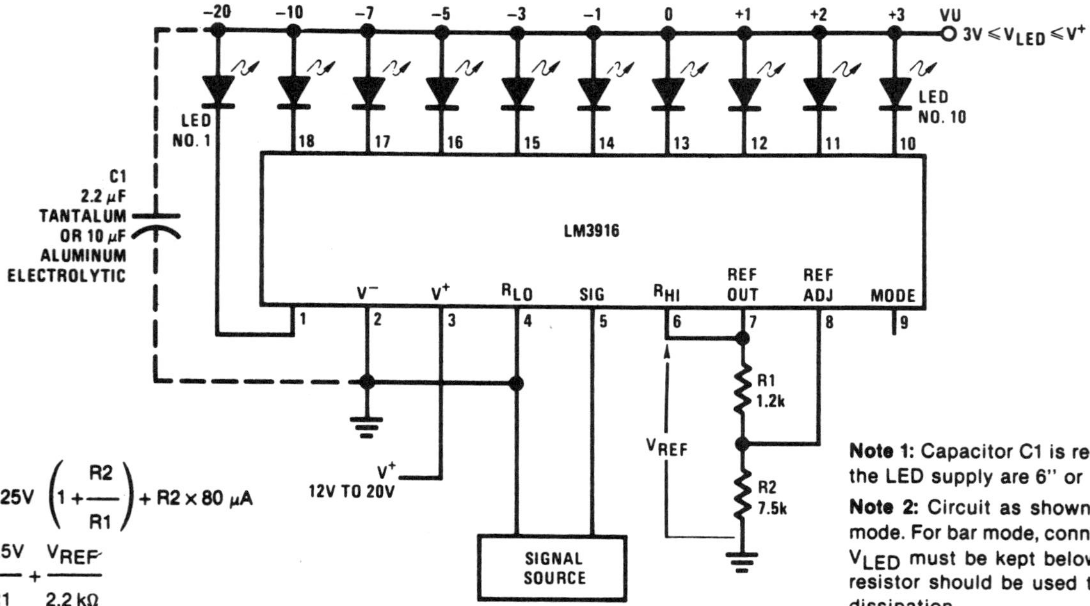

$$V_{REF} = 1.25V\left(1 + \frac{R2}{R1}\right) + R2 \times 80\ \mu A$$

$$I_{LED} = \frac{12.5V}{R1} + \frac{V_{REF}}{2.2\ k\Omega}$$

Note 1: Capacitor C1 is required if leads to the LED supply are 6" or longer.

Note 2: Circuit as shown is wired for dot mode. For bar mode, connect pin 9 to pin 3. V_{LED} must be kept below 7V or dropping resistor should be used to limit IC power dissipation.

0V to 10V VU meter (NS).

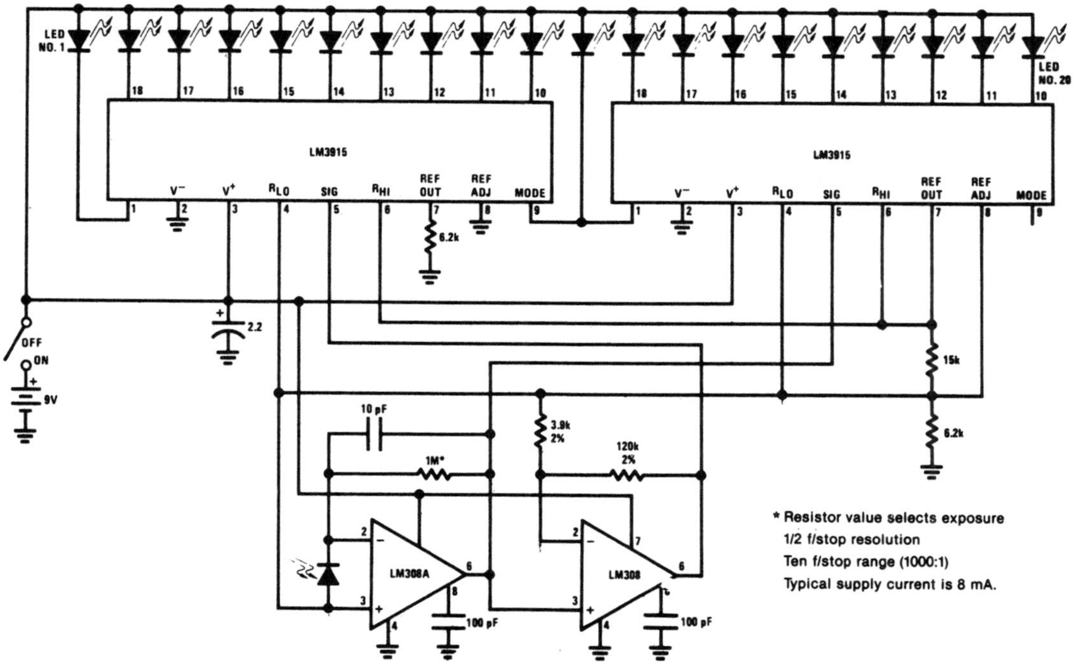

Light meter (NS).

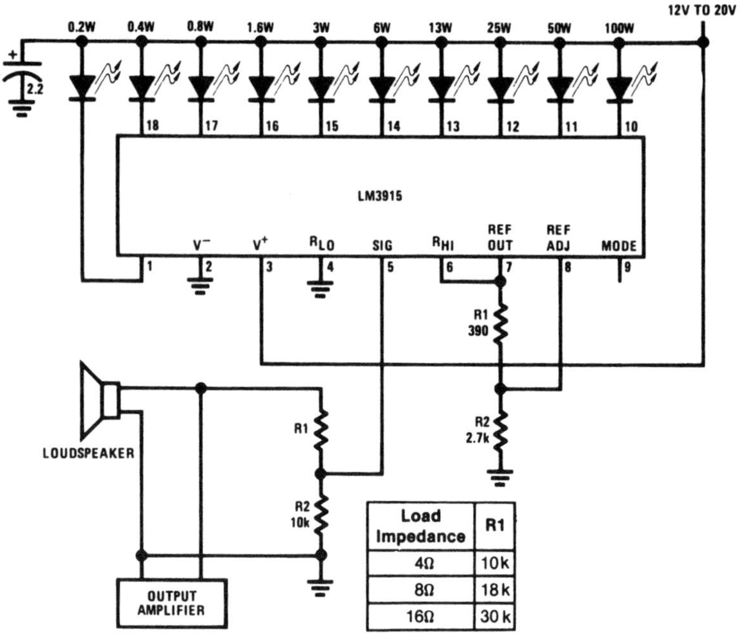

Load Impedance	R1
4Ω	10k
8Ω	18k
16Ω	30k

See Application Hints for optional Peak or Average Detector

Audio power meter (NS).

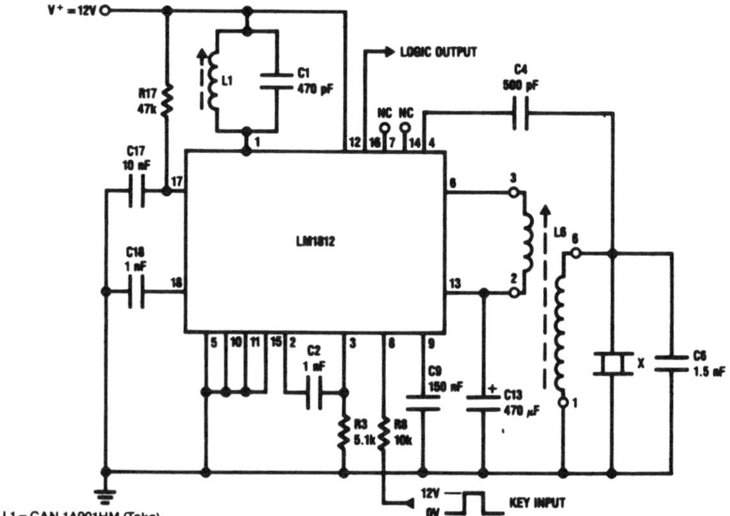

L1 = CAN-1A901HM (Toko)
L6 = 719VXA-A018YSU (Toko)
X = R283E (Massa Products)

200 kHz ultrasonic ranging system for 4 inches to 6 feet in air (NS).

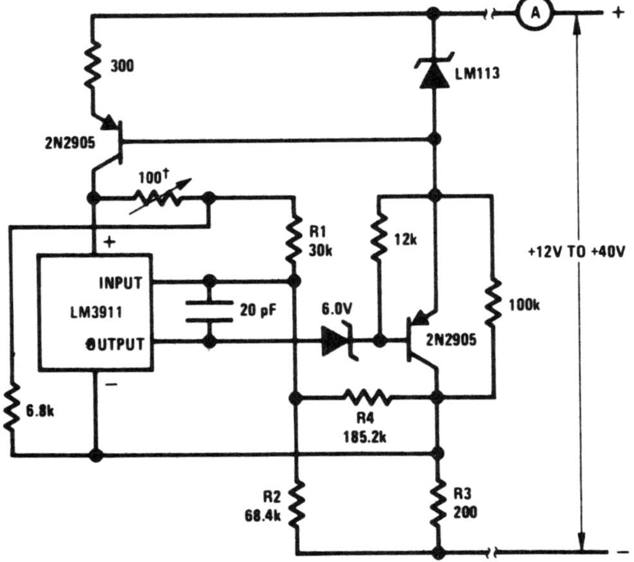

$$R2\,(\Omega) = \frac{\left(V_Z - 0.01\,T_L\right)\left(I_H - \frac{0.01\,T_H}{R1}\right) + \left(V_Z - 0.01\,T_H\right)\left(\frac{0.01\,T_L}{R1} - I_L\right)}{\frac{0.01}{R1\,R3}\left[T_H\,(V_Z - 0.01\,T_L) - T_L\,(V_Z - 0.01\,T_H)\right]}^{**}$$

$$R3\,(\Omega) \geqq \frac{V_Z\left(\frac{T_H}{T_L} - 1\right)}{I_H - \frac{I_L\,T_H}{T_L}}$$

$$\frac{1}{R4} = \frac{1}{(V_Z - 0.01\,T_L)(R2)}\left[\frac{(R2)(0.01\,T_L)}{R1} + \frac{\left(\frac{V_Z - 0.01\,T_L}{R2} - I_L\right)}{\frac{1}{R2} + \frac{1}{R3}}\right] - \frac{1}{R2}$$

T_L = Temperature for I_L (K)
T_H = Temperature for I_H (K)
V_Z = Zener voltage (V)
I_L = Low temperature output current (A)
I_H = High temperature output current (A)

*Values shown for I_{OUT} = 1 mA to 10 mA for 10°F to 100°F
†Set temperature

** The 0.01 in the above and following equations is in units of V/ X or V/ C, and is a result of the basic 0.01V/ K sensitivity of the transducer

Two terminal temperature to current transducer* (NS).

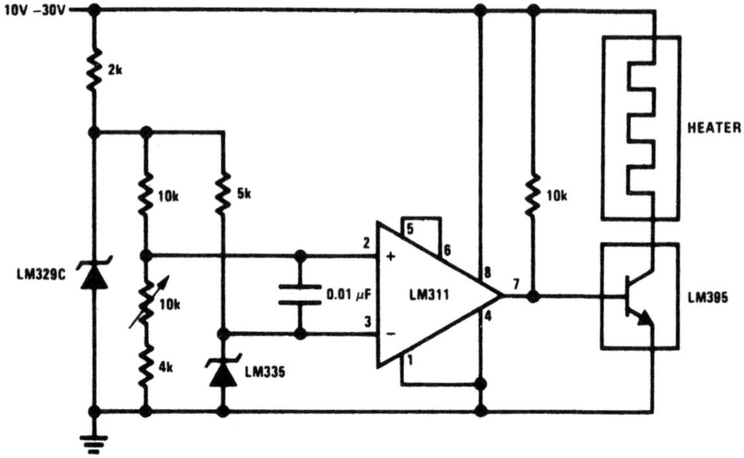

Simple temperature controller (NS).

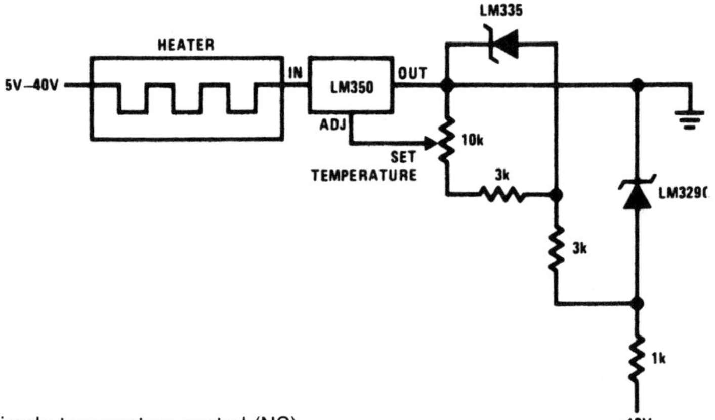

Simple temperature control (NS).

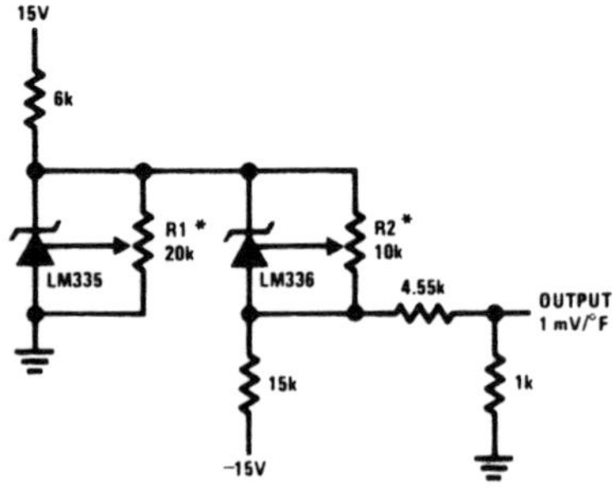

*Adjust R2 for 2.554V across LM336.
Adjust R1 for correct output.

Ground referred fahrenheit thermometer (NS).

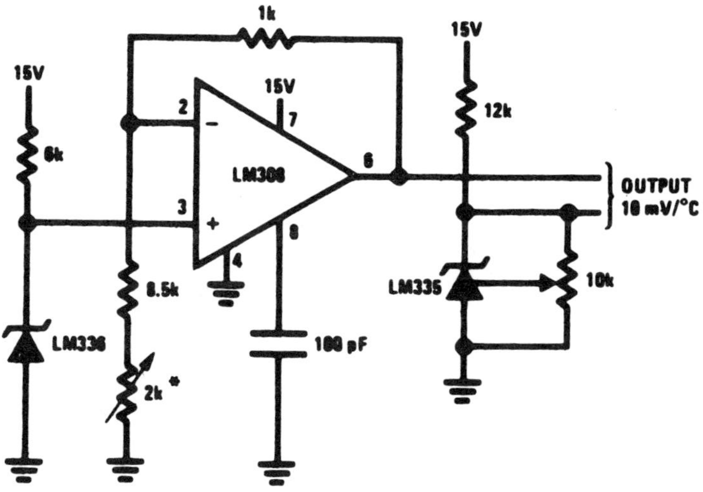

* Adjust for 2.73V at output of LM308

Centigrade thermometer (NS).

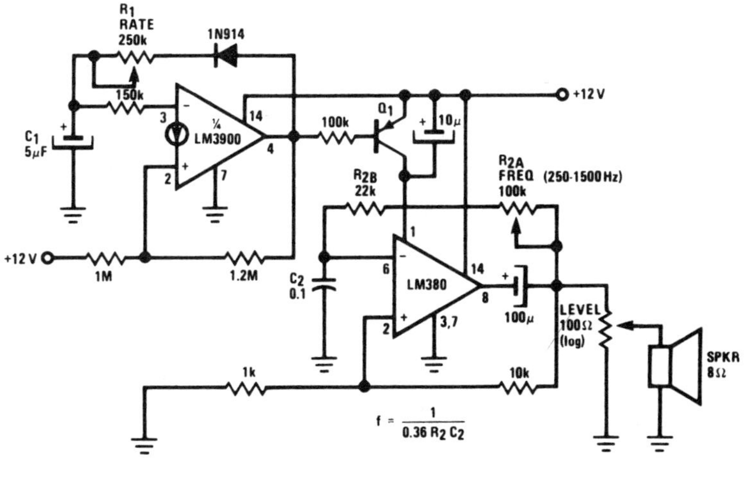

Siren with programmable frequency and rate adjustment (NS).

*To calibrate adjust R2 for 2.554V across LM336.
Adjust R1 for correct output.

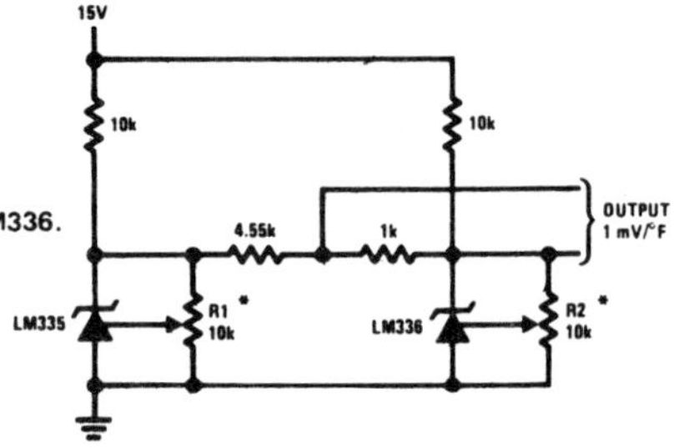

Fahrenheit thermometer (NS).

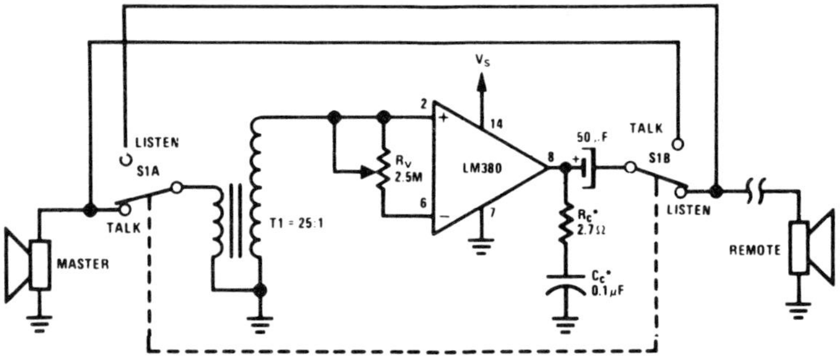

Intercom (NS).

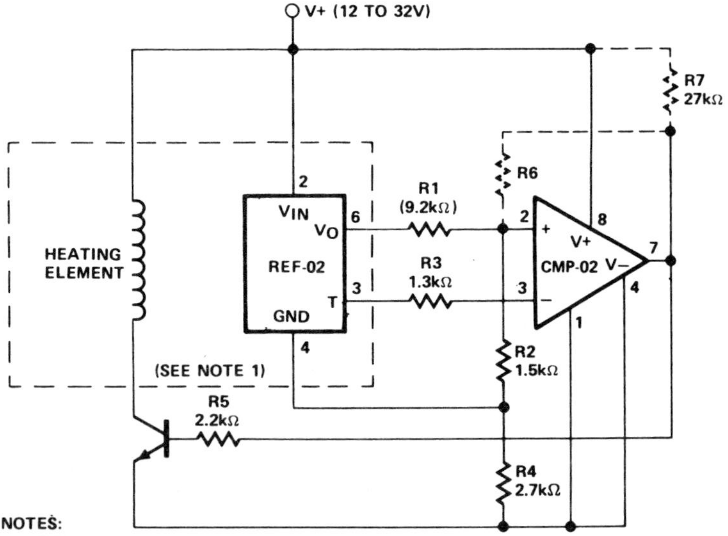

NOTES:

1. REF-02 SHOULD BE THERMALLY CONNECTED TO SUBSTANCE BEING HEATED.
2. NUMBERS IN PARENTHESES ARE FOR A SETPOINT TEMPERATURE OF 60°C.
3. R3 = R1||R2||R6

Temperature controller (PM).

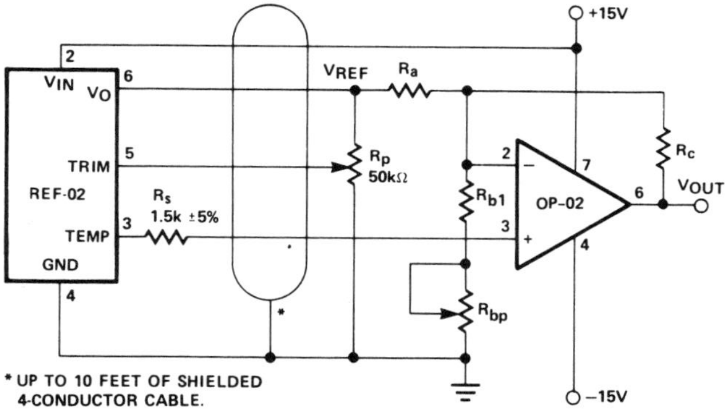

* UP TO 10 FEET OF SHIELDED 4-CONDUCTOR CABLE.

FOR THEORY OF OPERATION AND CALIBRATION PROCEDURE CONSULT APPLICATION NOTE 18, "THERMOMETER APPLICATIONS OF THE REF-02".

RESISTOR VALUES

TCV_{OUT} SLOPE (S)	10mV/°C	100mV/°C	10mV/°F
TEMPERATURE RANGE	−55°C to +125°C	−55°C to +125°C	−67°F to +257°C
OUTPUT VOLTAGE RANGE	−0.55V to +1.25V	−5.5V to +12.5V*	−0.67V to +2.57V
ZERO SCALE	0V@0°C	0V@0°C	0V@0°F
R_a (±1% resistor)	9.09kΩ	15kΩ	7.5kΩ
R_{b1} (±1% resistor)	1.5kΩ	1.82kΩ	1.21kΩ
R_{bp} (Potentiometer)	200Ω	500Ω	200Ω
R_c (±1% resistor)	5.11kΩ	84.5kΩ	8.25kΩ

* For 125°C operation, the op amp output must be able to swing to +12.5V, increase V_{IN} to +18V from +15V if this is a problem.

Precision temperature transducer with remote sensor (PM).

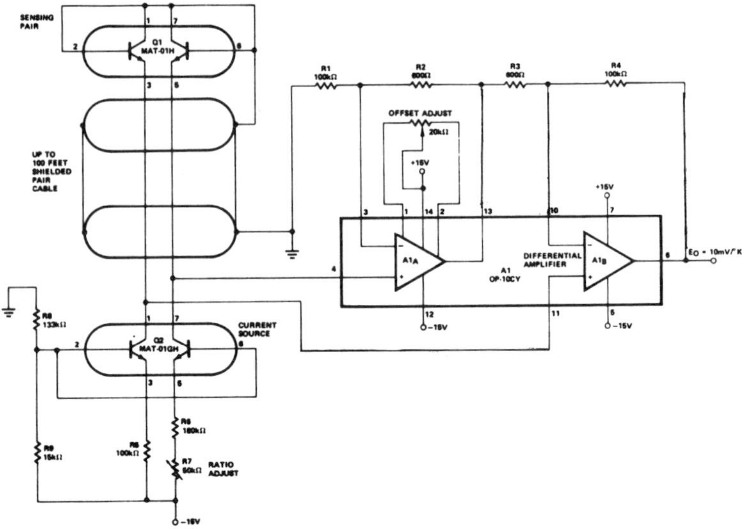

Complete schematic-temperature measurement system (PM).

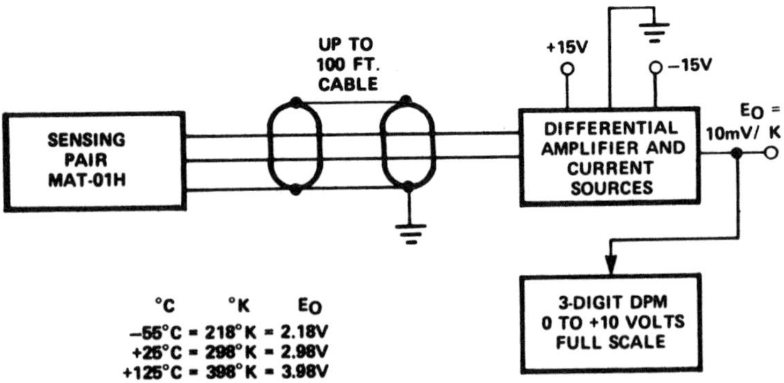

Basic digital thermometer readout in degrees Kelvin (°K) (PM).

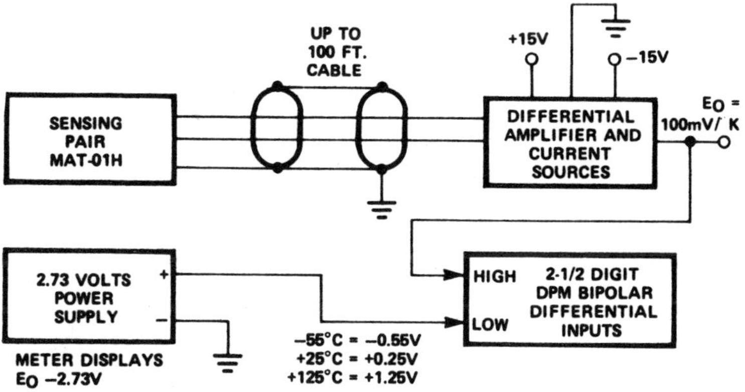

Digital thermometer with readout in °C (PM).

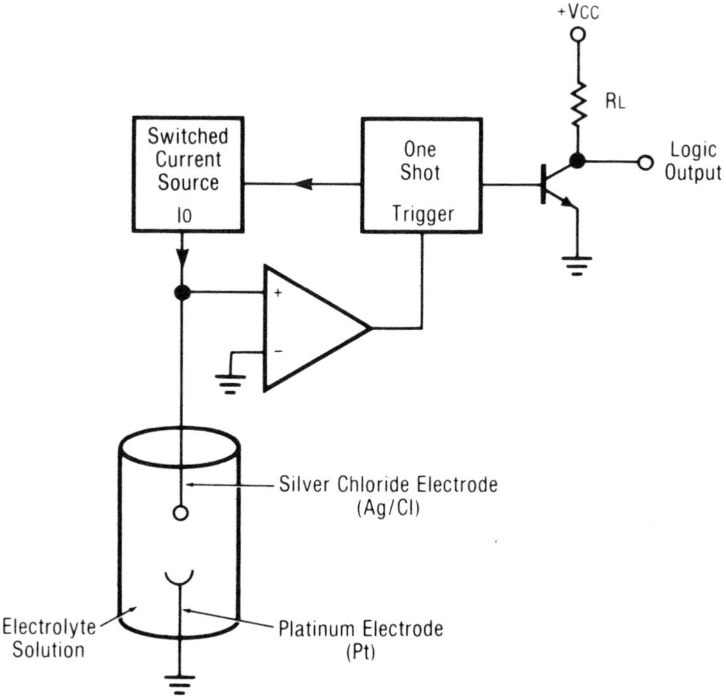

Chloride titrator (R).

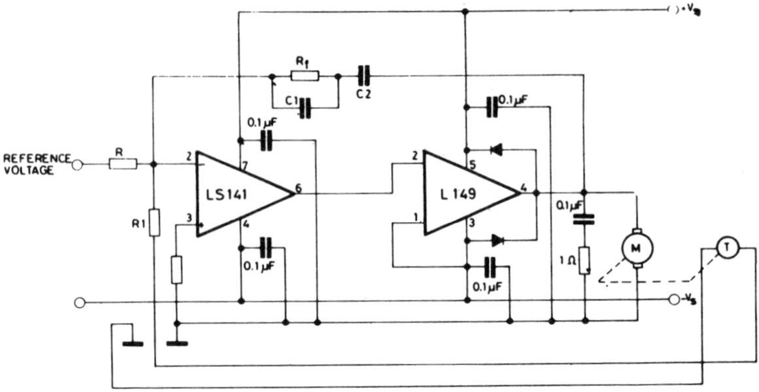

Bidirectional speed control of dc motor (SGS).

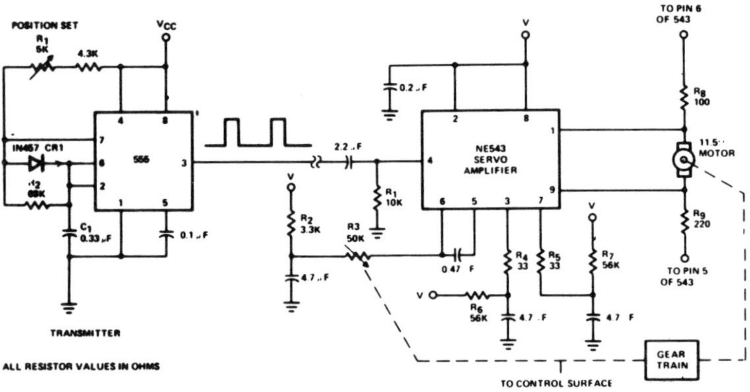

Servo system controller (S).

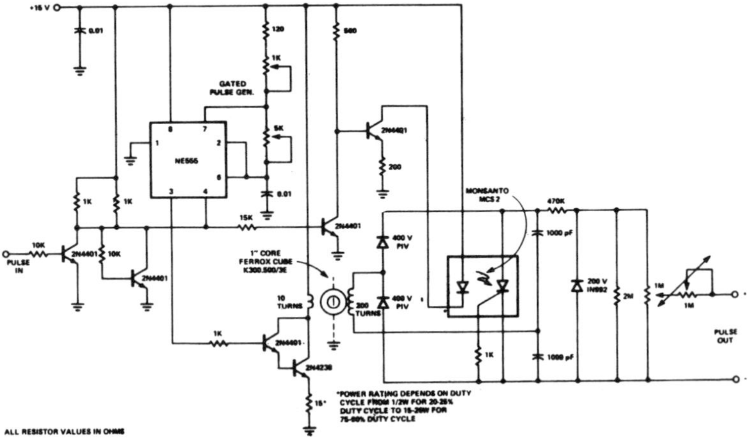

Stimulus isolator (S).

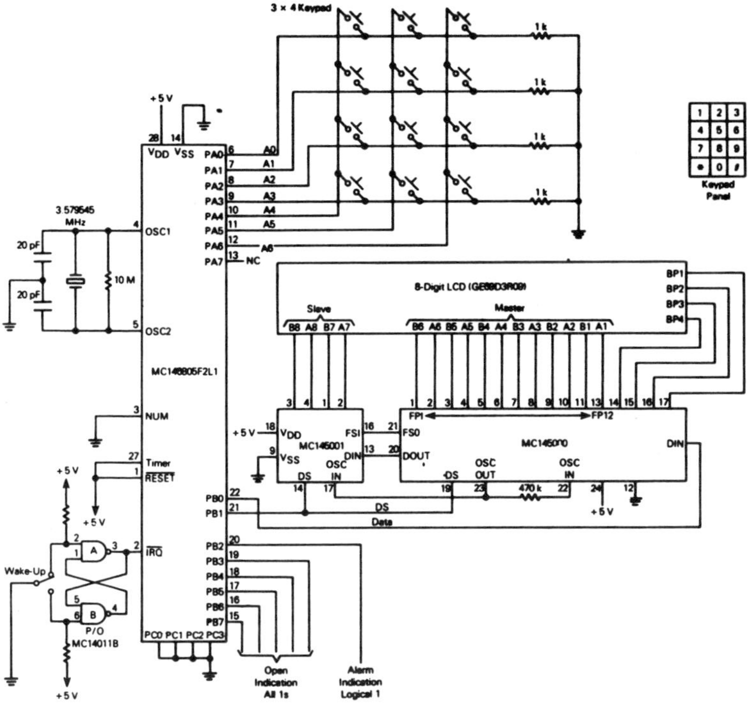

Digital lock system schematic diagram (M).

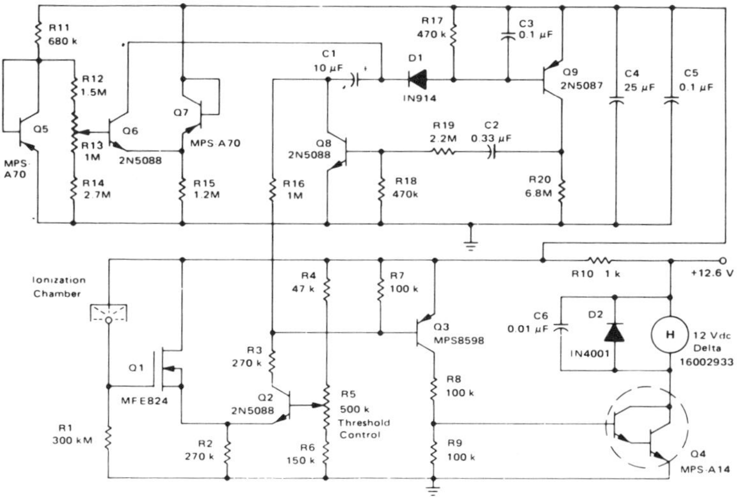

Ionization chamber smoke detector #2 - discrete version (M).

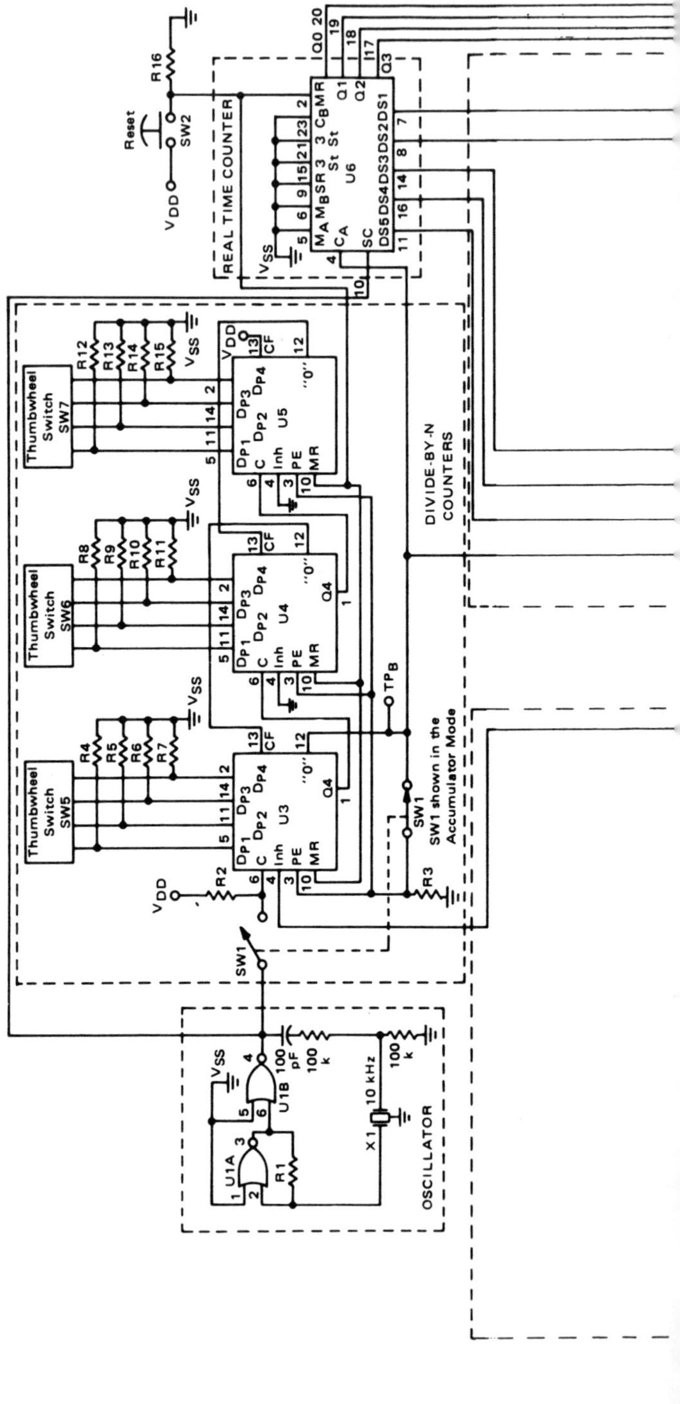
REAL TIME COUNTER
U6
Reset
SW2
R16
DIVIDE-BY-N COUNTERS
Thumbwheel Switch SW7
Thumbwheel Switch SW6
Thumbwheel Switch SW5
U5
U4
U3
SW1
SW1 shown in the Accumulator Mode
OSCILLATOR
U1A
U1B
R1
X1
10 kHz

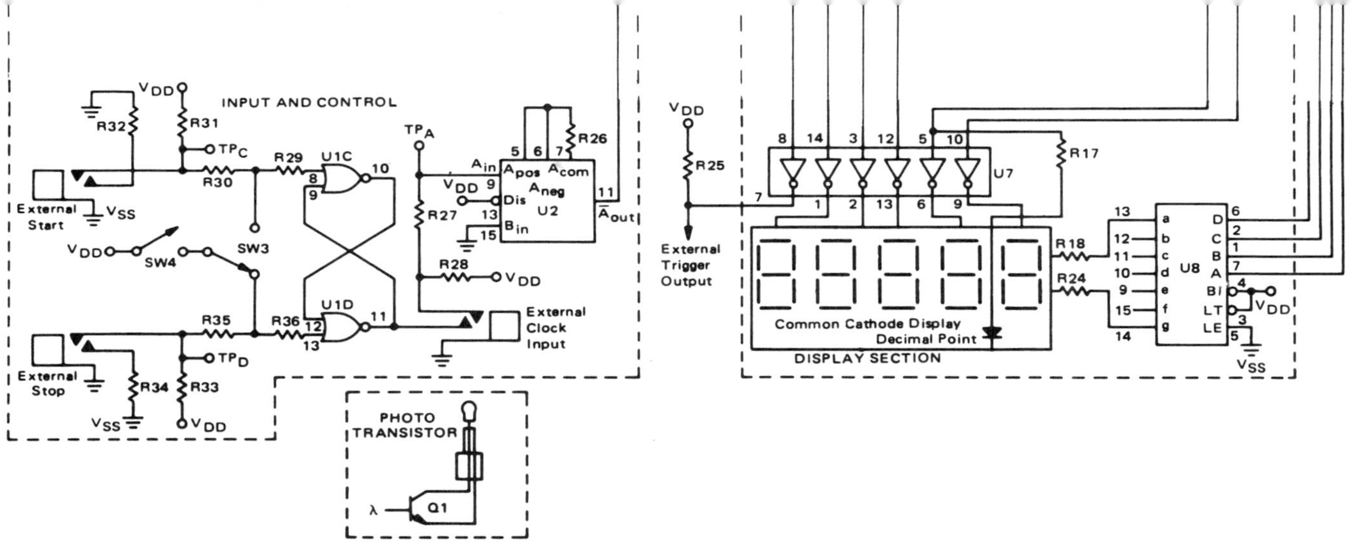

Parts List

U1	(1)	MC14001	SW1	(1)	DPDT Toggle Switch	R2-16,25,26	(17)	10 kΩ, ¼ W
U2	(1)	MC14583	SW2	(1)	SPST Pushbutton Switch	R17	(1)	470 Ω, ¼ W
U3-5	(3)	MC14522	SW3	(1)	SPDT Latching Push-button Switch	R18-24	(7)	390 Ω, ¼ W
U6	(1)	MC14534				R27	(1)	1 MΩ, ¼ W
U7	(1)	MC75492	SW4	(1)	SPST Toggle Switch	R28,31,33	(3)	10 kΩ, ¼ W
U8	(1)	MC14511	SW5-7	(3)	BCD Thumbwheel Switches	R29,30,35,36	(4)	470 kΩ, ¼ W
X1	(1)	STATEK Crystal, 10 kHz, SX1H10K	Q1	(1)	MRD3055	R32,34	(2)	560 Ω, ¼ W
			R1	(1)	5.1 MΩ, ¼W			

1 Display, HP5082-7432
1 Display, HP5082-7433
P.C. Board, DCE Circuits

Five digit accumulator/elapsed time indicator (M).

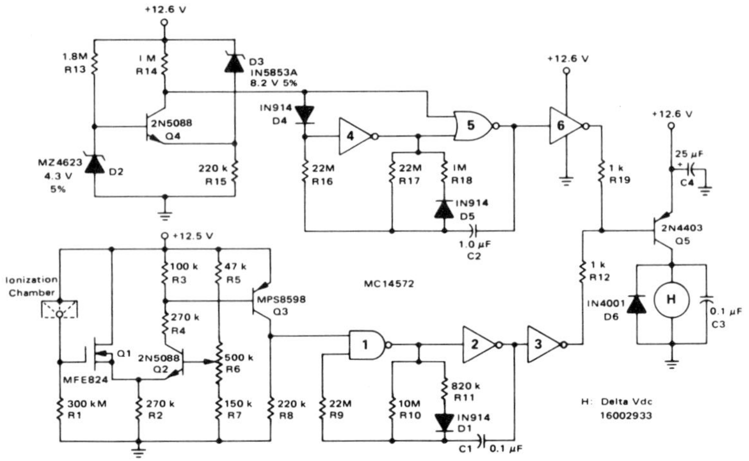

Ionization chamber smoke detector #1 - McMOS oscillators (M).

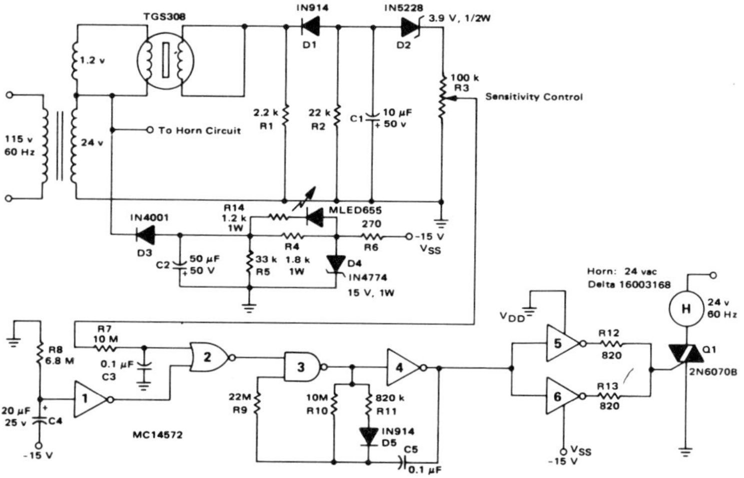

TGS gas/smoke detector, McMOS-triac full-wave control (M).

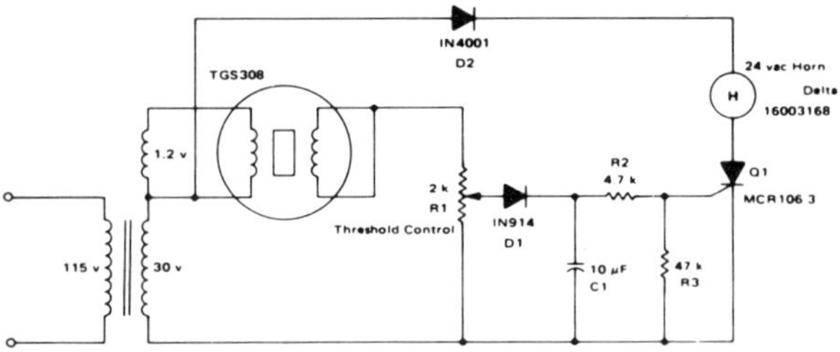

Simple SCR gas/smoke detector - half-wave control (M).

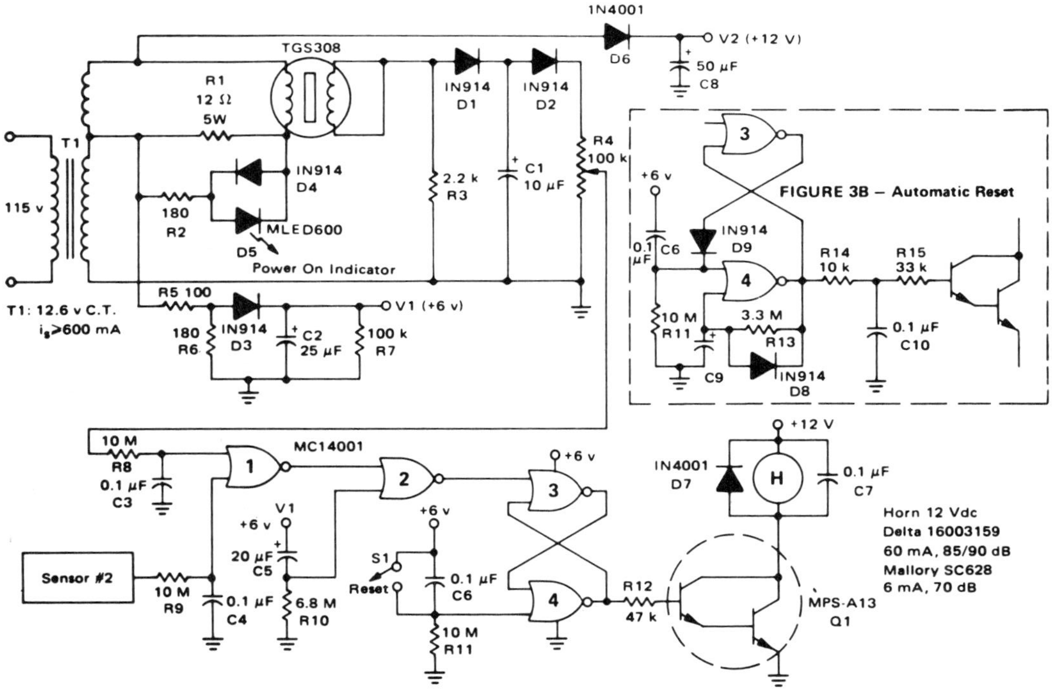

TGS gas/smoke detector - McMOS latch (M).

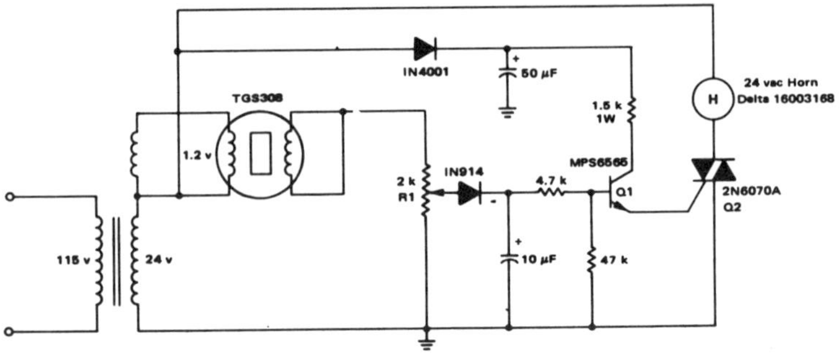

TGS gas/smoke detector—triac full-wave control (M).

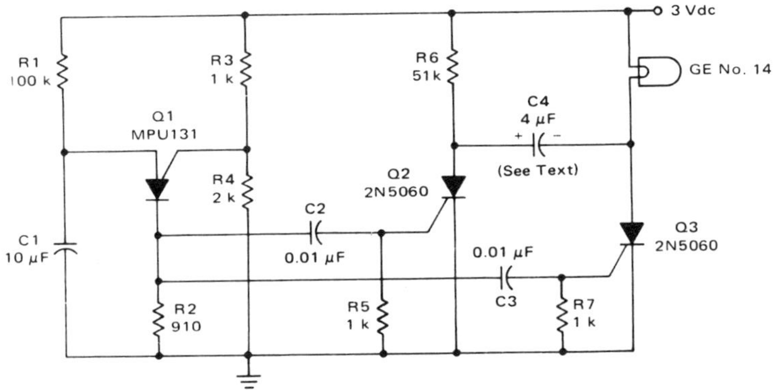

Low voltage lamp flasher (M).

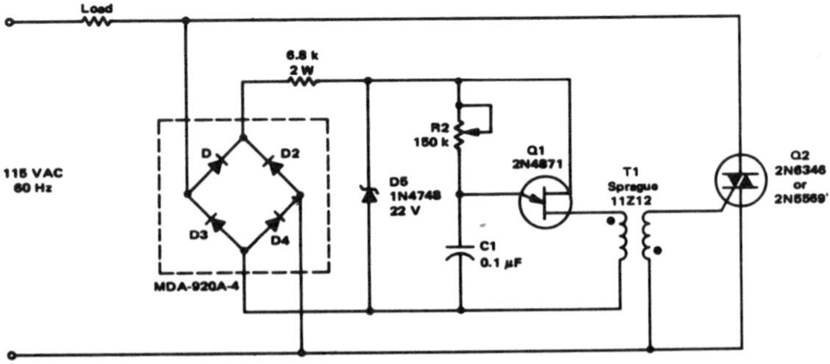

800 W triac light dimmer (M).

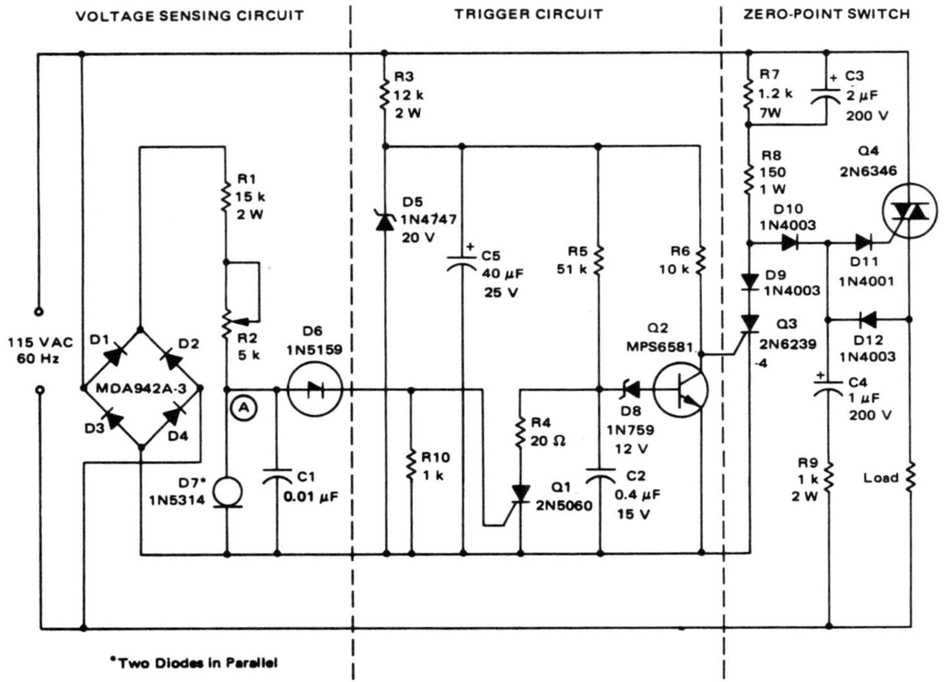

Overvoltage protection with automatic reset (M).

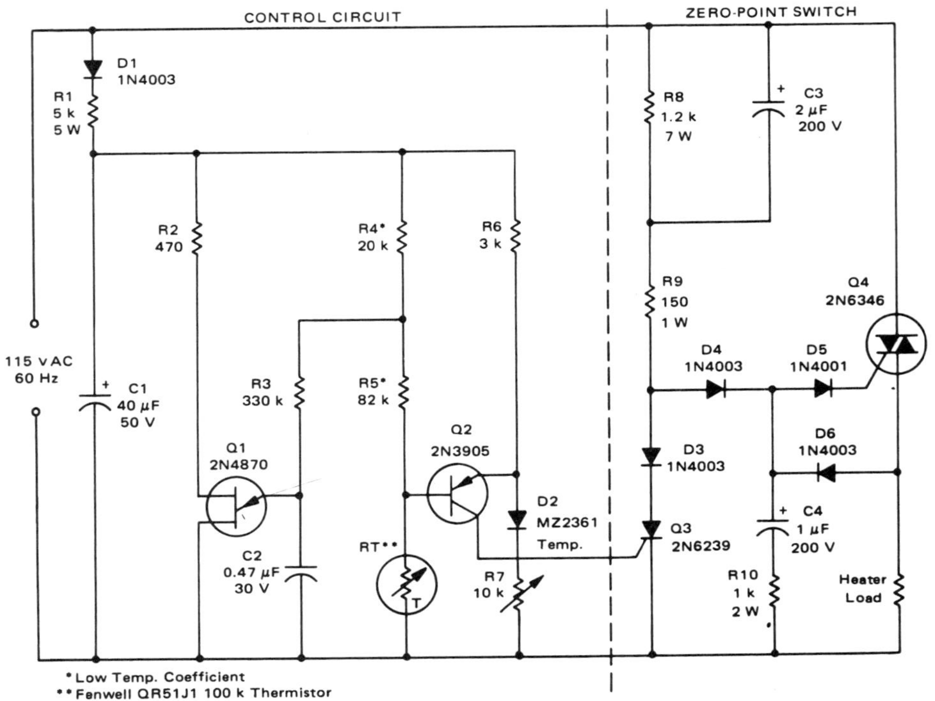

*Low Temp. Coefficient
**Fenwell QR51J1 100 k Thermistor

Temperature controller (M).

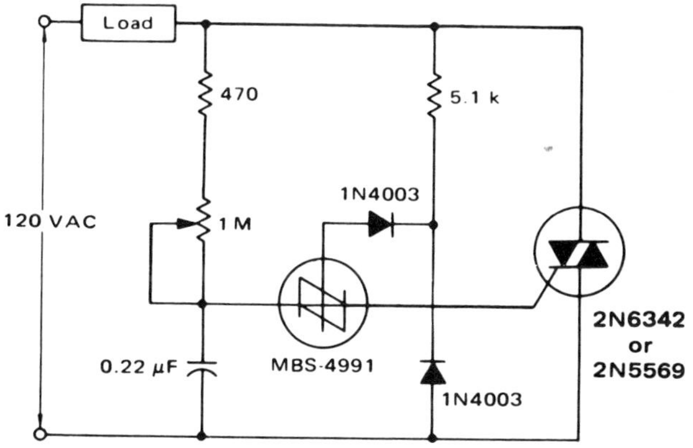

Hysteresis-free power controller (M).

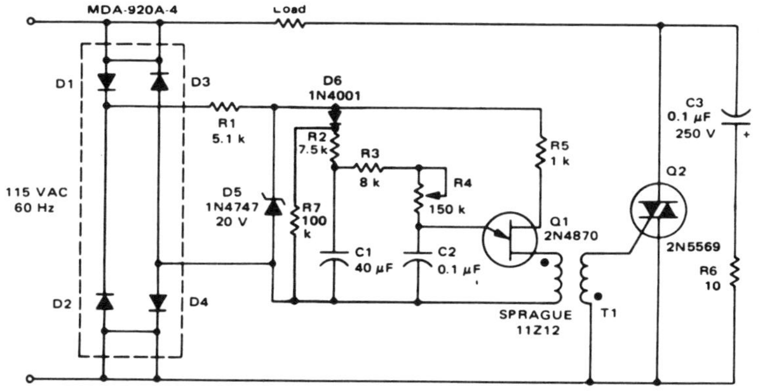

800 W soft-start light dimmer (M).

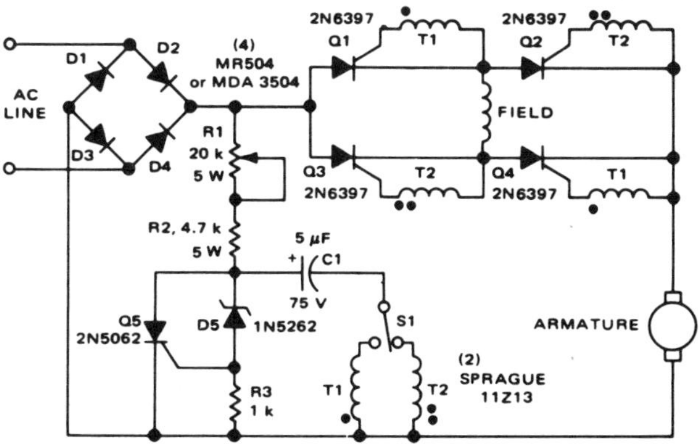

Direction and speed control for series-wound or universal motor (M).

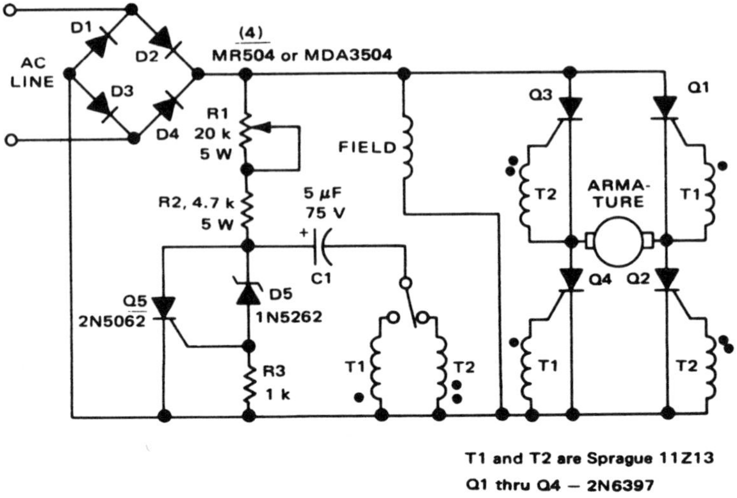

Direction and speed control for shunt-wound motor (M).

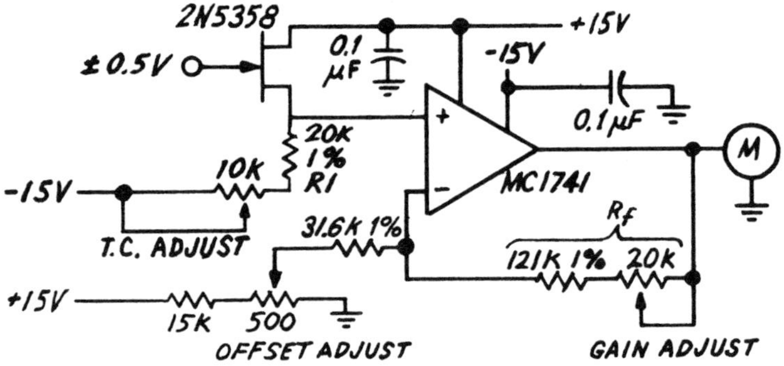

Signal conditioner (M).

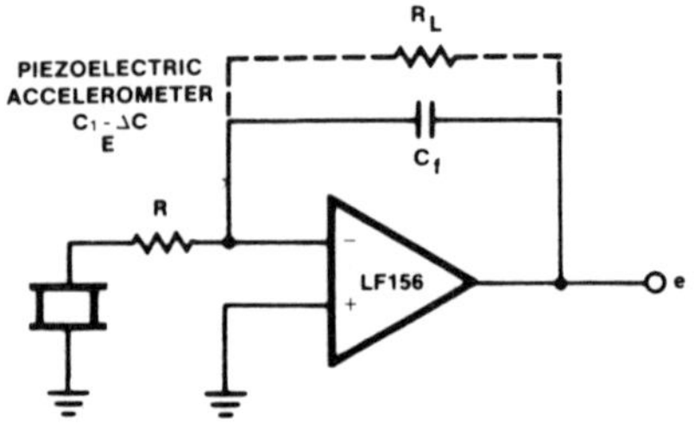

Piezoelectric charge amplifier (S).

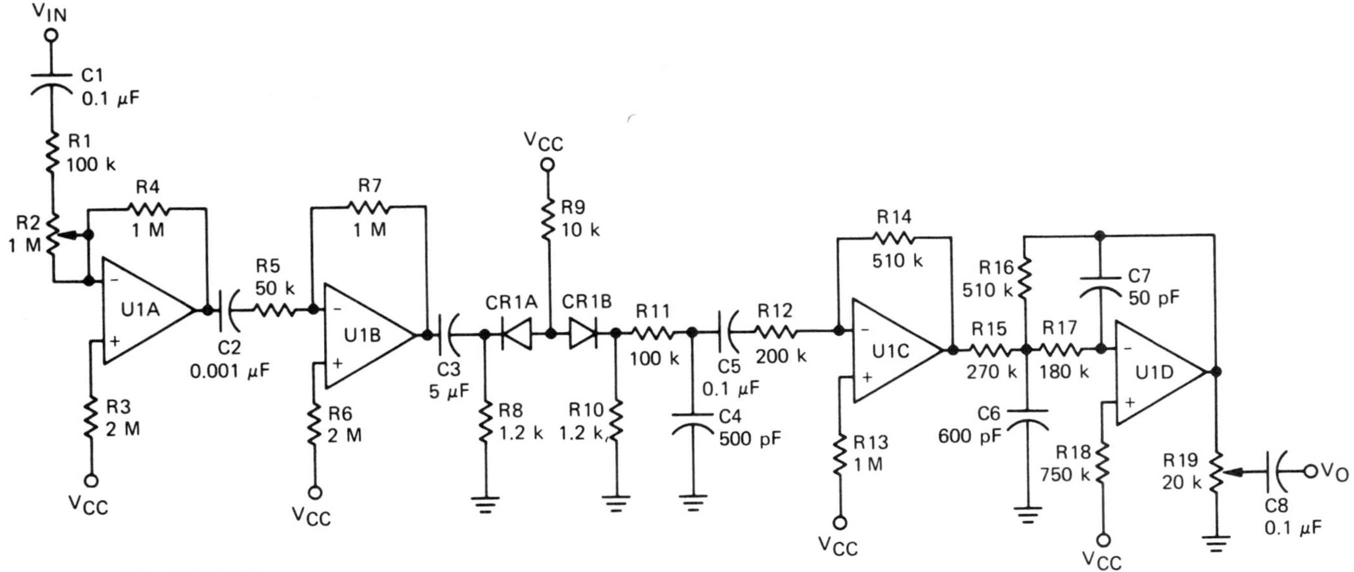

U1 = MC3401
CR1 = MSD6150
V_{CC} = 9 Vdc

Voice processor schematic diagram (M).

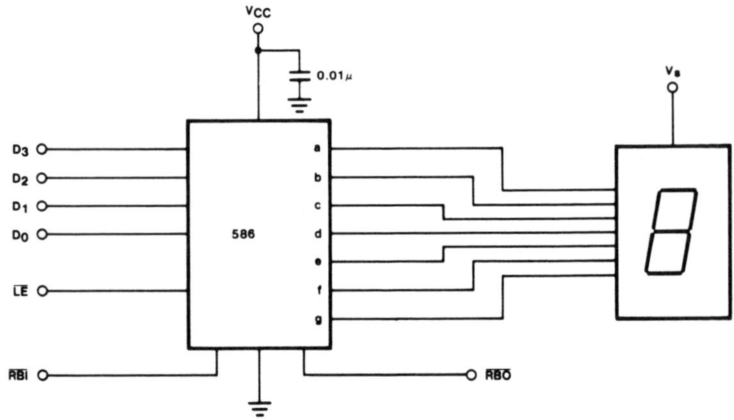

NOTE
Decoupling capacitor on V_{CC} should be 0.01μF ceramic

Driving a single digit (S).

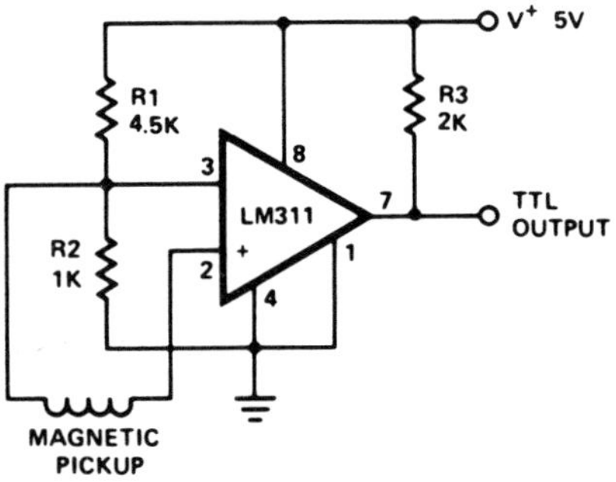

Detector for magnetic transducer (S).

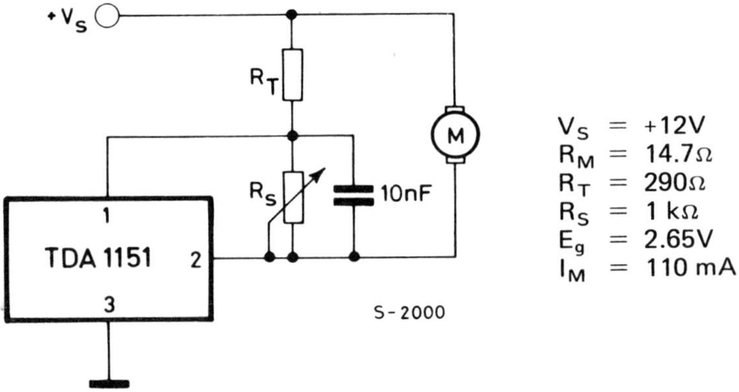

Low cost application circuit for motor speed regulation (SGS).

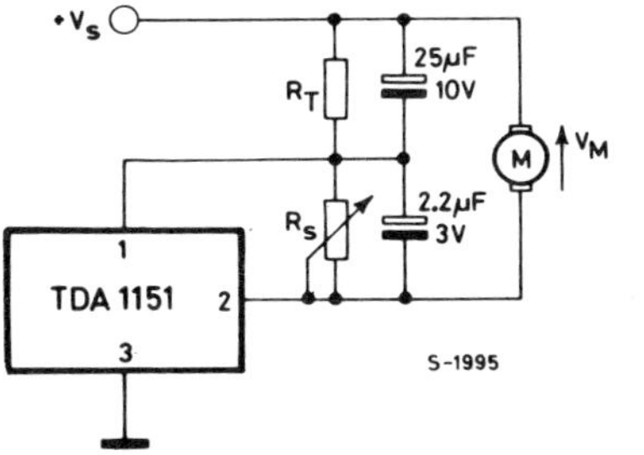

$V_s = +9V$
$R_M = 14.2\Omega$
$R_T = 280\Omega$
$R_S = 1\ k\Omega$
$E_g = 2.9V$
$I_M = 150\ mA$
$V_M = R_M \cdot I_M + E_g = 5.03V$

Note: A ceramic capacitor of 10 nF between pins, 1 and 2 improves stability in some applications.

Motor speed regulator (SGS).

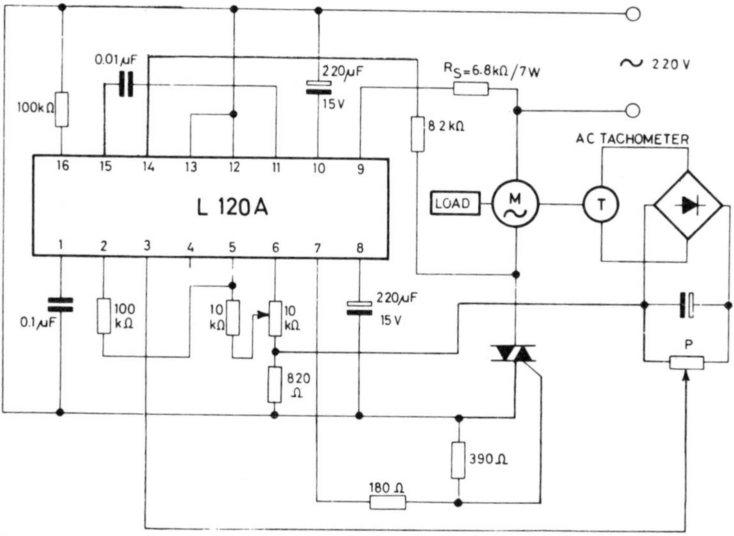

Application circuit for ac motor speed regulators (SGS).

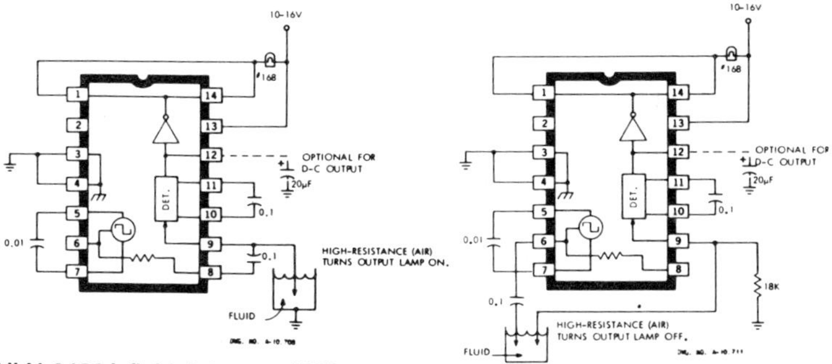

ULN-2429A fluid detectors (SE).

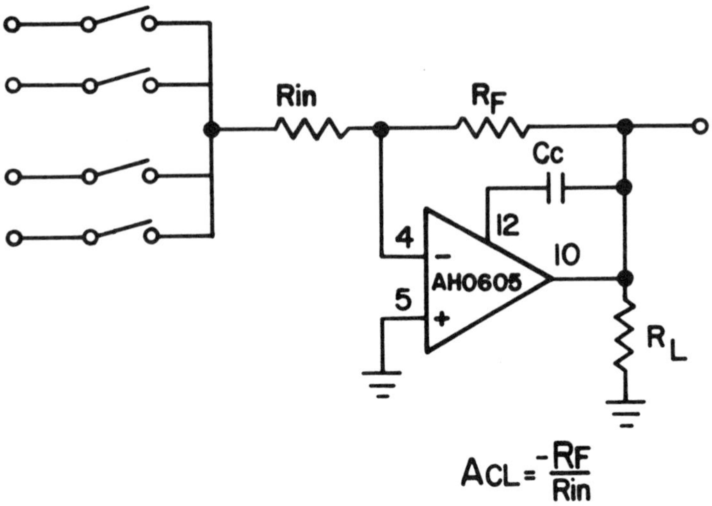

Multiplexed high speed line driver (OE).

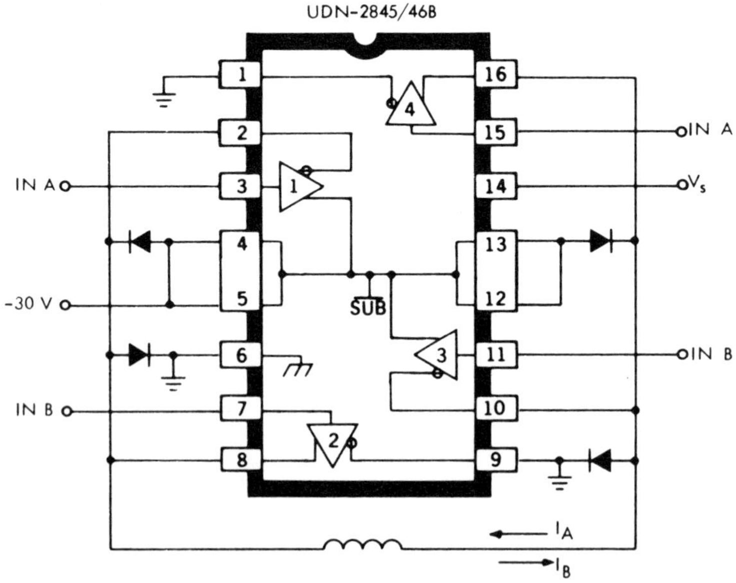

Typical bipolar motor drive application (SE).

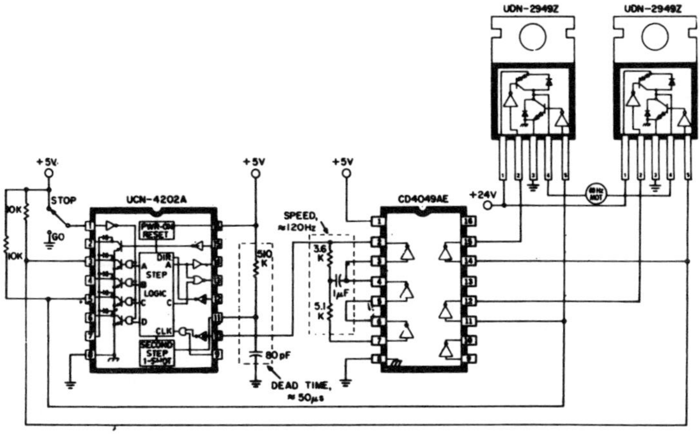

A-C motor drive circuit (SE).

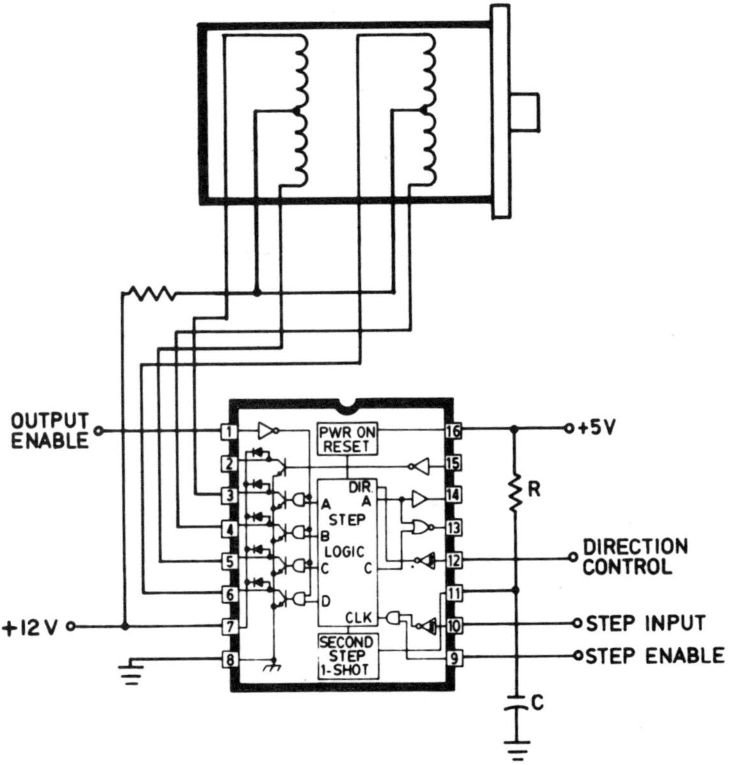

L/R drive circuit used to drive A 12-volt 500 mA unipolar stepper motor (double-step mode) (SE).

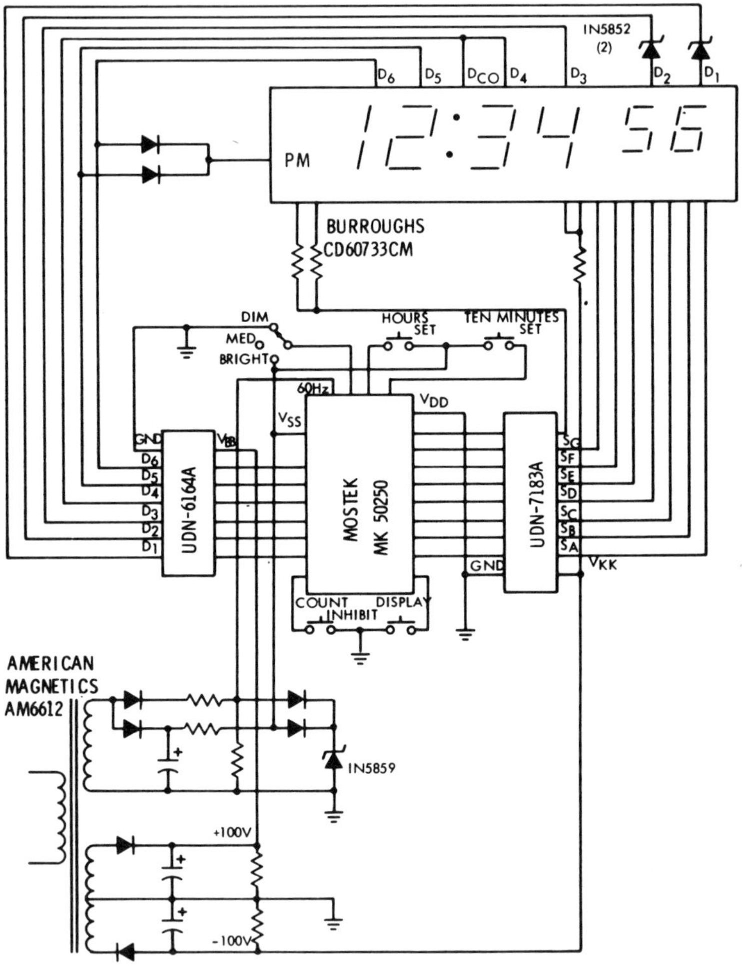

Typical six-digit clock (SE).

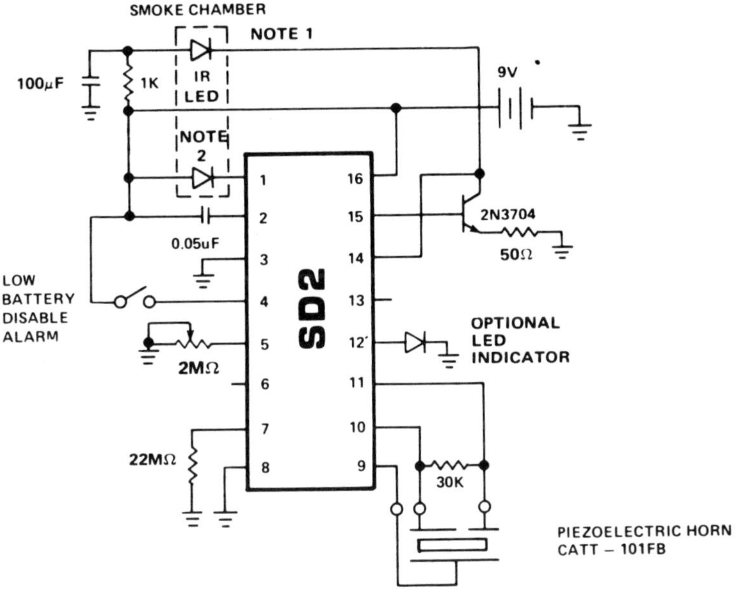

Notes: 1. IR Diode RCA Type SG 1010A or Spectronics Type SE 5455-4
Clairex Type CLED-1
2. IR Photo detectors Vactec VTS4085

CMOS photoelectric smoke detector (SU).

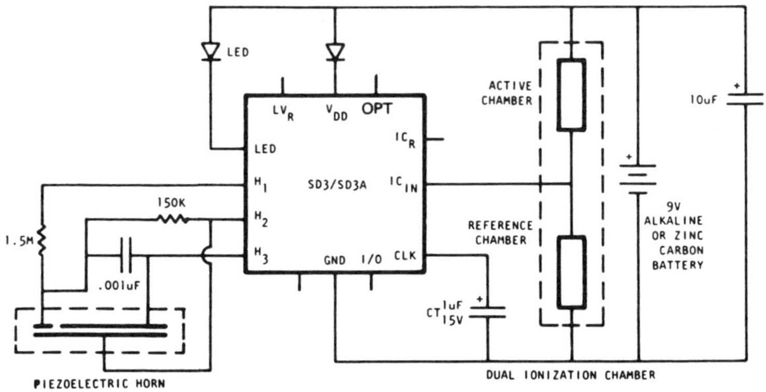

SD3/SD3A with a dual ionization chamber and piezoelectric horn together with an LED as battery connection indicator (SU).

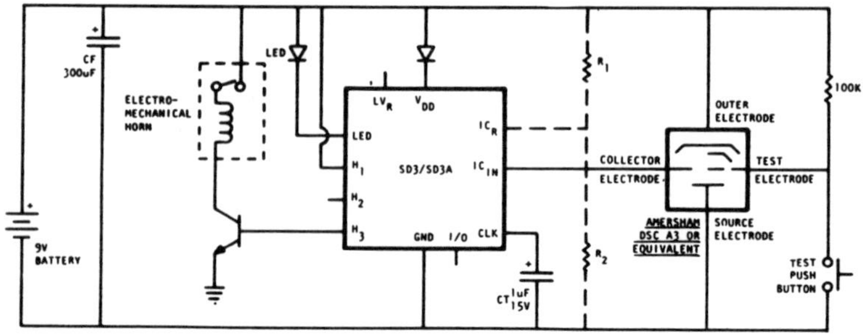

SD3/SD3A with an Amersham DSC A3 concentric chamber and an electromechanical horn. Special features are optional R1/R2 resistor network for adjusting comparator trip voltage and built-in test electrode for in-circuit alarm test (SU).

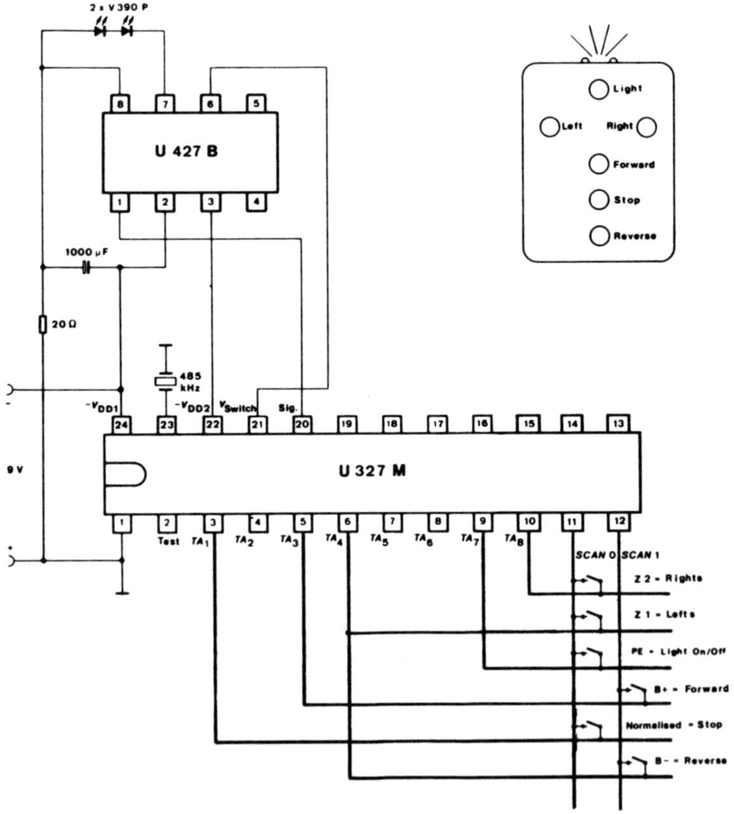

Model car transmission circuit (AEG).

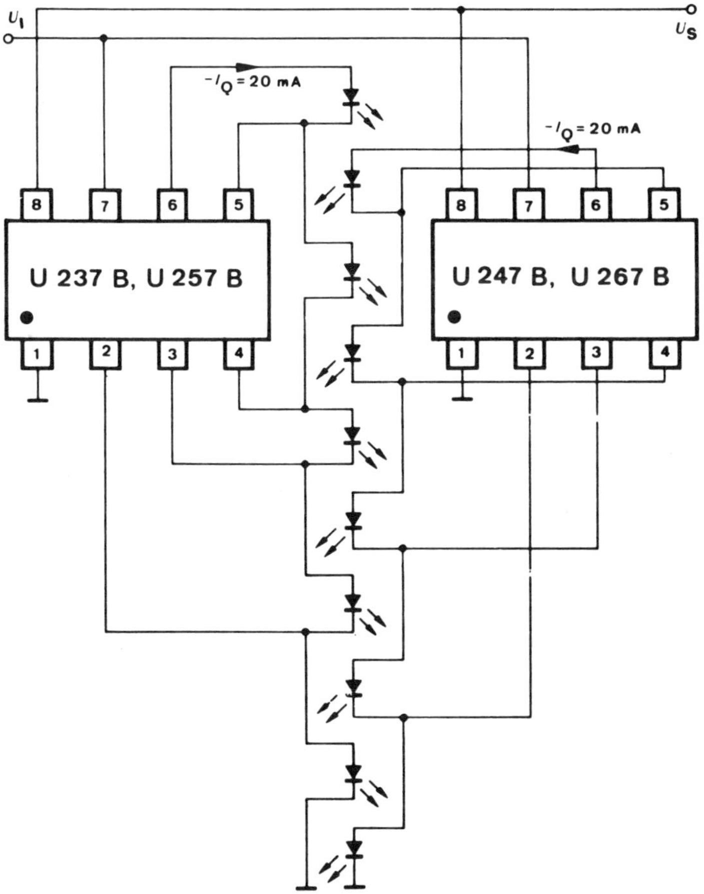

10 LED bar display (AEG).

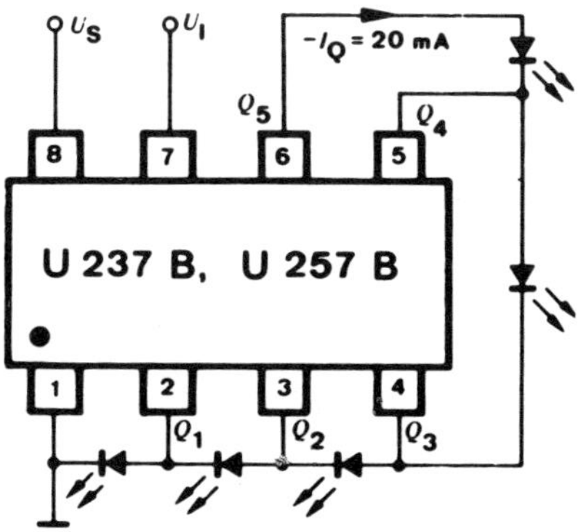

5 LED bar display (AEG).

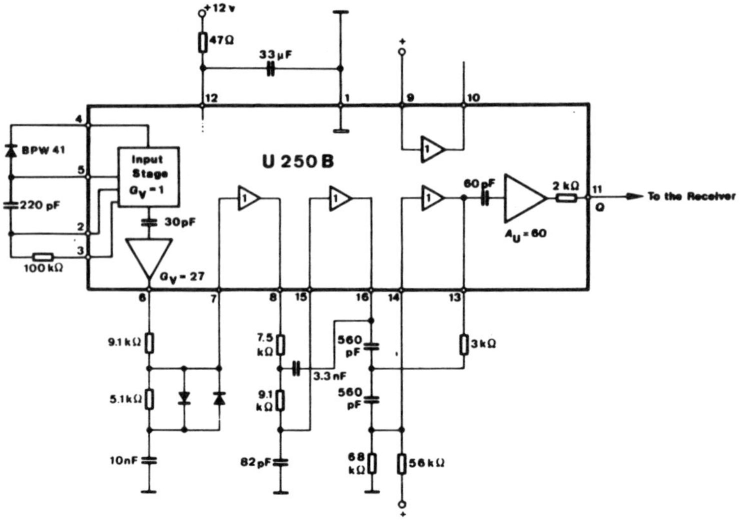

Preamplifier circuit (AEG).

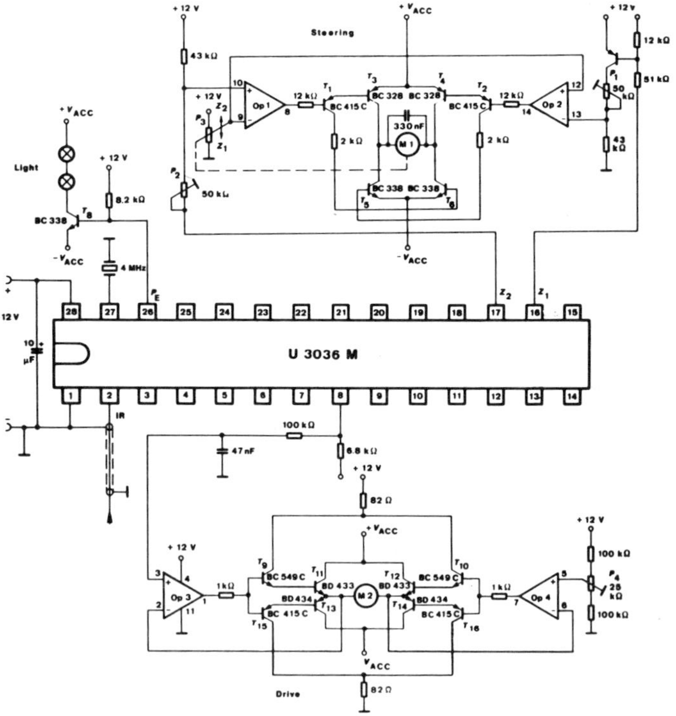

Receiver and the control circuitry for the drive, steering and lights (AEG).

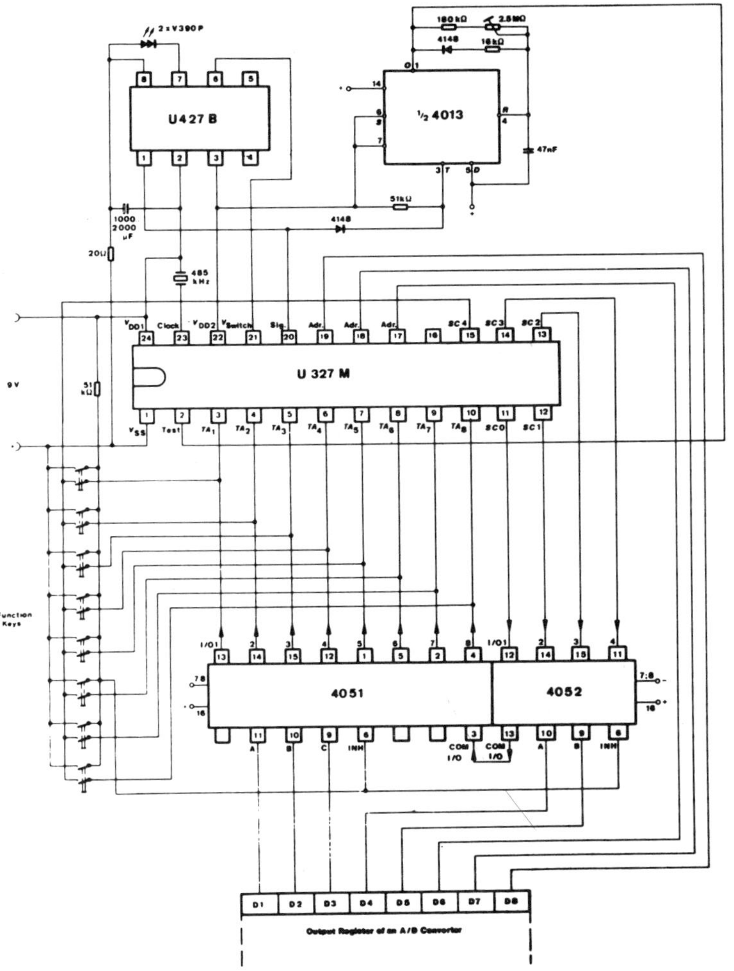

"Wire-less" measurement transmission (AEG).

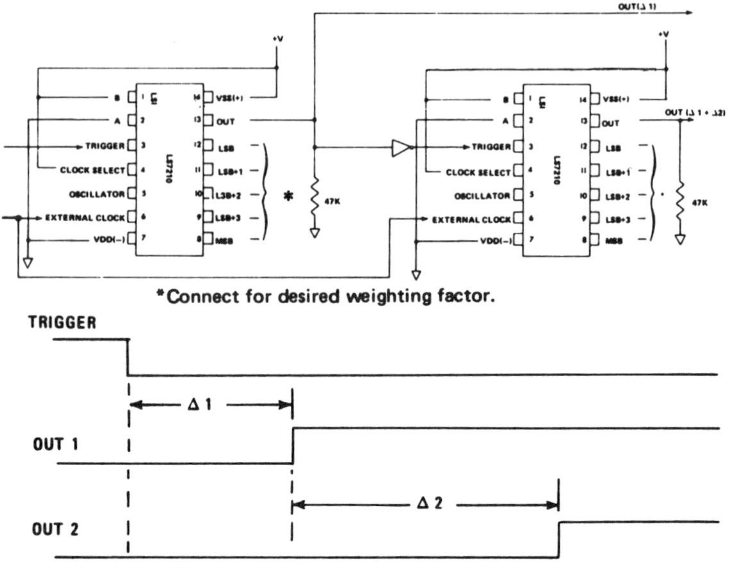

*NOTE:
Output of LS7210 is open drain FET. Some load to ground is required to cause output to go negative.

Sequential turn on (LSI).

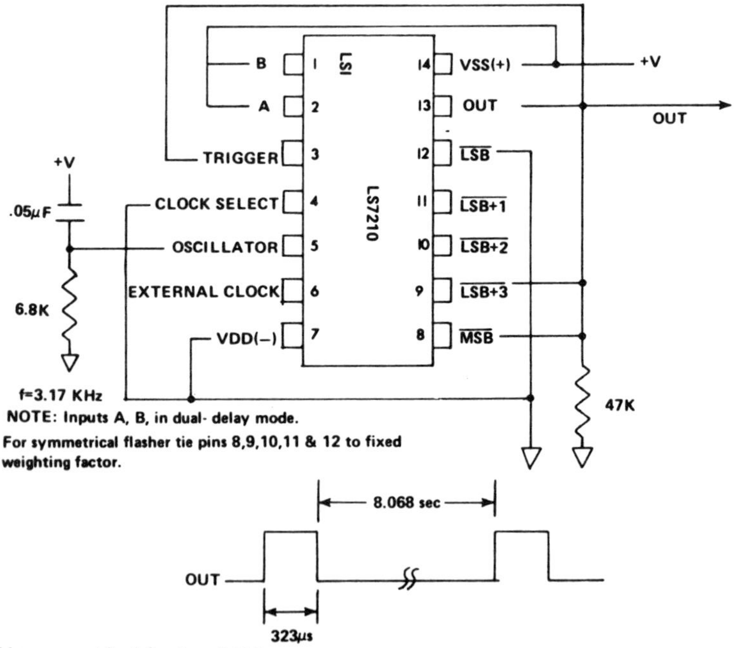

Unsymmetrical flasher (LSI).

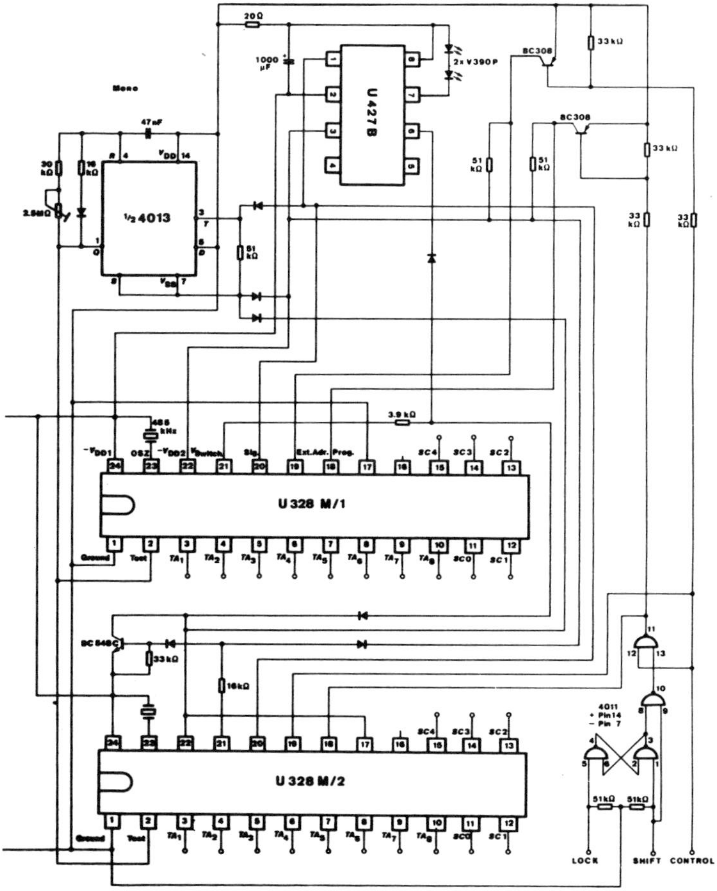

Fast keyboard connection (AEG).

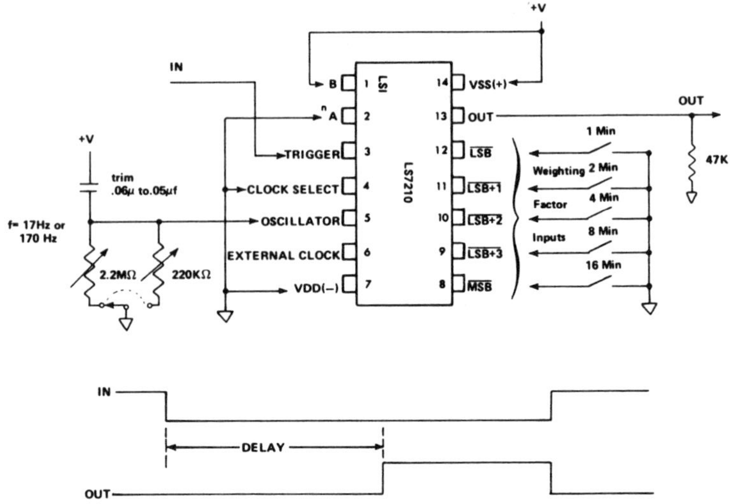

LS7210 in delayed operate mode to achieve one to 31 minute delay (LSI).

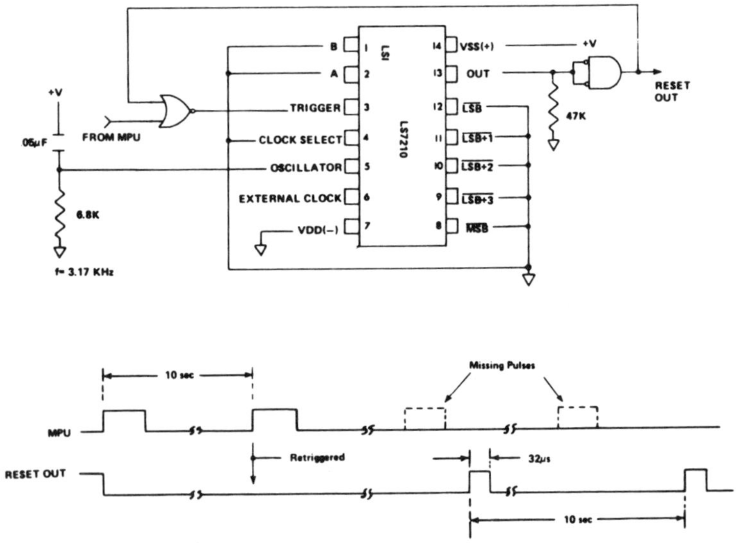

NOTE: Inputs A, B are in One Shot mode.
In this application an output is generated whenever the perodic sampling signal from the MPU is interrupted.

Auto reset watchdog circuit (LSI).

Chapter 18

Timers and Switches

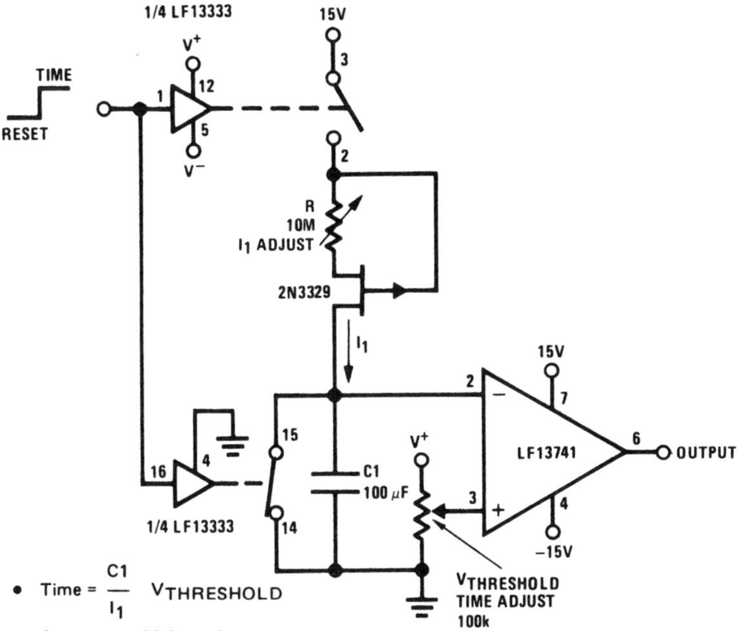

- Time = $\frac{C1}{I_1}$ $V_{THRESHOLD}$
- Output goes high on time out
- Reverse op amp inputs for output low on time out
- C1 low leakage capacitor

Long time timer (NS).

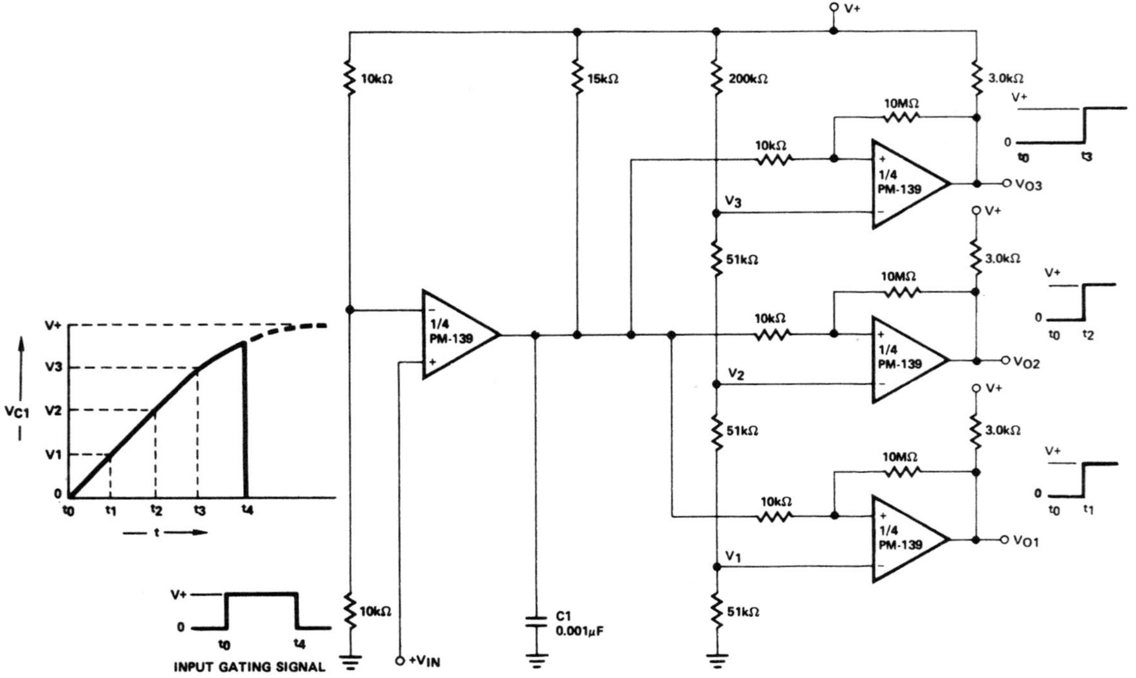

Time delay generator (PM).

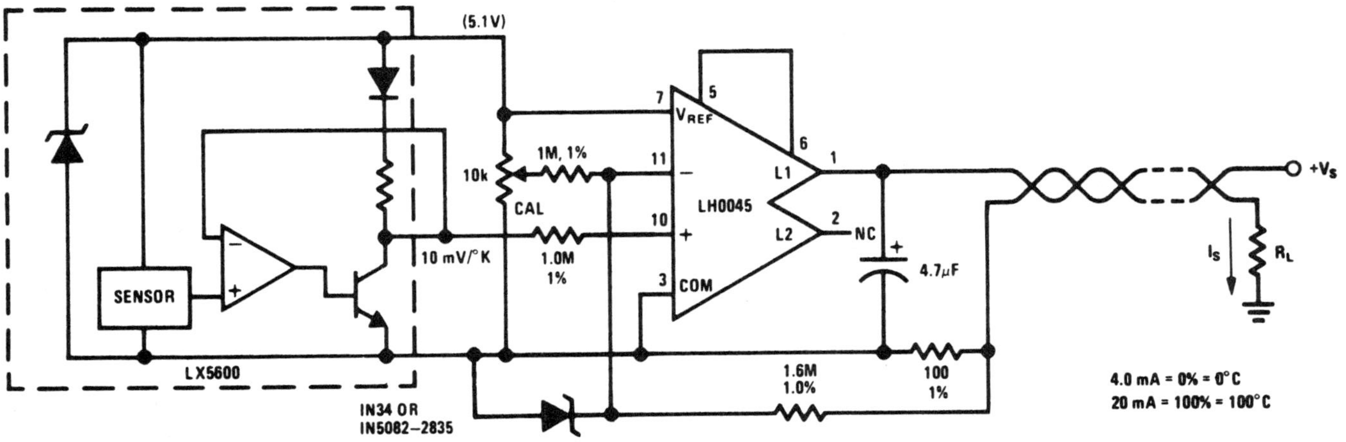

Electronic temperature sensor (NS).

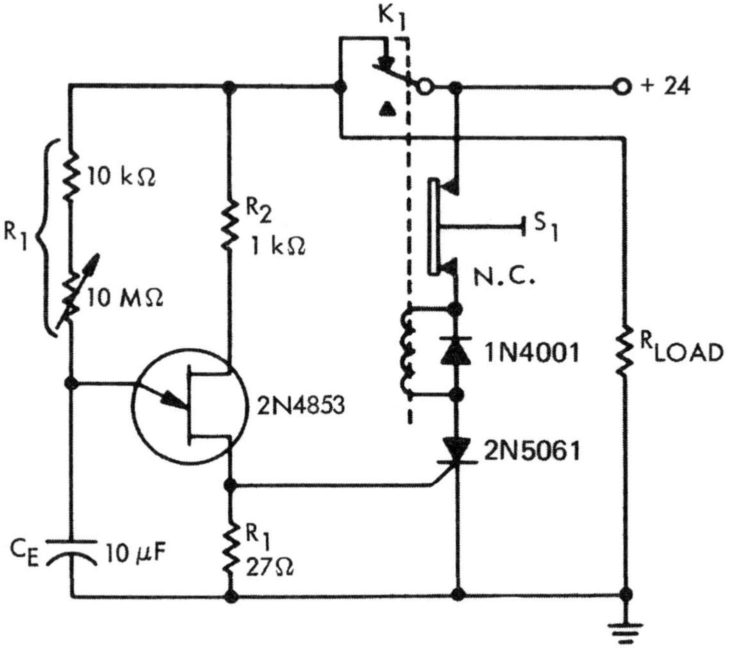

A simple time delay circuit (M).

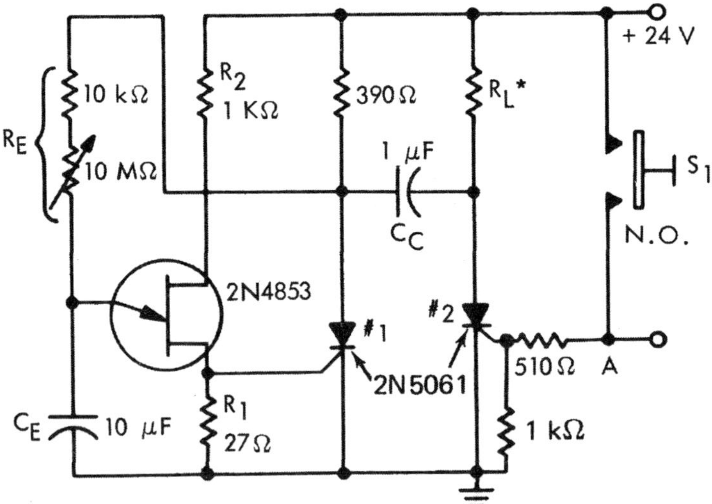

*Value of R_L must be low enough to allow hold current to flow in the SCR.

A simple time delay circuit using two SCR's (M).

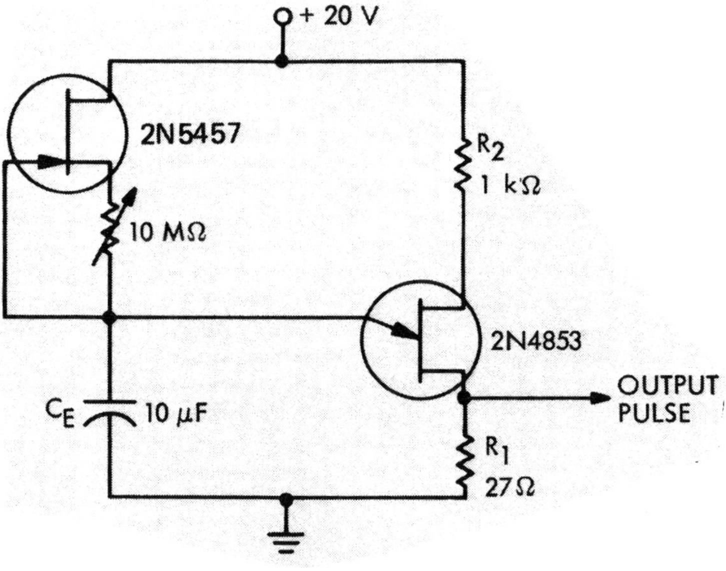

A time delay circuit featuring constant current charging (M).

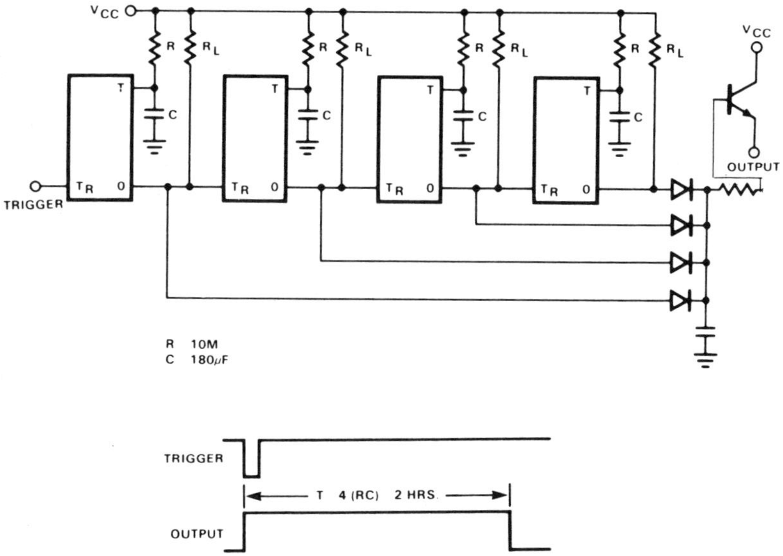

558 two hour timer (S).

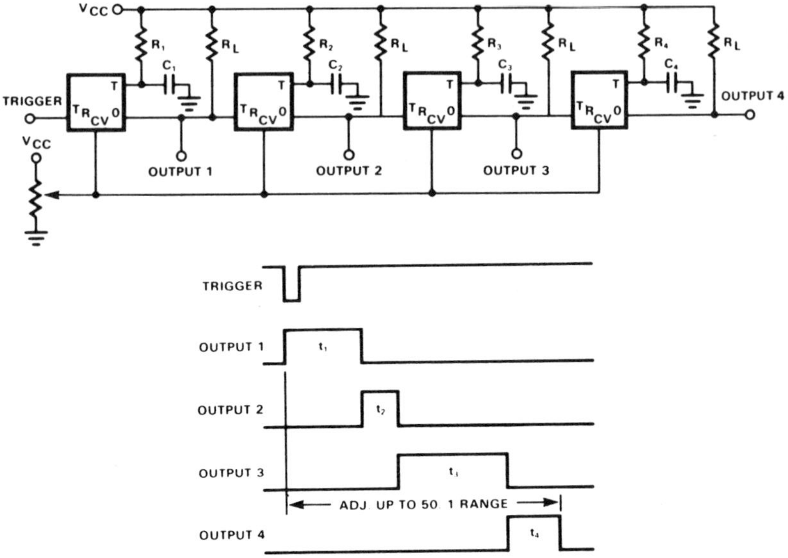

NOTE
t_1, t_2, t_3, t_4 remain proportional over entire adj. range

558 sequential timer with voltage controlled cycle time (50:1 range) (S).

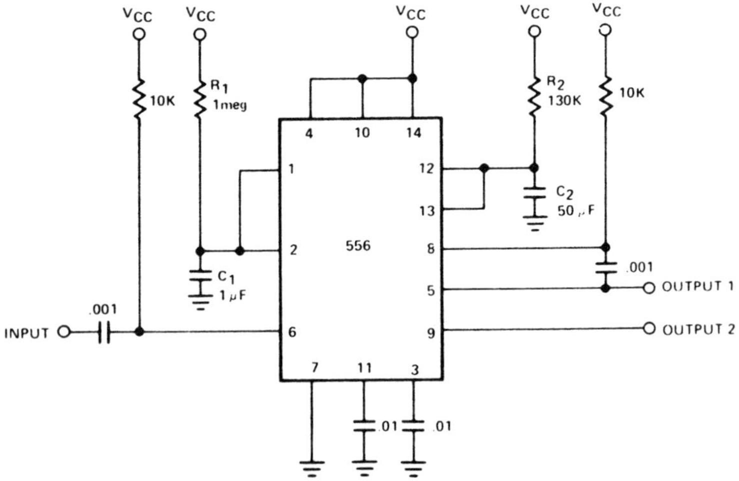

ALL RESISTOR VALUES ARE IN OHMS

Sequential timer (S).

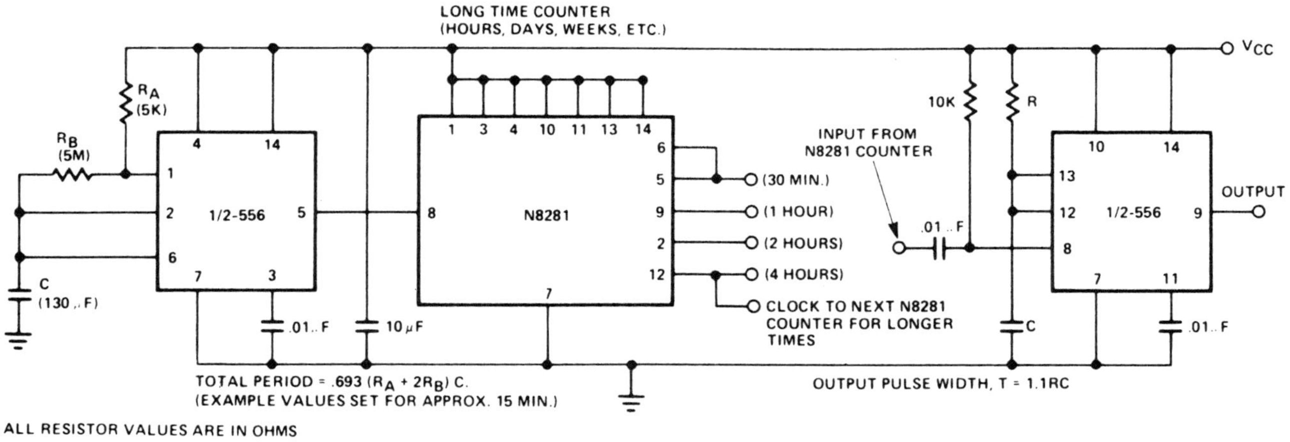

Method of achieving long time delays (S).

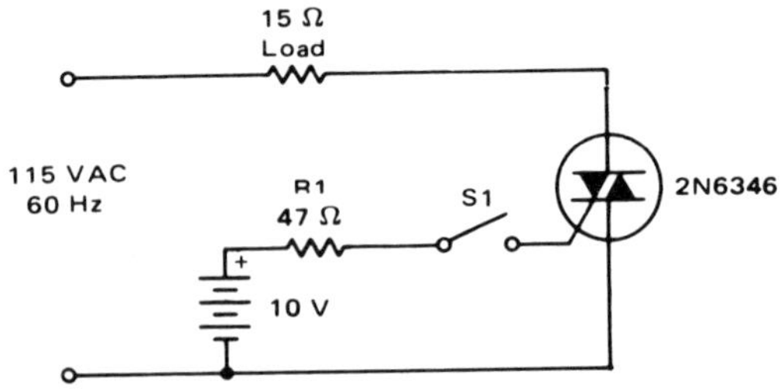

Low voltage controlled triac switch (M).

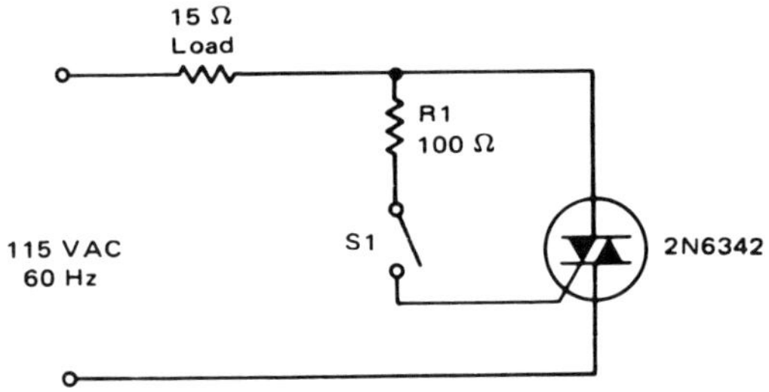

Triac ac static contactor (M).

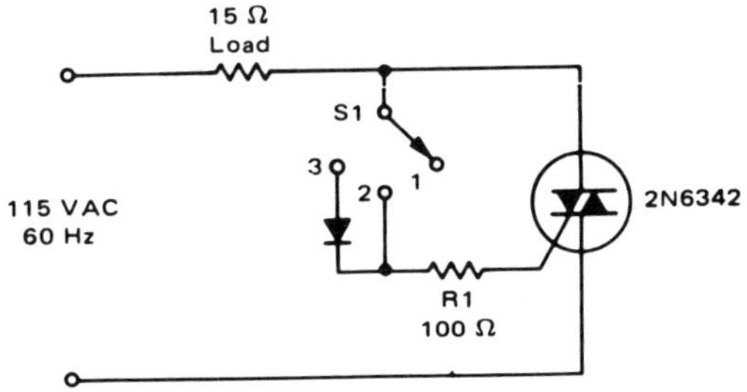

3 position static switch (M).

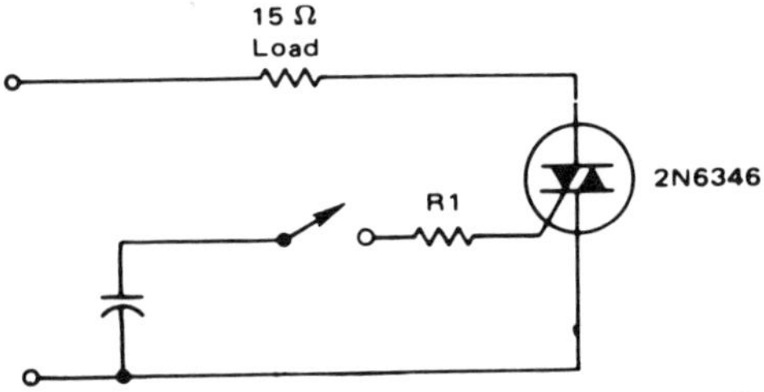

Ac controlled triac switch (M).

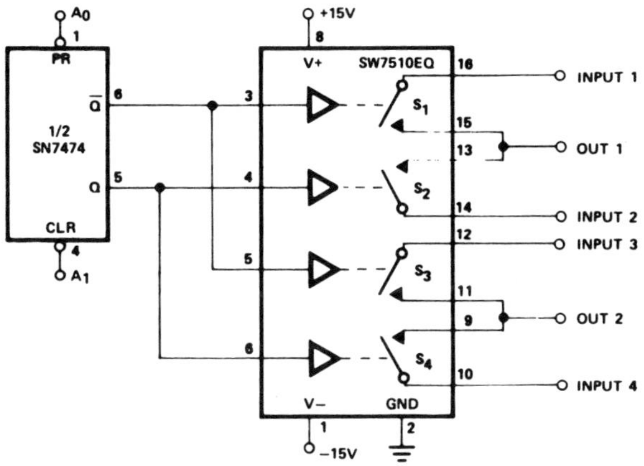

Truth Table

Command		State of Switches After Command	
A_0	A_1	S2 and S3	S1 and S3
1	1	same	same
0	1	on	off
1	0	off	on
0	0	INDETERMINATE	

Typical applications latching DPDT switch (PM).

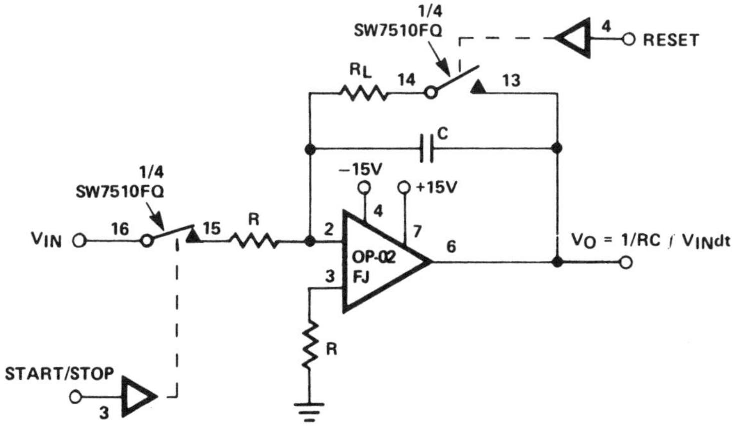

Integrator with analog reset and start/stop capability (PM).

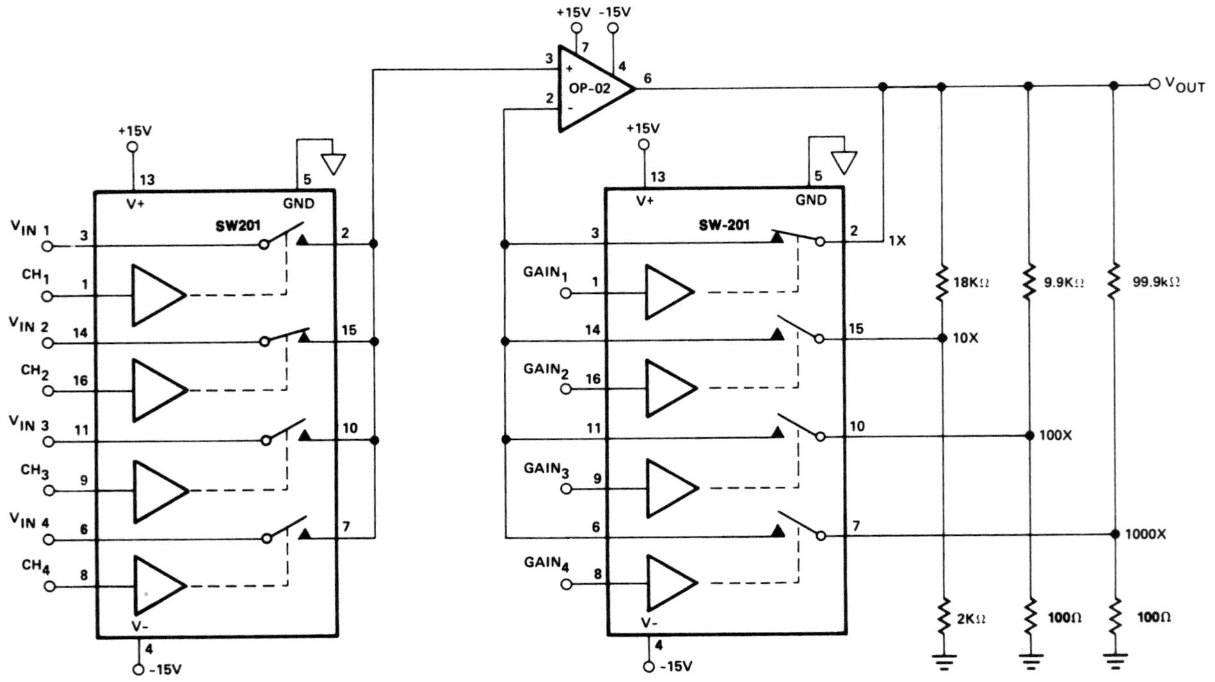

Programmable gain non-inverting amplifier with selectable inputs (PM).

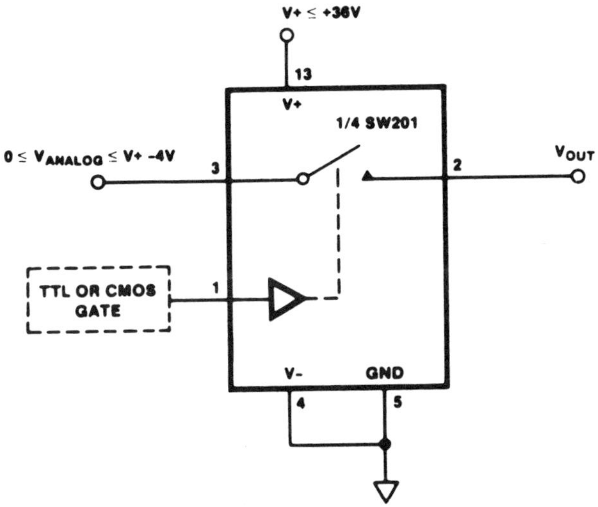

Operation from single positive power supply (PM).

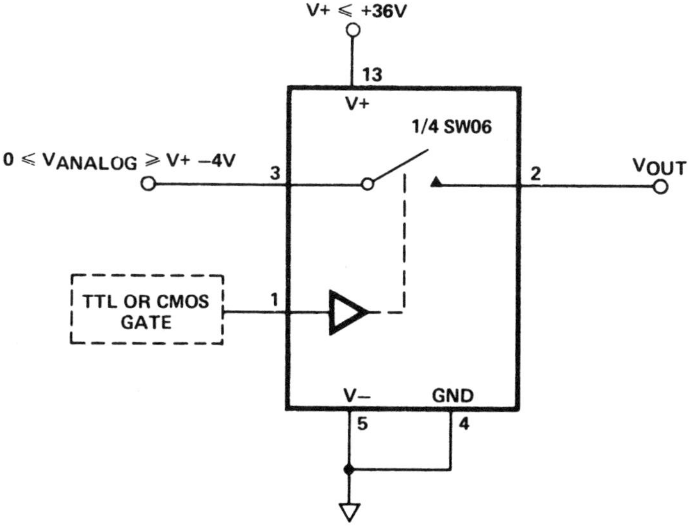

Operation from single positive power supply (PM).

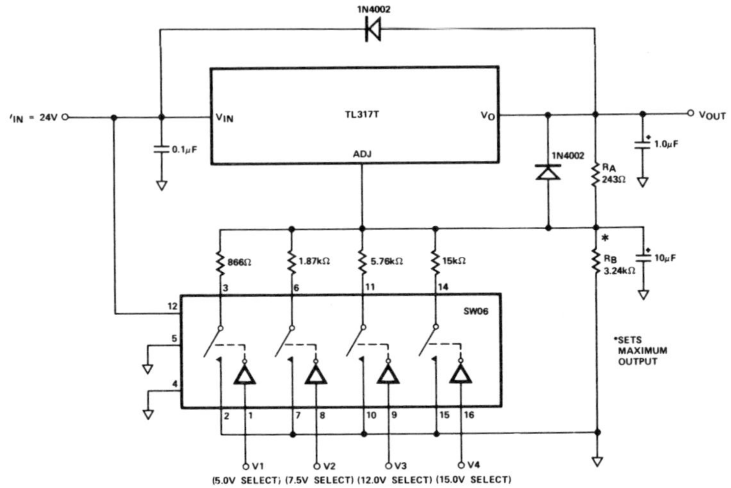

V_O	R_i	NOTE
5.0	866Ω	$V_O \approx 1.25 \left(1 + \frac{R_B \parallel R_i}{R_A}\right)$
7.5	1.87 kΩ	
12.0	5.76 kΩ	
15.0	15.0 kΩ	
SWITCH $R_{ON} \approx 60\Omega$ R_B = 3.24 kΩ		

Programmable voltage supply (PM).

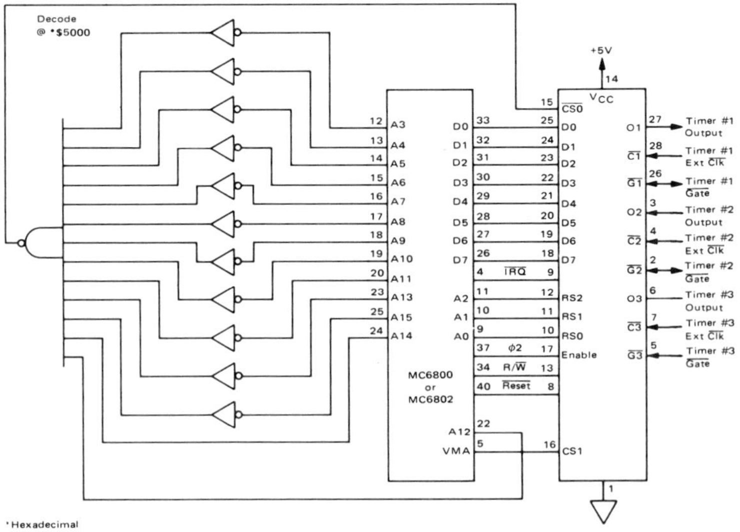

* Hexadecimal

PTM interfaced to the MPU (M).

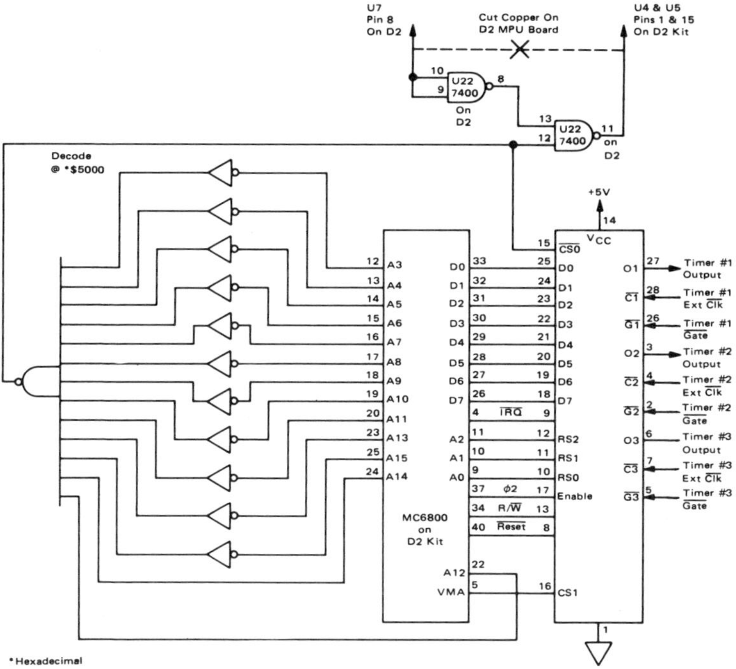

*Hexadecimal

PTM interfaced to MEK6800D2 (M).

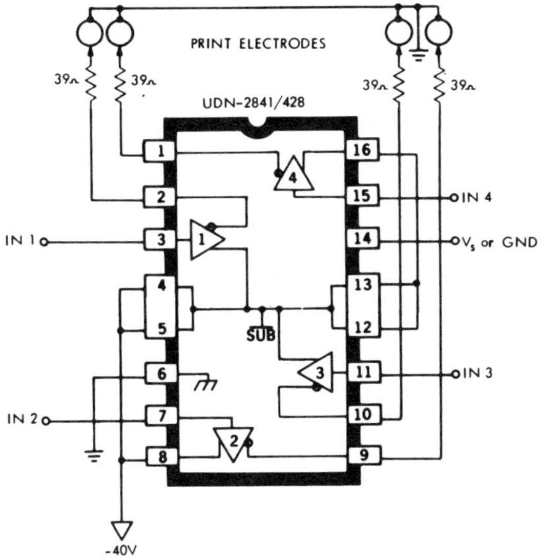

Typical electrosensitive printer application (SE).

Chapter 19

Automotive Circuits

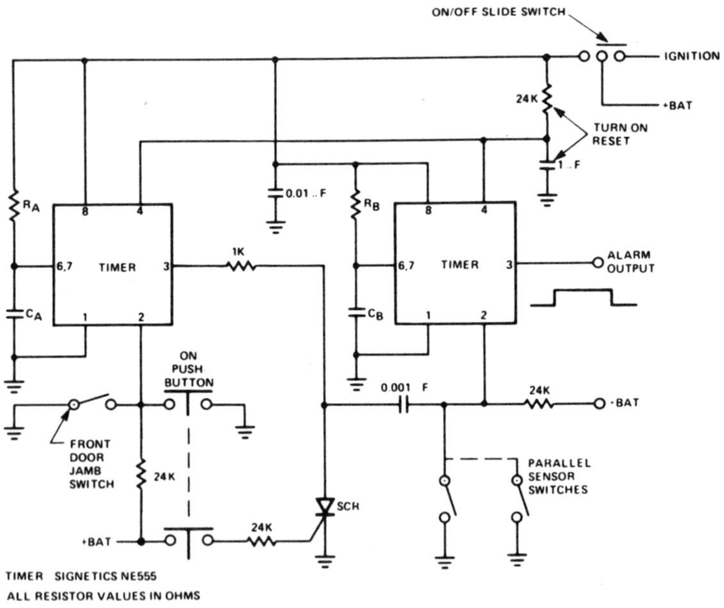

Auto burglar alarm (S).

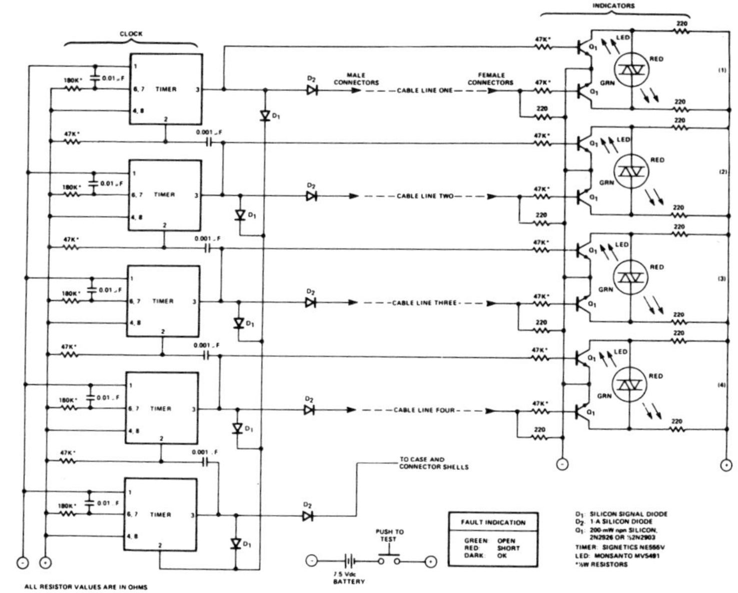

Cable tester (S).

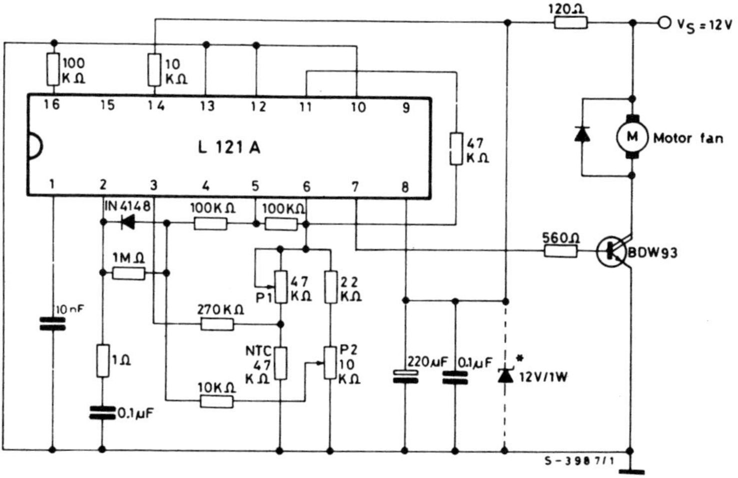

* Protection against overvoltages.
P_1 : system hysteresis setting
P_2 : temperature setting

Climate control for cars (SGS).

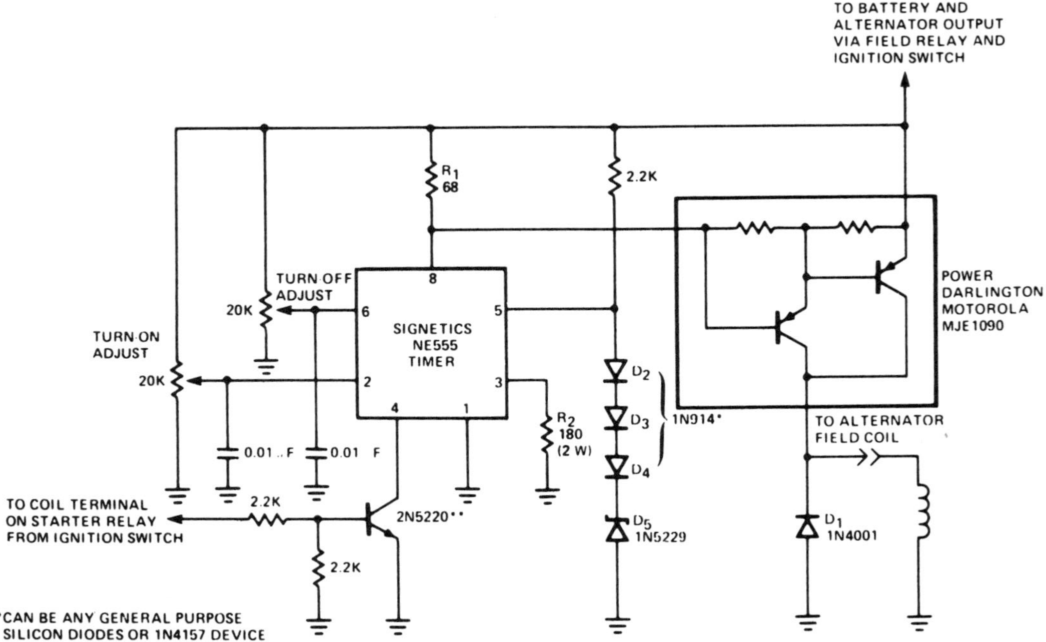

*CAN BE ANY GENERAL PURPOSE SILICON DIODES OR 1N4157 DEVICE

**CAN BE ANY GENERAL PURPOSE SILICON TRANSISTOR

ALL RESISTOR VALUES ARE IN OHMS

Automobile voltage regulator (S).

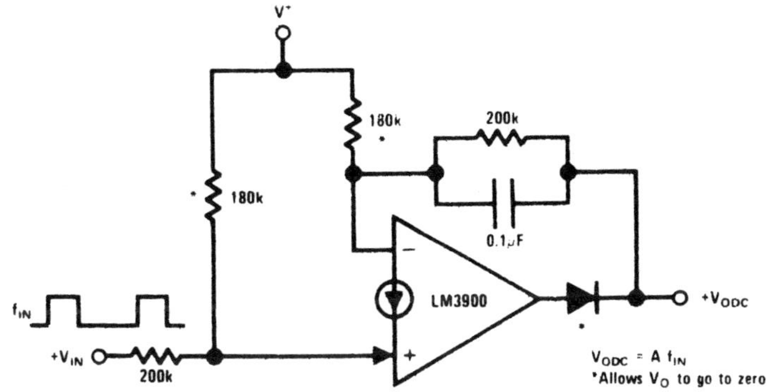

Tachometer (NS).

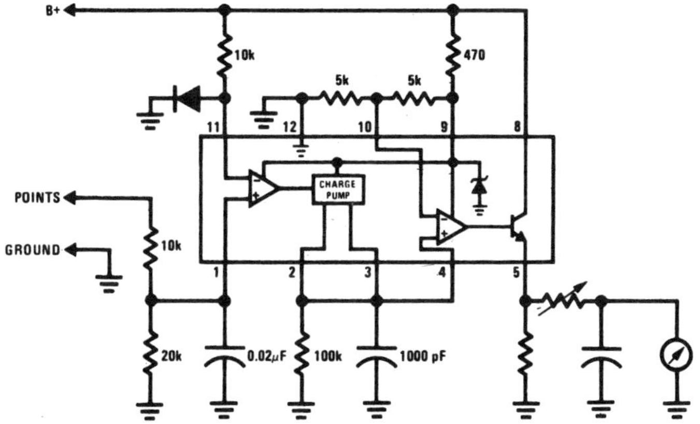

Breaker point dwell meter (NS).

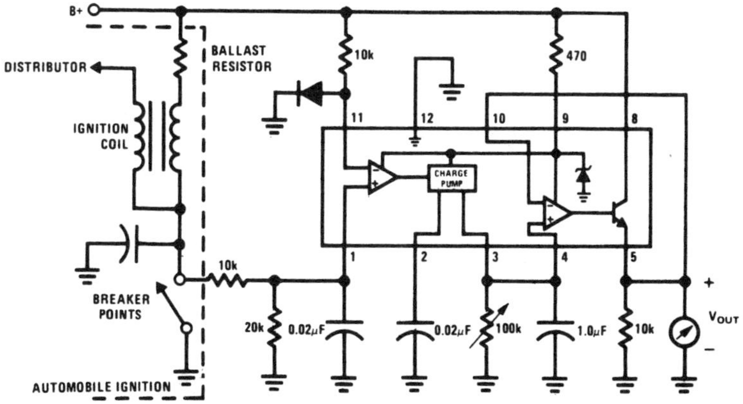

Voltage driven meter indicating engine RPM $V_o = 6V@400$ Hz or 6000 ERPM (8 cylinder engine) (NS).

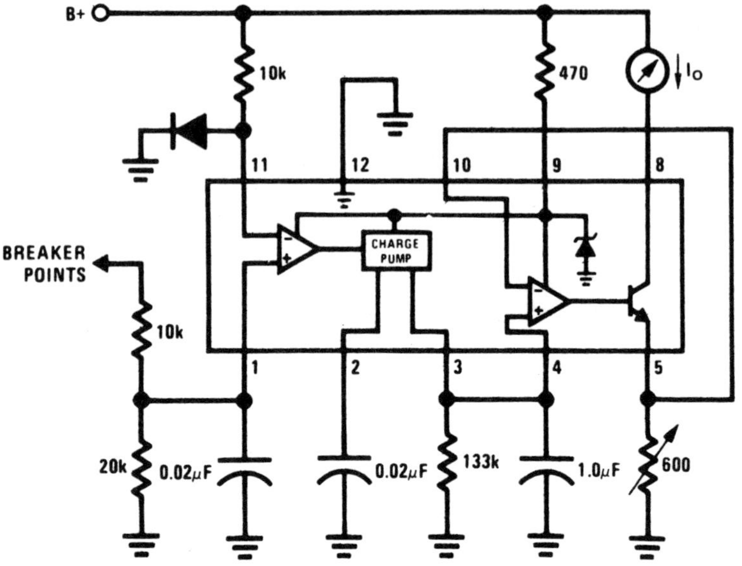

Current driven meter indicating engine RPM I_o = 10 mA@300 Hz or 6000 ERPM (6 cylinder engine) (NS).

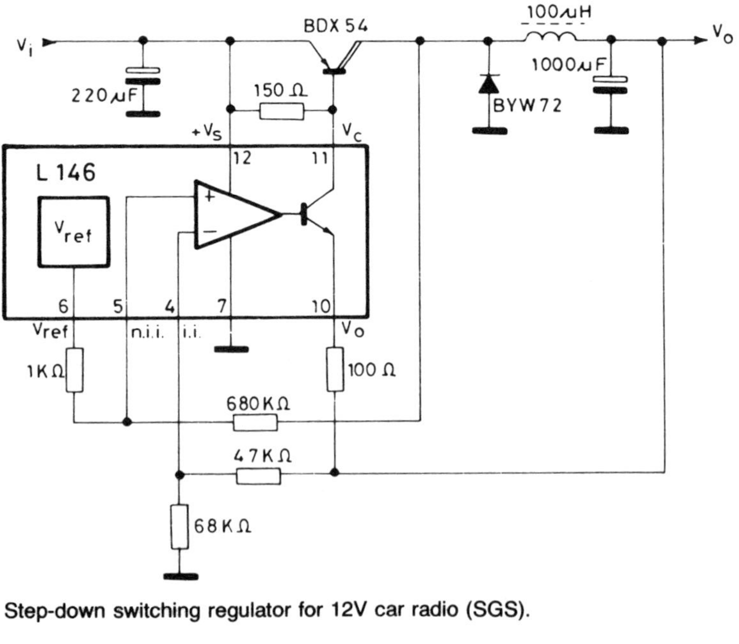

Step-down switching regulator for 12V car radio (SGS).

Chapter 20

Logic Circuits and Counters

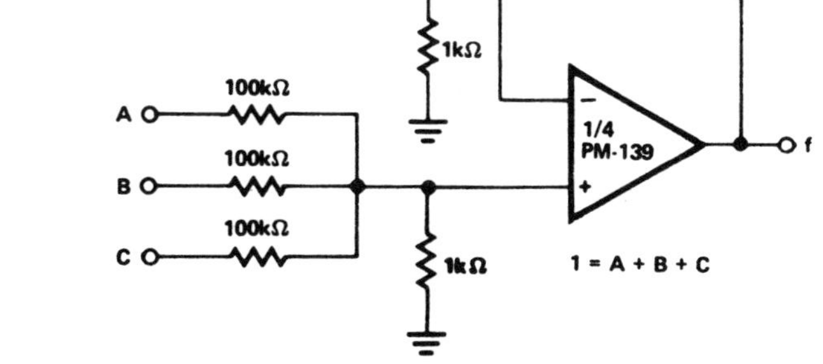

OR GATE (PM).

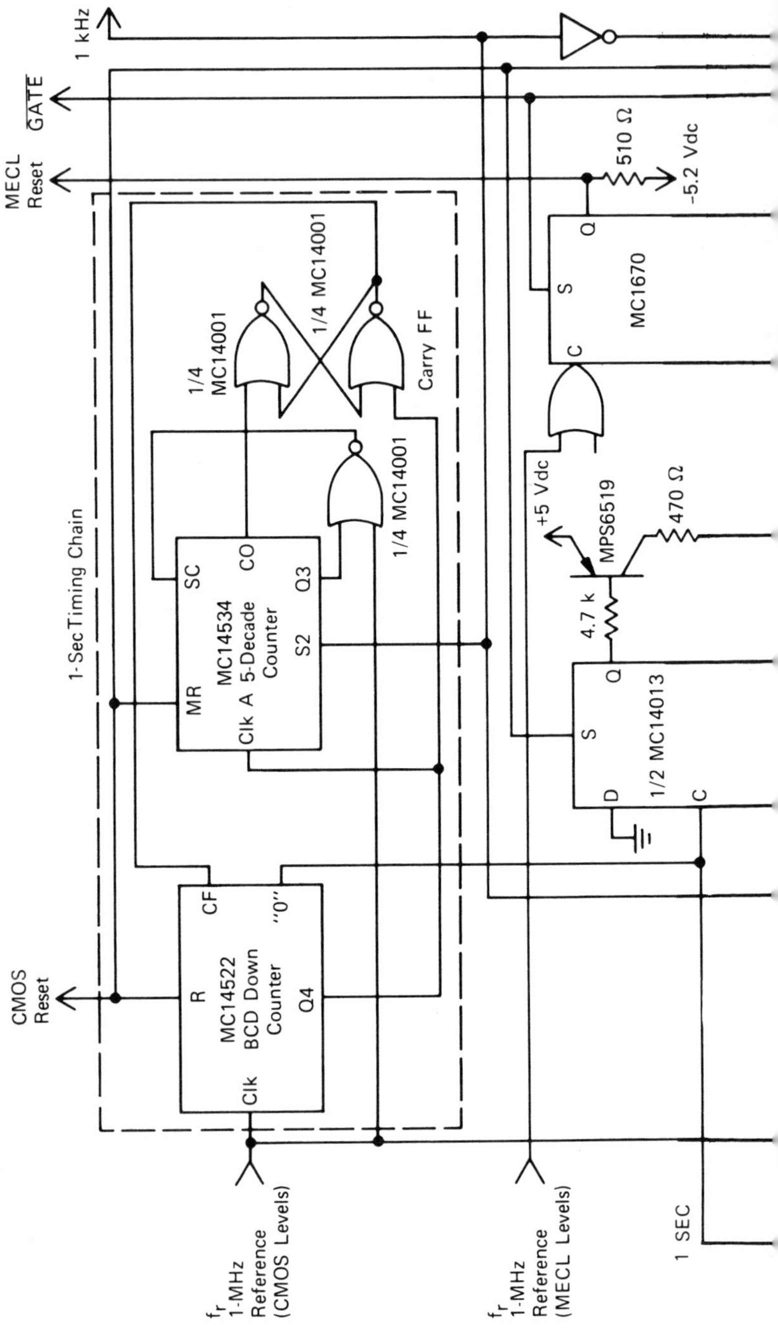
1 kHz
GATE
MECL Reset
CMOS Reset
510 Ω
−5.2 Vdc
MC1670
S
C
Q
1/4 MC14001
1/4 MC14001
Carry FF
1/4 MC14001
+5 Vdc
MPS6519
470 Ω
4.7 k
1-Sec Timing Chain
MC14534 5-Decade Counter
SC
CO
Q3
S2
MR
Clk A
1/2 MC14013
D
S
C
Q
MC14522 BCD Down Counter
CF
"0"
R
Q4
Clk
f_r 1-MHz Reference (CMOS Levels)
f_r 1-MHz Reference (MECL Levels)
1 SEC

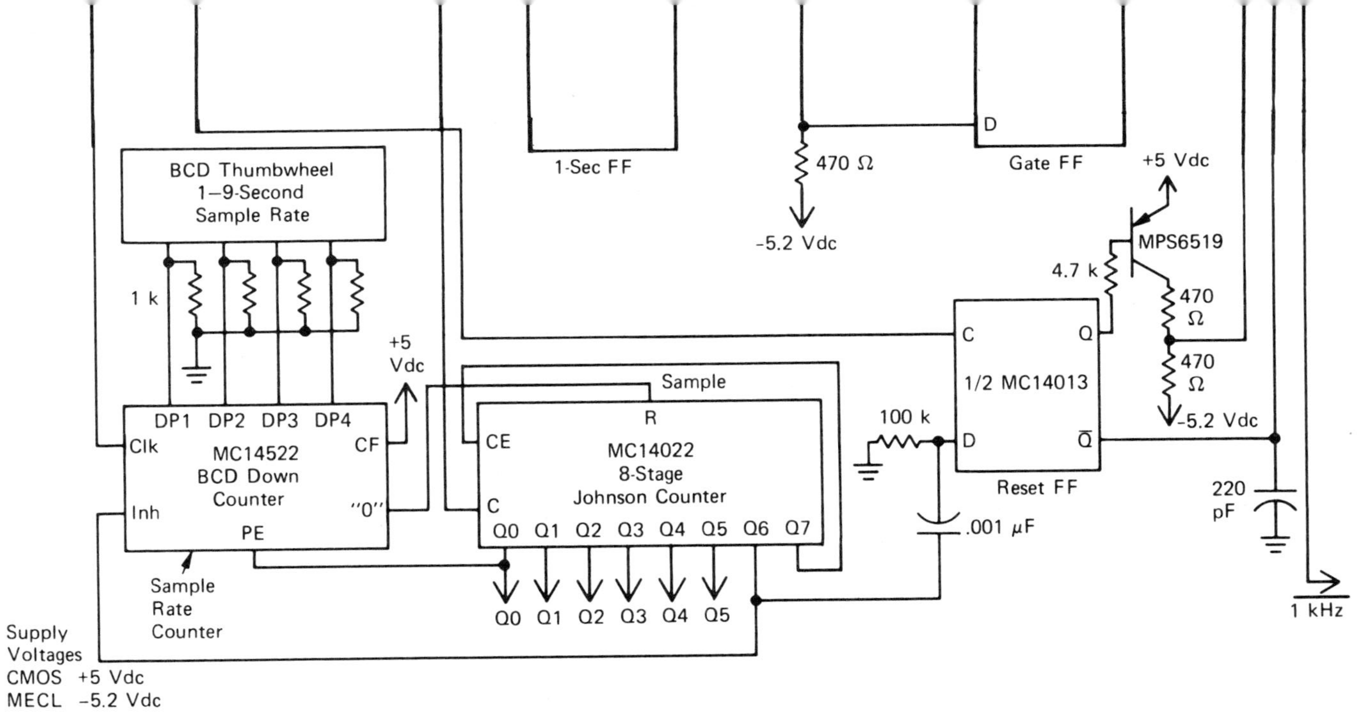

1.2-GHz counter: time base and control (M).

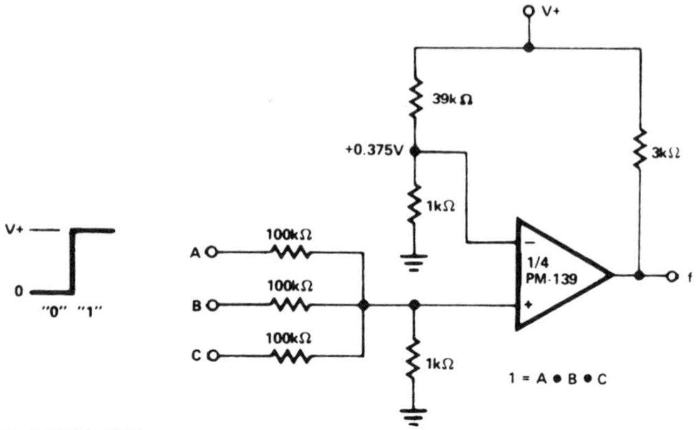

AND GATE (PM).

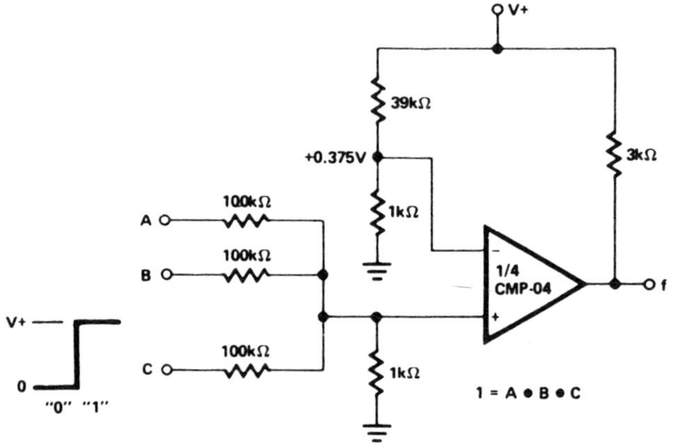

AND GATE (PM).

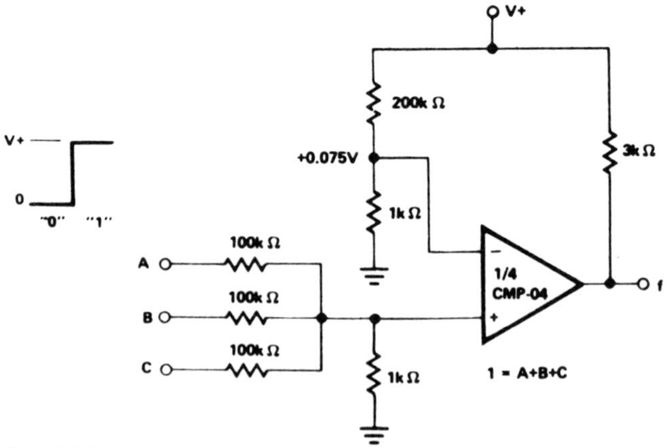

OR GATE (PM).

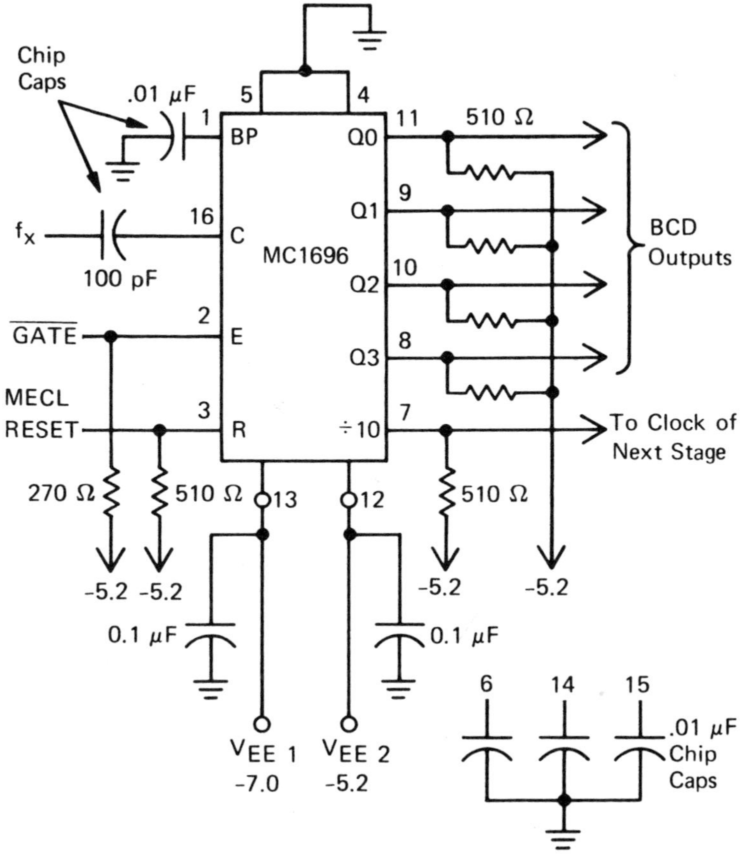

Typical configuration of MC1696 1.2-gigahertz decade counter for ac-coupled fx (M).

Chapter 21

Solar and Optoelectronic Circuits

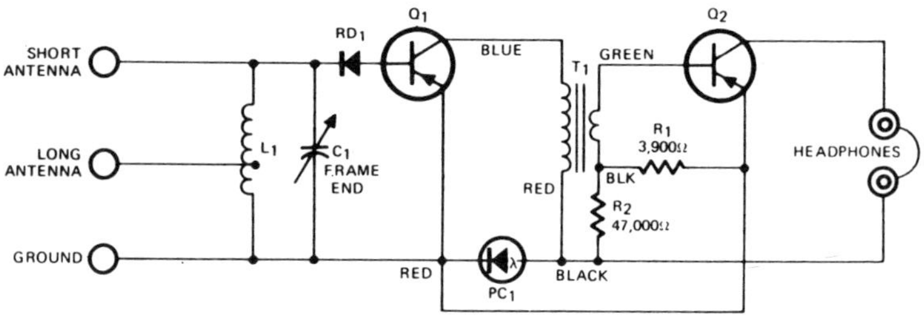

Q_1, Q_2	PNP Transistors (WEP254)
PC_1	Selenium or Silicon solar WEP8525, WEP8526, WEP8500
T_1	10,000 Ohm to 2,000 Ohm Interstage Transformer (Stancor TA-35, or equivalent)
L_1	Transistor Loopstick Antenna (J.W. Miller No. 2001, or equivalent)
C_1	365pf Variable Capacitor (WEP F70-452)
R_1	3,900 Ohm Resistor WEP22-2110
R_2	47K Ohm resistor WEP22-2136
RD_1	Germanium diode (WEP134)
Headphones	
	Transistor Socket (WEP8050 or WEP8051)
	Printed Circuit Board

Two-stage solar radio (W).

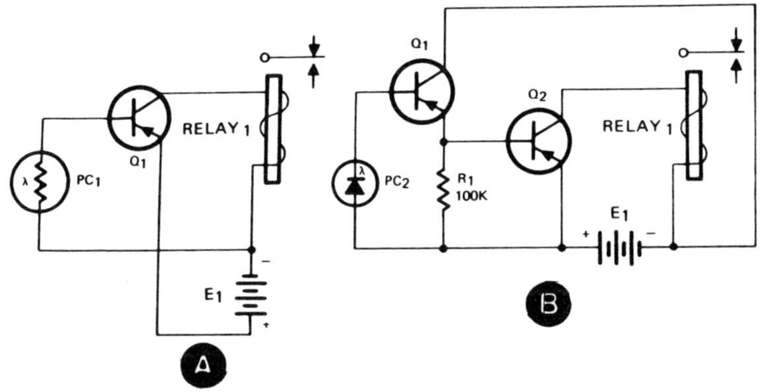

Q_1, Q_2	PNP General Purpose Transistor WEP254
PC_1	Cadmium Sulfide Photocell
PC_2	Selenium or Silicon Photocell WEP8525, WEP8526 or WEP8500
E_1	9 Volt Battery
R_1	100K Ohm, ½ Watt Resistor
$Relay_1$	5,000 Ohm Relay
	Transistor Socket WEP8050
	Printed Circuit Board
	PC Connector Path
	PC Patterns

Amplified solar relay (W).

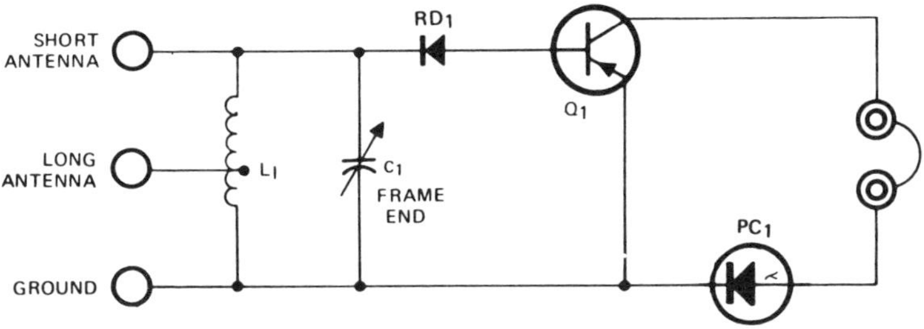

Q_1	PNP Transistor WEP254
C_1	365pf Variable Capacitor WEP F70-452
L_1	Transistor Loopstick Antenna (J.W. Miller No. 2001, or equivalent)
RD_1	Germanium Diode WEP134
PC_1	Selenium or Silicon Solar Cell WEP8525, WEP8526 or WEP8500
Headphones	2,000 or 4,000 Ohms WEP D70-092
	Transistor Socket WEP8050 or WEP8051
	Printed Circuit Board

Single stage solar radio (W).

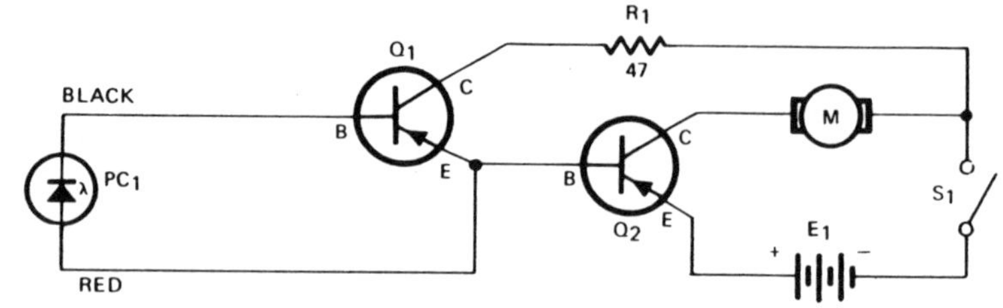

Q_1	PNP Transistor WEP254
Q_2	PNP Transistor (WEP230)
M	Small Motor (WEP8550)
R_1	47 Ohm, ½ Watt Resistor WEP22-2064
E_1	1.5 Volt Battery
PC_1	Silicon Photocell WEP8500
S_1	SPST Switch WEP TS101
	Heat Exchanger for Q_2 WEP8018 or 8019
	Transistor Socket WEP8050
	Printed Circuit Board

Solar model control (W).

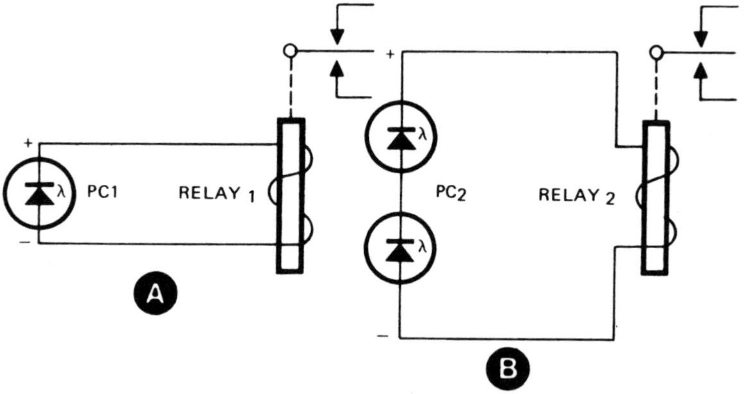

PC_1	WEP Photocell	1-WEP8525 or 1-WEP8526	1-WEP8500	2-WEP8500
$Relay_1$	Ultra-sensitive Relay	AYLZ7303-100 (Barber Coleman)	AYLZ7325-10 (Barber Coleman)	5F-1655-PAL (Sigma)

PC_2	Silicon photocells (6 WEP-WEP8500 cells)
$Relay_2$	50 Ohm model airplane radio control relay

Solar relays (W).

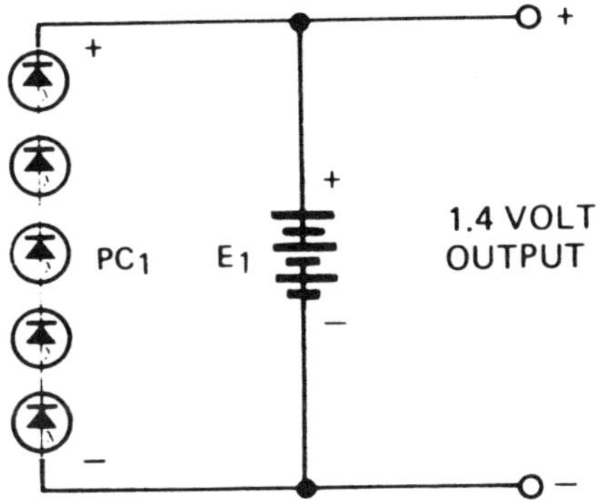

E1	1.5 Volt AA battery
PC1	Selenium or Silicon photocells (WEP-Five WEP8526, or Five WEP8502 connected in series [red to black]).

Solar batteries (W).

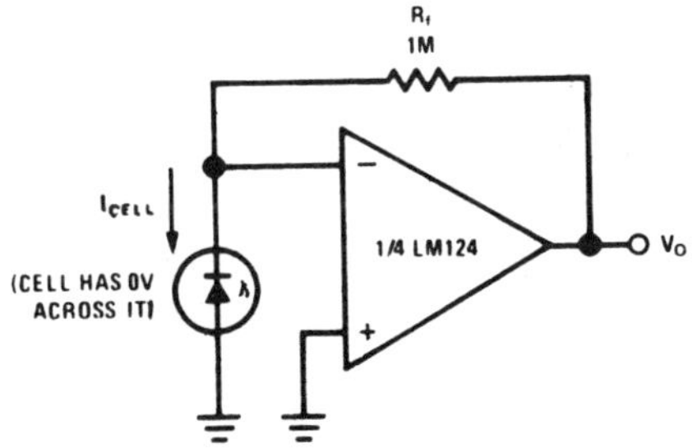

Photovoltaic-cell amplifier (NS).

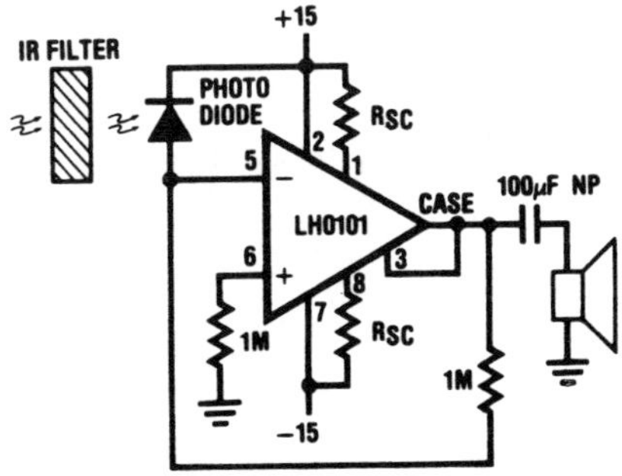

Remote loudspeaker via infrared link (NS).

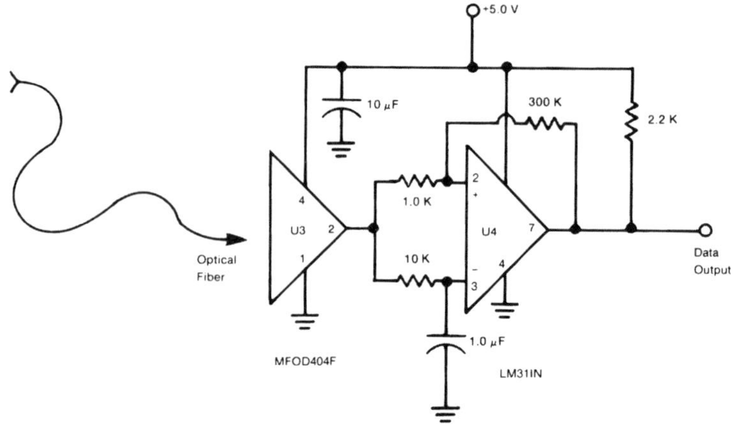

Simple F/O data receiver using IDP and a voltage comparator (M).

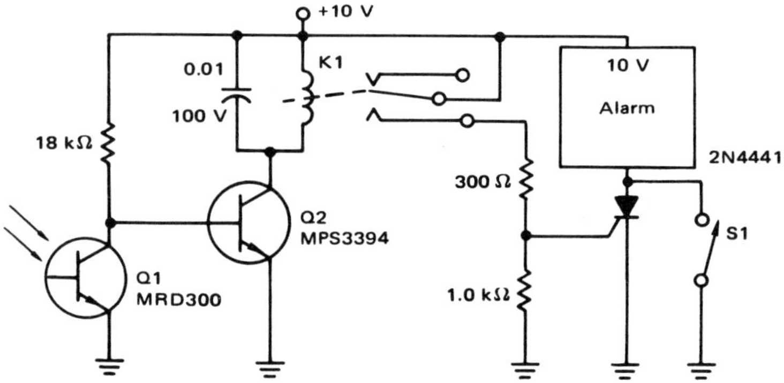

Light-relay operated SCR alarm circuit (M).

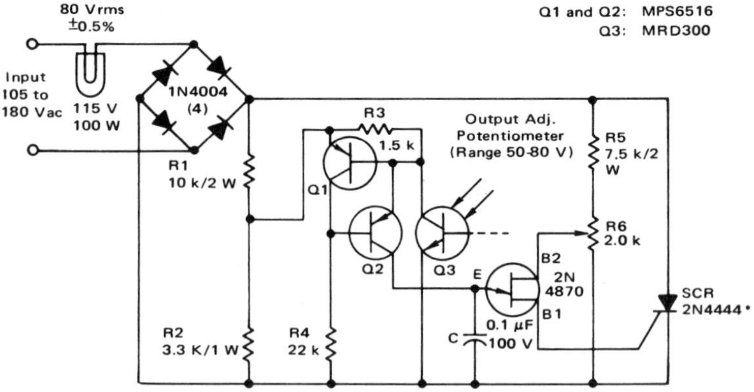

Circuit diagram of voltage regulator for projection lamp (M).

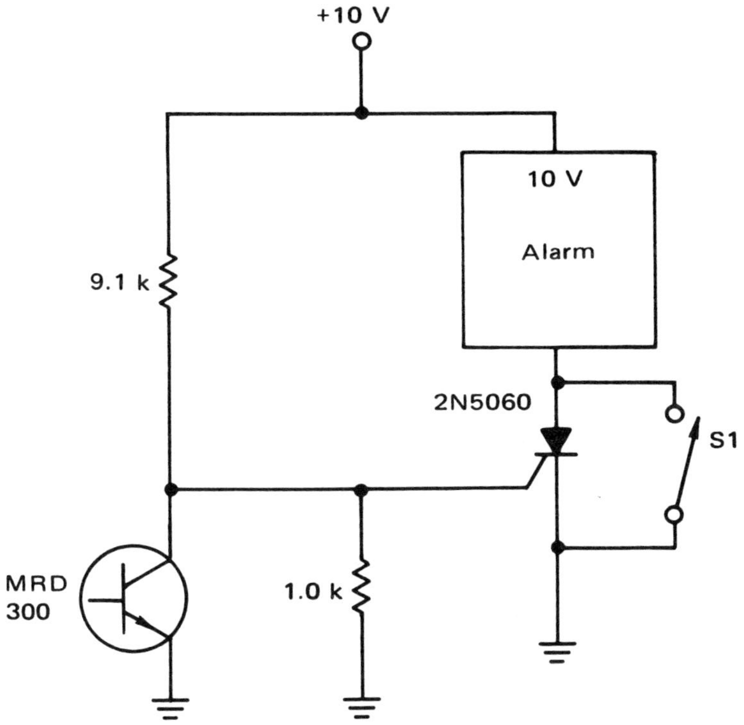

Light operated SCR alarm using sensitive-gate SCR (M).

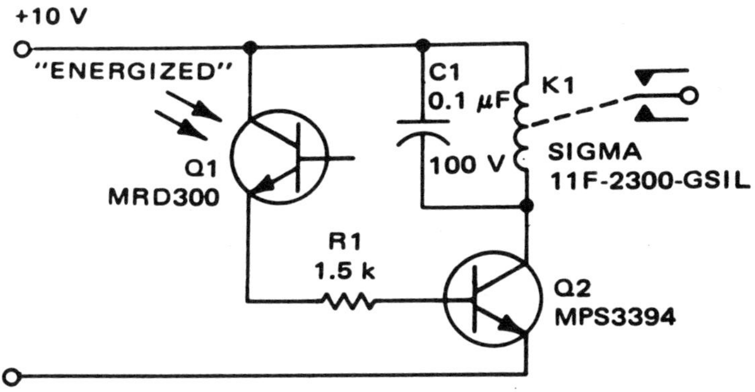

Light-operated relay (M).

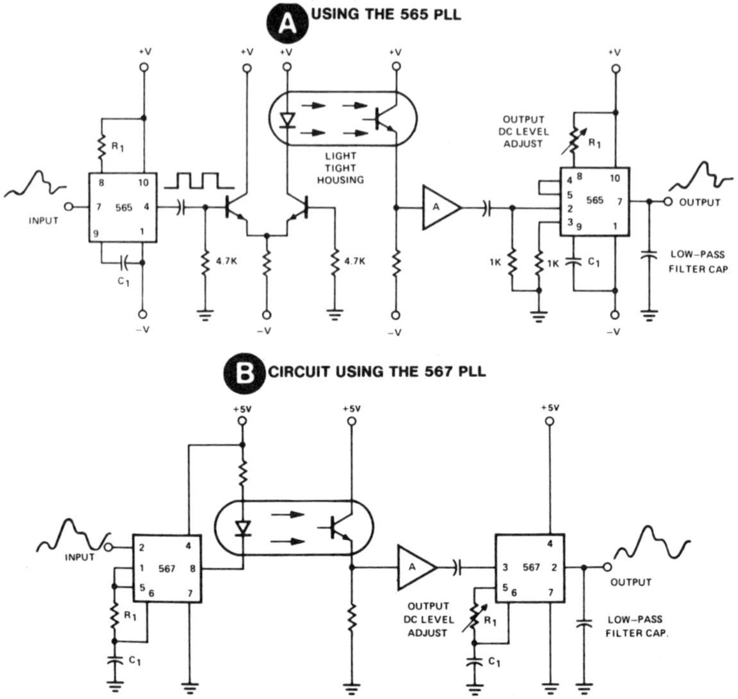

Light-coupled analog isolators (S).

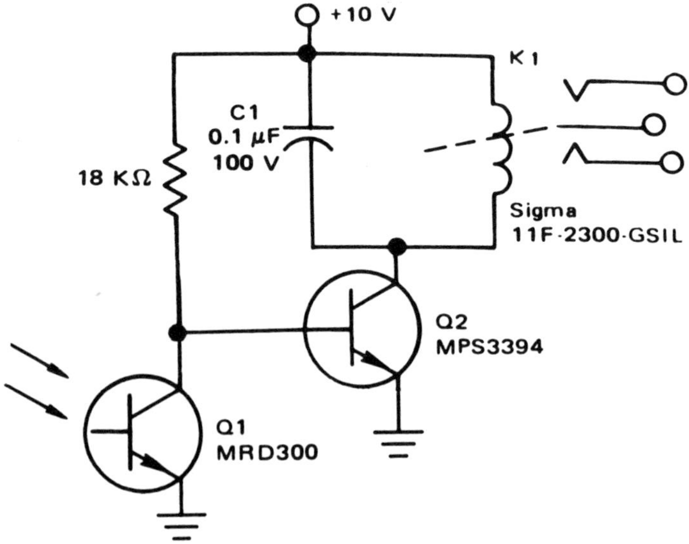

Light de-energized relay (M).

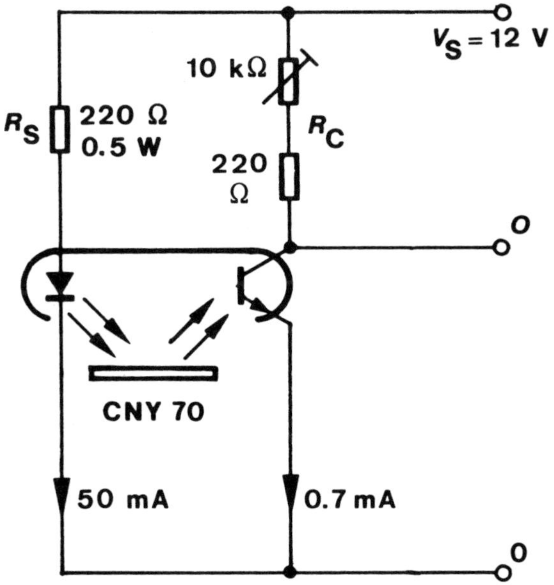

Simple circuit for analog applications of reflective optocouplers (AEG).

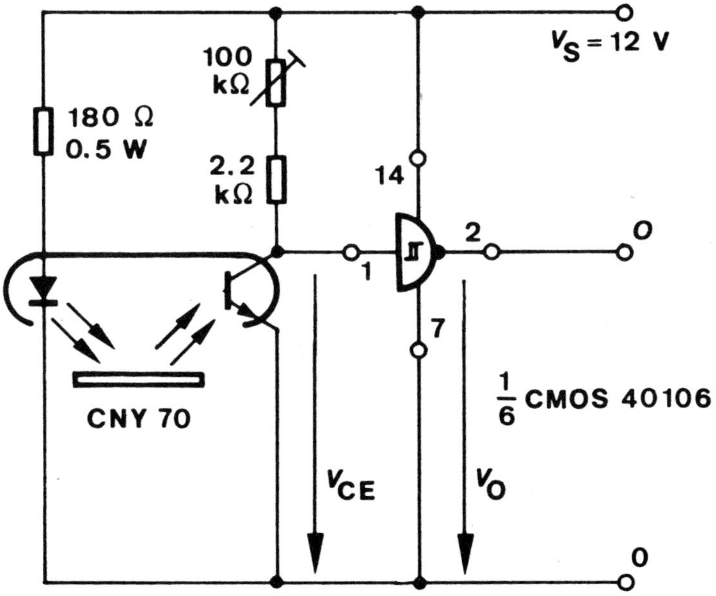

A digital application of a reflective optocoupler (AEG).

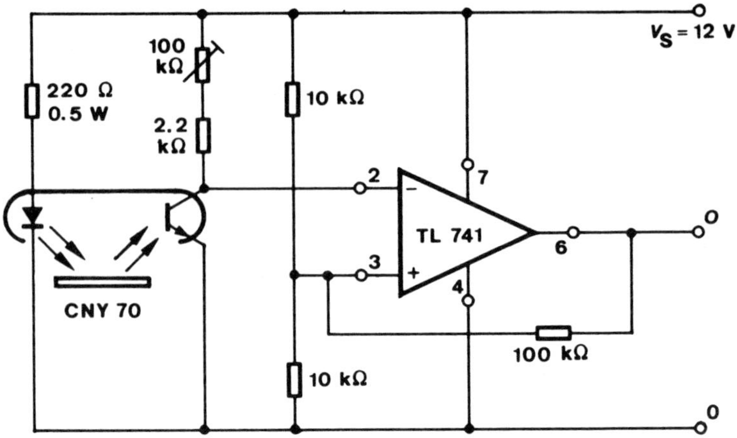

Improved digital reflective optocoupler (AEG).

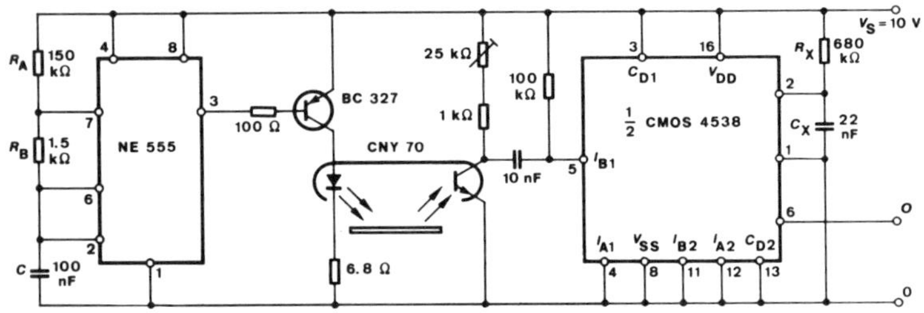

Proximity detector using a reflective optocoupler (AEG).

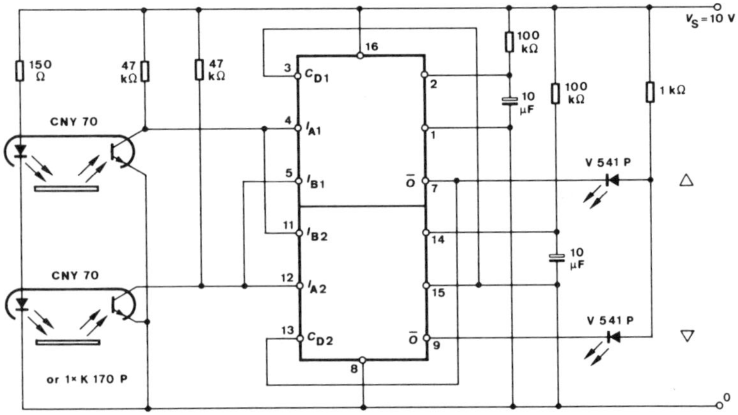

Motion detector sensor using reflective optocouplers (AEG).

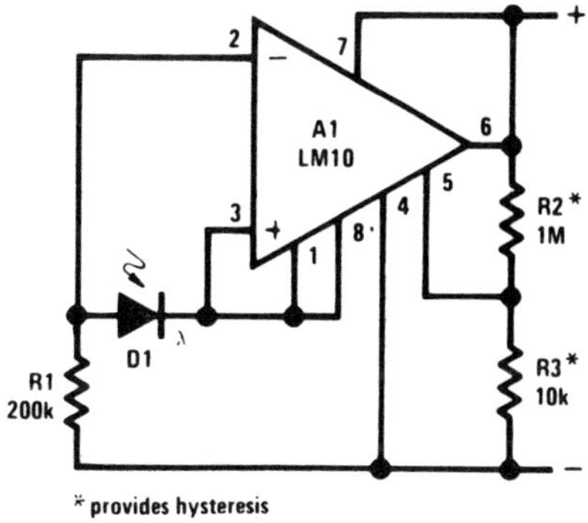

Light level sensor (NS).

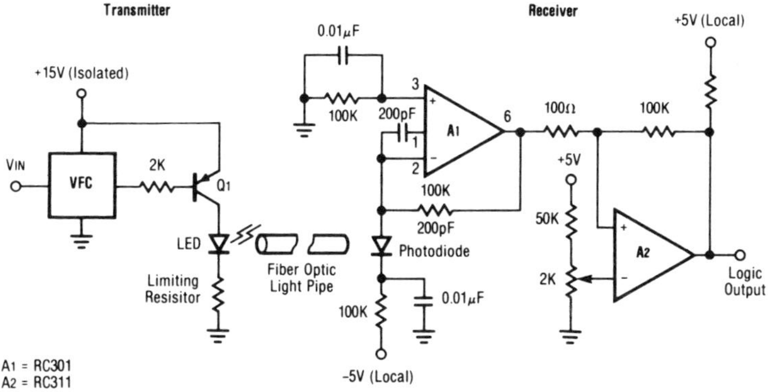

Fiberoptic transmission (R).

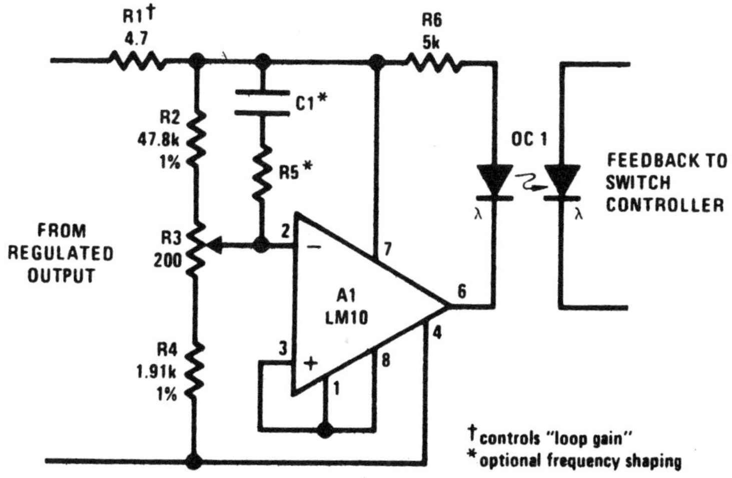

Isolated voltage sensor (NS).

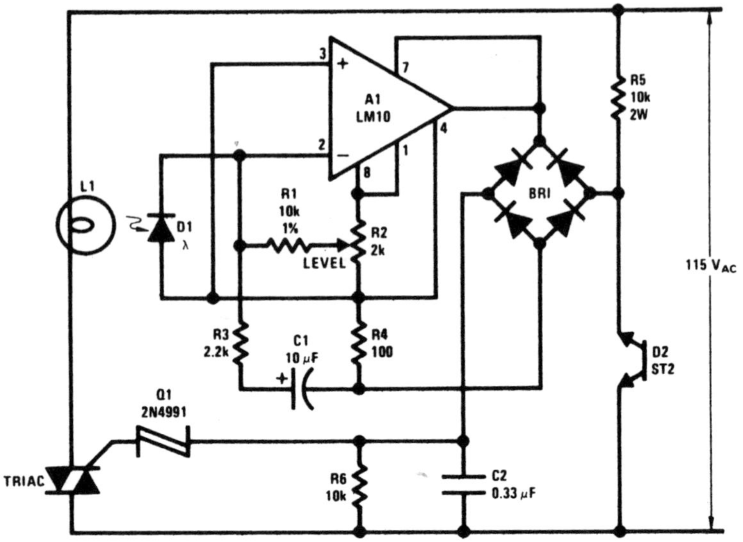

Light-level controller (NS).

Chapter 22

Interfacing and Computer Circuits

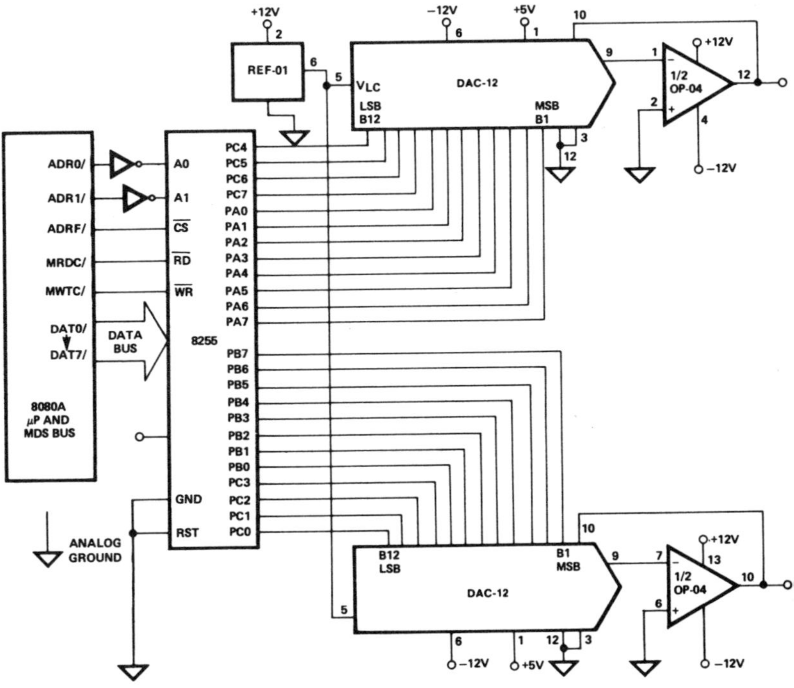

12-bit DAC to 8-bit interface (PM).

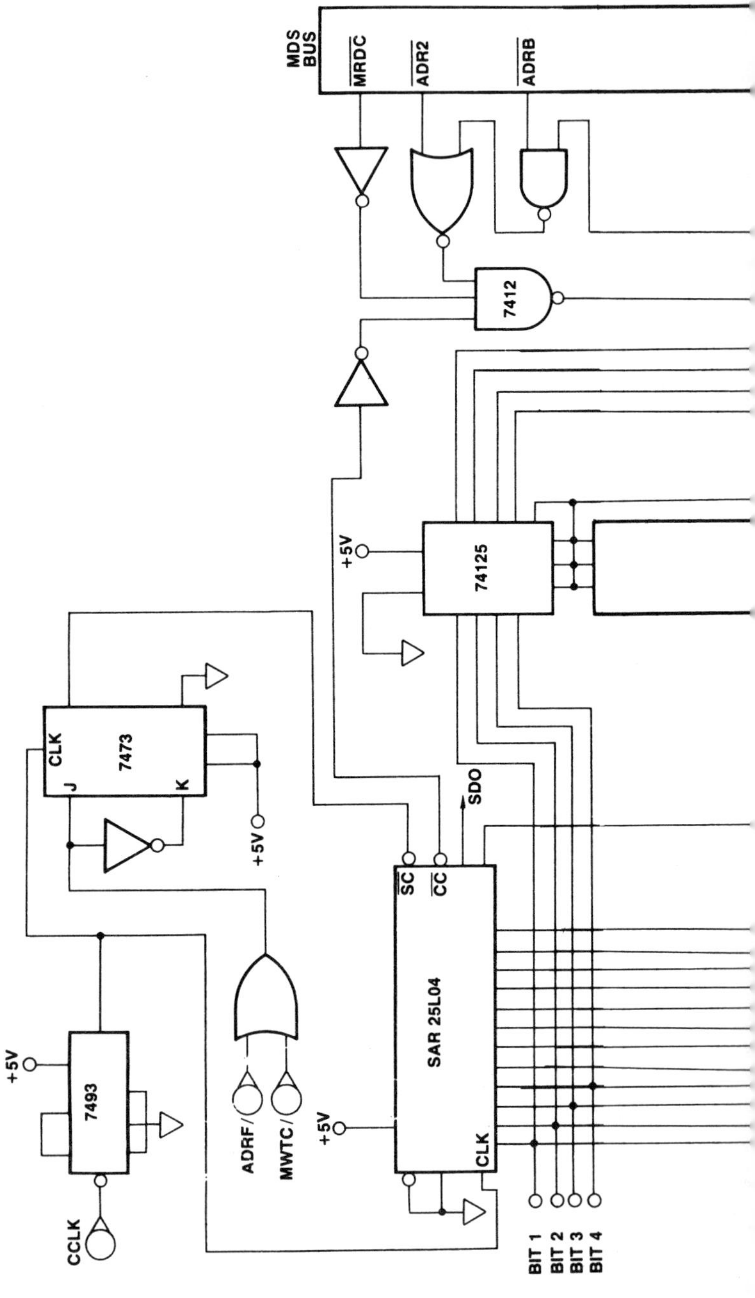
MDS BUS
MRDC
ADR2
ADRB
7412
+5V
74125
CLK
7473
J
K
+5V
SDO
SC
CC
SAR 25L04
CLK
+5V
7493
+5V
ADRF
MWTC
CCLK
BIT 1
BIT 2
BIT 3
BIT 4

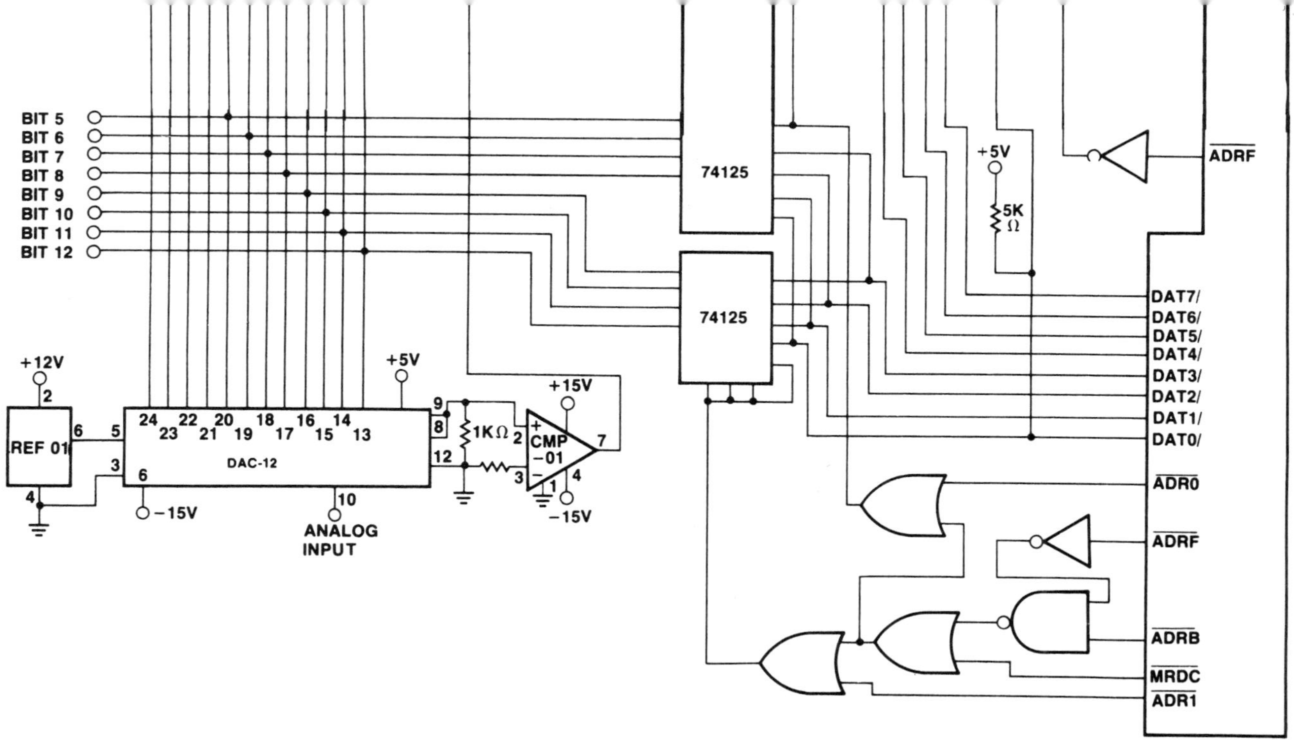

Complete 12-bit successive approximation ADC interface (PM).

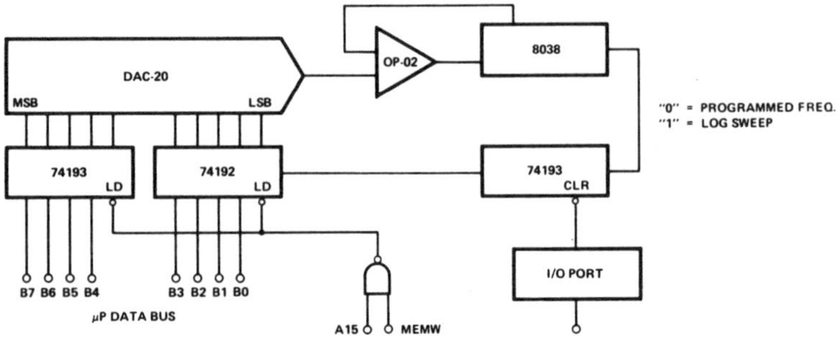

Microprocessor interface to function generator with software selectable sweep (PM).

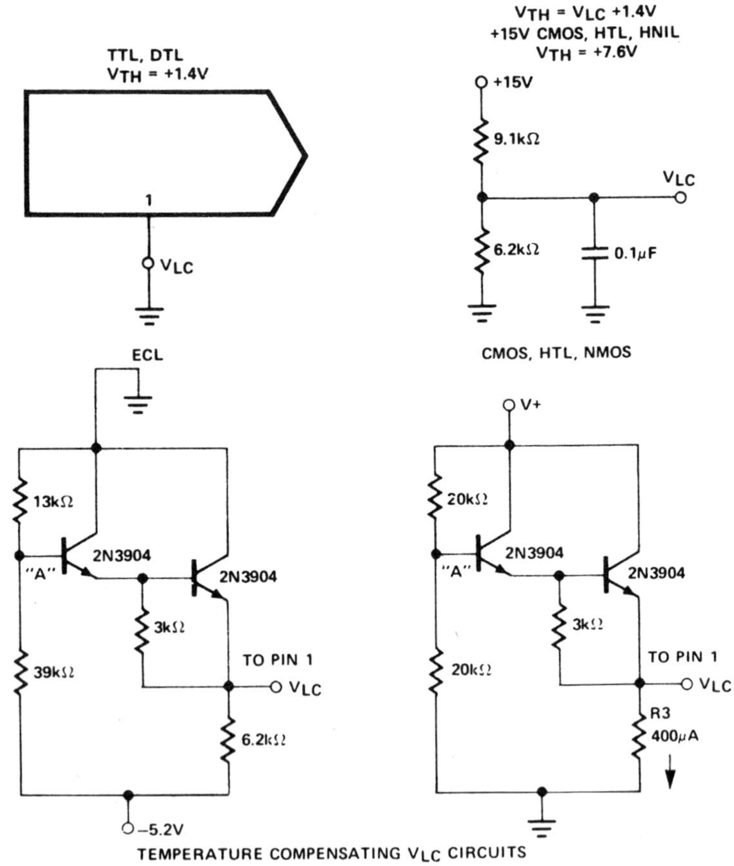

Interfacing with various logic families ($V_{TH}=V_{LC}+1.4V$) (PM).

Driving CMOS

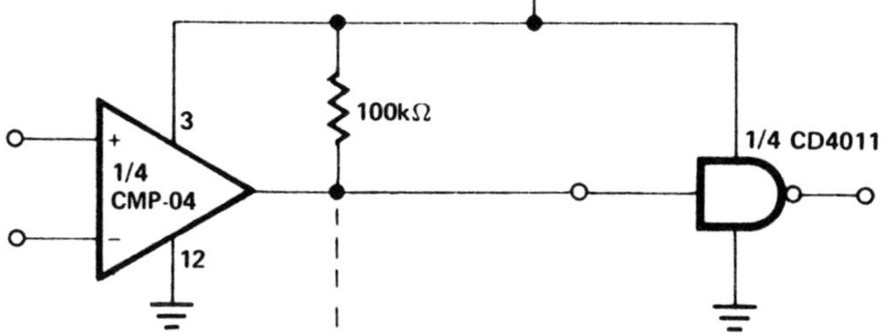

Driving TTL

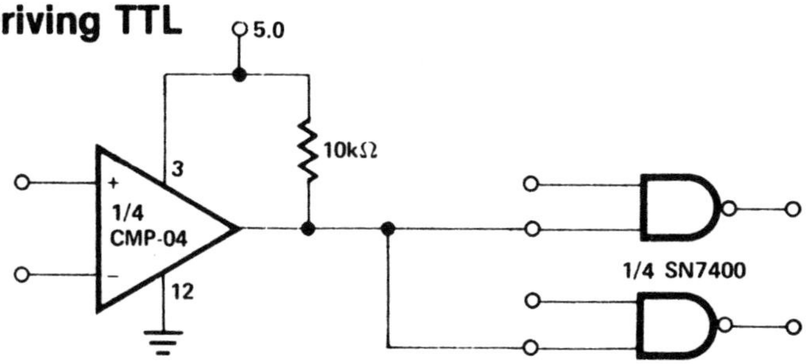

Typical interface (PM).

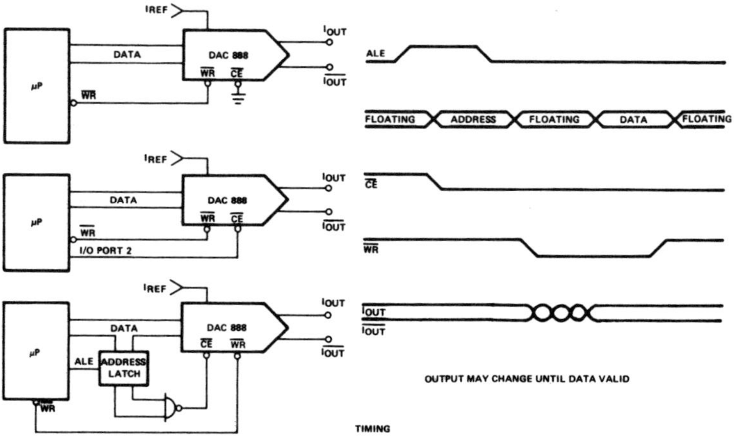

8048 Interface (PM).

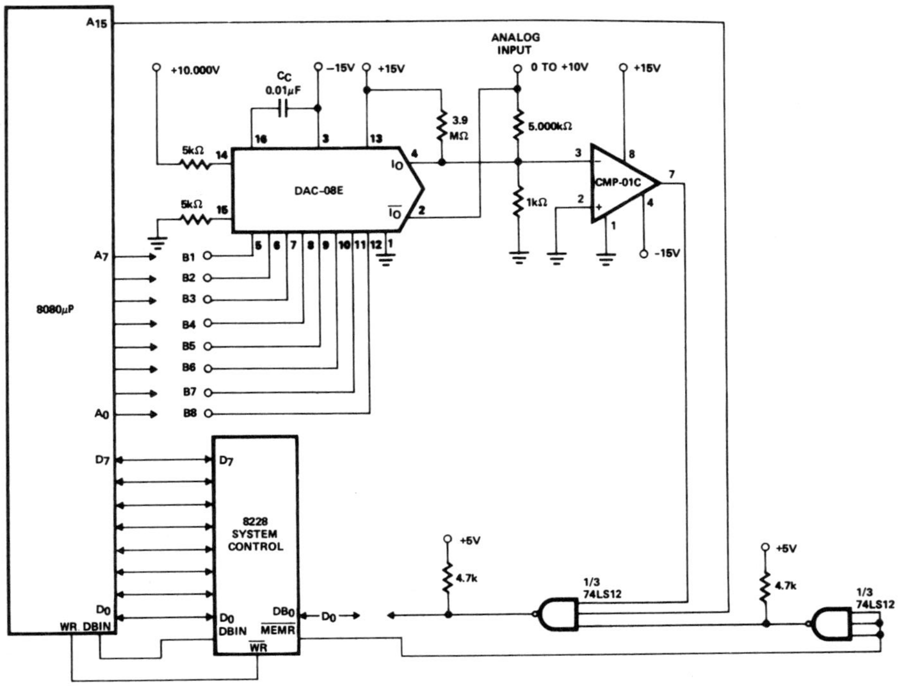

Complete schematic of DAC-08/8080A interface (PM).

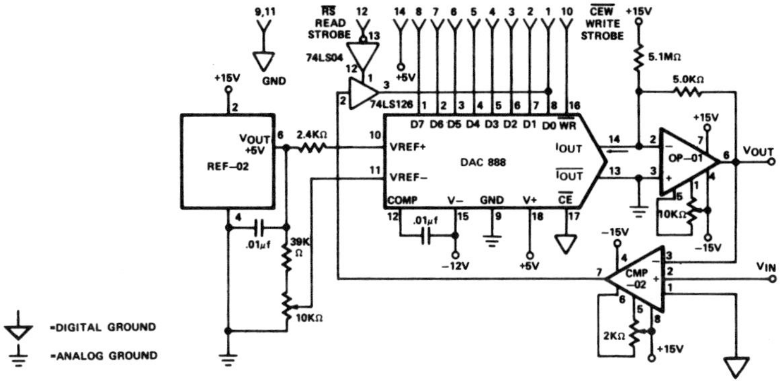

'Software SAR'A/D converter (with 6502 micro processor) (PM).

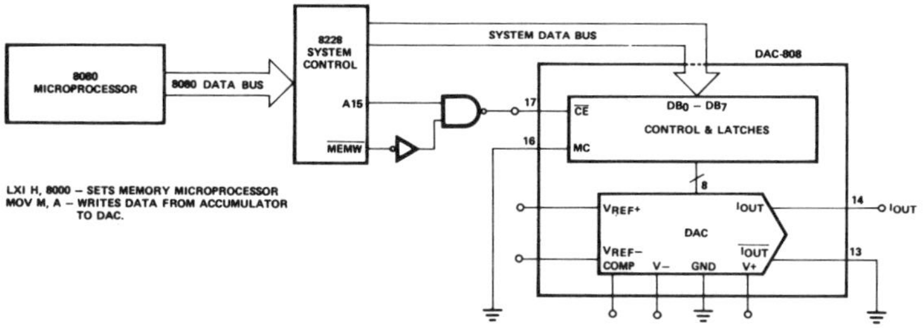

8080 Microprocessor interface—8-bit transfer (PM).

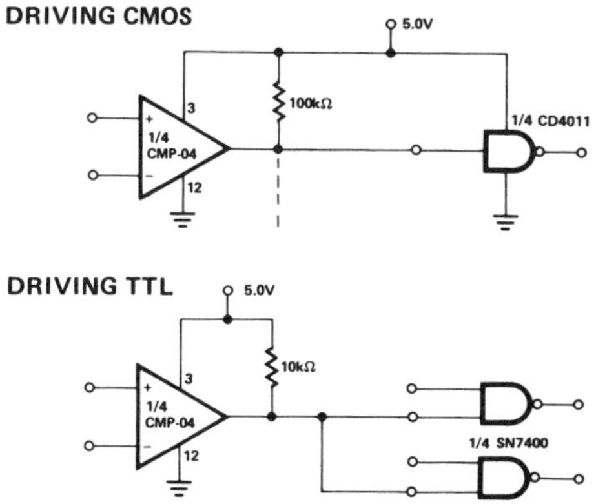

Typical interface (PM).

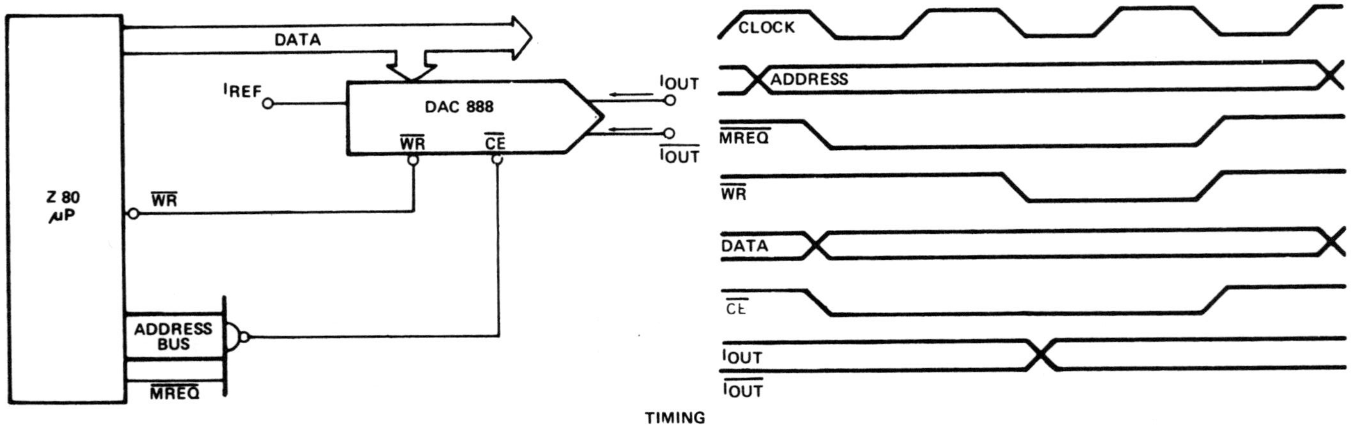

Z-80 Interface (PM).

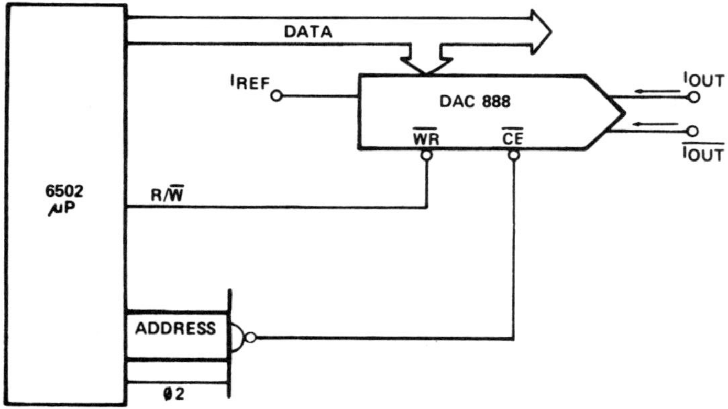

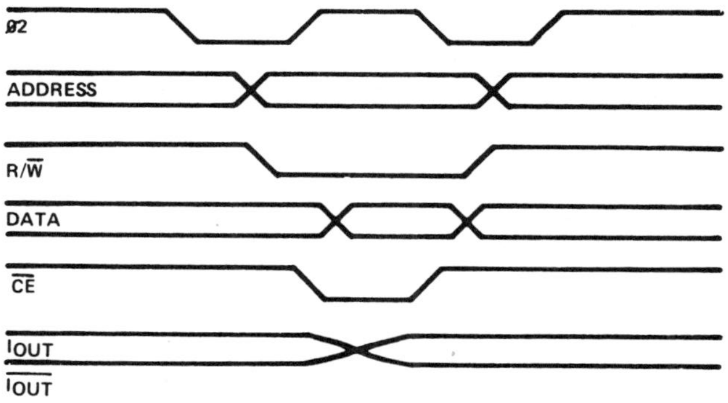

TIMING

6502 Interface (PM).

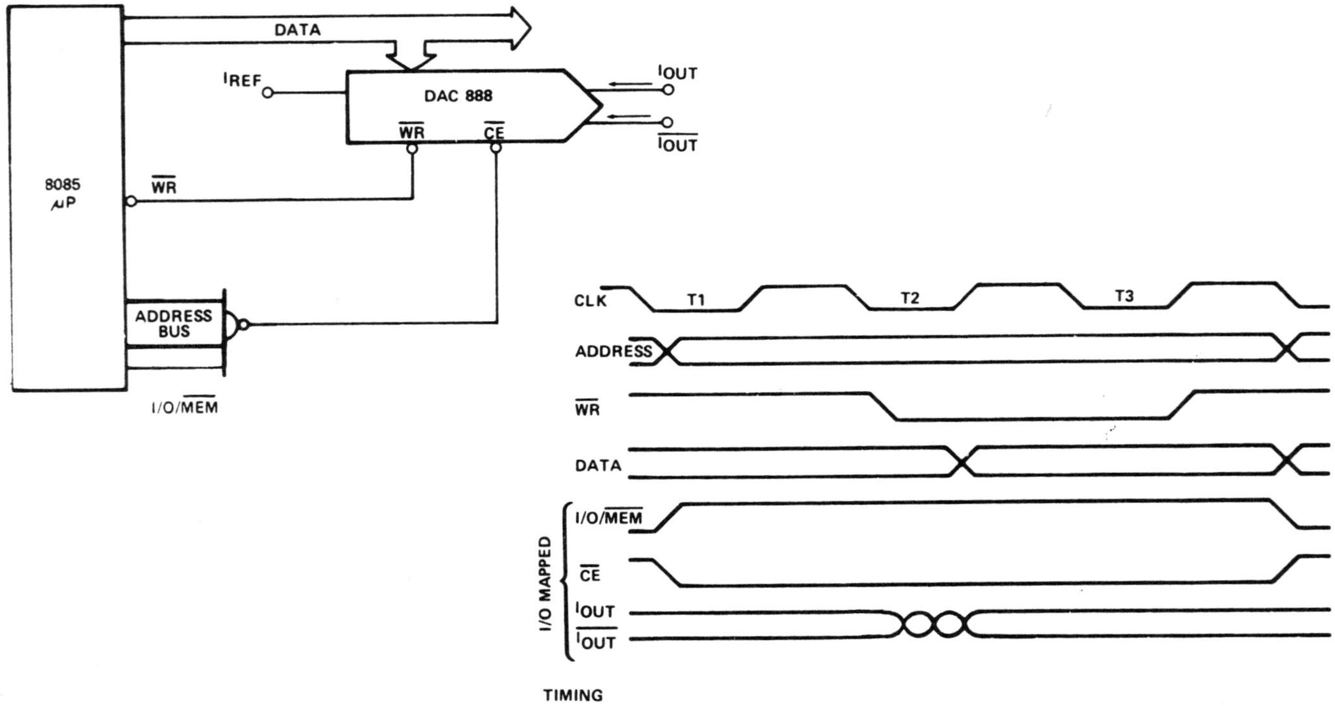

8080 Interface (PM).

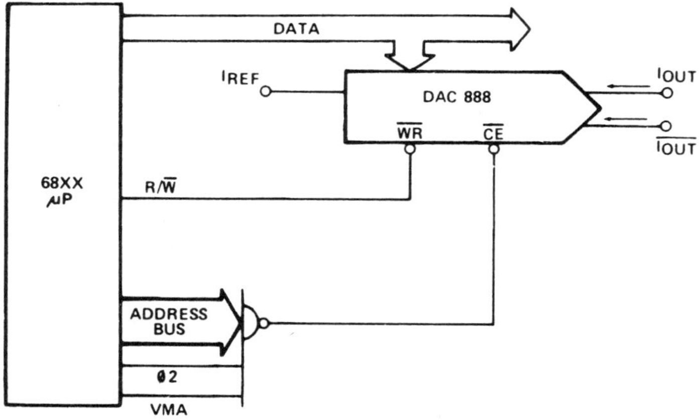

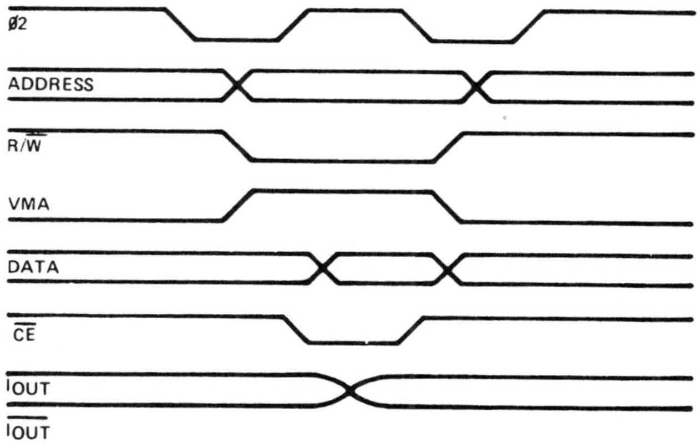

TIMING

6800, 6801, 6809 Interface (PM).

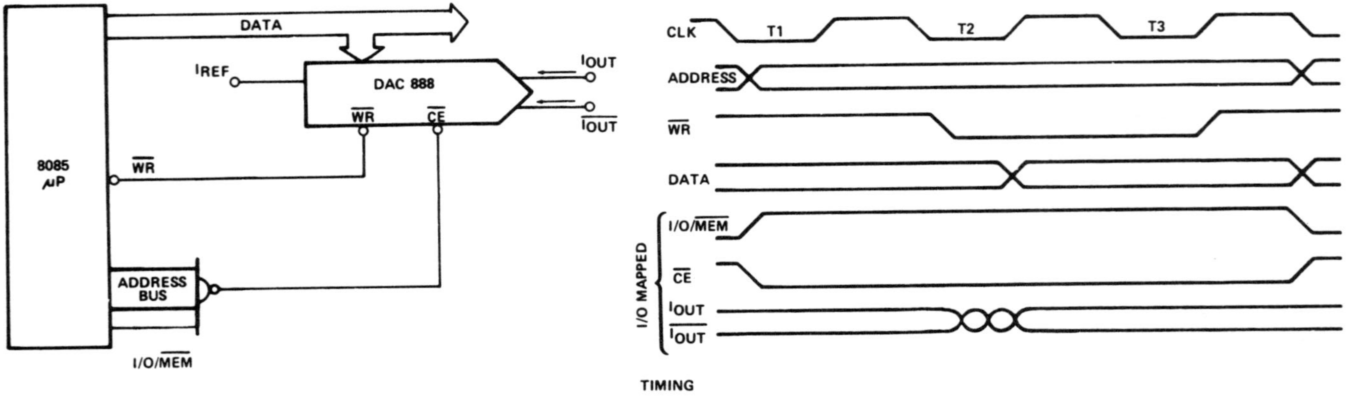

8085 Interface (PM).

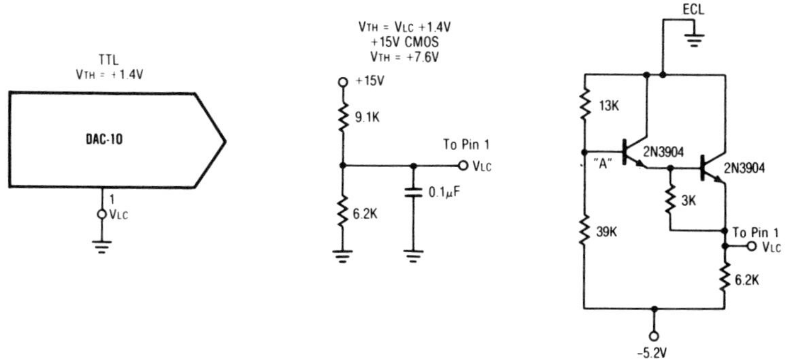

Interfacing with various logic families (R).

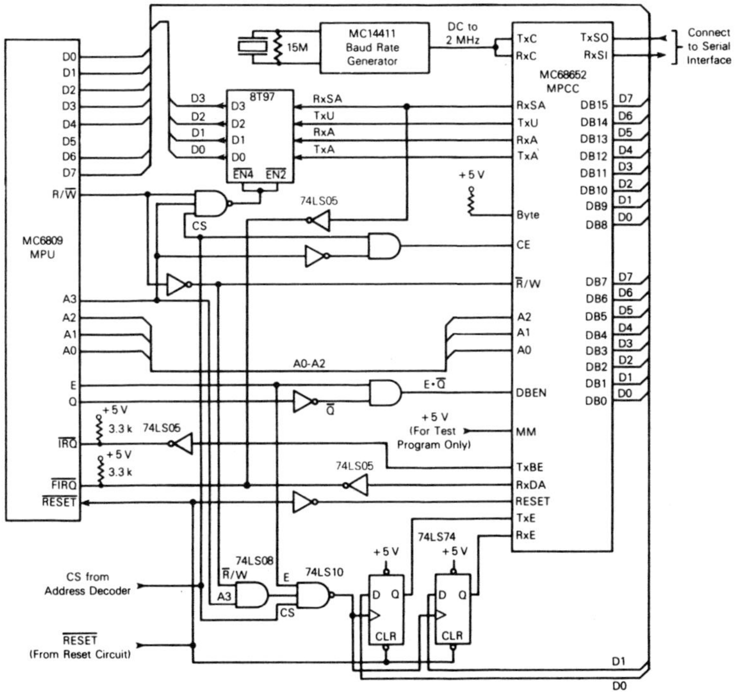

MC6809 to MC68652 interface connection, schematic diagram (M).

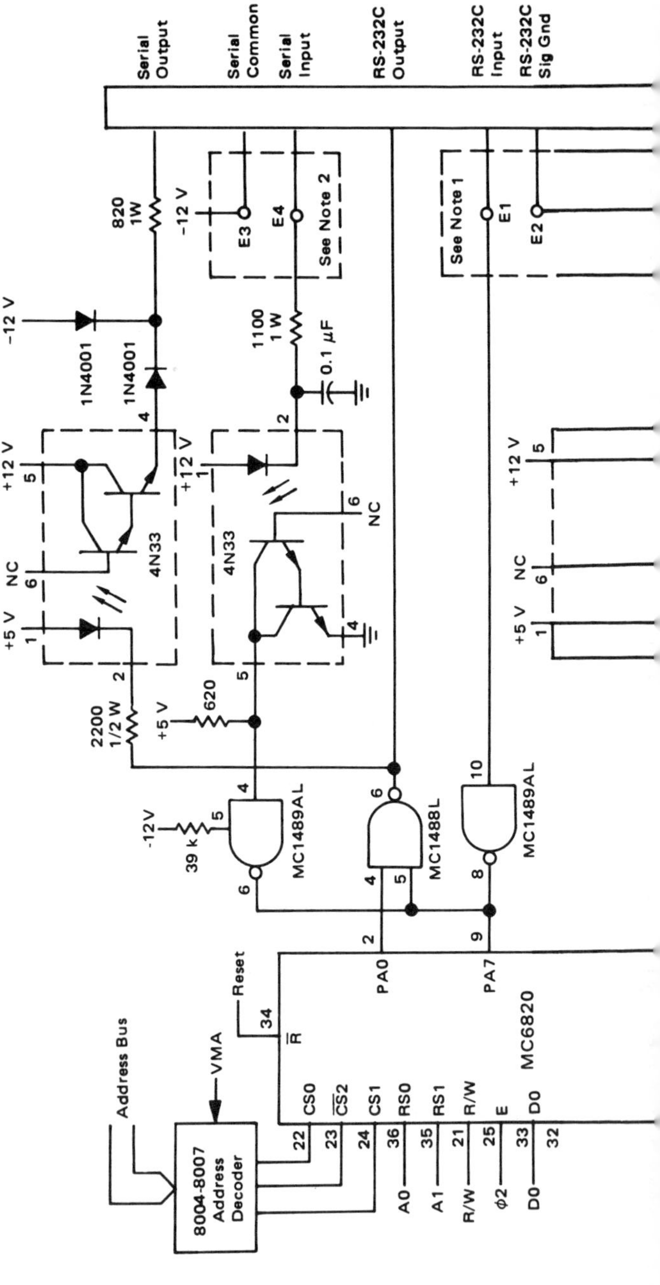

Serial Output
Serial Common
Serial Input
RS-232C Output
RS-232C Input
RS-232C Sig Gnd
820 1W
-12 V
E3
E4
See Note 2
See Note 1
E1
E2
-12 V
1N4001
1N4001
1100 1 W
0.1 μF
+12 V
4N33
4N33
NC
+5 V
2200 1/2 W
+5 V
620
MC1489AL
MC1488L
MC1489AL
-12V
39 k
Reset
Address Bus
VMA
8004-8007 Address Decoder
MC6820
PA0
PA7
CS0
CS2
CS1
RS0
RS1
R/W
E
D0
A0
A1
R/W
φ2
D0

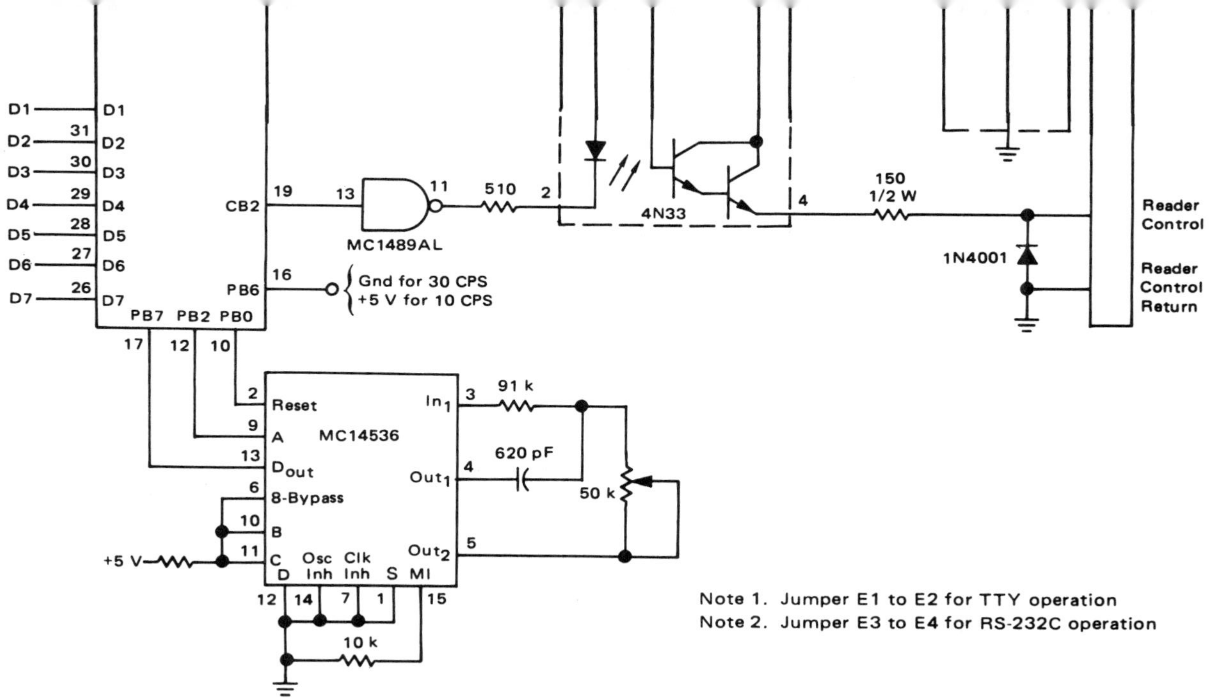

Note 1. Jumper E1 to E2 for TTY operation
Note 2. Jumper E3 to E4 for RS-232C operation

TTY/RS-232C interface used with MIKBUG ROM (M).

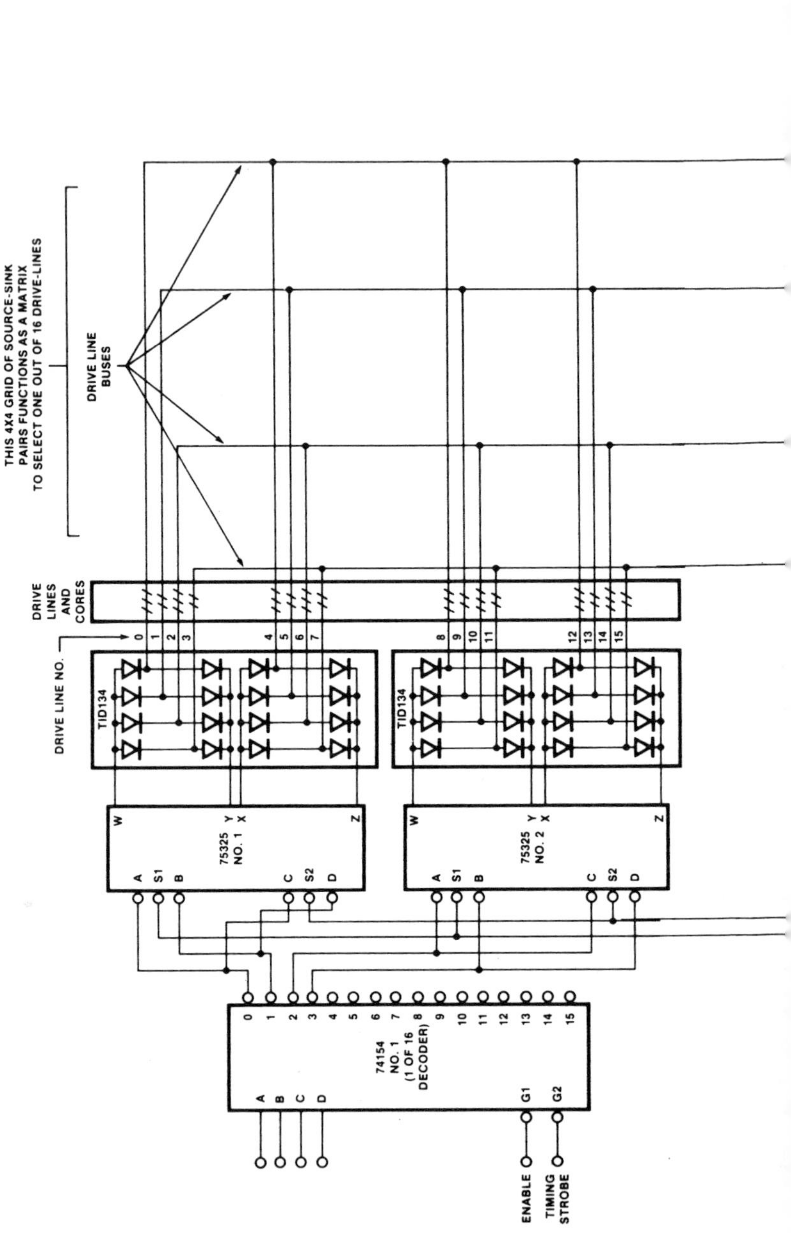

THIS 4X4 GRID OF SOURCE-SINK PAIRS FUNCTIONS AS A MATRIX TO SELECT ONE OUT OF 16 DRIVE-LINES
DRIVE LINE BUSES
DRIVE LINES AND CORES
DRIVE LINE NO.
TID134
75325 NO. 1
75325 NO. 2
W
Y
X
Z
A
S1
B
C
S2
D
74154 NO. 1 (1 OF 16 DECODER)
G1
G2
ENABLE
TIMING STROBE

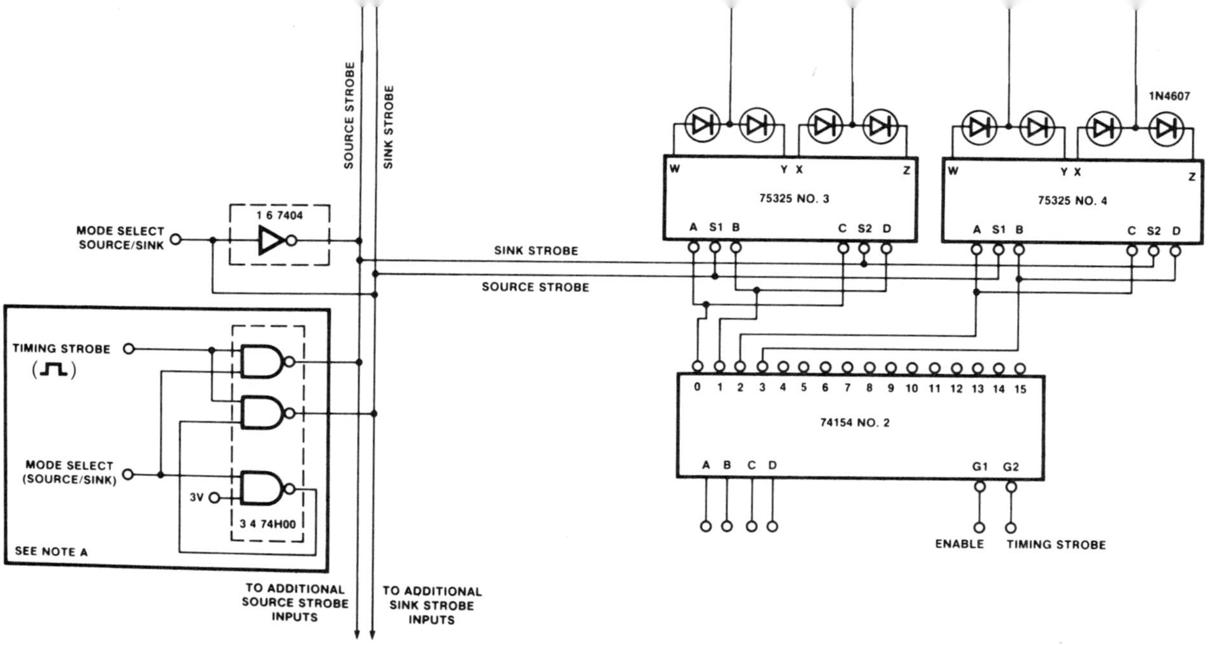

NOTE A This optional mode select and timing strobe technique can be used in place of the 7440 mode select and 74154 timing strobe when minimum time skew is desired.

75325 used as a memory driver to select one of sixteen drive lines (S).

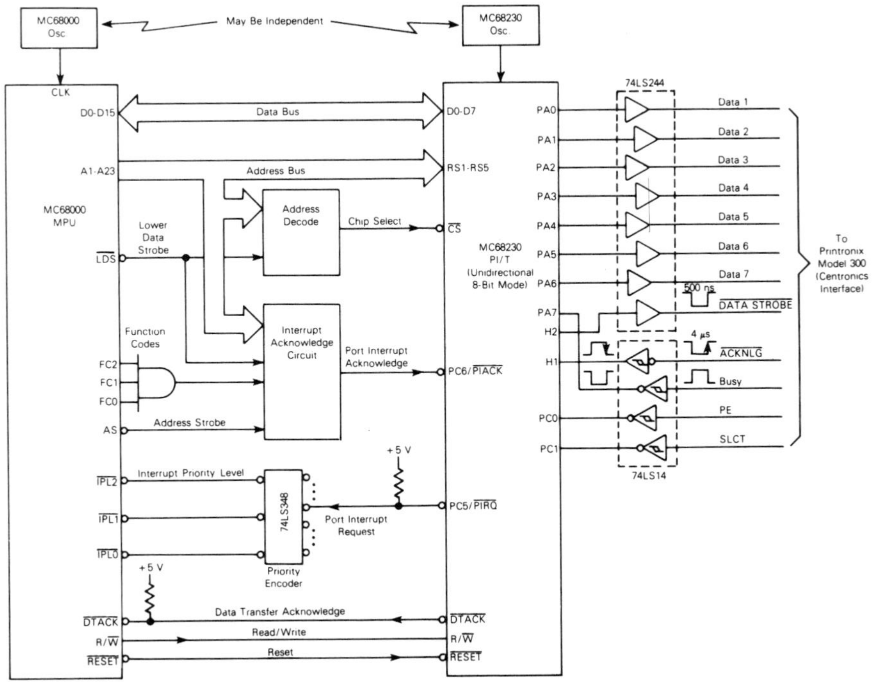

MC68000-PI/T-printer interface schematic diagram (M).

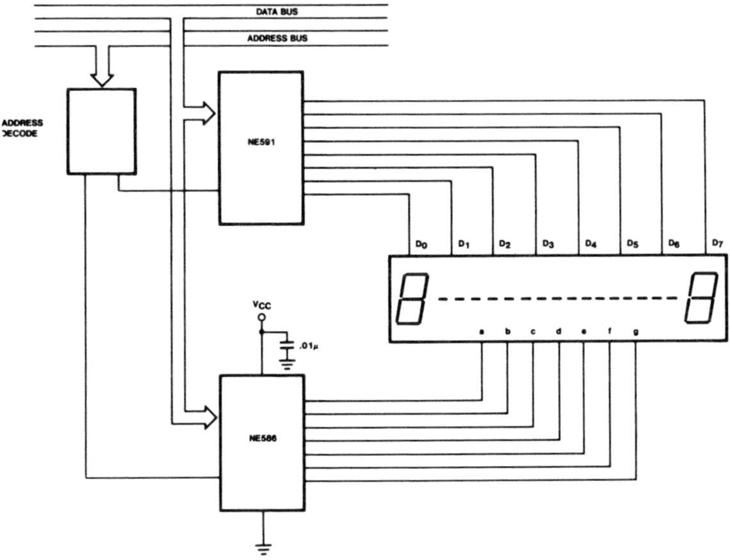

Interfacing 8-digit LED display with μP bus (S).

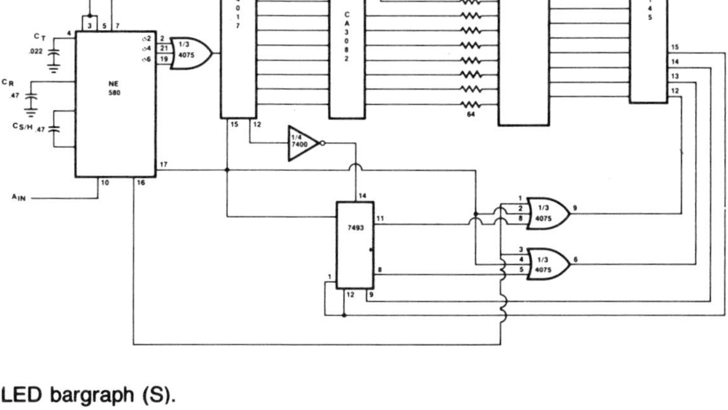

LED bargraph (S).

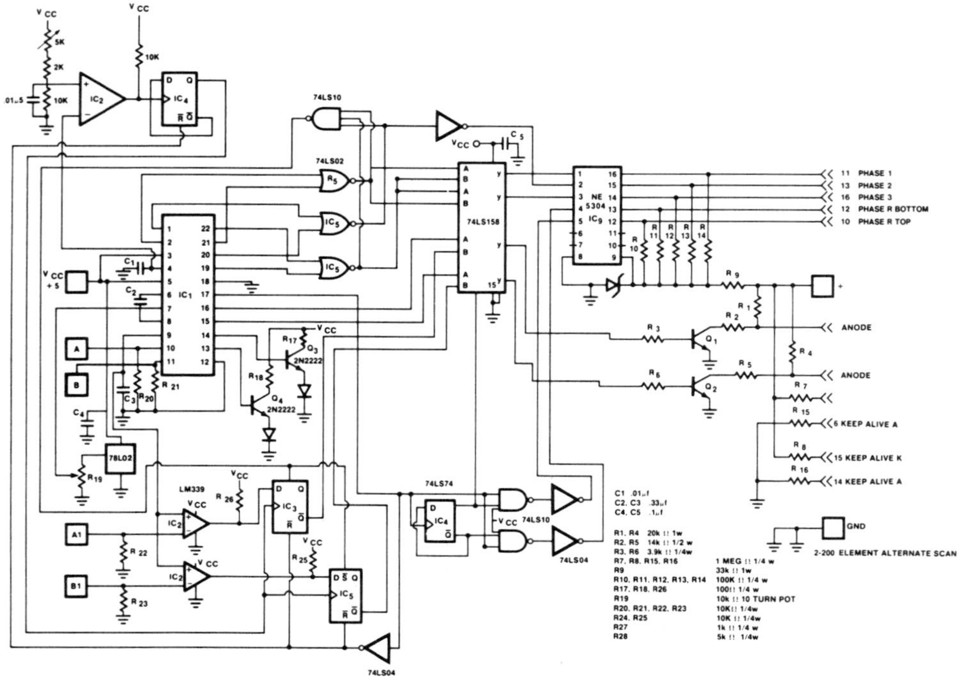

Dual resettable bargraph (S).

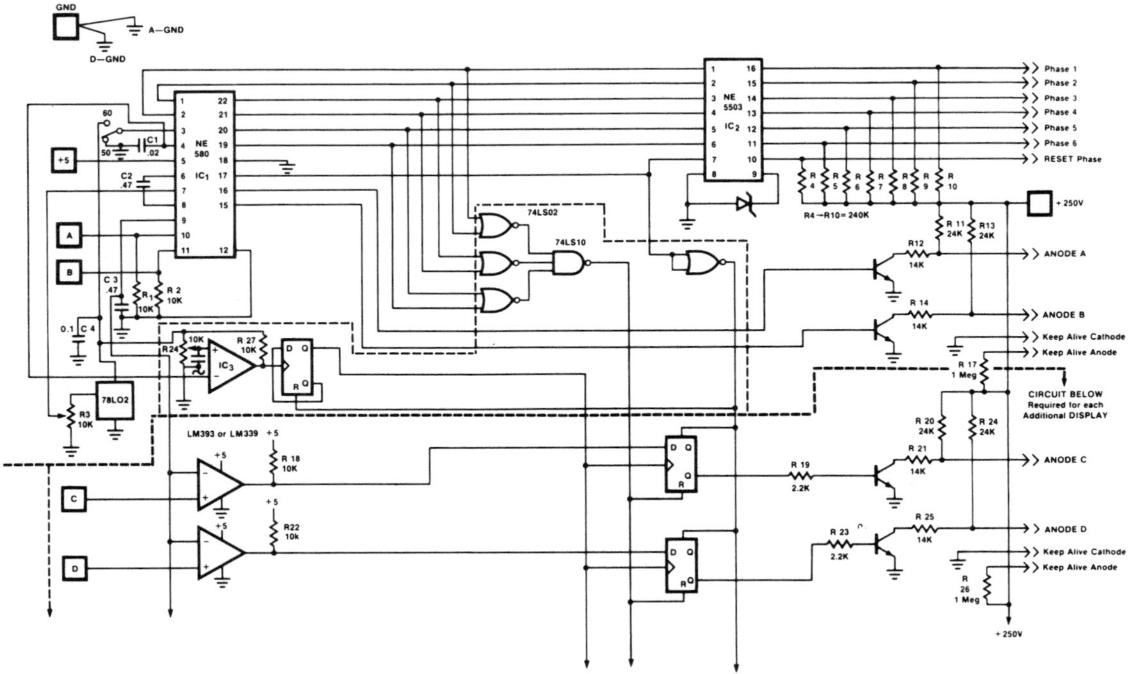

Driving multiple displays with common cathode phasing (S).

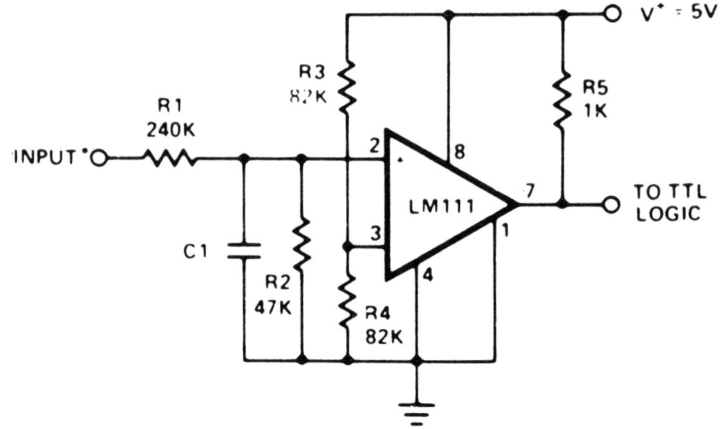

*Values shown are for a 0 to 30V logic swing and a 15V threshold.

†May be added to control speed and reduce susceptability to noise spikes.

TTL interface with high level logic (S).

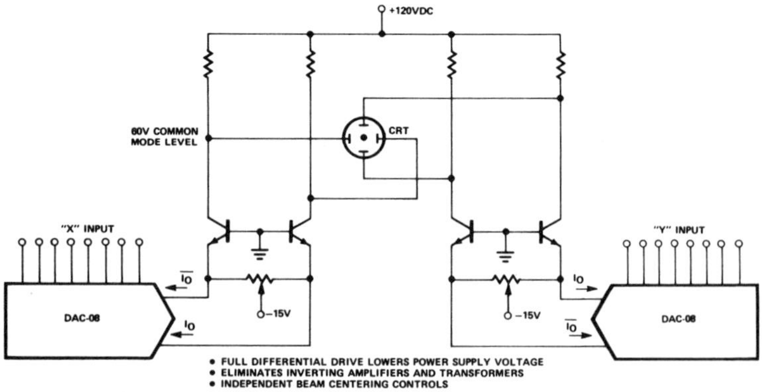

CRT display driver (PM).

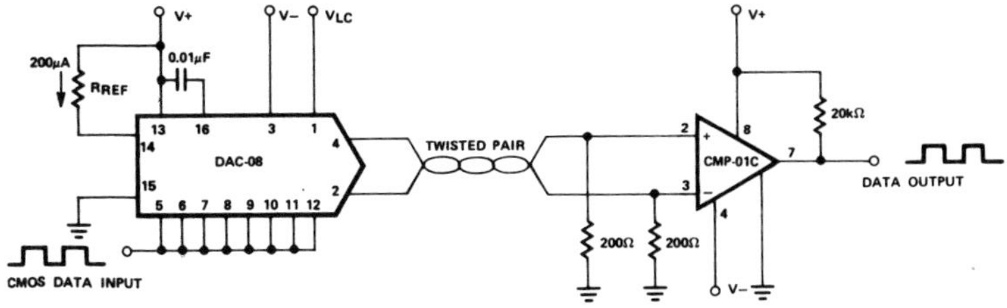

CMOS differential line driver/receiver (PM).

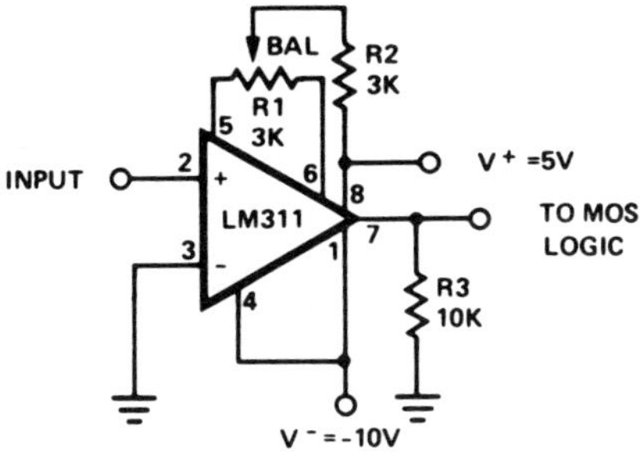

Zero crossing detector driving MOS logic (S).

Chapter 23

Miscellaneous Circuits

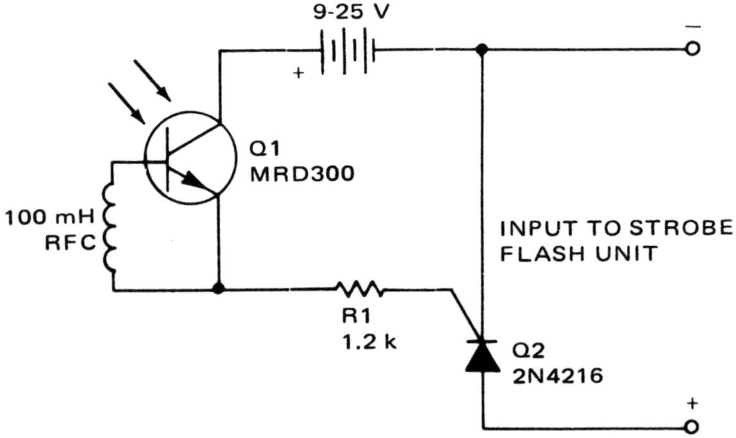

Strobeflash slave adapter (M).

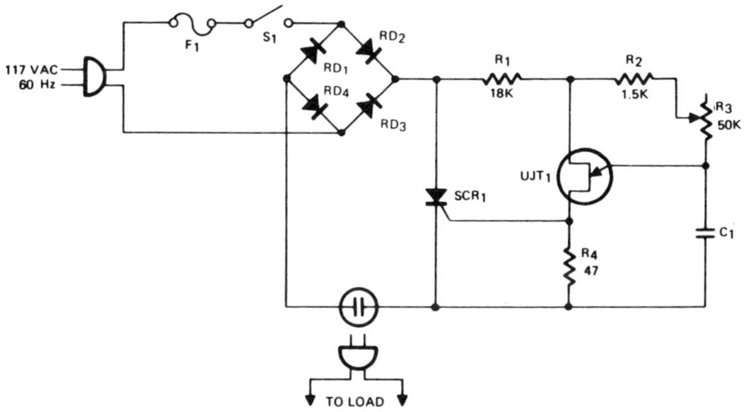

UJT1	Unijunction Transistor WEP 310		
R1	18K Ohm, 2 Watt Resistor, WEP22-4126	C1	0.1 μF, 150V Capacitor
R2	1.5K Ohm, 1 Watt Resistor WEP22-3100	S1	Switch WEP TS101
R3	50K Ohm Linear Potentiometer		Transistor Sockets WEP8051
R4	47 Ohm, 1 Watt Resistor WEP22-3136		Printed Circuit Board
	Socket and Plug		
	Heat Exchanger WEP8022		

RD1-4 SCR1 F1	Motors up to 2 Amps Lamps up to 400 Watts or one WEP6253	Motors up to 5 Amps Lamps up to 1,000 Watts or one WEP6273
	5 Amp Fuse WEP FG5-2	10 Amp Fuse

SCR light dimmer (W).

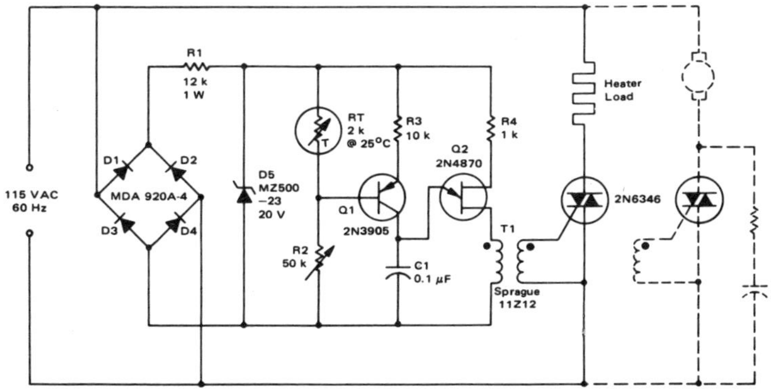

Temperature-sensitive heater control (M).

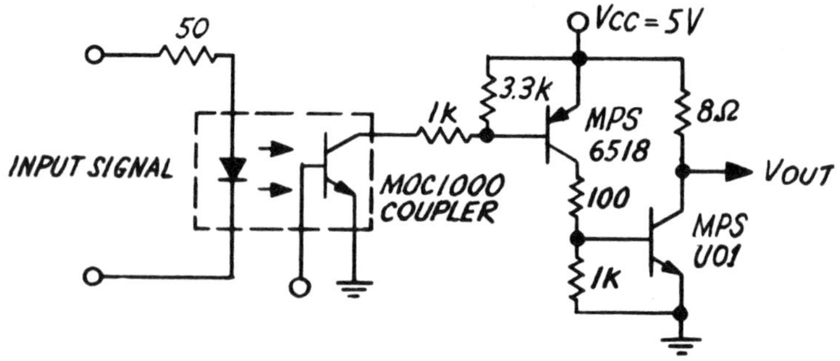

Pulse amplifier (M).

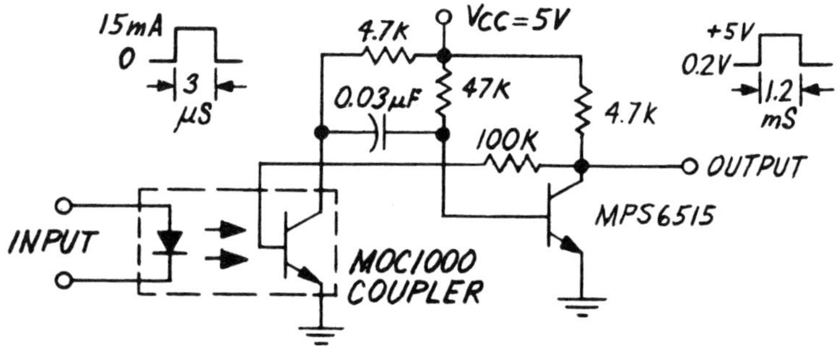

Pulse stretcher (M).

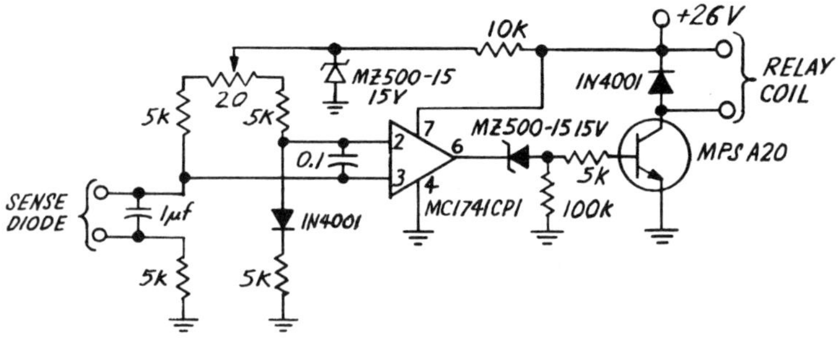

Temperature sensor (M).

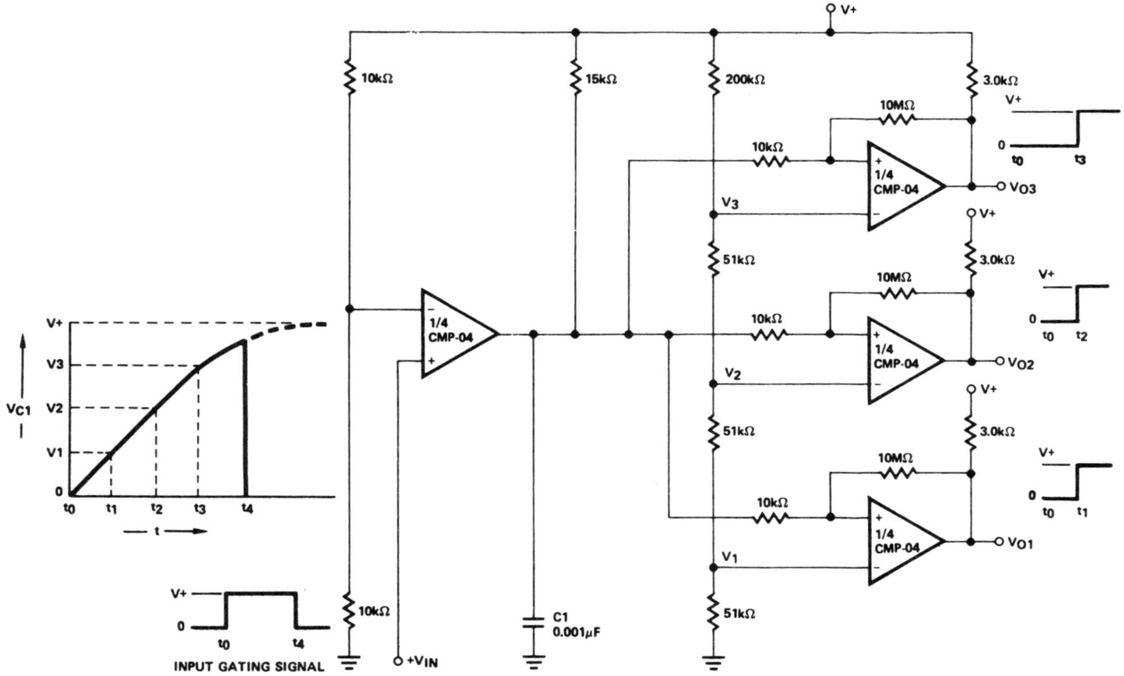

Time delay generator (PM).

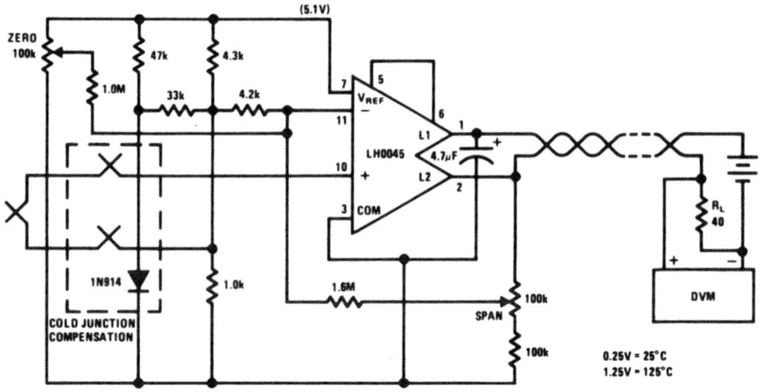

Remote sensing digital thermometer (NS).

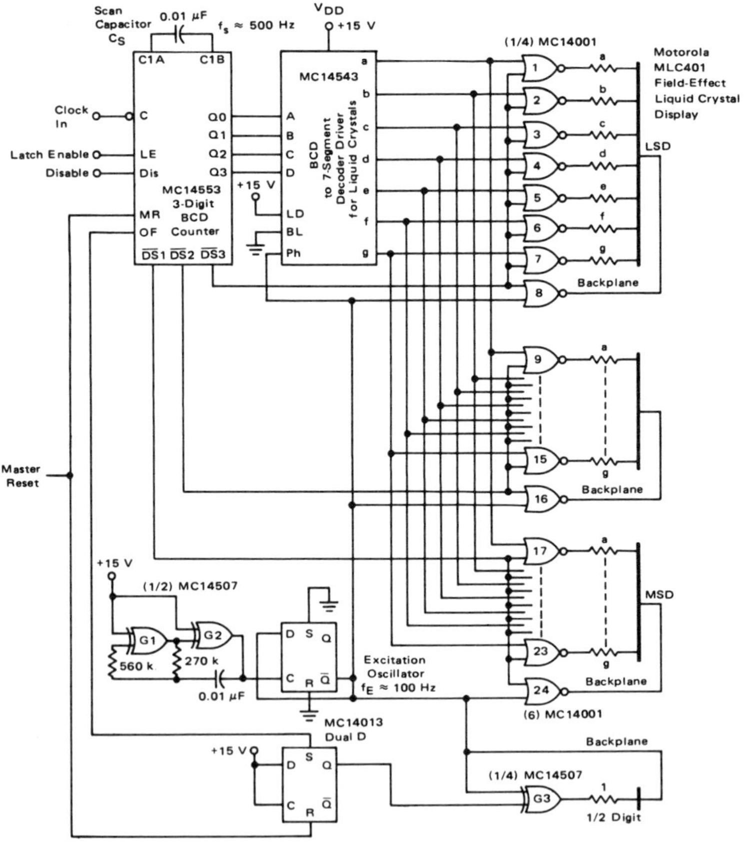

3½-digit multiplexed MLC401 field-effect liquid crystal display (M).

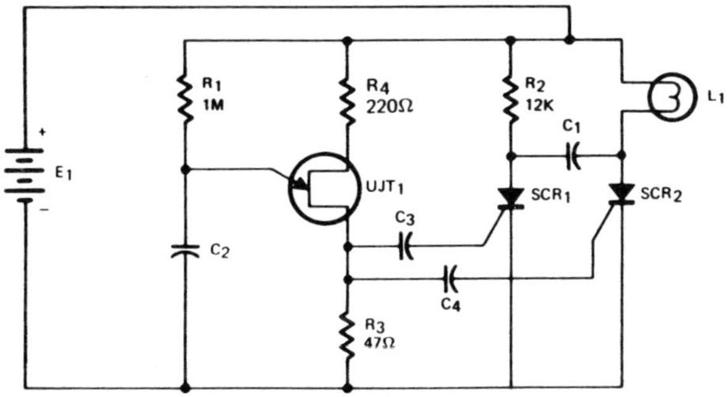

A

RD_{1-4}	20 Amp, 200 Volt Silicon Rectifier WEP4085
SCR_{1-2}	9 Amp, 200 Volt Silicon Controlled Rectifier WEP6273
C_1	50 μF, 160 Volt Electrolytic Capacitor
C_2	0.05 μF Capacitor
R_1	1 K Ohm, 1 Watt Resistor WEP22-3096
R_2	33 K Ohm, ½ Watt Resistor WEP22-2132
R_3	1 M Ohm, ½ Watt Resistor WEP22-2096
S_1	Toggle Switch WEP TS-101
L_1	Neon Bulb
L_2	Lamp
	Printed Circuit Board
	Line Cord WEP33-140
	Lamp Socket

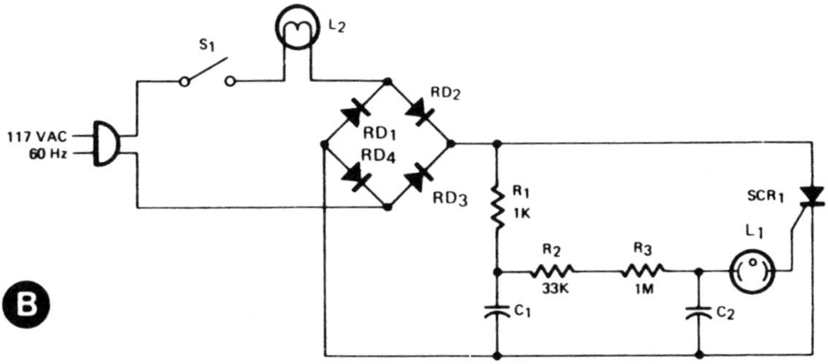

B

SCR_1, SCR_2	5 Amp, 50 Volt Silicon Controlled Rectifier WEP6241
UJT_1	Unijunction Transistor WEP310
R_1	1 M Ohm, ½ Watt Resistor (for 60 flashes per minute) 500 K Ohm & Potentiometer (for variable frequency)
R_2	12 K Ohm, 2 Watt Resistor WEP22-4122
R_3	47 Ohm, ½ Watt Resistor WEP22-2064
R_4	220 Ohm, ½ Watt Resistor WEP22-2080
C_1	2 μF, 50 Volt Electrolytic Capacitor WEP AX2-50
C_2	0.47 μF, 20 Volt Capacitor
$C_{3,4}$	0.1 μF, 50 Volt Capacitor
E_1	24 Volt Battery
L_1	24 Volt Lamp
	Transistor/SCR Sockets WEP8050
	Printed Circuit board

SCR light flasher circuits (W).

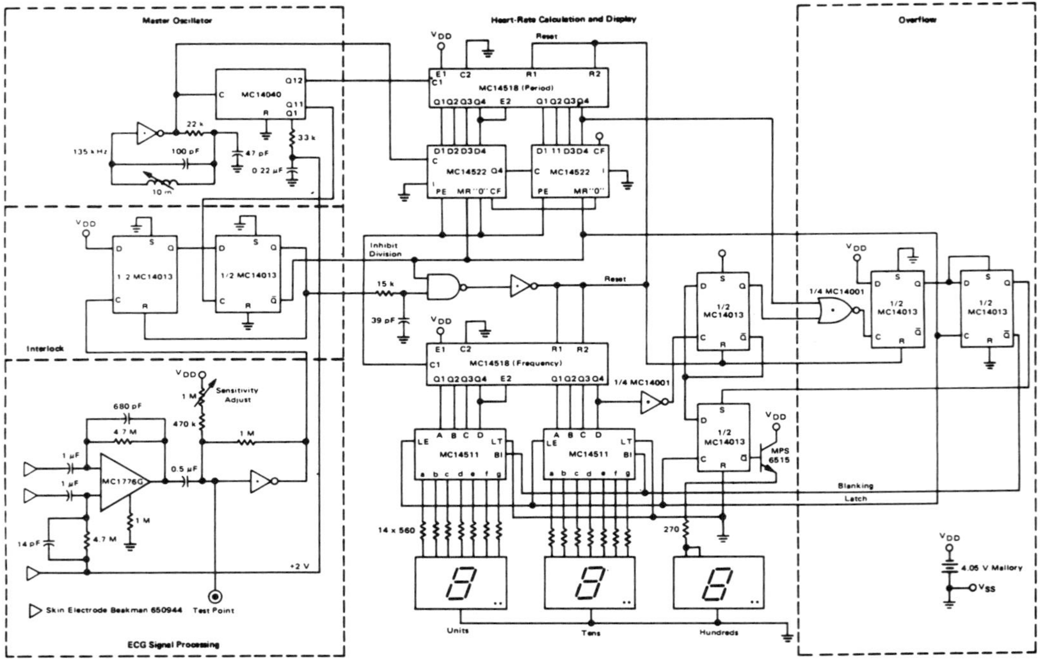

*MC14572 Hex Functional Gate **Common Cathode Display HP5082-7740

Circuit diagram of the heart-rate monitor (M).

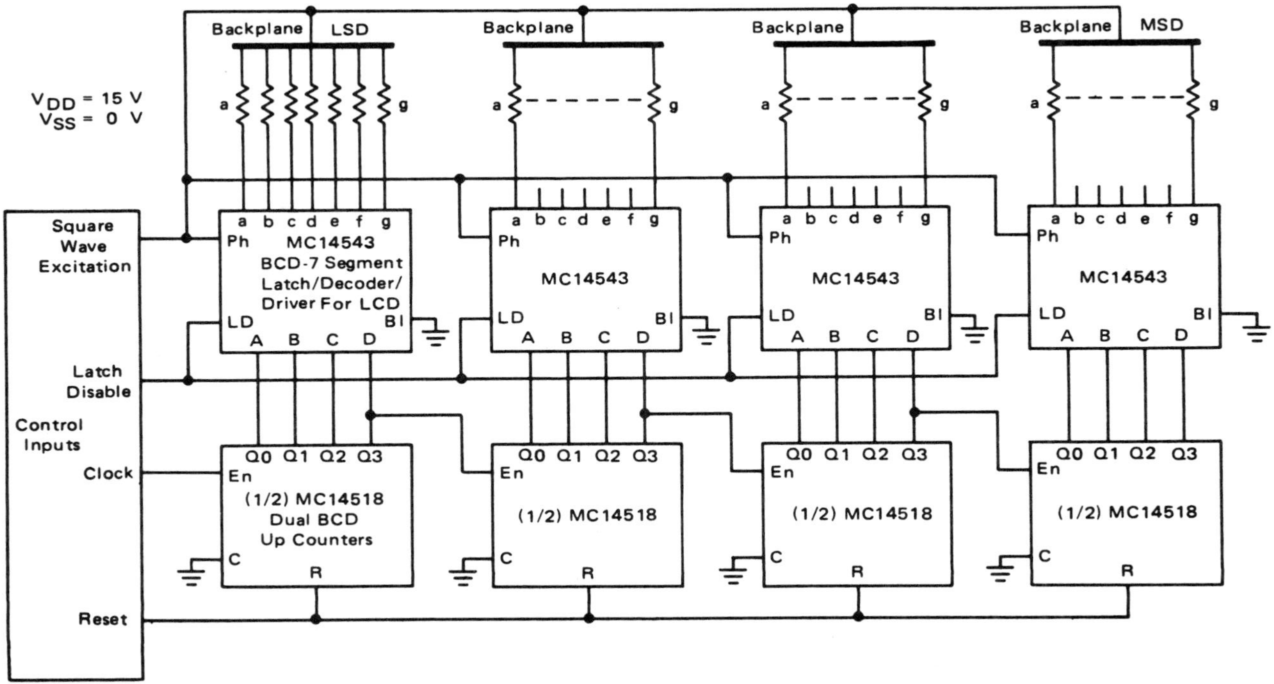

4-digit direct drive liquid crystal display (LCD) (M).

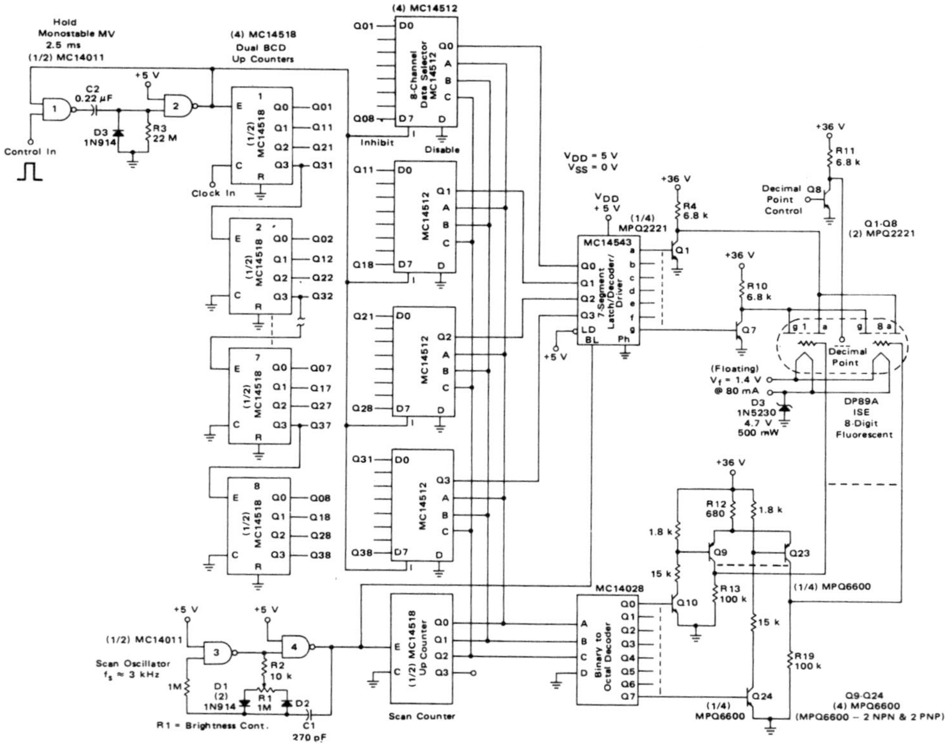

8-digit fluorescent triode display system (M).

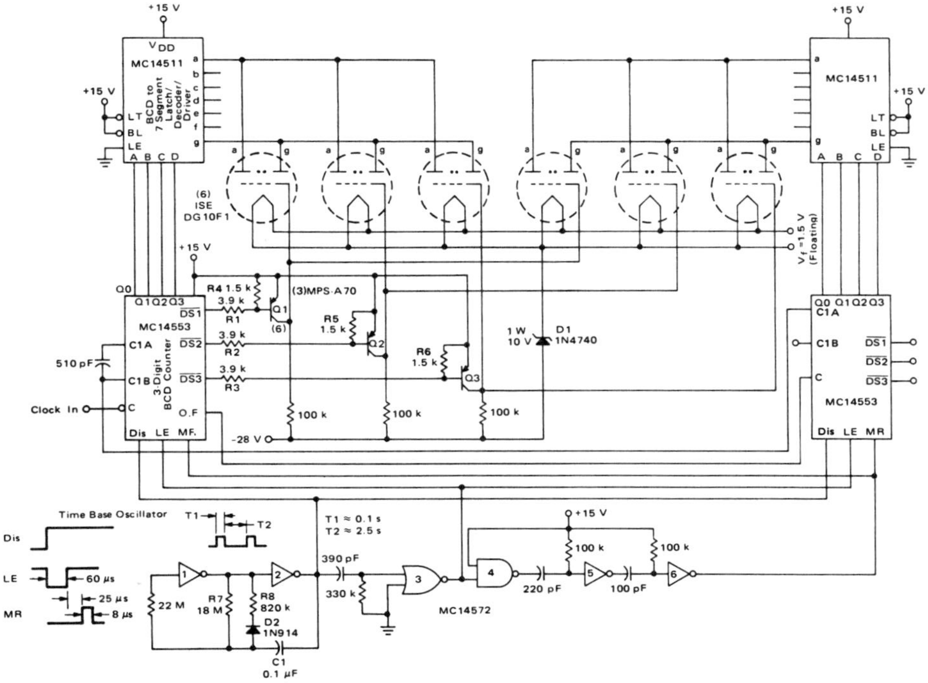

6-digit fluorescent triode display (M).

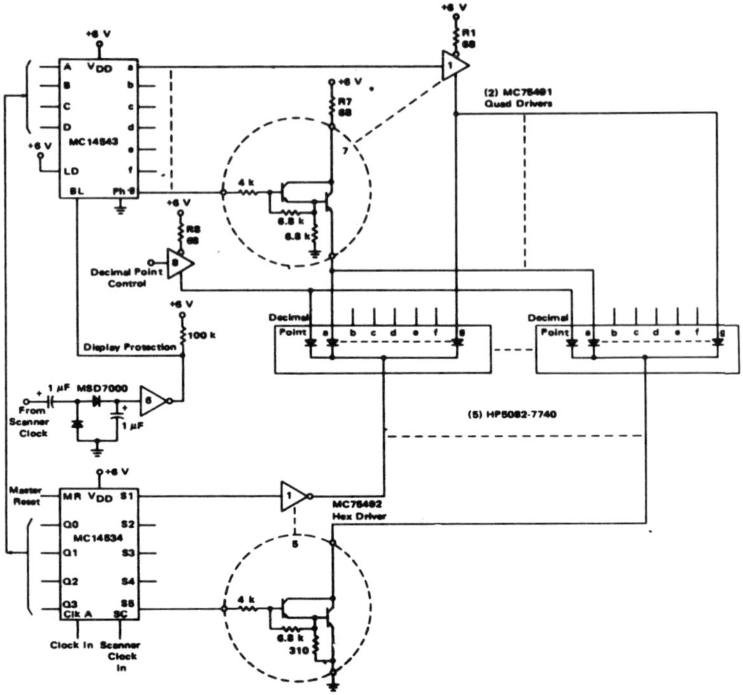

5-digit LED display (M).

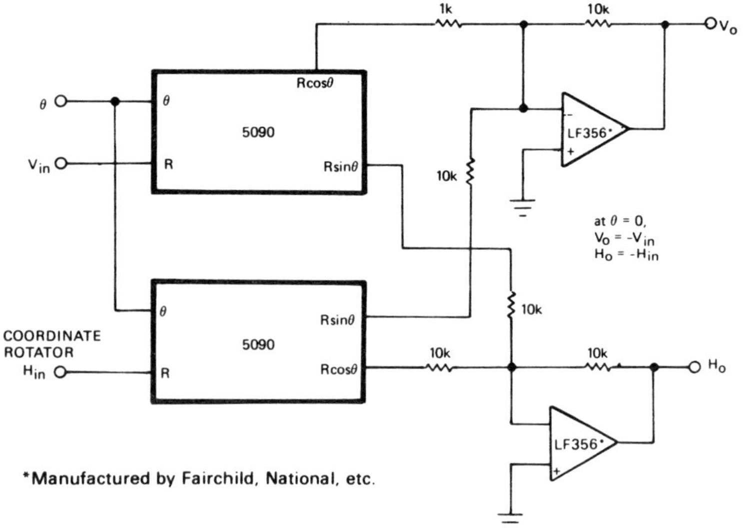

Coordinate rotator (OE).

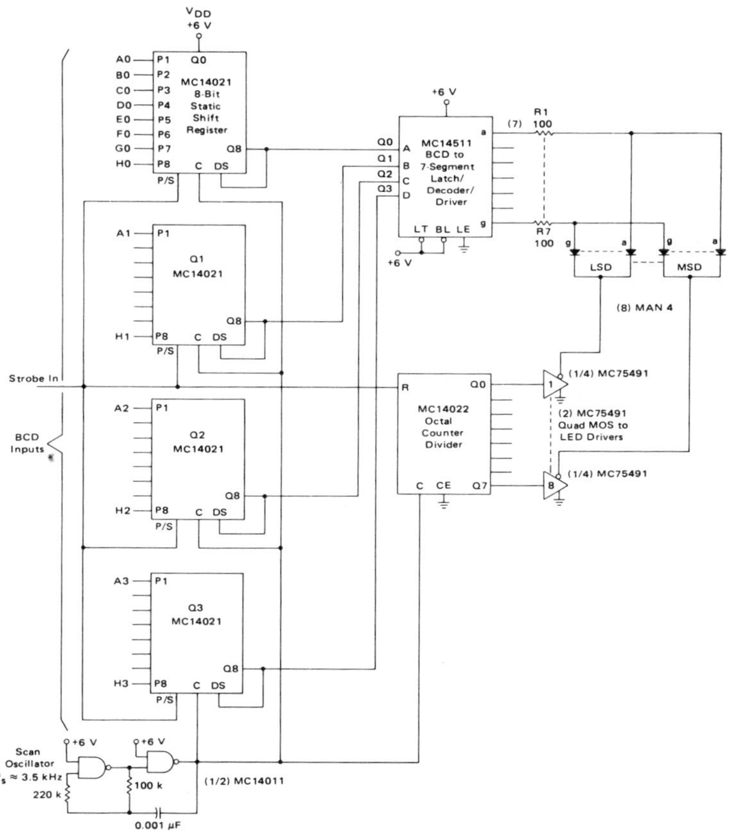

8-digit multiplexed LED display (M).

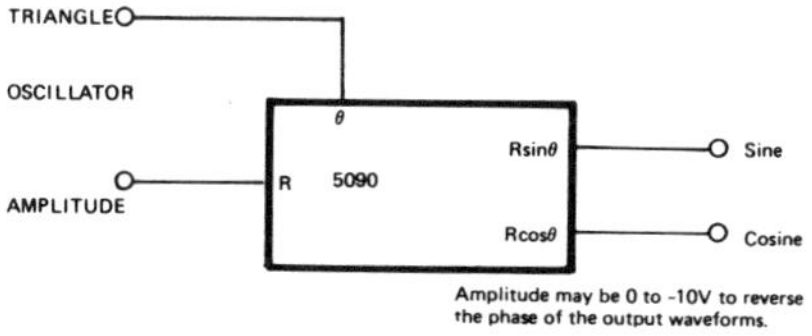

Sine and cosine function generation/circle generator (OE).

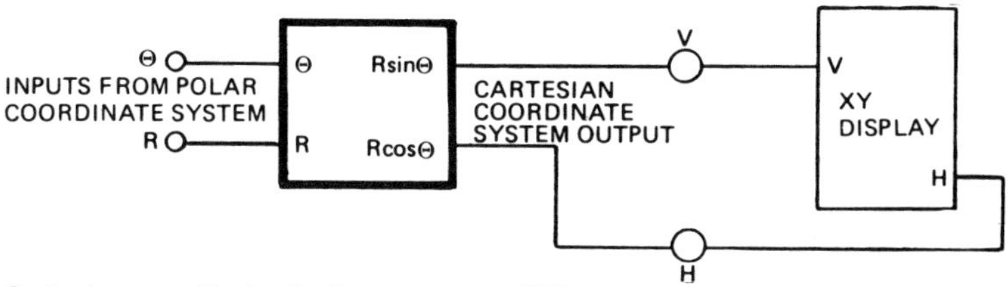

Cartesian coordinate display generator (OE).

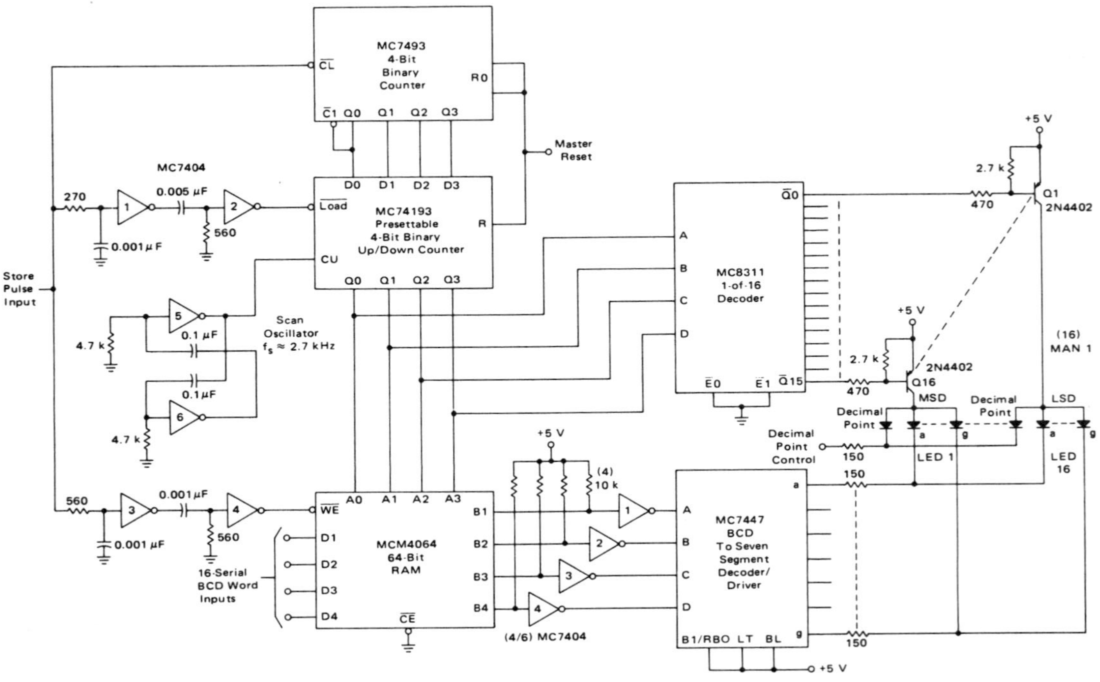

16-digit multiplexed display for LEDs (M).

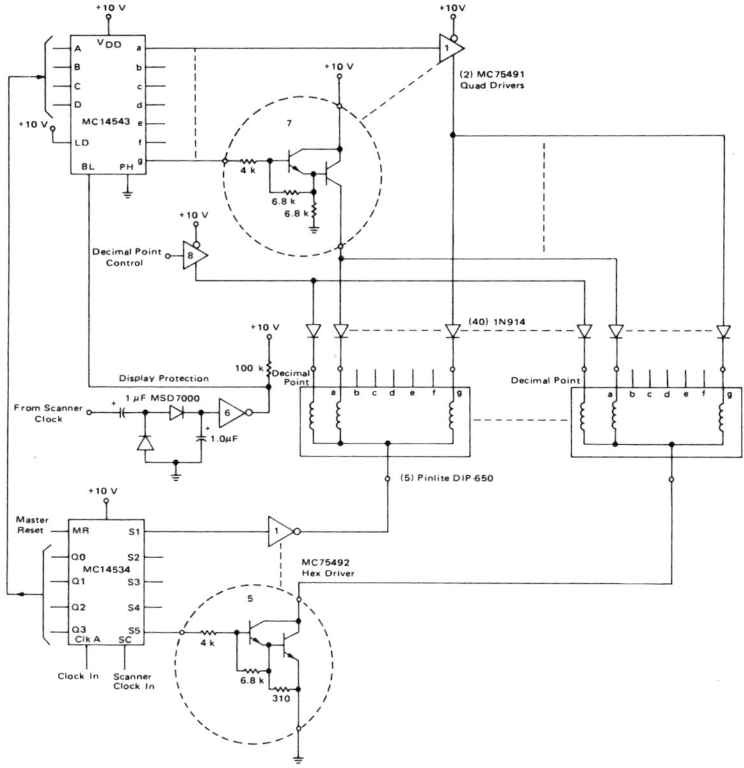

5-digit incandescent display (M).

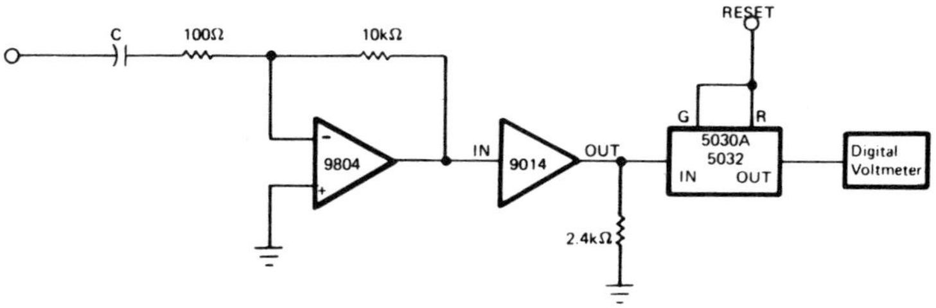

Peak-to-peak noise measurement, lower -3dB frequency; $f = 1/[2\pi C(10^2)]$ (OE).

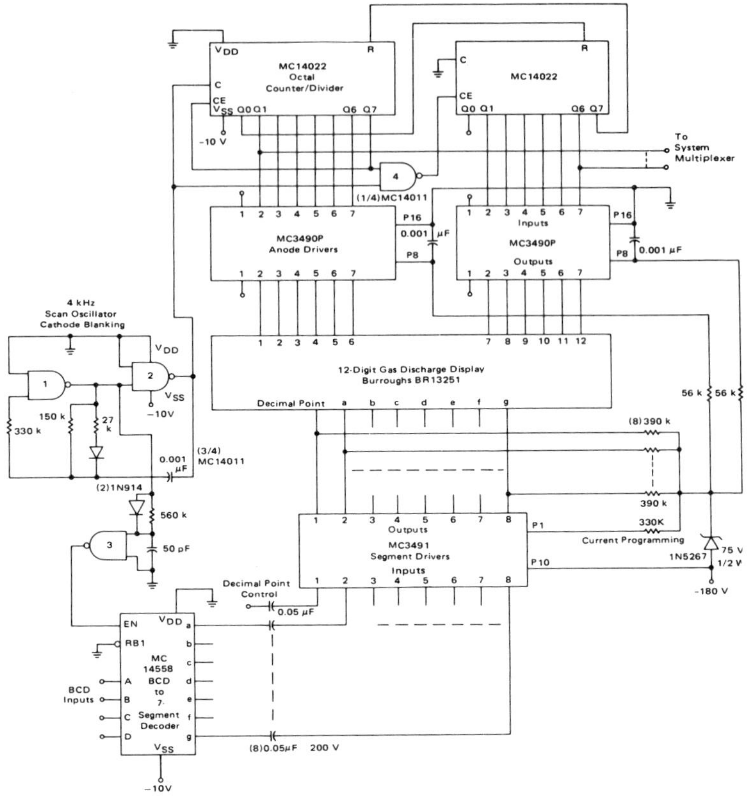

12-digit CMOS gas discharge display system (M).

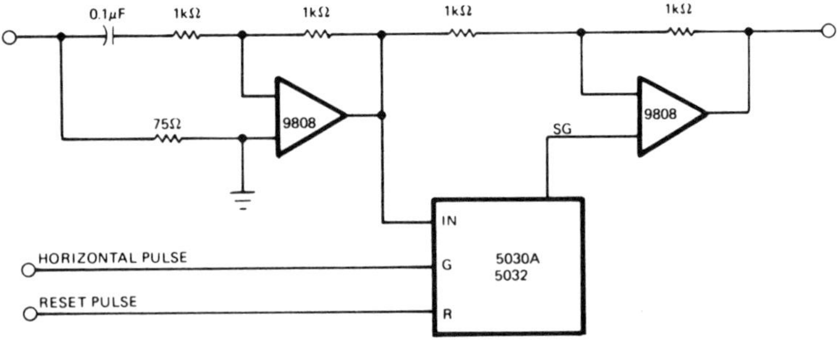

Video peak sense and hold (keyed video dc restorer) (OE).

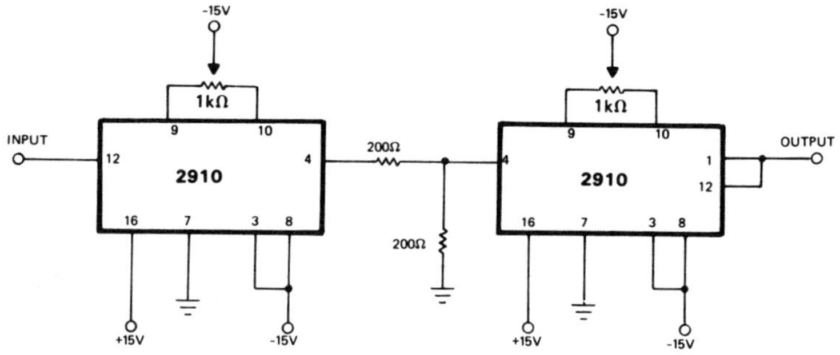

Wide-band unipolar square rooter (OE).

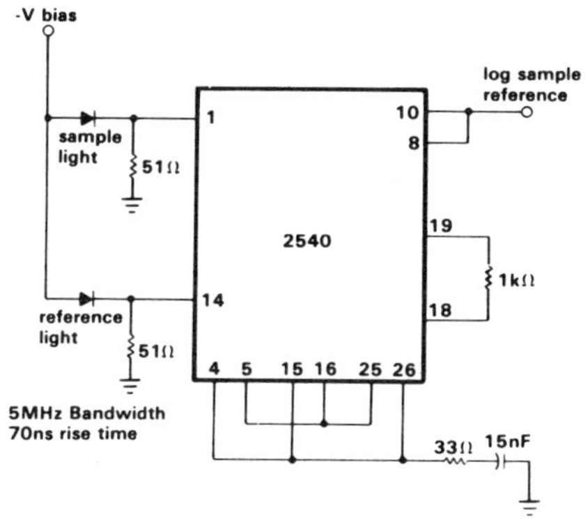

High speed logarithmic light ratio (OE).

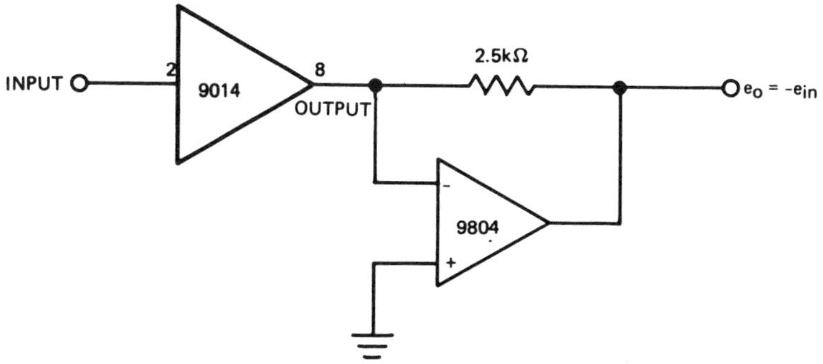

Input inverting and averaging circuit (OE).

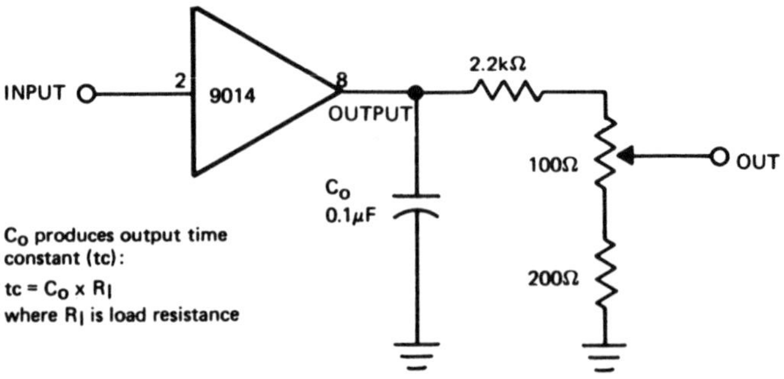

0-1 Volt output circuit (OE).

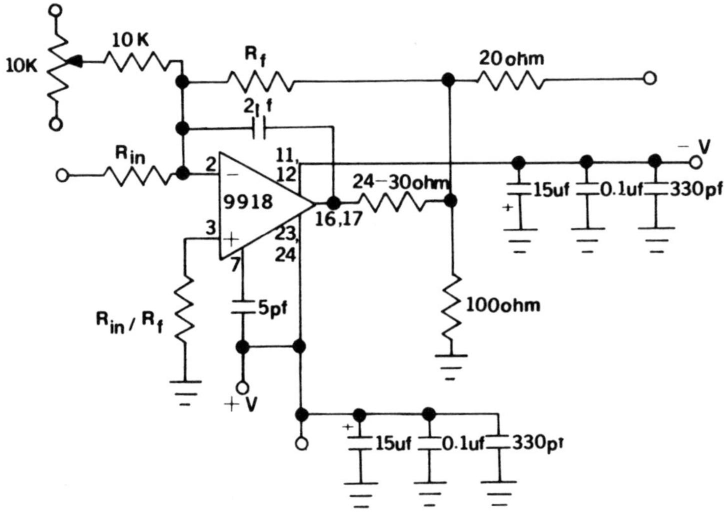

9918 offset adjust with power supply bypass capacitors (OE).

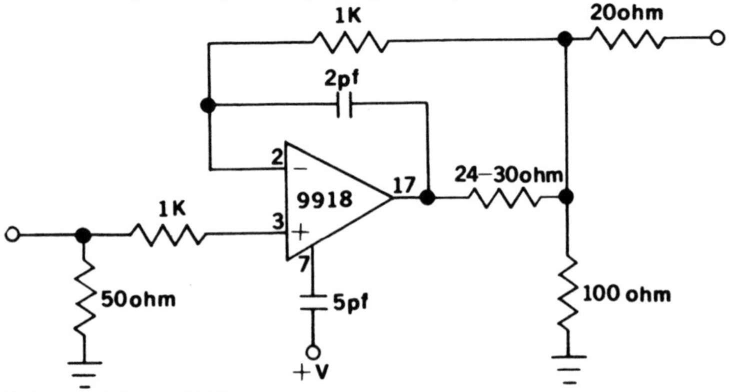

Unity gain follower (OE).

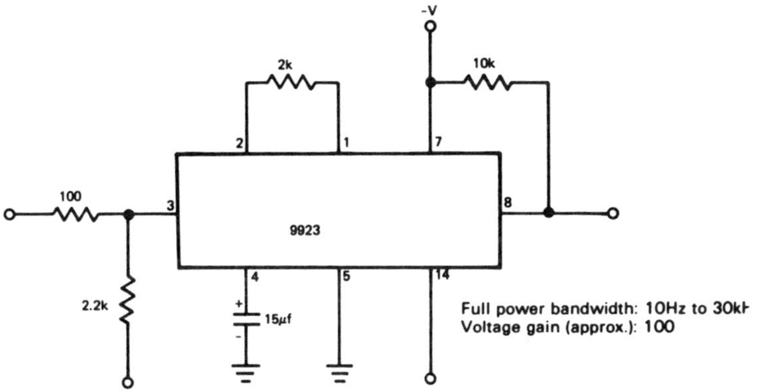

Open loop current booster (OE).

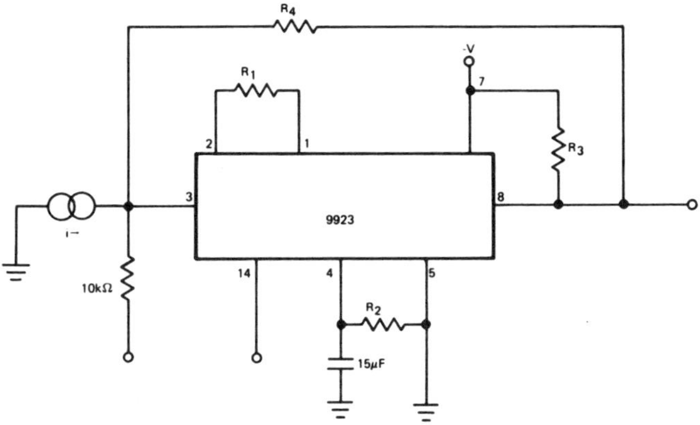

TABLE OF BANDWIDTH, GAIN, AND RESISTANCE VALUES				
3-dB bandwidth	100kHz	200kHz	300kHz	1MHz
Voltage gain (approx.)	39	30	20	10
R_1, Ω	20k	10k	6.8k	2k
R_2, Ω	300	300	300	300
R_3, Ω	10k	10k	10k	5.1k
R_4, Ω	39k	30k	20k	10k

Current booster with feedback loop (OE).

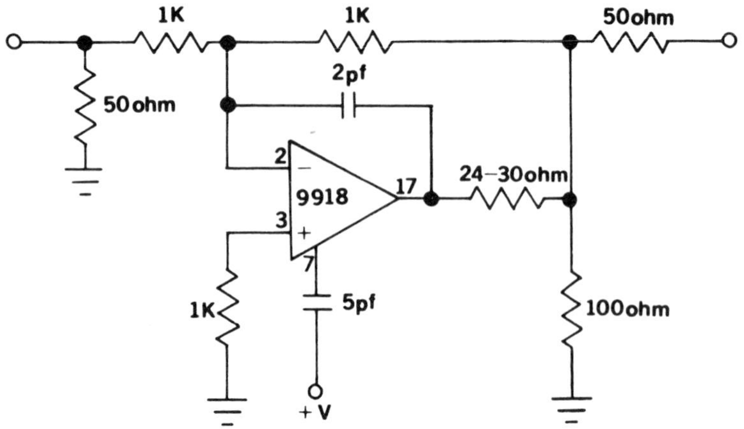

Fast settling unit gain inverting (OE).

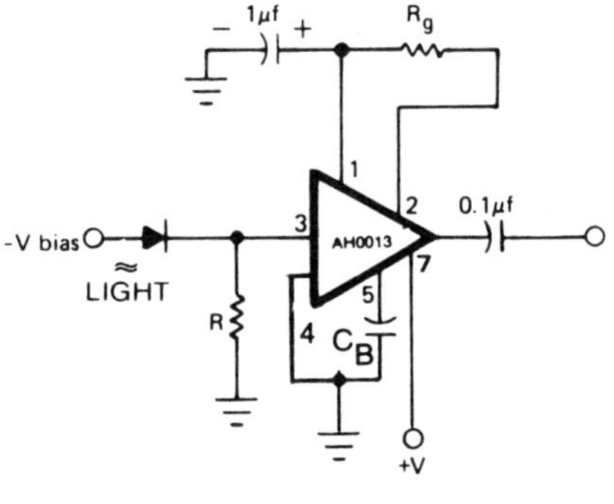

Low level light sensor (OE).

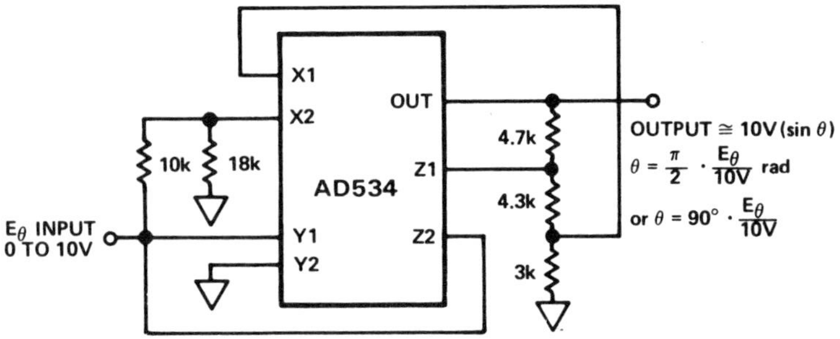

A simple sine-of-a-voltage circuit using the AD534 (AD).

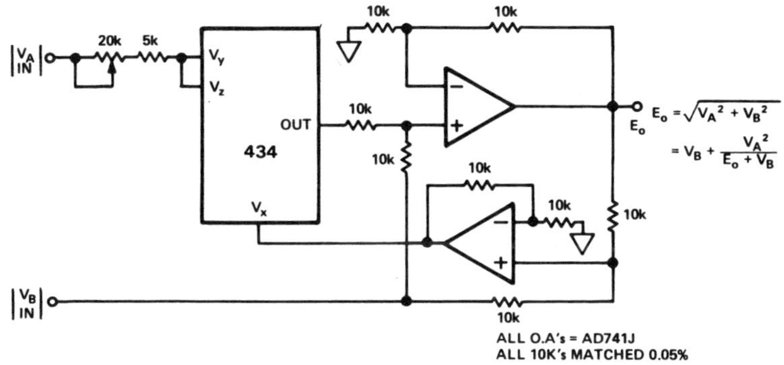

Square-root of the sum-of-the-squares (AD).

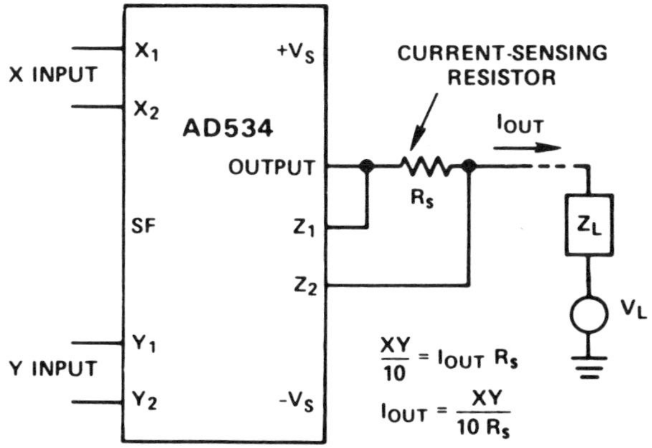

Output-to-current conversion (AD).

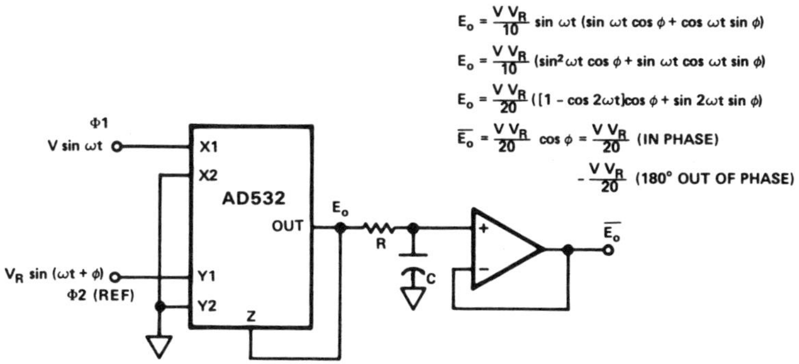

Phase-sensitive detector for sinusoidal signals. Measures magnitude of in-phase or 180°-out-of-phase input with proper polarity, depending on phase relationship to reference, with less than 1% nominal error for phase shift between signal and reference of less than 0.14 radian (8°) (AD).

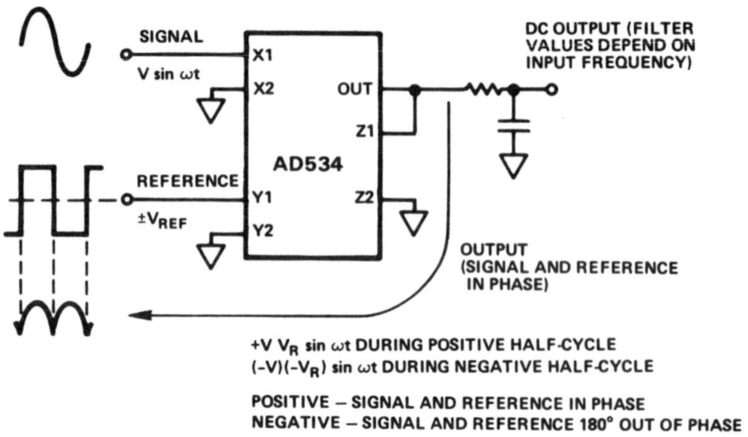

Phase-sensitive detector, squarewave reference.

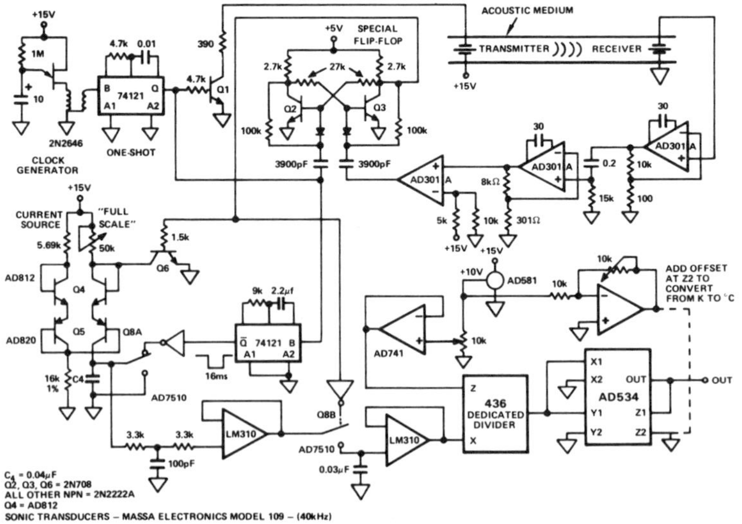

Acoustic-thermometer analog output (AD).

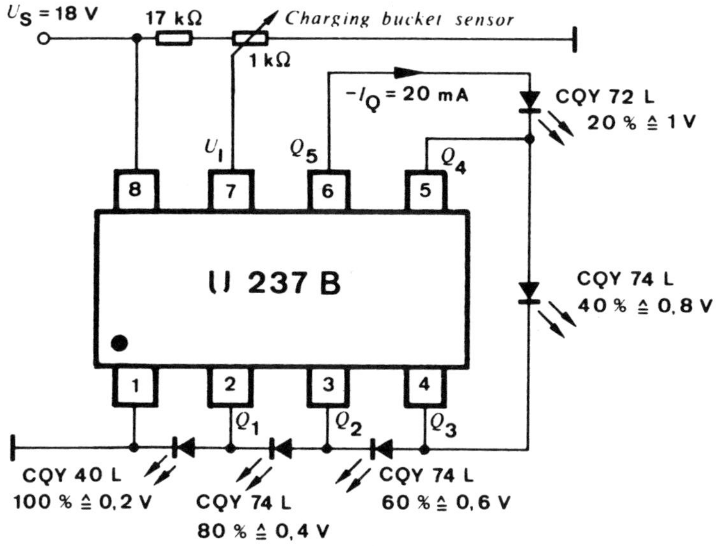

Liquid level display with linear resistance pickup (AEG).

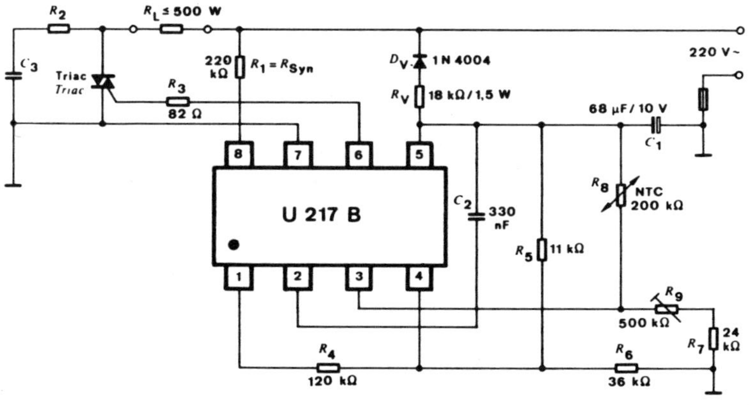

Circuit of the temperature controller (AEG).

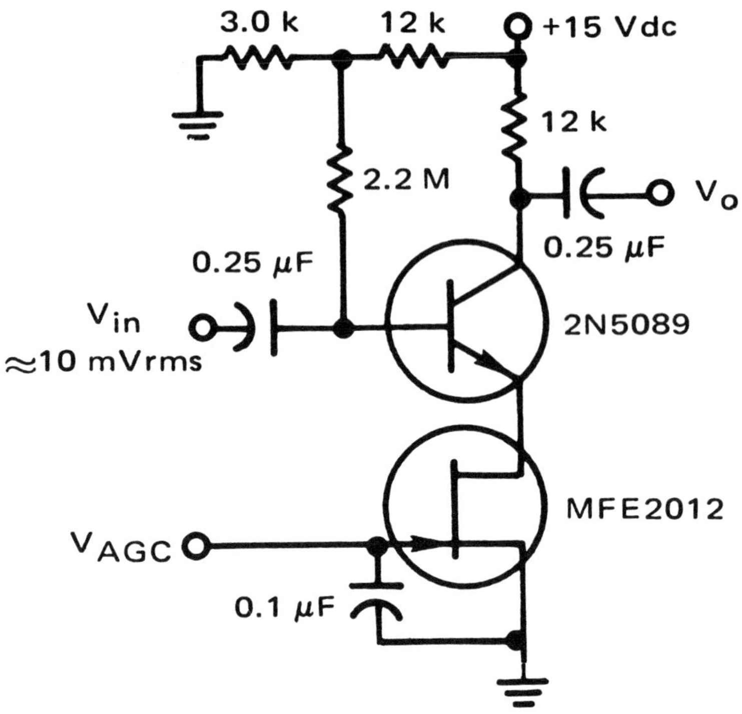

Simple AGC circuit (M).

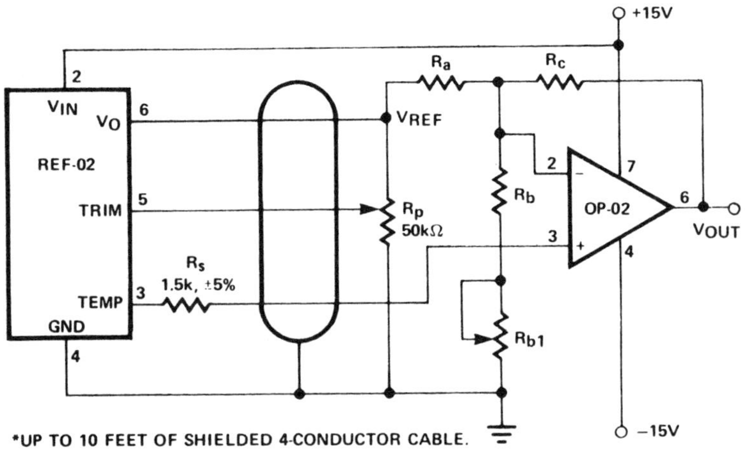

Precision temperature transducer with remote sensor (PM).

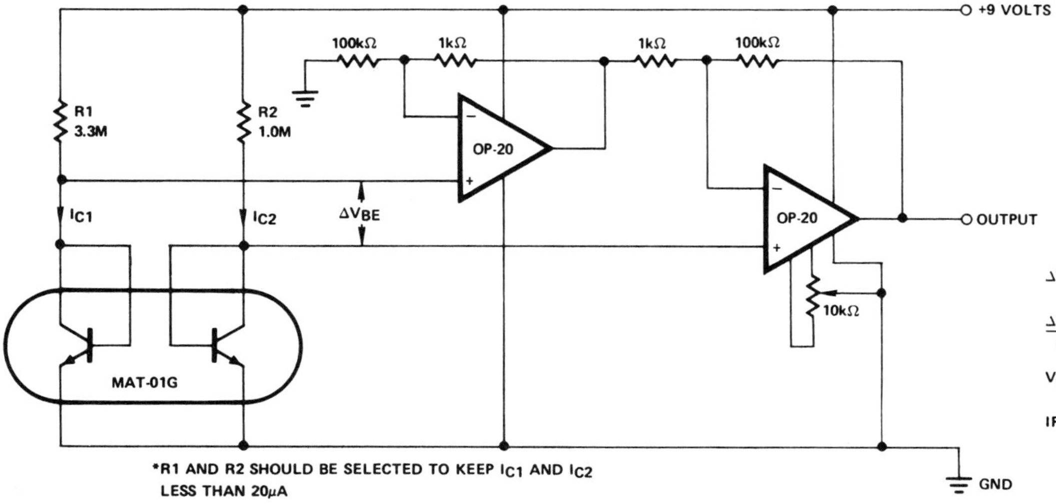

DESIGN EQUATIONS

$$\Delta V_{BE} = \frac{kT}{q} \ln \left(\frac{I_{C2}}{I_{C1}} \right)$$

$$\frac{\Delta V_{BE}}{\Delta T} = 85.8 \ln \left(\frac{I_{C2}}{I_{C1}} \right) \ |\mu V/ K|$$

$$V_{OUT} = 101(\Delta V_{BE})$$

$$\text{IF } \frac{R1}{R2} = 3.2, \text{ THEN } TCV_{OUT} = 10mV/ C$$

Temperature sensor (PM).

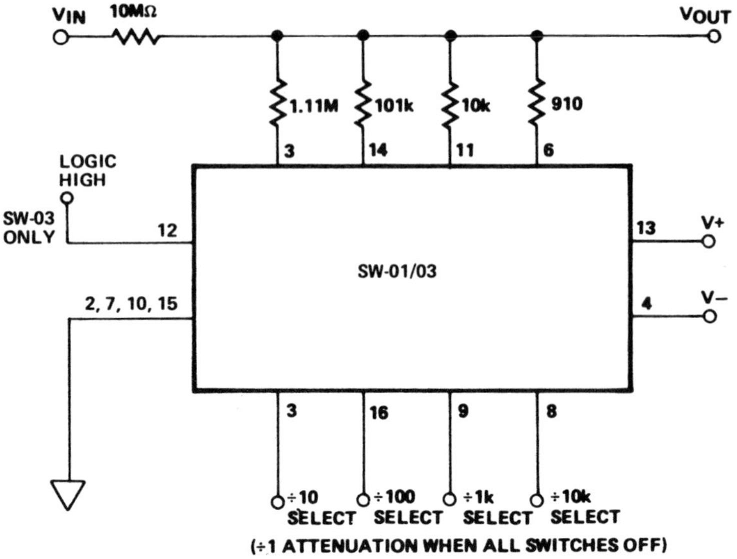

Programmable attenuator (1 to 0.0001) (PM).

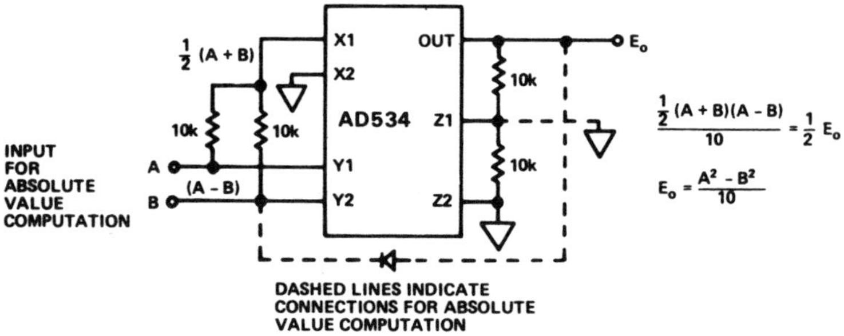

Difference-of-squares circuit (AD).

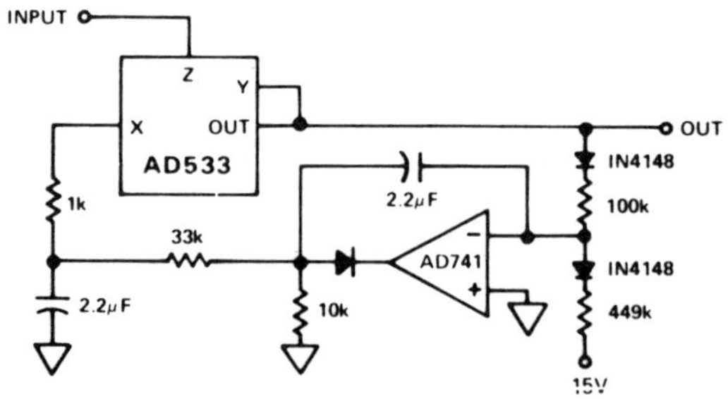

Automatic level control circuit (AD).

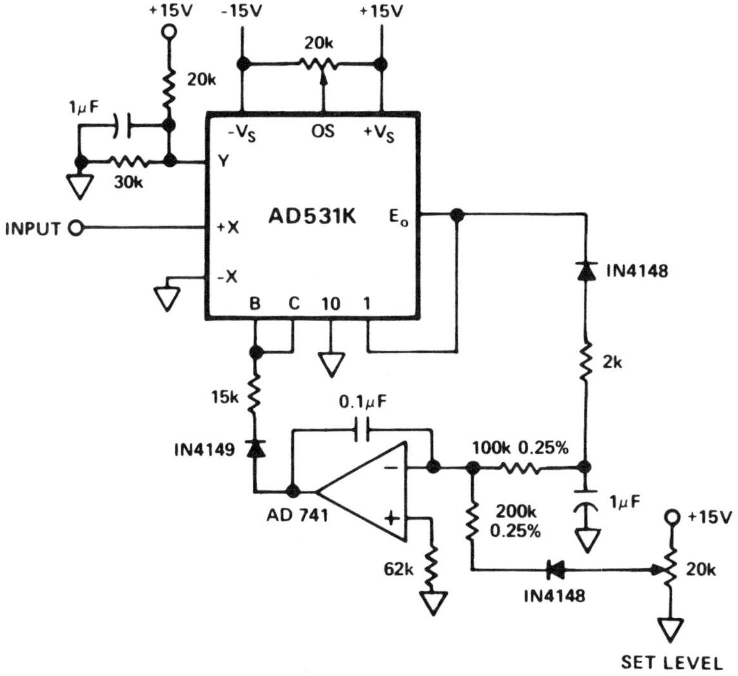

Automatic gain control circuit (AD).

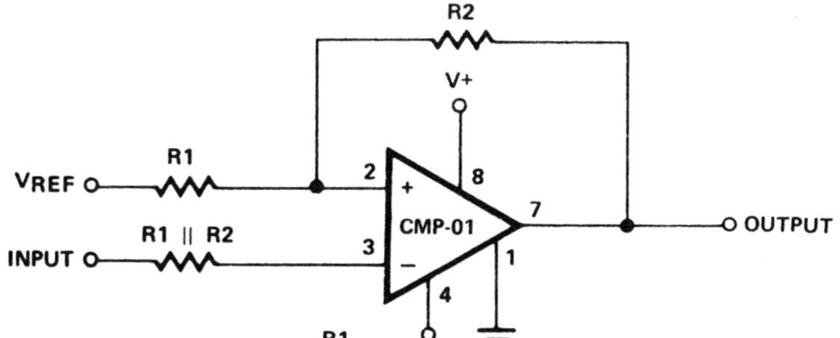

HYSTERESIS WIDTH $\leqslant$ 4V $\frac{R1}{R1 + R2}$

Level detector with hysteresis (Positive feedback) (PM).

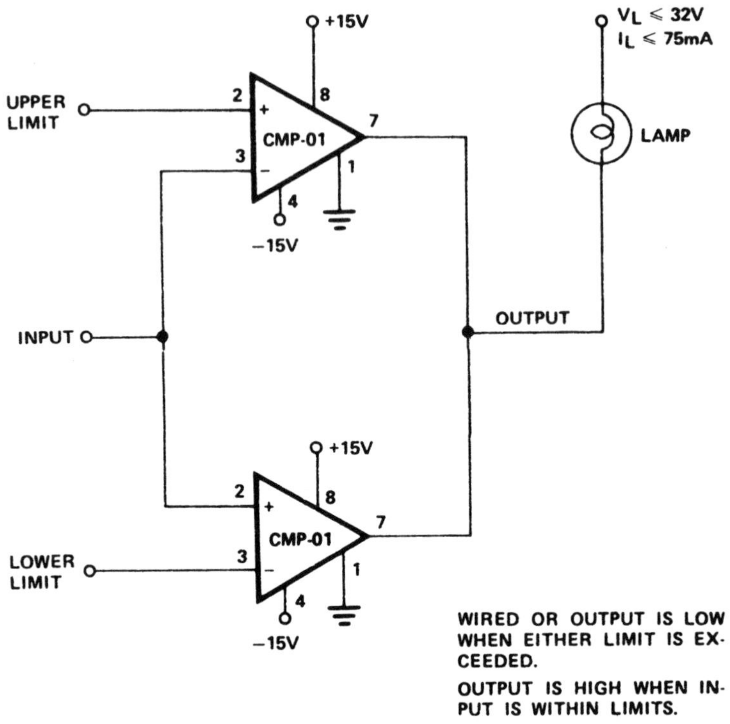

Precision, dual limit, go/no go tester (PM).

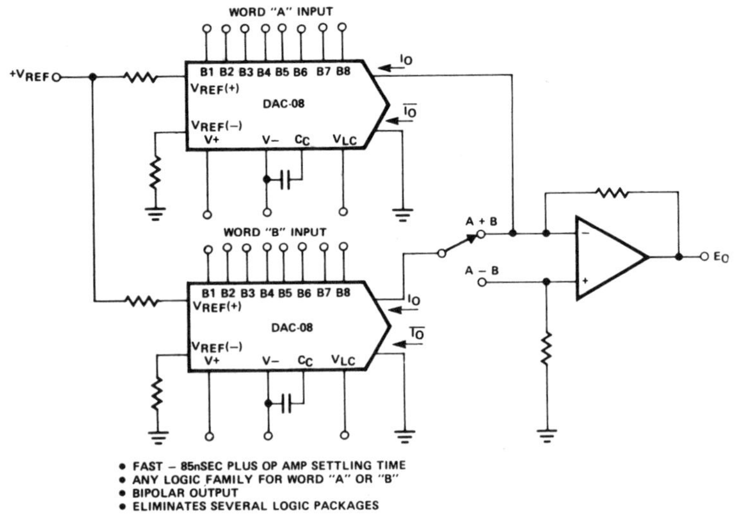

Digital addition or subtraction with analog output (PM).

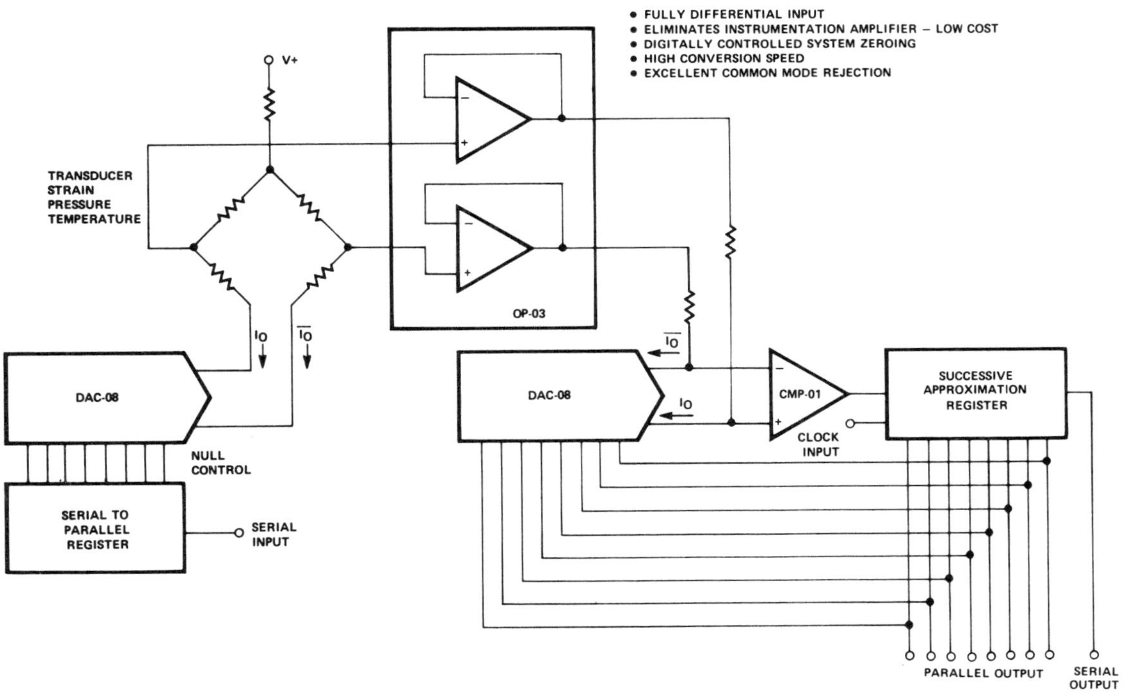

Bridge transducer control system with full differential input (PM).

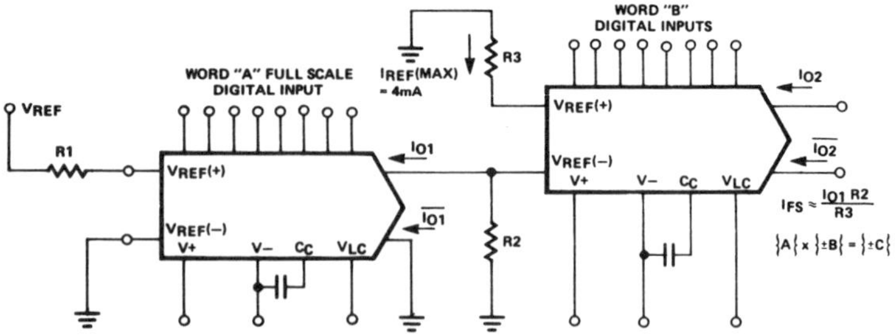

- IFS IS THE PRODUCT OF TWO DIGITAL INPUT WORDS
- MAY BE USED AS AN 8 x 8 DIGITAL MULTIPLIER WITH ANALOG OUTPUT
- ELIMINATES DAC AFTER DIGITAL MULTIPLICATION
- FUNCTIONS WITH ANY LOGIC FAMILY
- NOTE: LIMIT WORD "B" INPUT RISE AND FALL TIMES TO 200nSEC MINIMUM

Digitally controlled full-scale calibration (Multiplier) (PM).

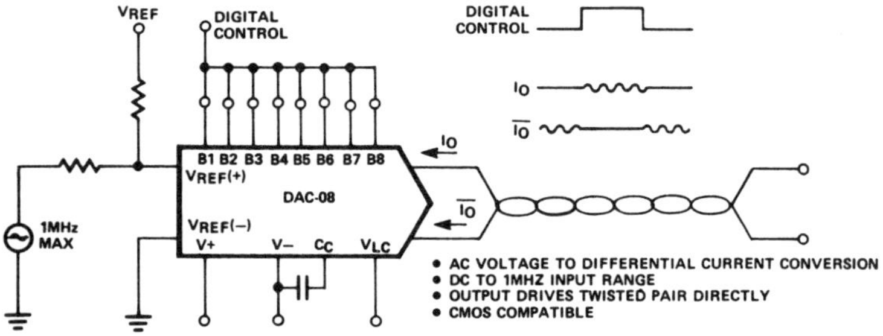

Modem transmitter (PM).

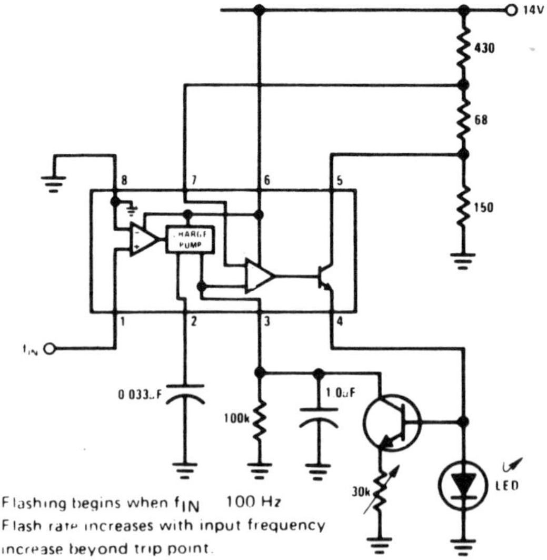

Flashing LED indicates overspeed (NS).

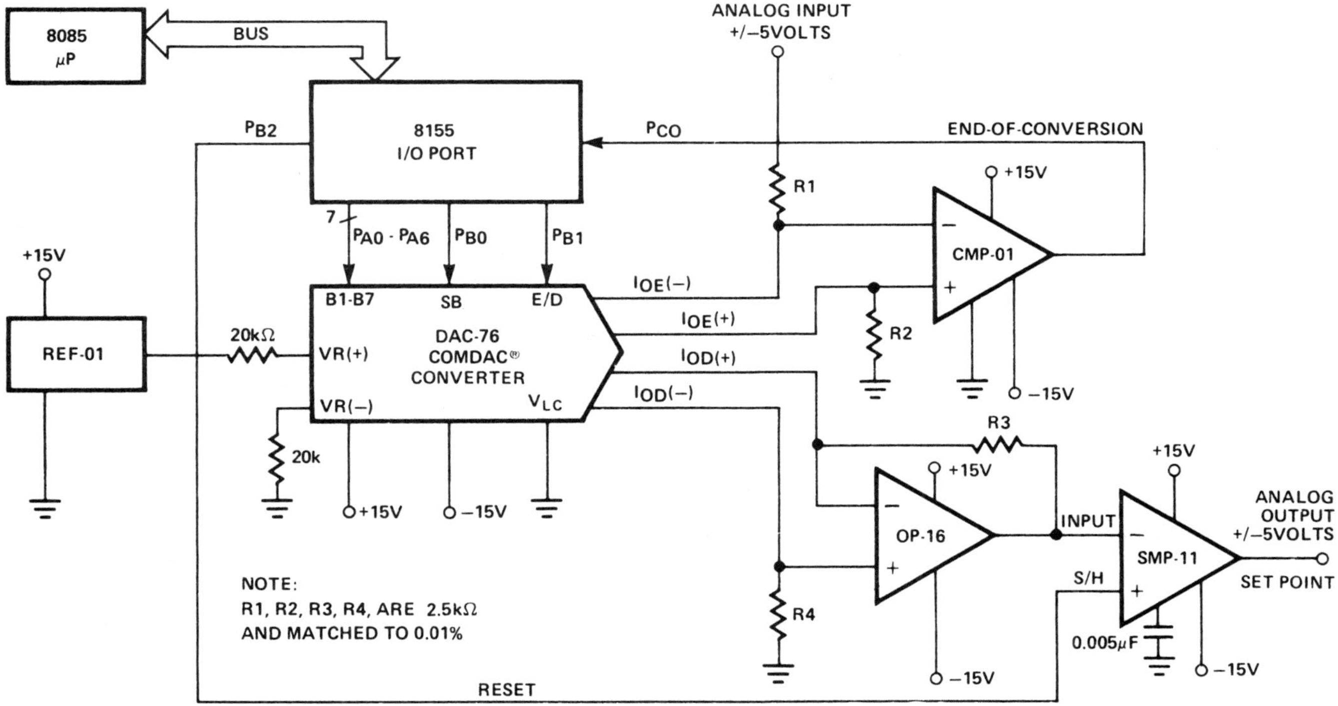

COMDAC® controller showing analog input/output (PM).

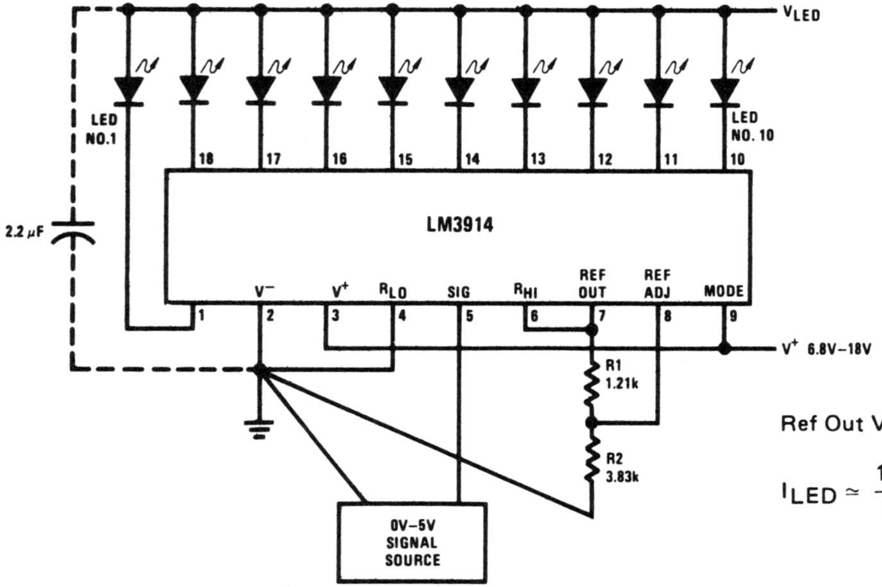

Note 1: Grounding method is typical of *all* uses. The 2.2 μF tantalum or 10 μF aluminum electrolytic capacitor is needed if leads to the LED supply are 6″ or longer.

$$\text{Ref Out V} = 1.25 \left(1 + \frac{R2}{R1}\right)$$

$$I_{LED} \simeq \frac{12.5}{R1}$$

0V to 5V bargraph meter (NS).

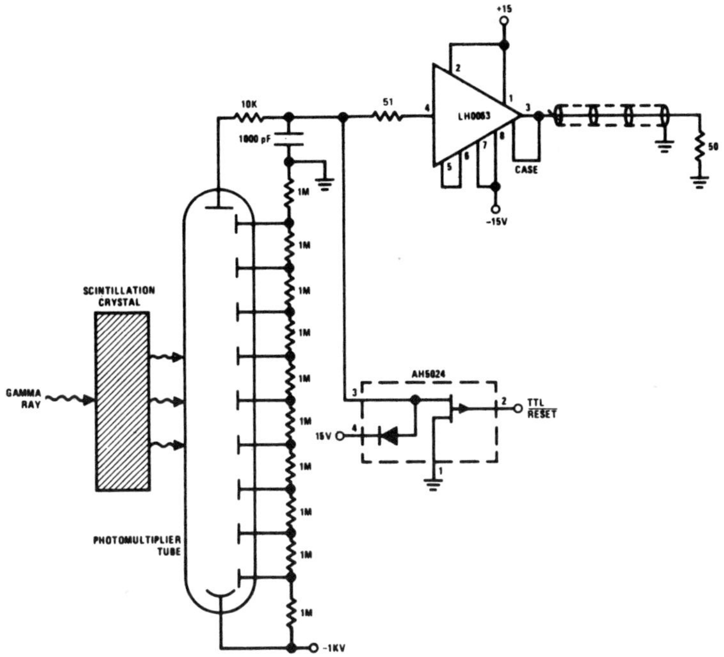

Gamma ray pulse integrator (NS).

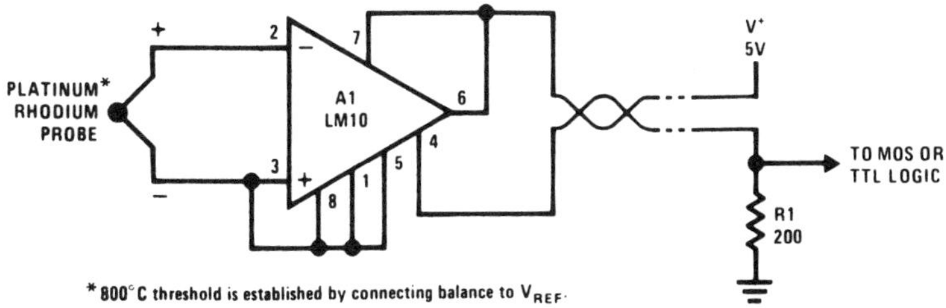

Flame detector (NS).

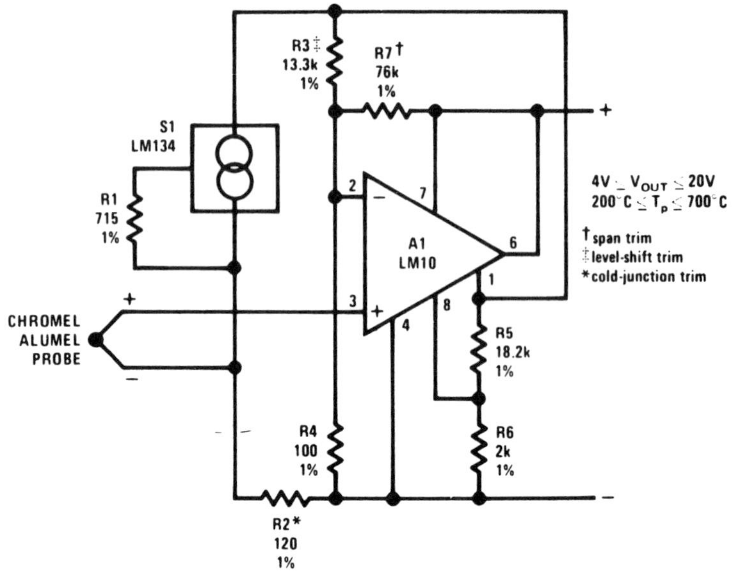

Remote thermocouple amplifier (NS).

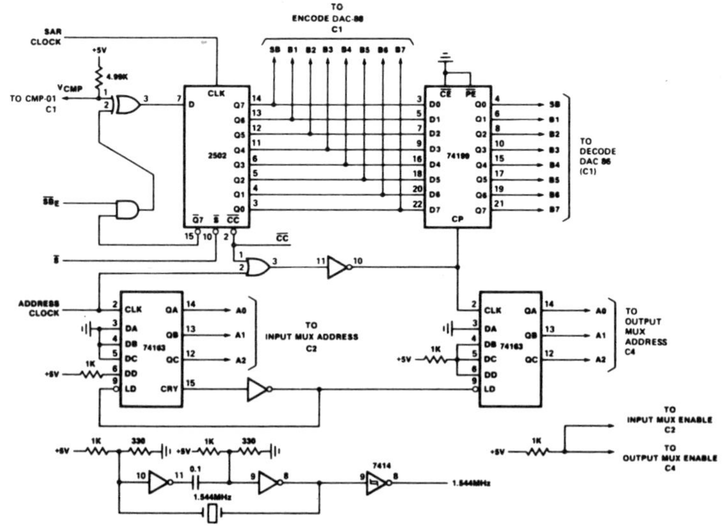

Encode/decode controller (PM).

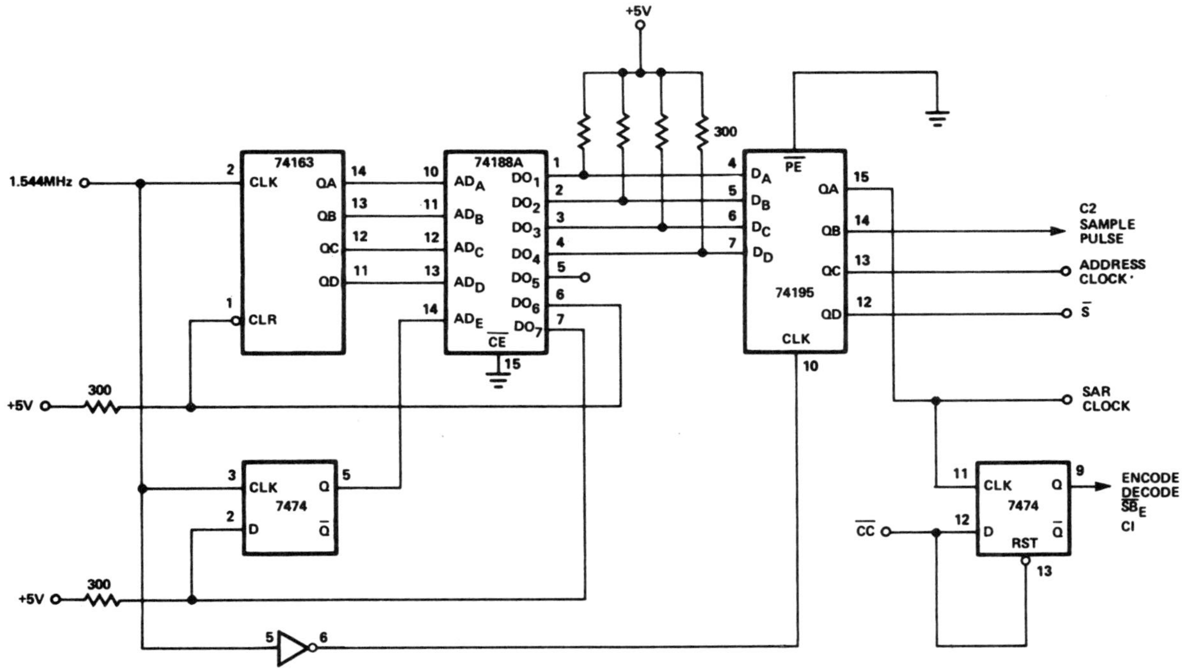

Encode clock (PM).

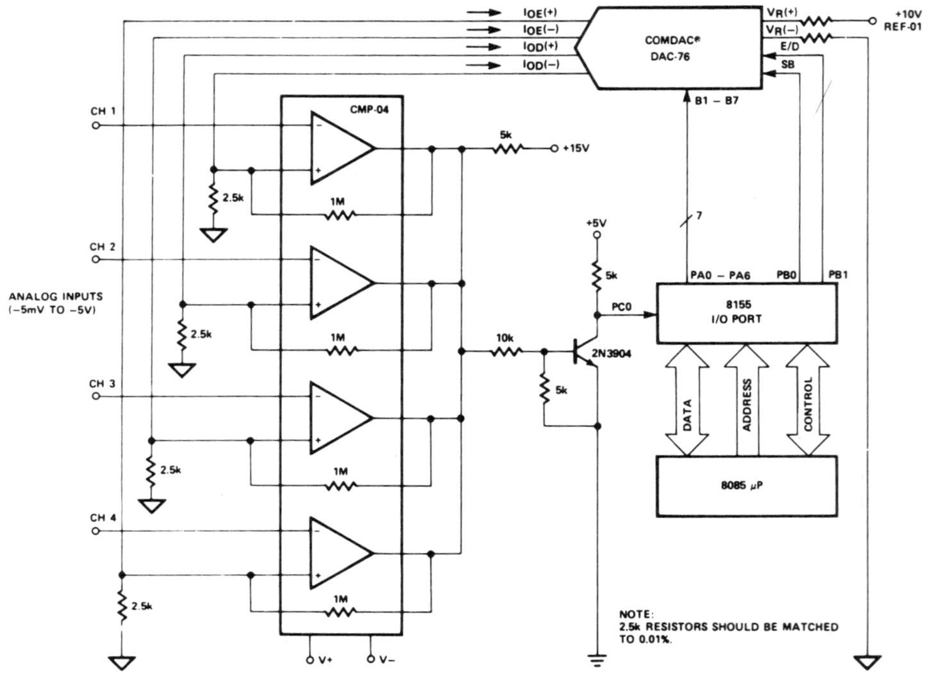

Four channel data acquisition system (PM).

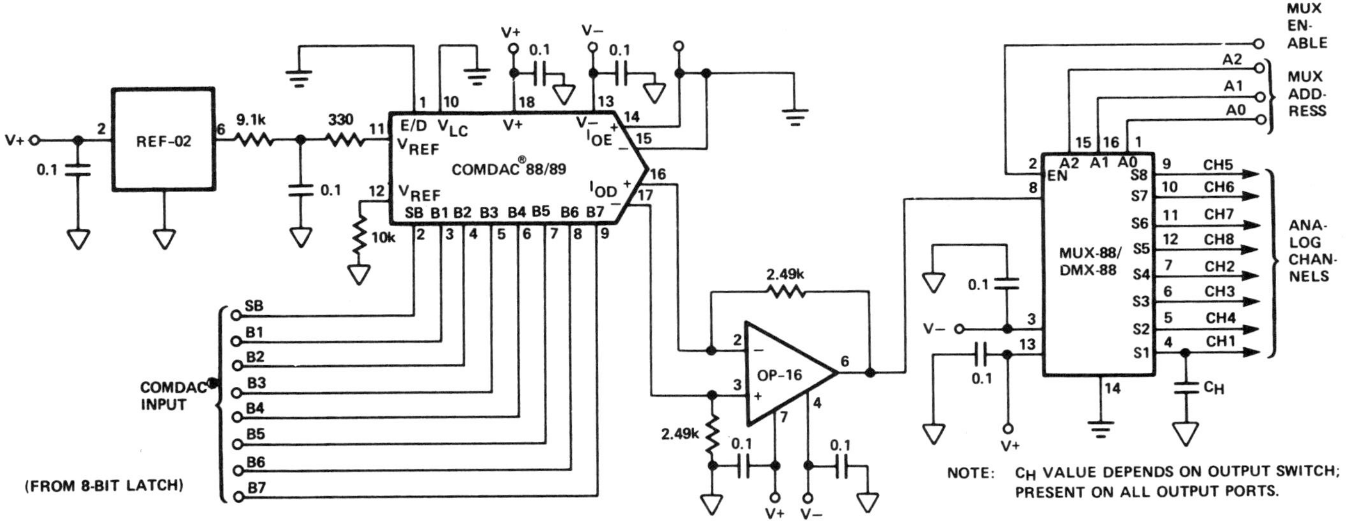

Demo system decode board (PM).

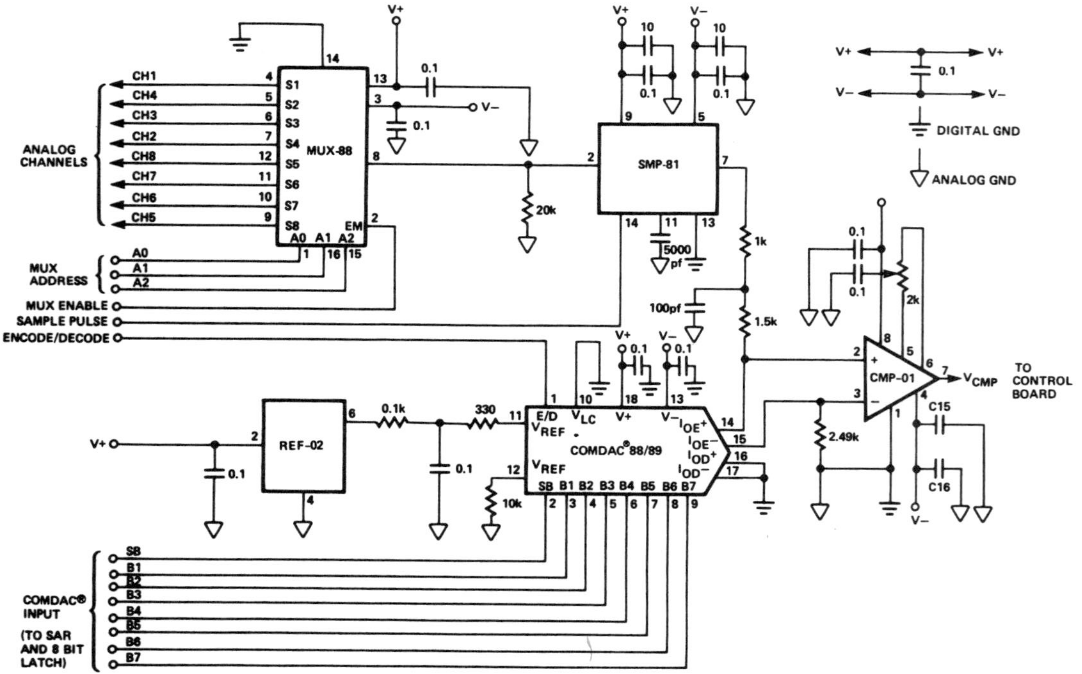

Demo system encode board (PM).

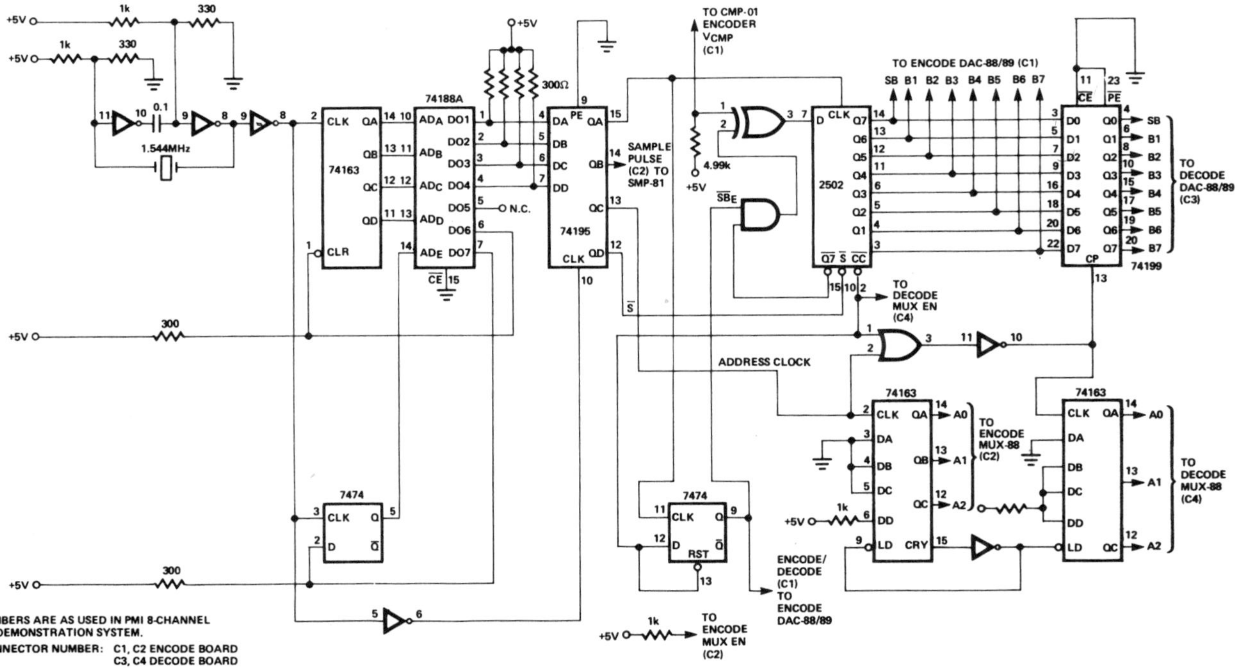

PIN NUMBERS ARE AS USED IN PMI 8-CHANNEL CODEC DEMONSTRATION SYSTEM.

(C1) CONNECTOR NUMBER: C1, C2 ENCODE BOARD
C3, C4 DECODE BOARD

(REFER TO 16 AND 14 PIN CONNECTORS AS SHOWN IN PMI AN-37)

CODEC demonstration system digital control board (PM).

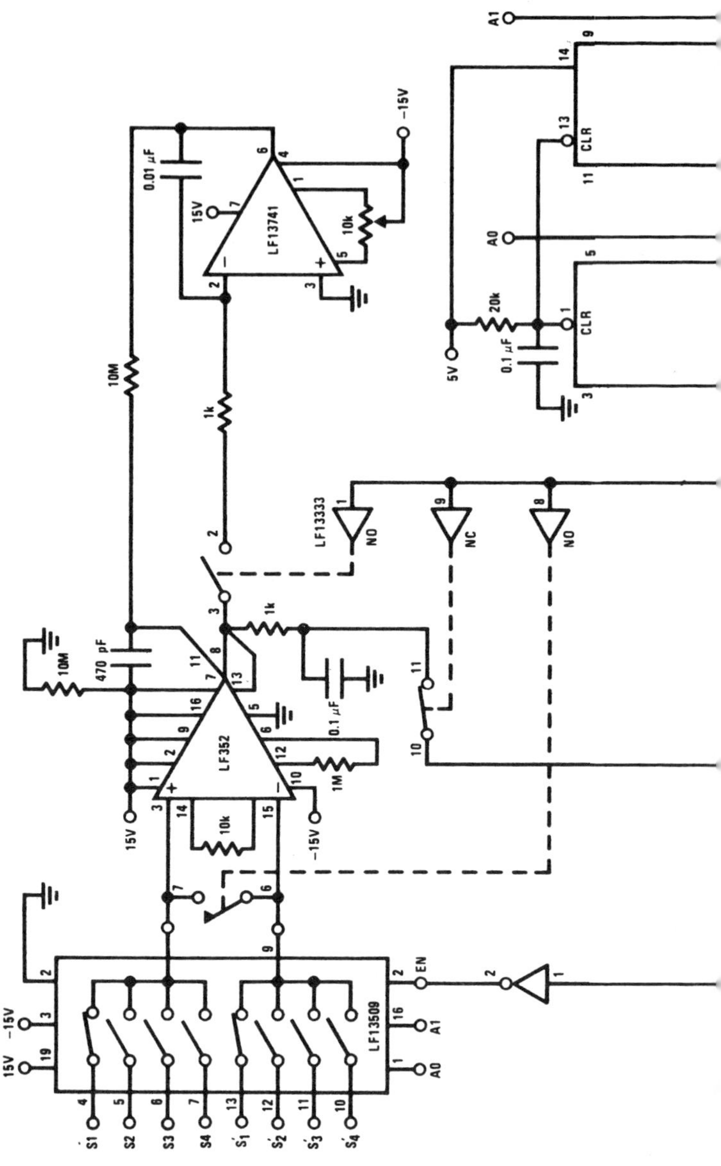
LF13741
LF352
LF13509
LF13333
0.01 μF
10M
470 pF
1k
10k
1M
0.1 μF
20k
15V
−15V
5V
A0
A1
CLR
EN
NO
NC
S1
S2
S3
S4

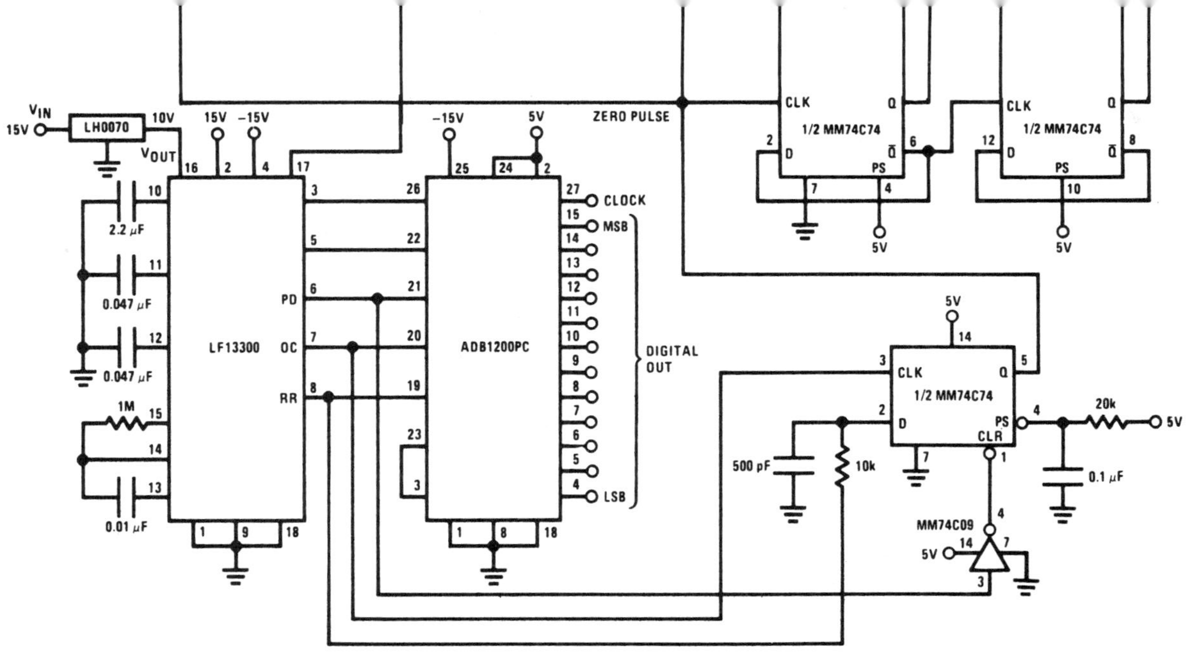

- f_{CLOCK} max = 200 kHz
- The LF352 instrumentation amplifier is auto zeroed during offset correction cycle of the LF13300 A/D
- The system accuracy will mostly depend on the instrumentation amplifier gain linearity

4-channel differential multiplexer with auto zeroed instrumentation amplifier and 12-bit A/D converter (NS).

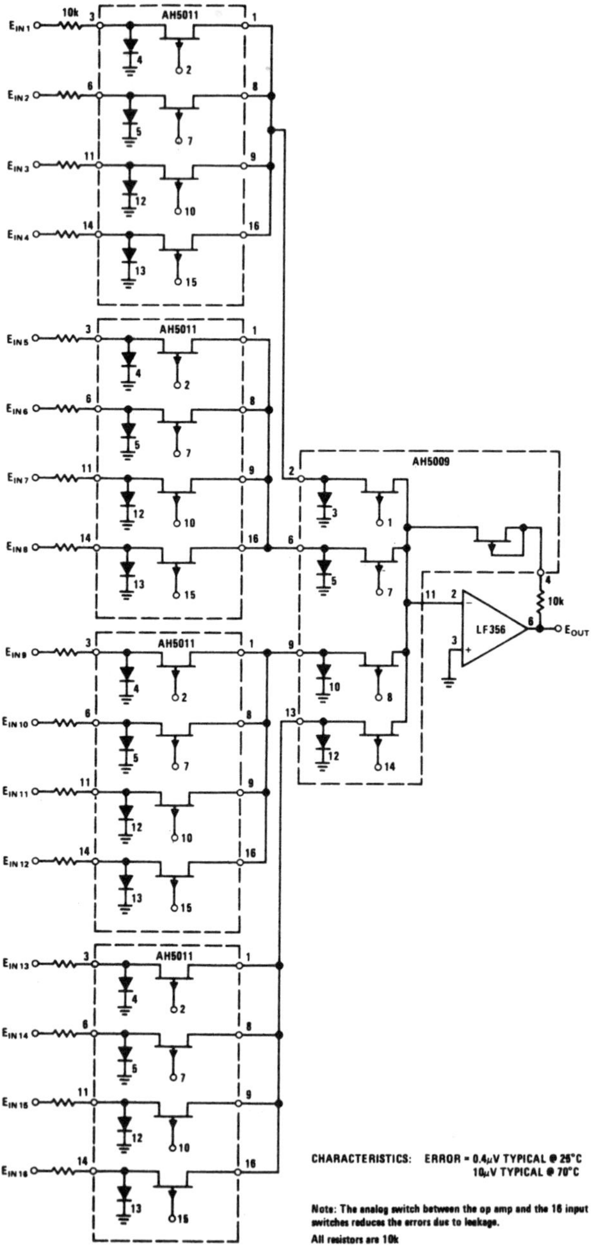

16-channel multiplexer (NS).

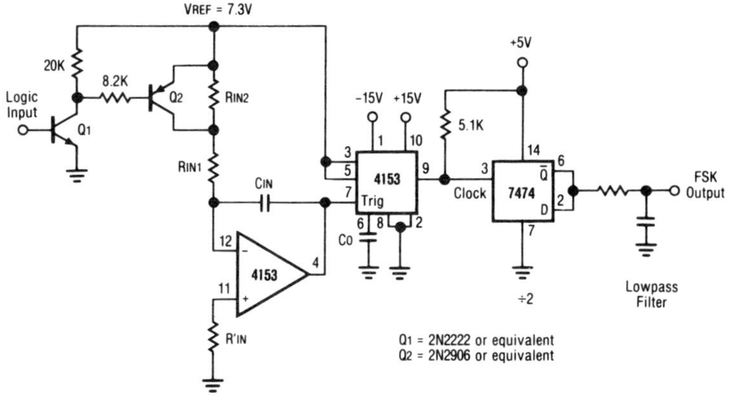

4153 FSK modulator (R).

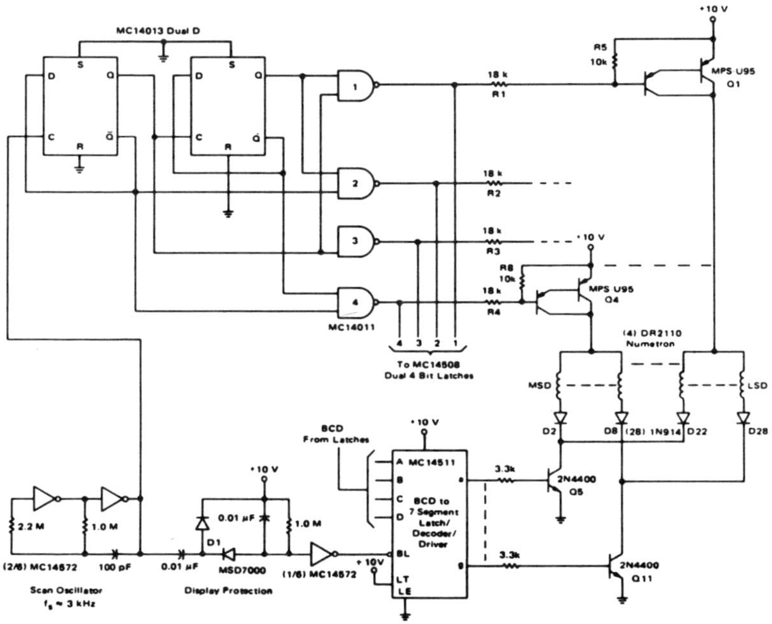

Interfacing CMOS to 4-digit incandescent display (M).

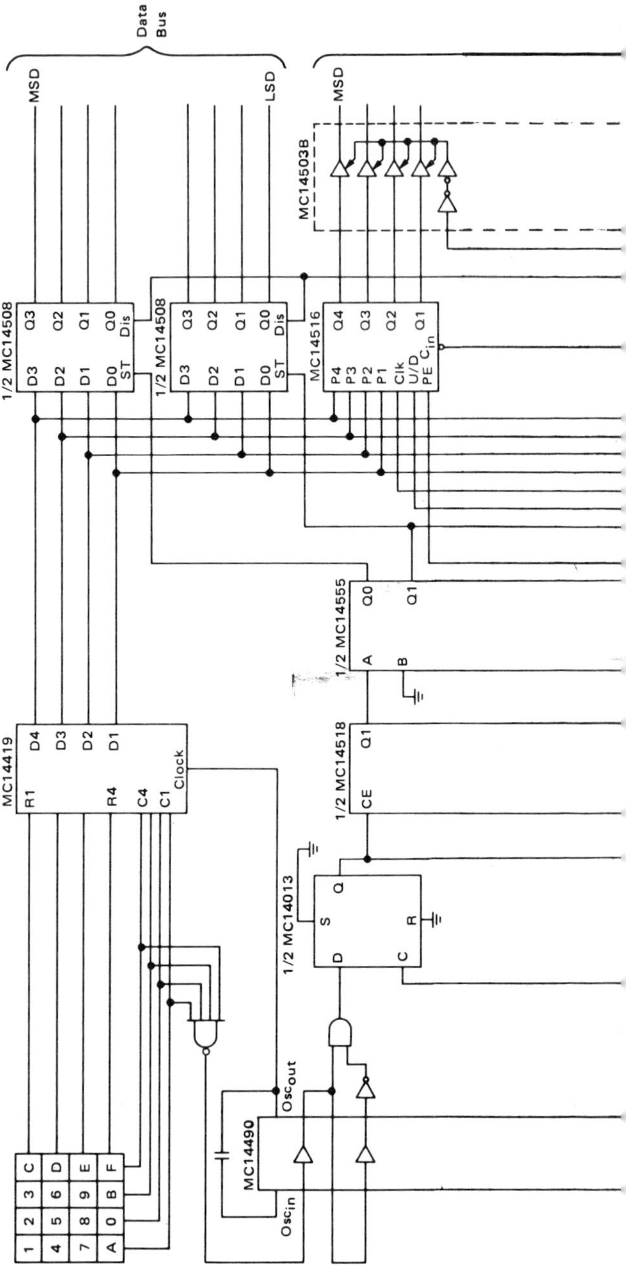

Data Bus
MSD
LSD
MSD
MC14503B
1/2 MC14508
D3 D2 D1 D0 ST
Q3 Q2 Q1 Q0 Dis
1/2 MC14508
D3 D2 D1 D0 ST
Q3 Q2 Q1 Q0 Dis
MC14516
P4 P3 P2 P1 Clk U/D PE
Q4 Q3 Q2 Q1 C_in
1/2 MC14555
A B
Q0 Q1
MC14419
R1 R4 C4 C1 Clock
D4 D3 D2 D1
1/2 MC14518
CE Q1
1/2 MC14013
S D Q R C
MC14490
Osc_in
Osc_out
1 2 3 C
4 5 6 D
7 8 9 E
A 0 B F

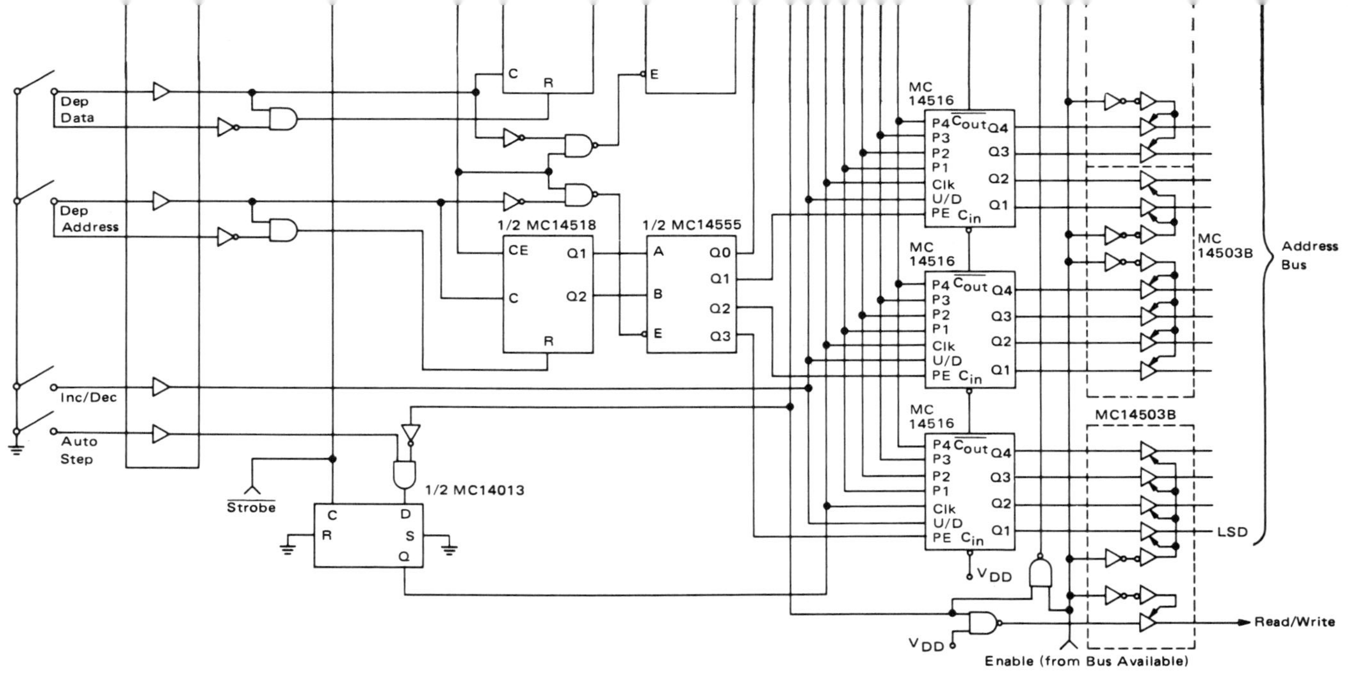

CMOS keyboard entry system (M).

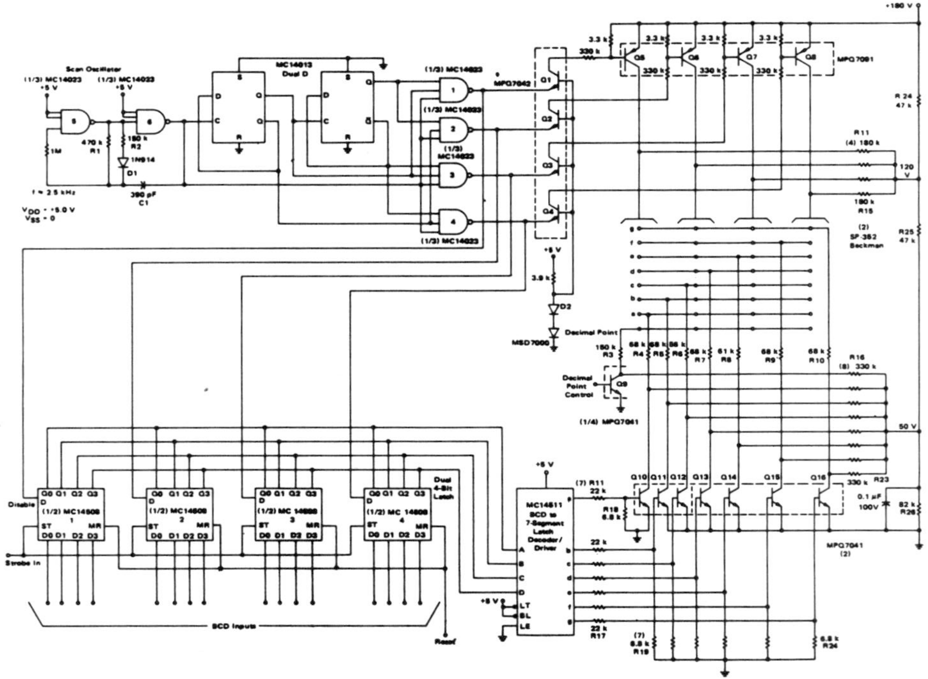

4-digit CMOS gas discharge display (M).

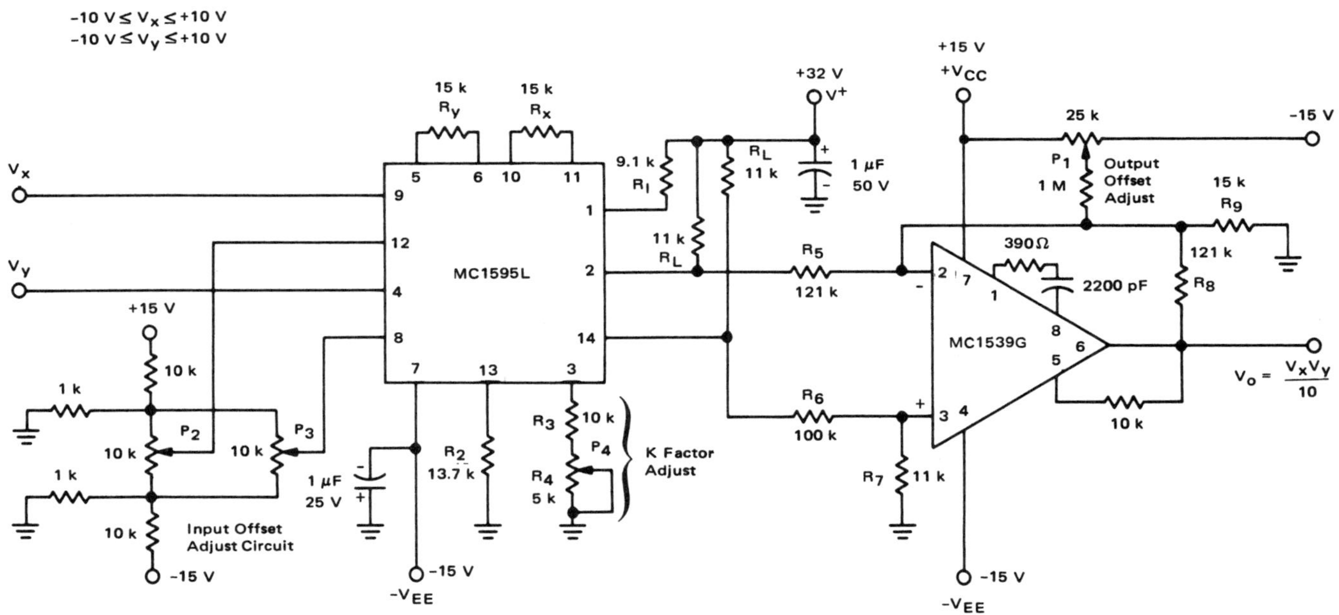

Multiply with op amp level shift (M).

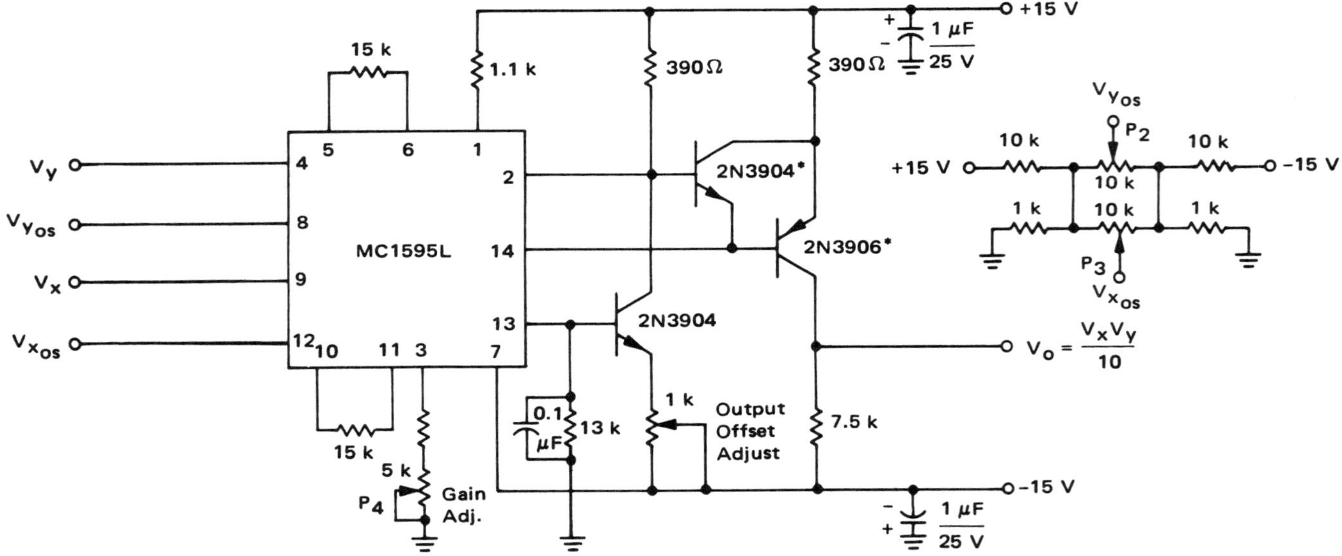

Discrete level shifting circuit (M).

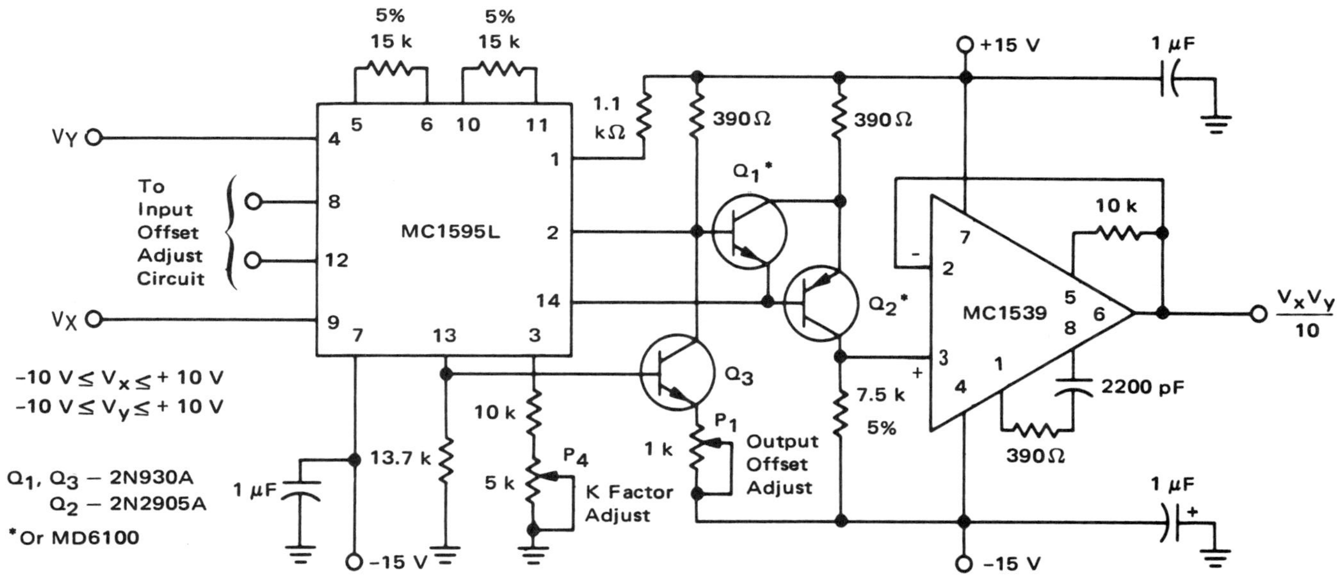

Multiply with discrete level shift (M).

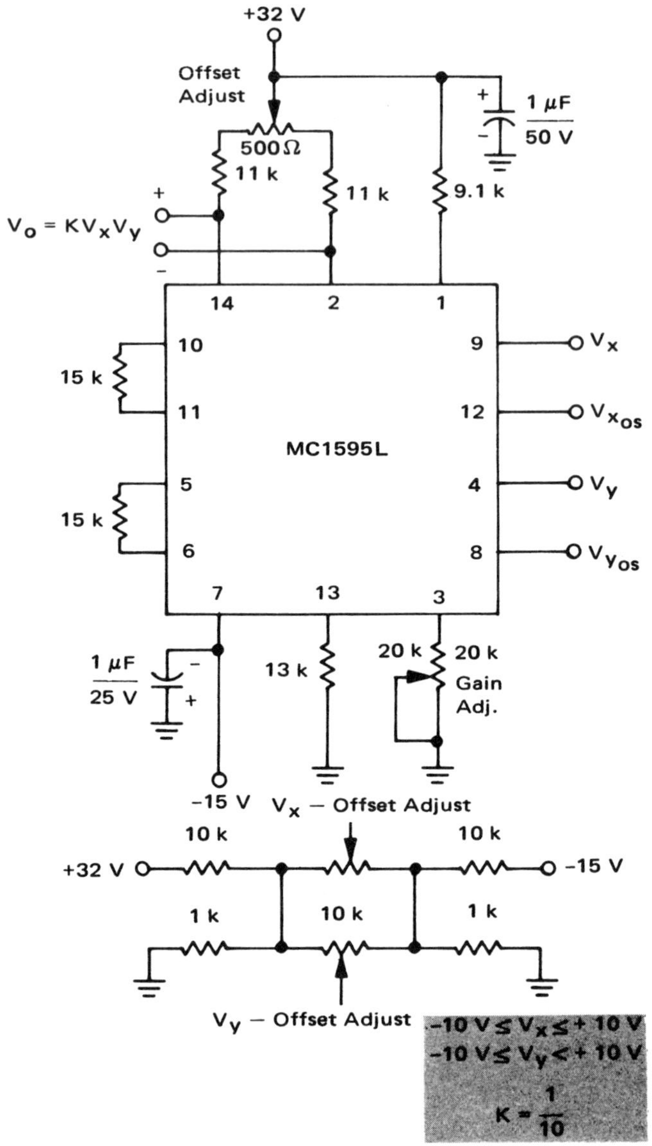

Basic multiply circuit (M).

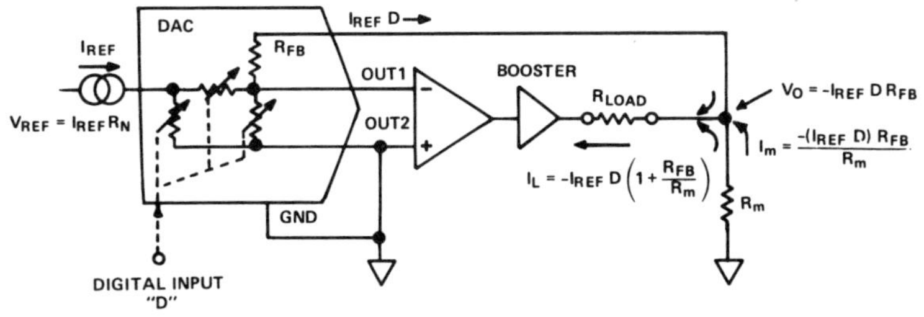

Current gain (floating load) (AD).

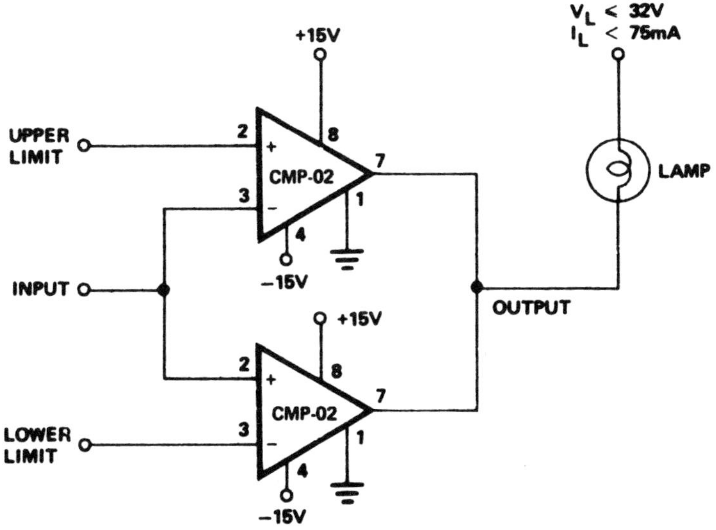

Precision dual limit go/no go tester (PM).

Chapter 24

Special Transistor Circuits

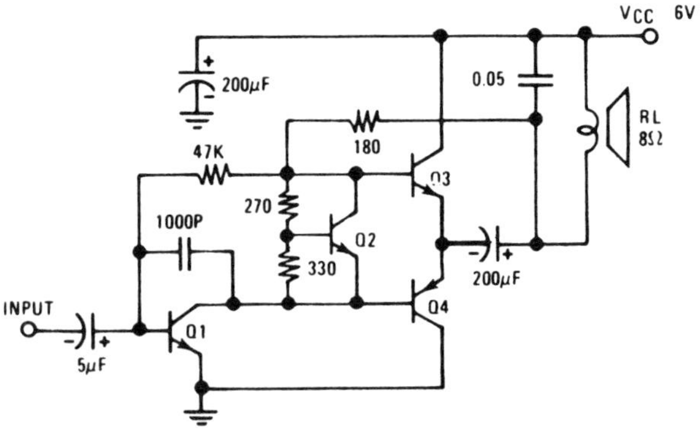

Q1 NB111EH/J
Q2 NR001E
Q3 NA11EG/J
Q4 NA12EG/J

380mW 6V/8Ω OTL amplifier (NS).

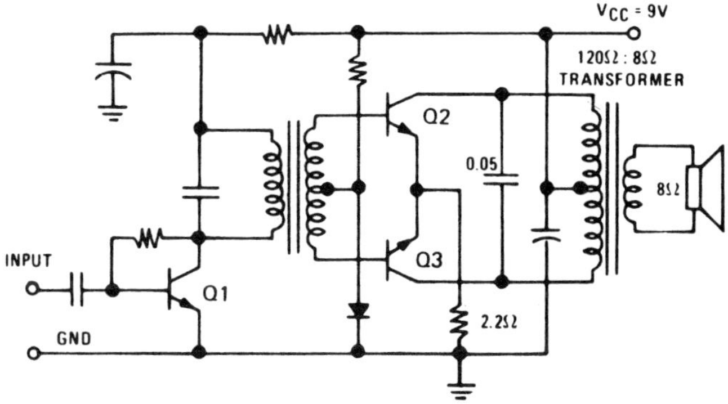

Q1 NB111EH/J Q2 NA11EG/J Q3 NA11EG/J

1.2W audio amplifier (NS).

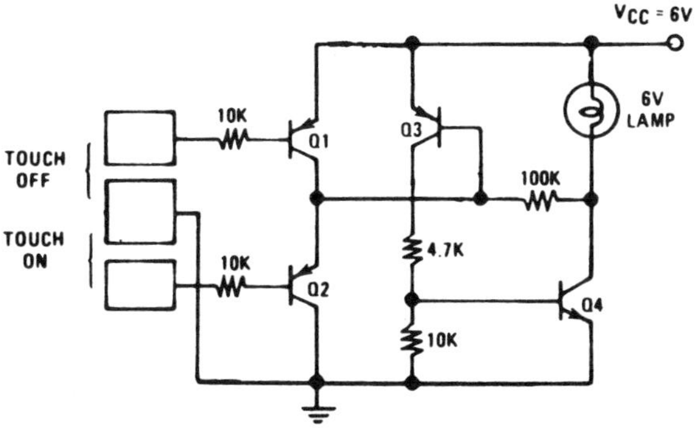

Q1 NB021EY Q3 NB021EY
Q2 NB021EY Q4 NA11EX

Touch-on/touch-off electronic switch (NS).

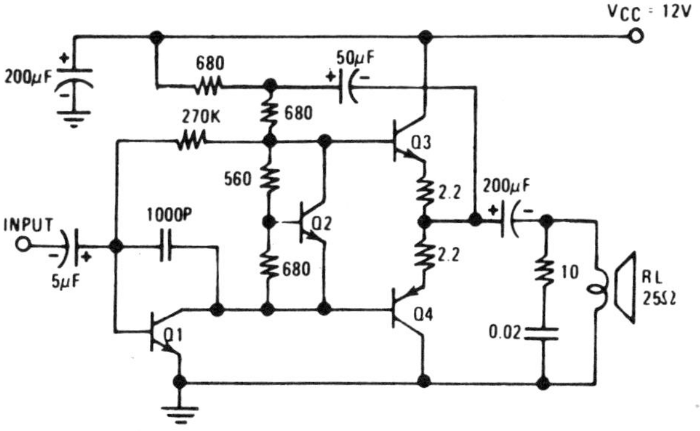

Q1 NB111EH/J Q3 NA11EG/J
Q2 NR001E Q4 NA12EG/J

650mW 12V/25Ω OTL amplifier (NS).

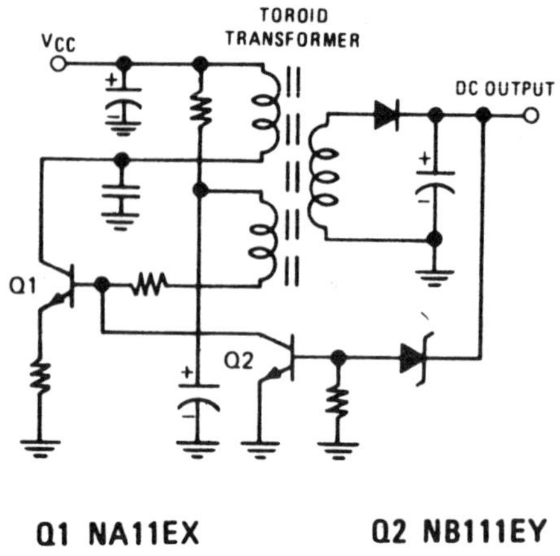

Typical converter circuit (NS).

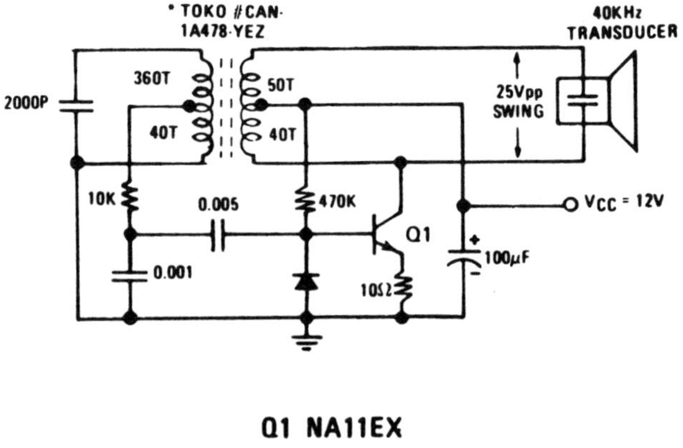

40kHz ultrasonic transmitter (NS).

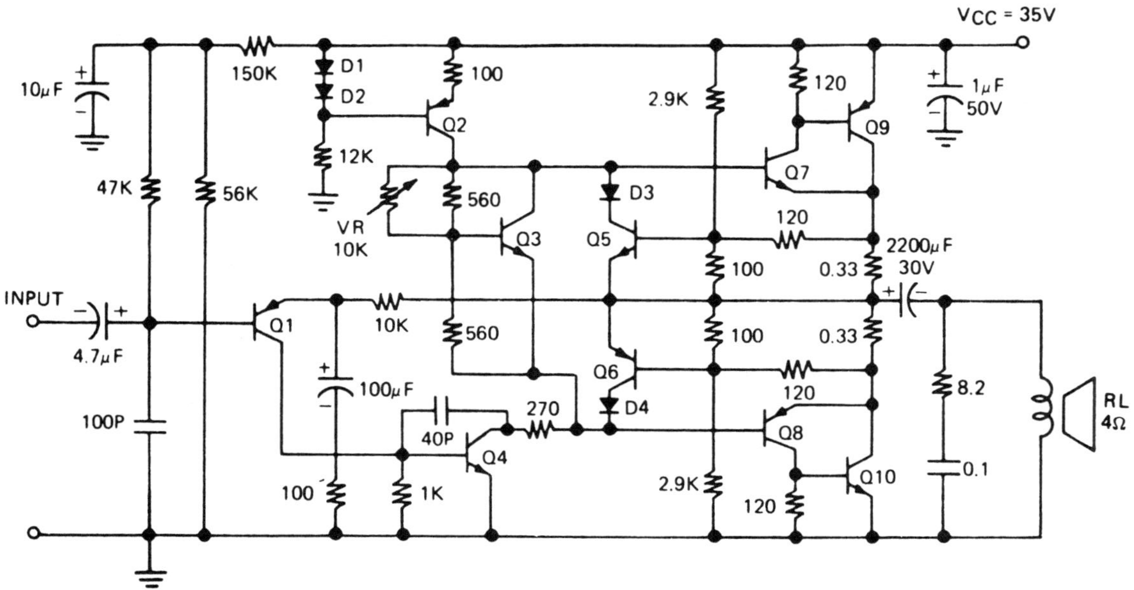

Q1	NB022EY
Q2	NB123EY
Q3	NR001E
Q4	NB113EY
Q5	NB111EY
Q6	NB121EY
Q7	NB313Y
Q8	NB323Y
Q9	NA62W
Q10	NA61W

25 watt OTL amplifier (NS).

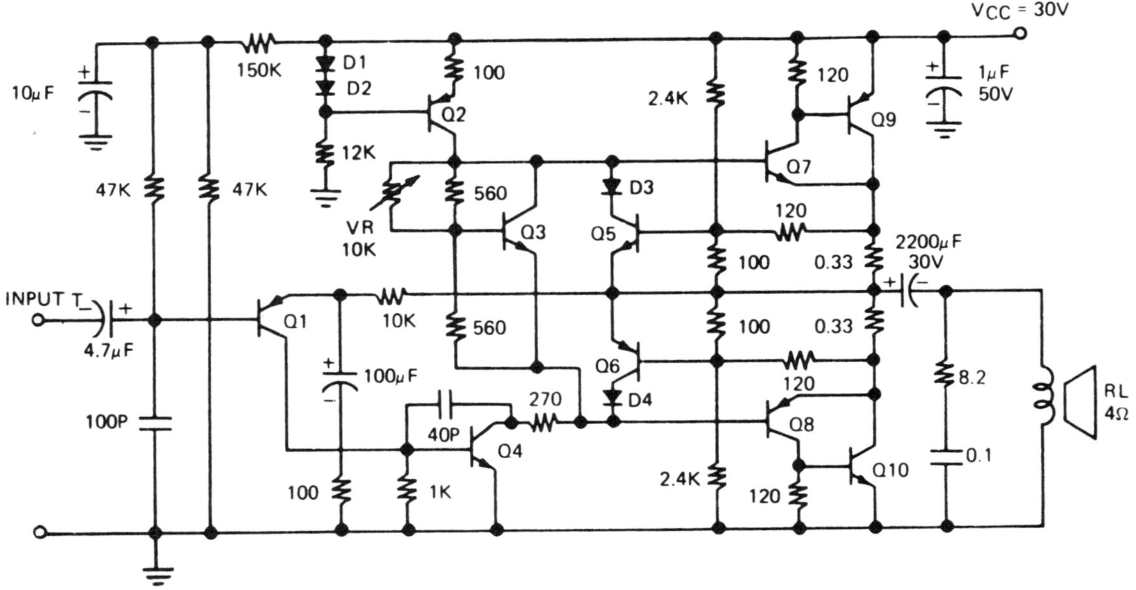

Q1	NB022EY
Q2	NB122EY
Q3	NR001E
Q4	NB112EY
Q5	NB111EY
Q6	NB121EY
Q7	NB313Y
Q8	NB323Y
Q9	NA62W
Q10	NA61W

18 watt OTL amplifier (NS).

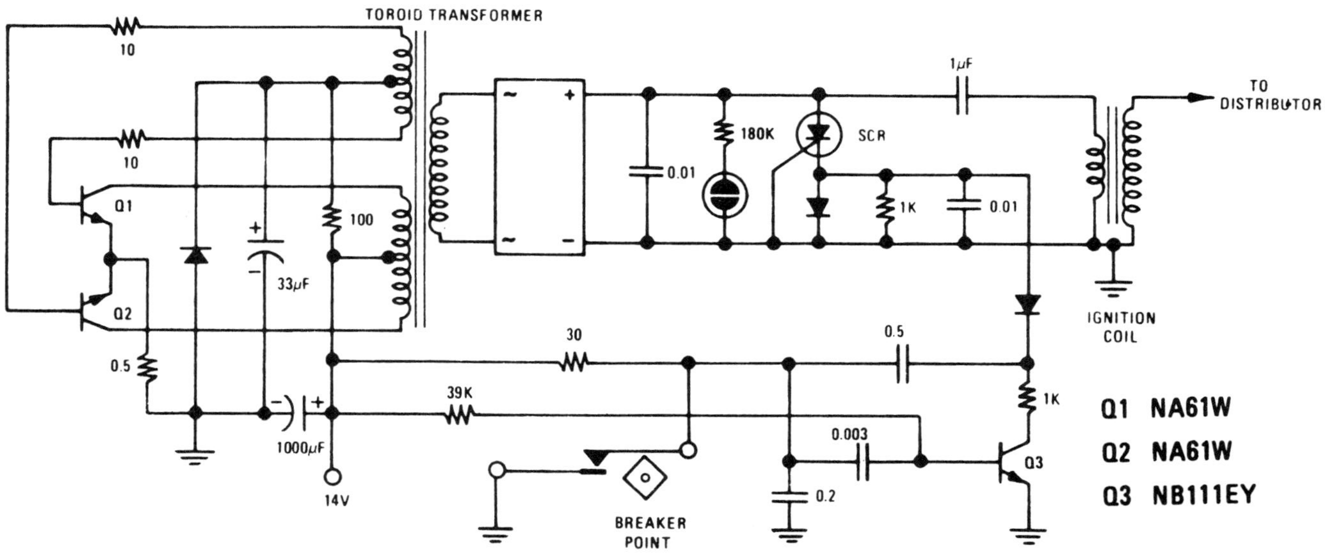

Capacitor discharge ignition system (NS).

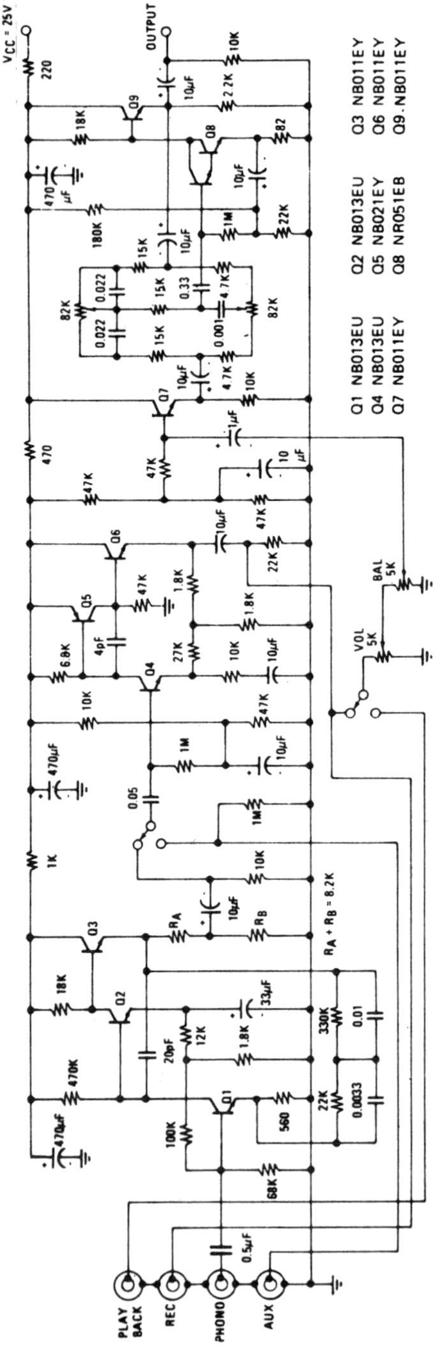

High quality preamplifier with tone control circuit (NS).

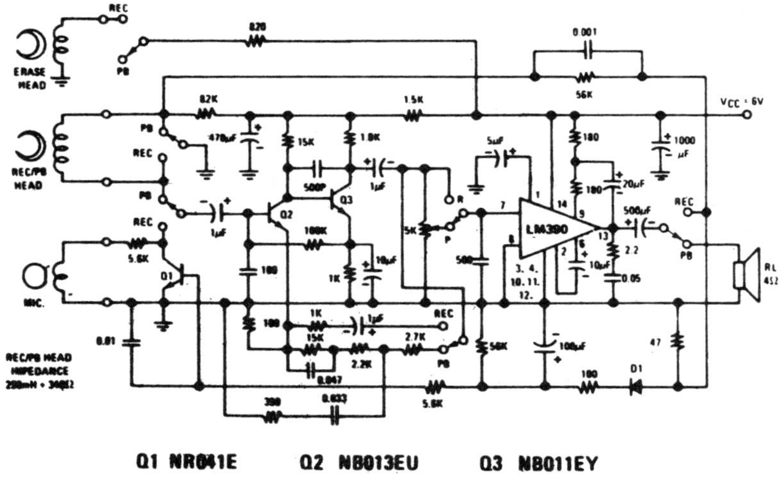

Battery operated recording/playback cassette circuit (NS).

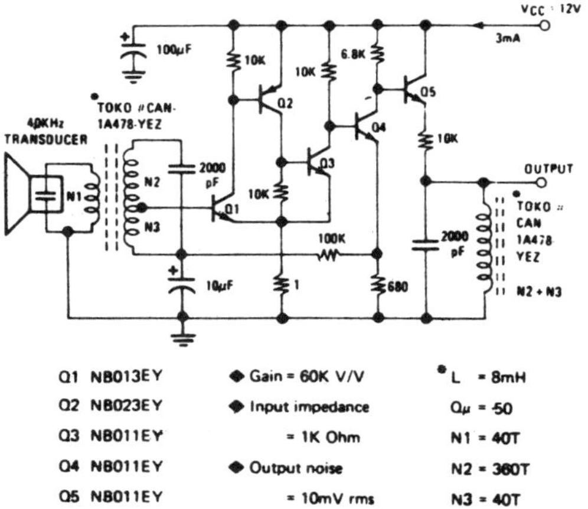

High gain ultrasonic amplifier (NS).

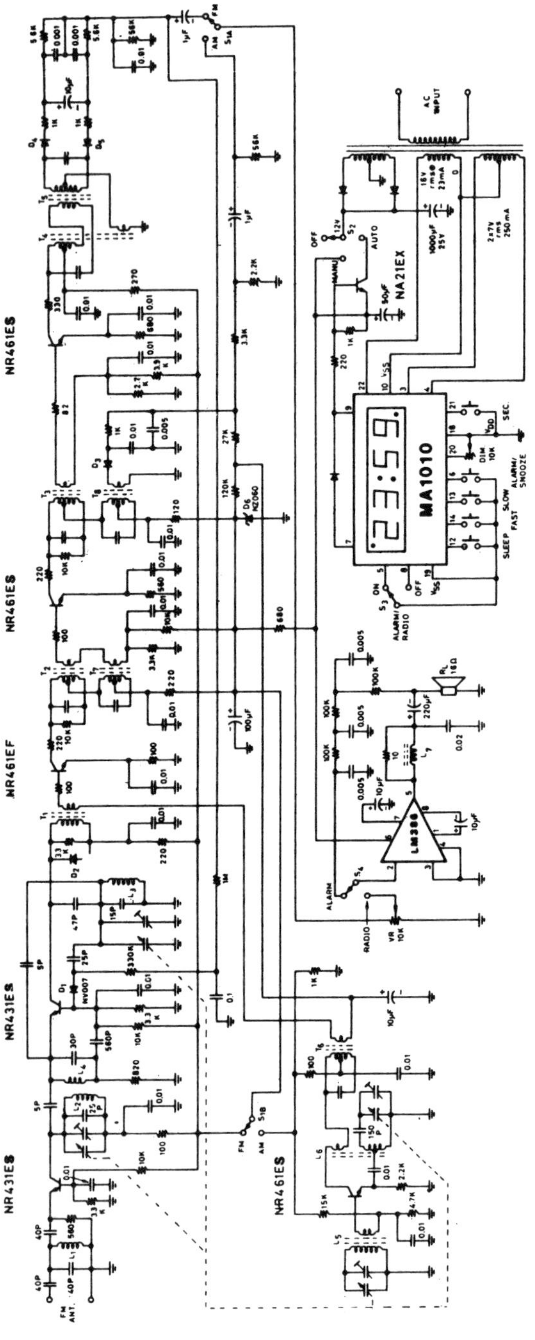
NR431ES
NR431ES
NR461EF
NR461ES
NR461ES
NR461ES
MA1010
NA21EX
LM386
FM ANT.
AC INPUT

L1 SWG#22, N = 5.5T, Dia = 4mm
L2 SWG#22, N = 4T, Dia = 4mm
L3 SWG#22, N = 4T, Dia = 3mm

L4 SWG#28, N = 20T, Dia = 3mm L = 0.75μH
L5 95T : 8T, L = 600μH, Qμ = 300
L7 N = 1.5T, PHILIPS #4312–020–34401

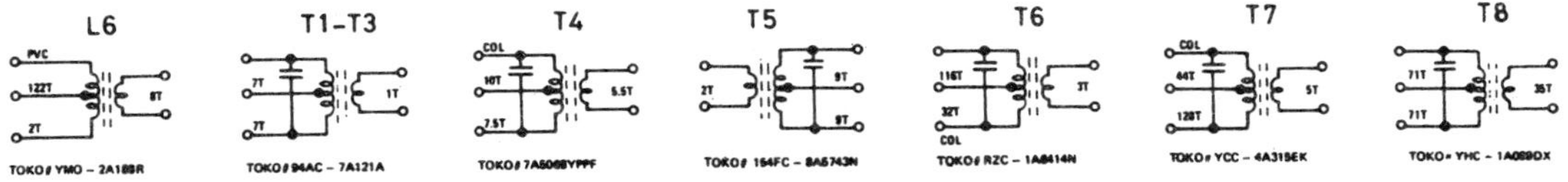

FM performance (88–108 MHz)

- 30dB quieting sensitivity: 5μV
- limiting sensitivity: 20μV
- AM rejection: 40dB
- AFC holding range: 800KHz
- Bandwidth: 180KHz

AM performance (525–1650 KHz)

- maximum sensitivity: 100μV/M
- 20dB quieting sensitivity: 280μV/M
- selectivity ± 10KHz: –28dB
- AGC figure of merit: 40dB
- overload distortion: 6%

AUDIO performance

- gain at 1 KHz: 200
- 10% THD output power: 900mW
- frequency response: 70Hz – 12KHz
- typical system dist: 0.8%
- alarm tone frequency: 600Hz

AM/FM clock radio (NS).

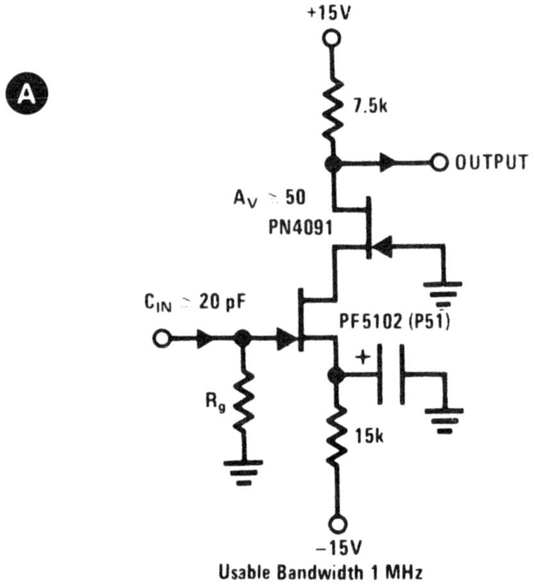

Wide band, low input capacitance, very low noise preamplifier (NS).

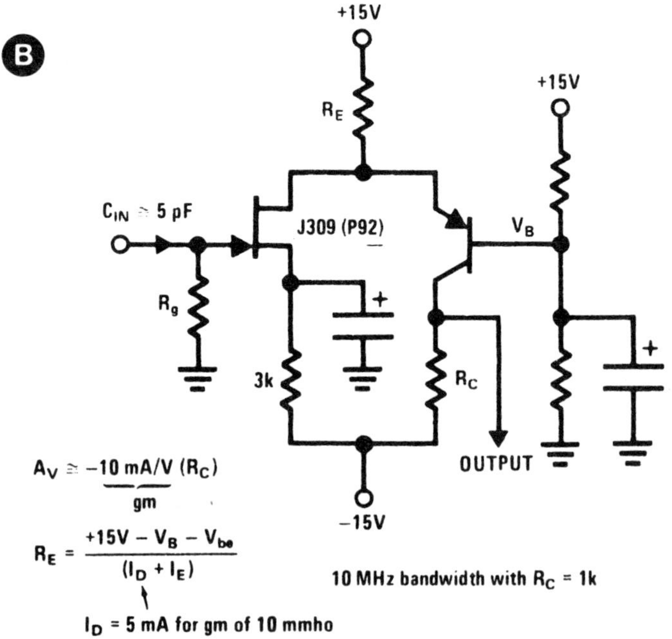

Low noise, very low input capacitance video amplifier (NS).

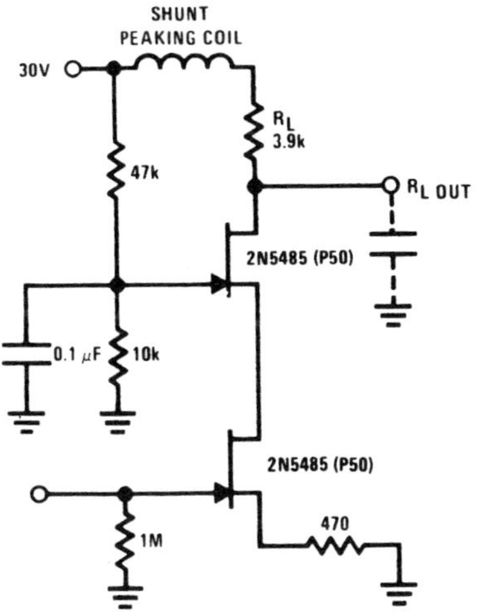

FET cascode video amplifier (NS).

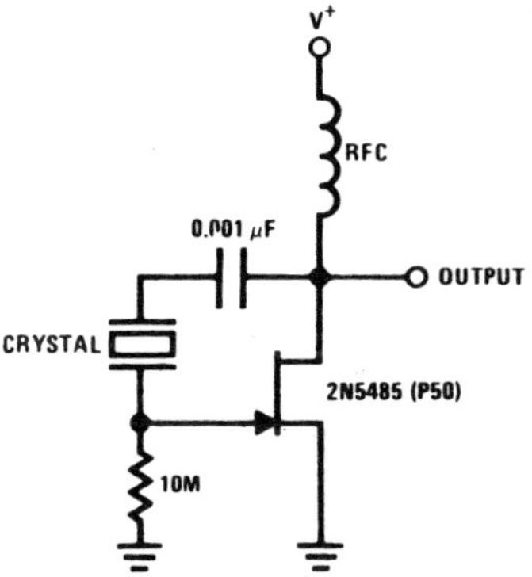

JFET Pierce crystal oscillator (NS).

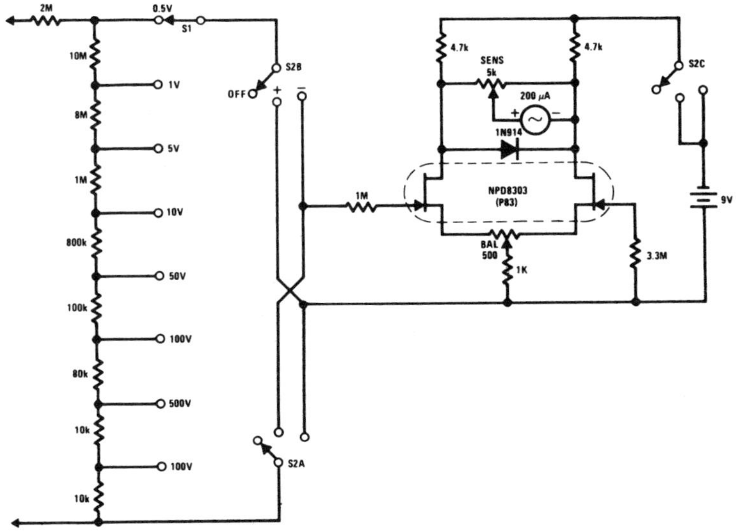

FET voltmeter (NS).

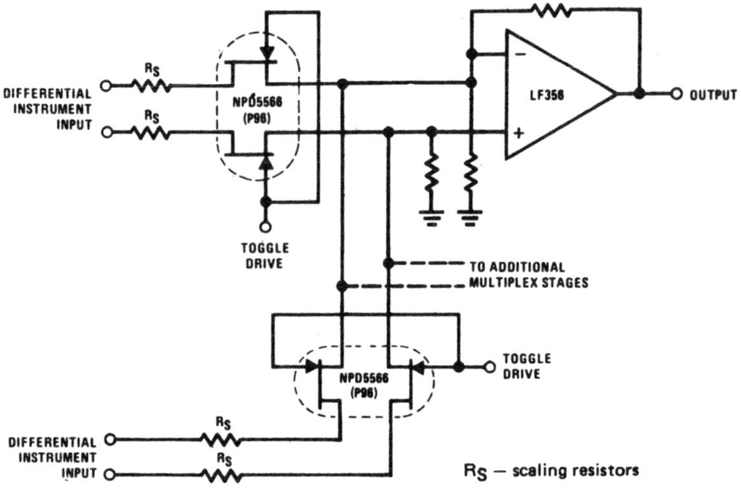

Differential analog switch (NS).

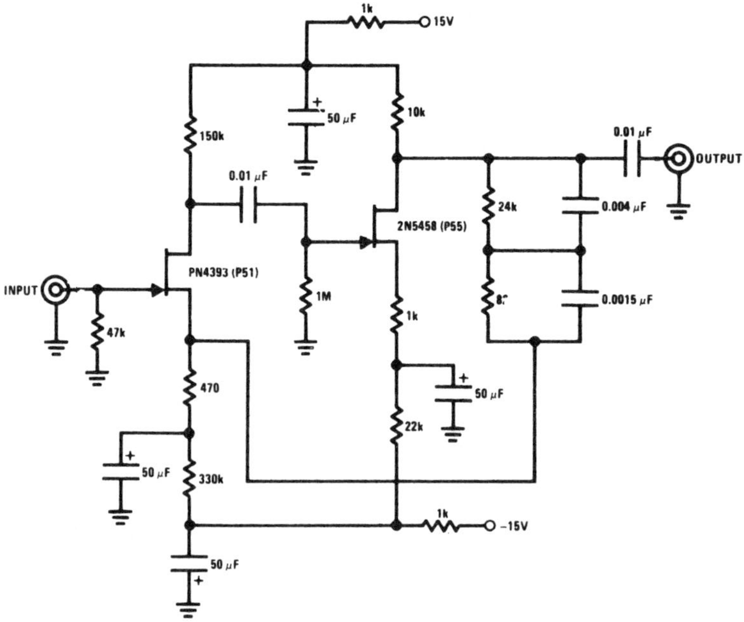

Magnetic pickup phono preamplifier (NS).

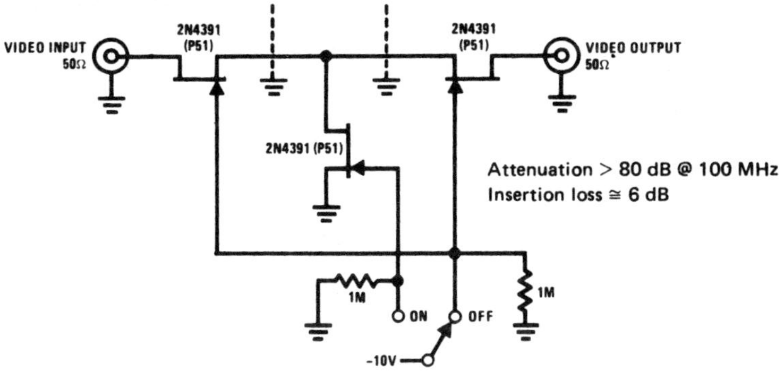

High frequency switch (NS).

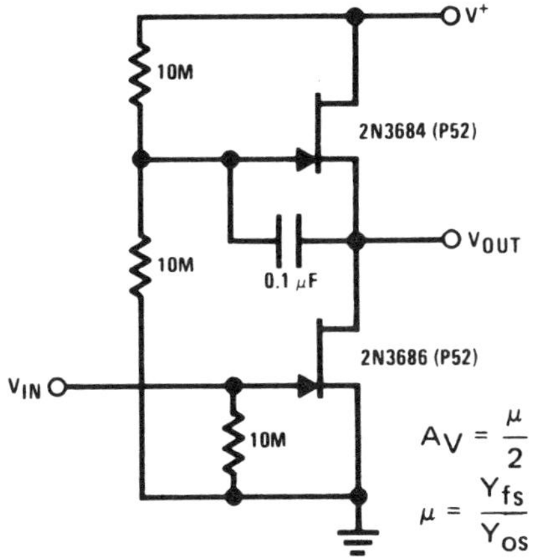

$$A_V = \frac{\mu}{2} = 500 \text{ typical}$$

$$\mu = \frac{Y_{fs}}{Y_{os}}$$

Ultra-high gain audio amplifier (NS).

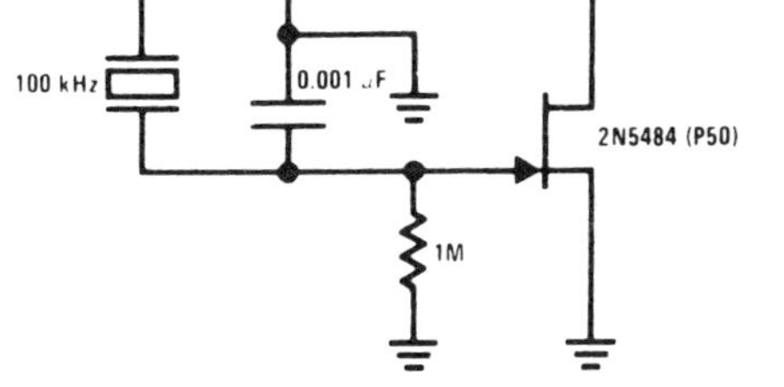

Stable low frequency crystal oscillator (NS).

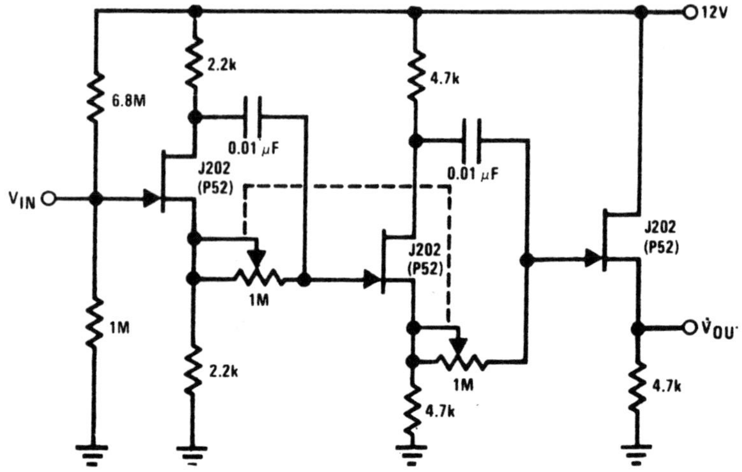

0 to 360° phase shifter (NS).

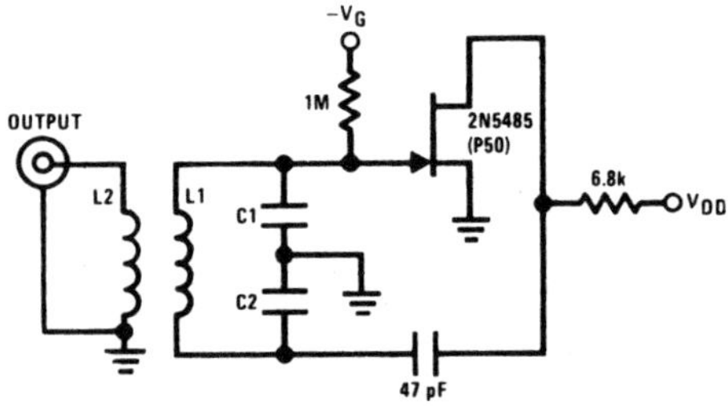

20 MHz oscillator values

C1 ≅ 700 pF	L1 = 1.3 μH
C2 = 75 pF	L2 = 10T 3/8" dia 3/4" long
V_{DD} = 16V	I_D = 1 mA

20 MHz oscillator performance

Low distortion 20 MHz osc
2nd harmonic − 60 dB
3rd harmonic > −70 dB

Low distortion oscillator (NS).

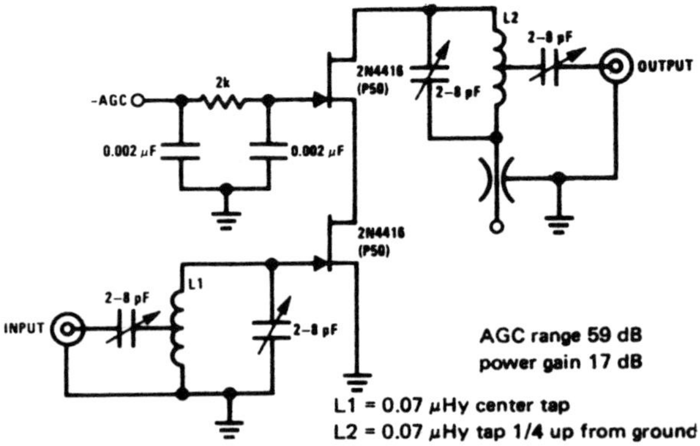

AGC range 59 dB
power gain 17 dB

L1 = 0.07 μHy center tap
L2 = 0.07 μHy tap 1/4 up from ground

200 MHz cascode amplifier (NS).

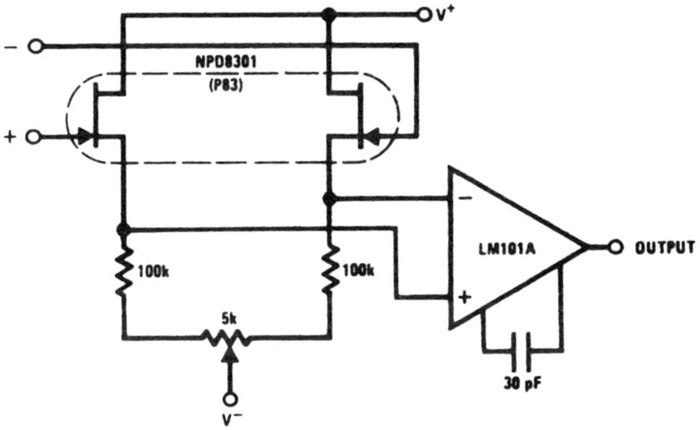

FET Op AMP (NS).

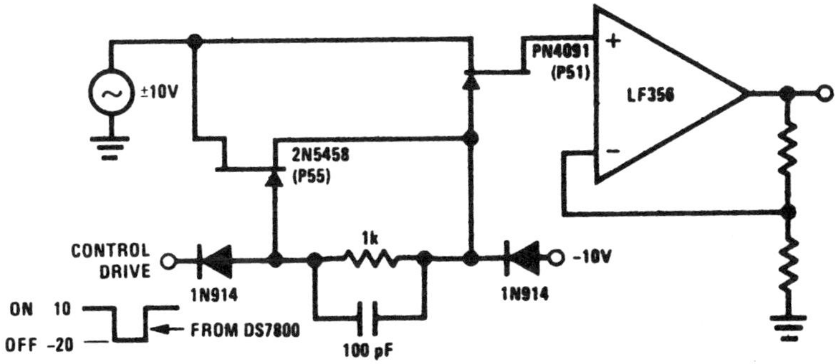

High toggle rate high frequency analog switch (NS).

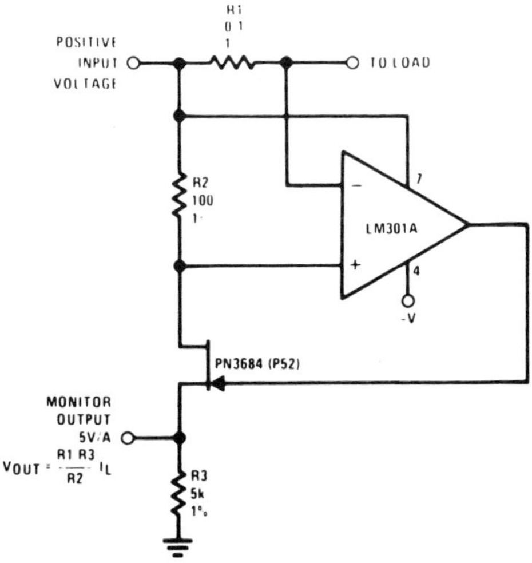

Current monitor (NS).

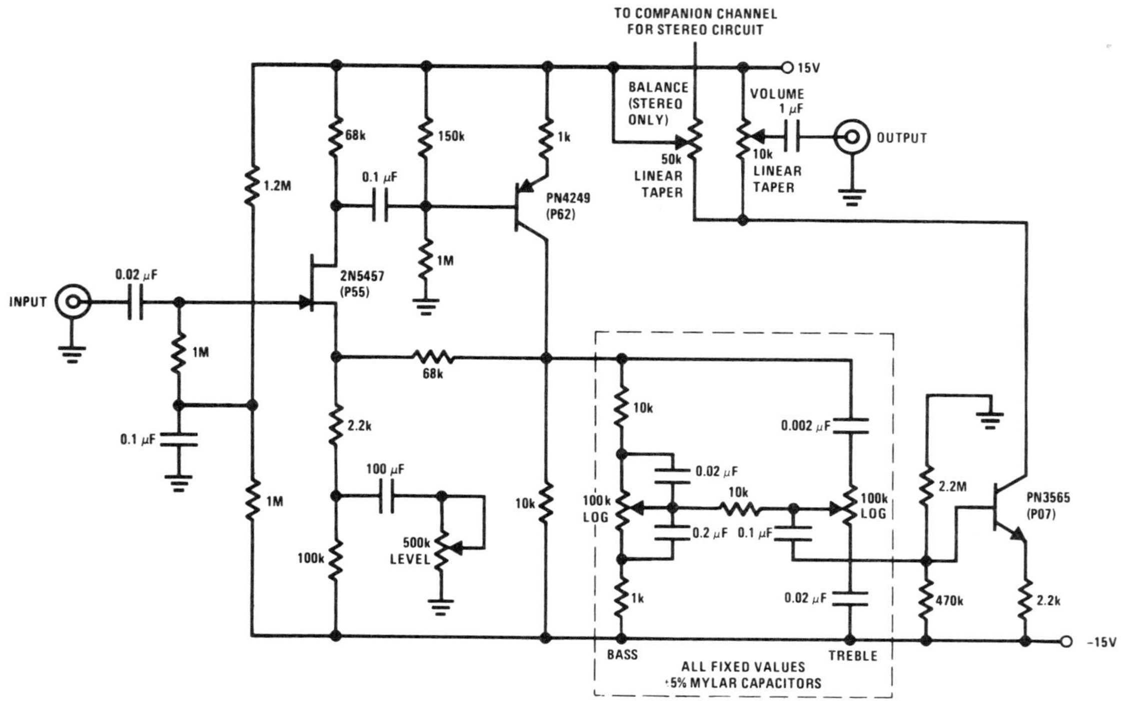

Low cost high level preamp and tone controlled circuit (NS).

Chapter 25

Bonus Circuits for Beginners

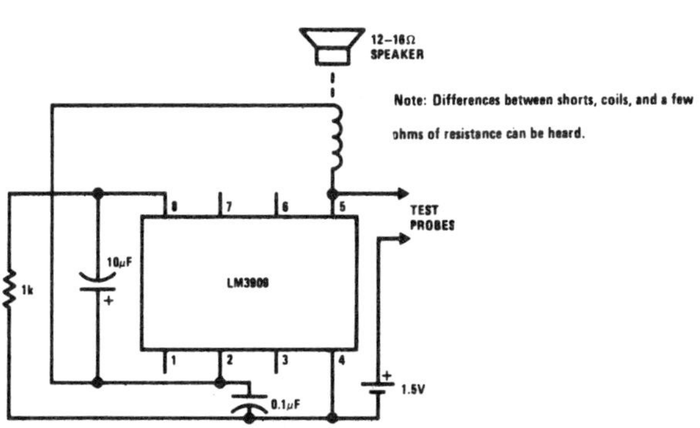

"Buzz box" continuity and coil checker (NS).

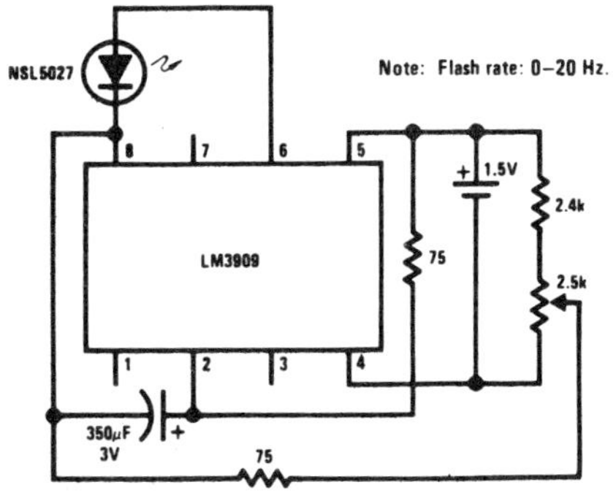

Variable flasher (NS).

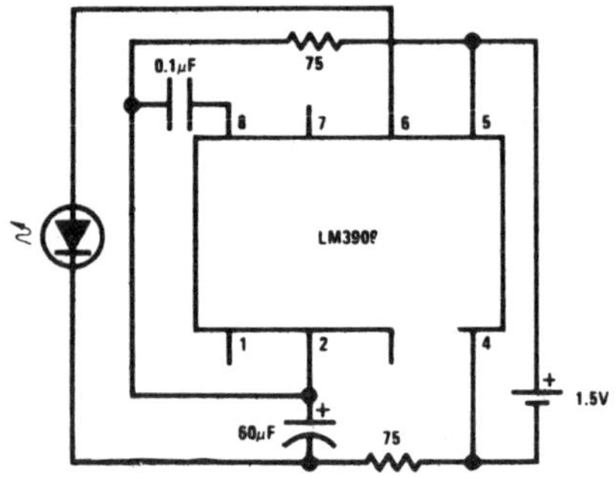

Note: High efficiency, 4 mA drain.

Note: Continuous appearing light obtained by supplying short, high current, pulses (2 kHz) to LEDs with higher than battery voltage available.

LED booster (NS).

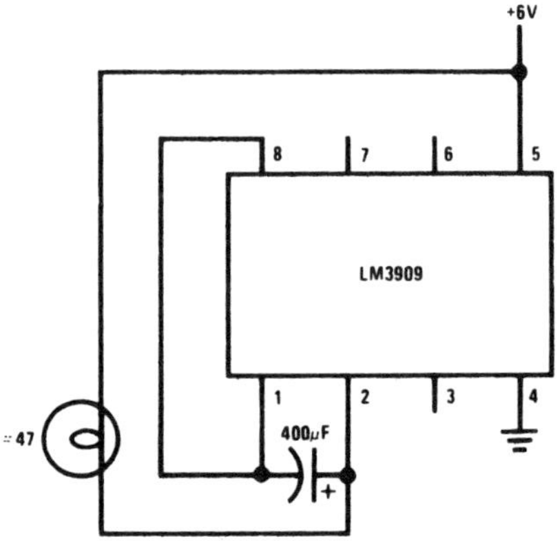

Note: Flash rate: 1.5 Hz.

Incandescent bulb flasher (NS).

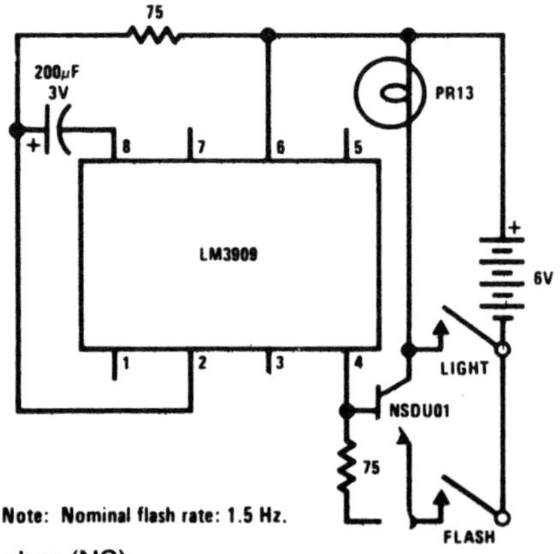

Note: Nominal flash rate: 1.5 Hz.

Emergency lantern/flasher (NS).

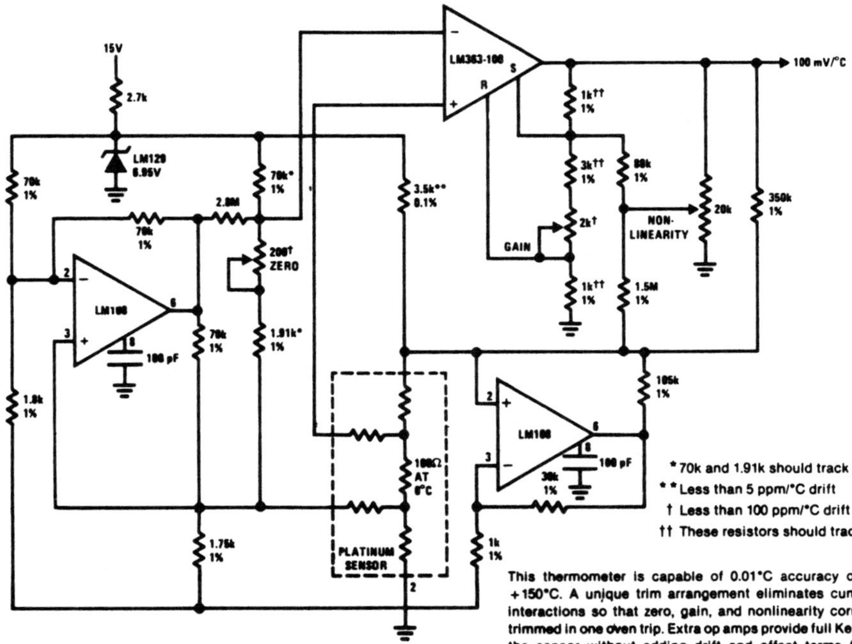

*70k and 1.91k should track to 5 ppm/°C

**Less than 5 ppm/°C drift

† Less than 100 ppm/°C drift

†† These resistors should track to 20 ppm/°C

This thermometer is capable of 0.01°C accuracy over −50°C to +150°C. A unique trim arrangement eliminates cumbersome trim interactions so that zero, gain, and nonlinearity correction can be trimmed in one oven trip. Extra op amps provide full Kelvin sensing on the sensor without adding drift and offset terms found in other designs.

Curvature corrected platinum RTD thermometer (NS).

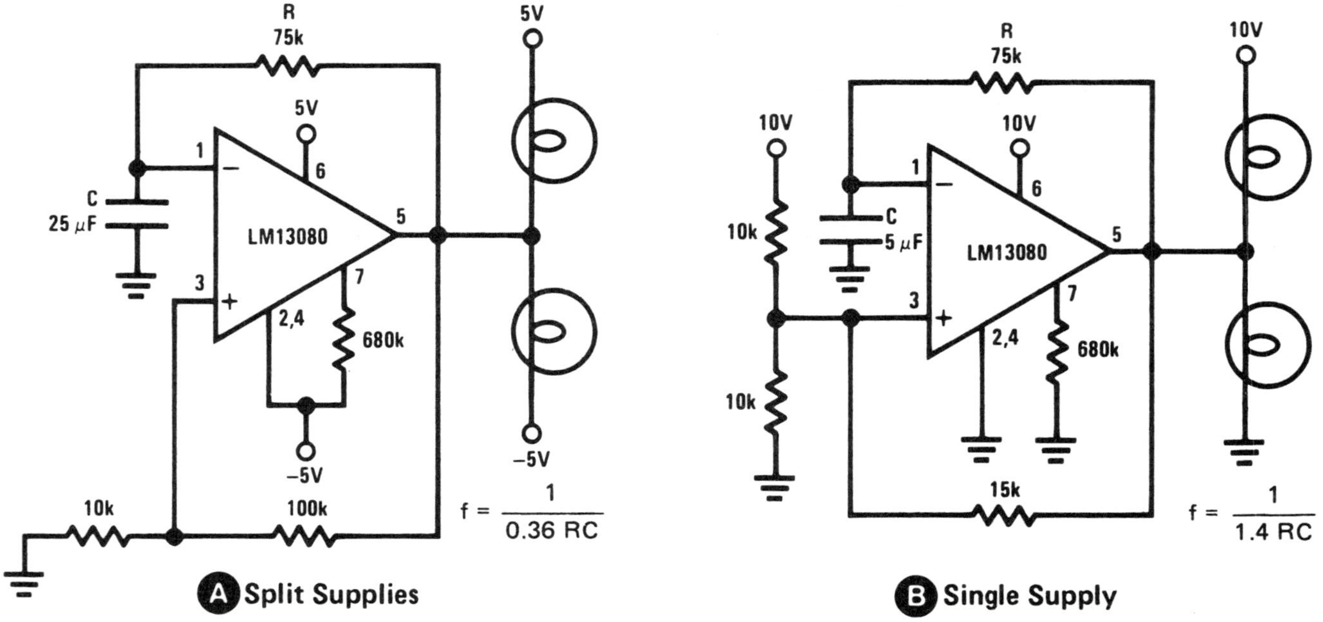

Low frequency lamp flasher/relay driver (NS).

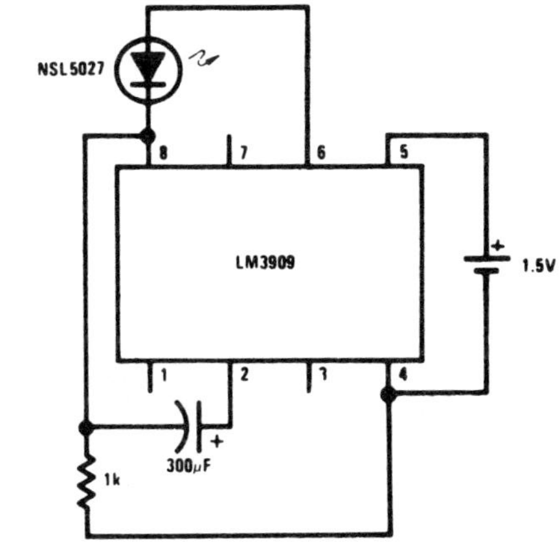

Note: Nominal flash rate: 2.6 Hz. Average I_{DRAIN} = 1.2 mA.

Fast blinker (NS).

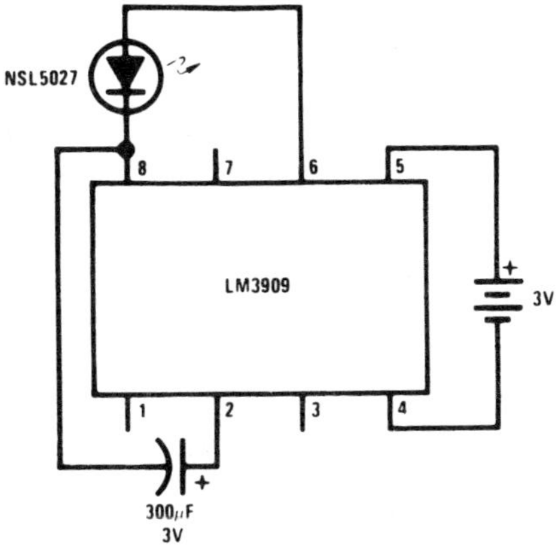

Note: Nominal flash rate: 1 Hz. Average I_{DRAIN} = 0.77 mA

3V flasher (NS).

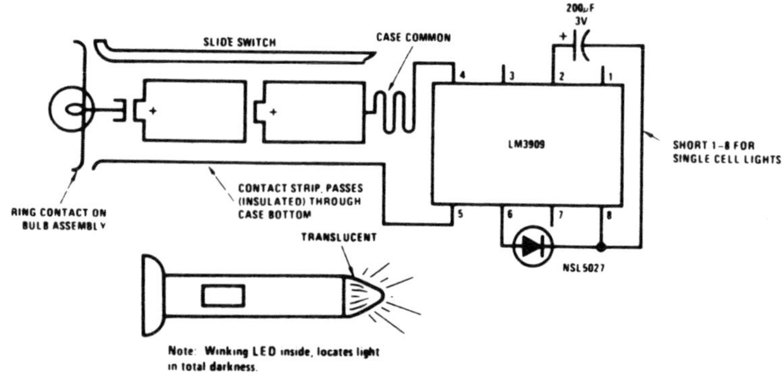

Note: LM3909, capacitor, and LED are installed in a white translucent cap on the flashlight's back end. Only one contact strip (in addition to the case connection) is needed for flasher power. Drawing current through the bulb simplifies wiring and causes negligible loss since bulb resistance cold is typically less than 2Ω.

Flashlight finder (NS).

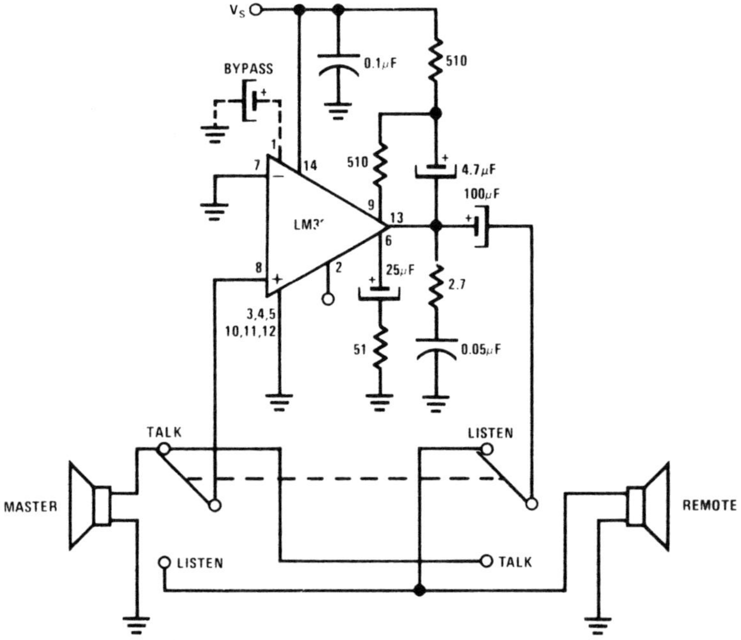

Intercom (NS).

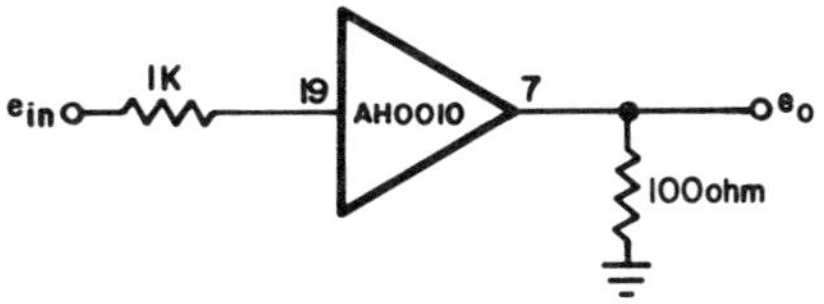

Voltage follower (OE).

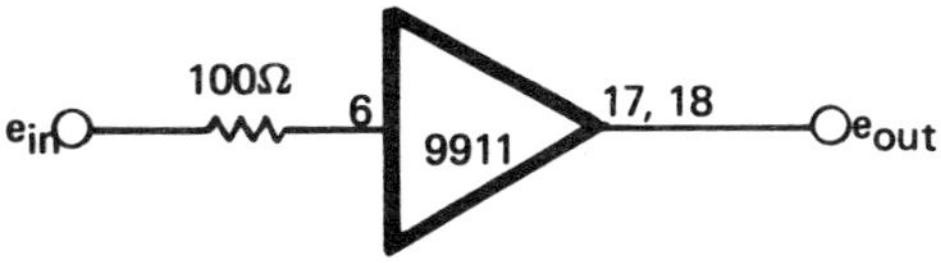

Voltage follower (OE).

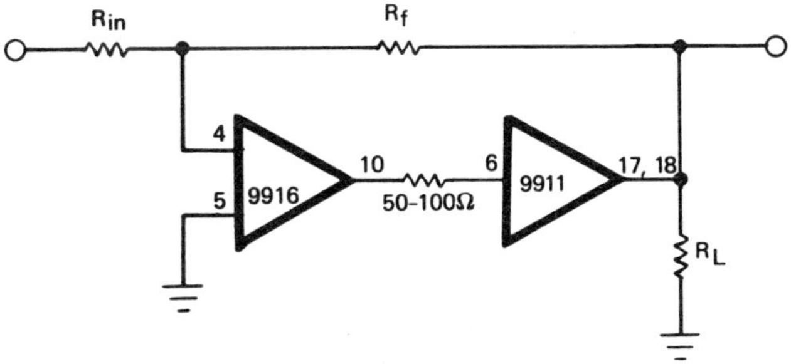

Current booster for video (OE).

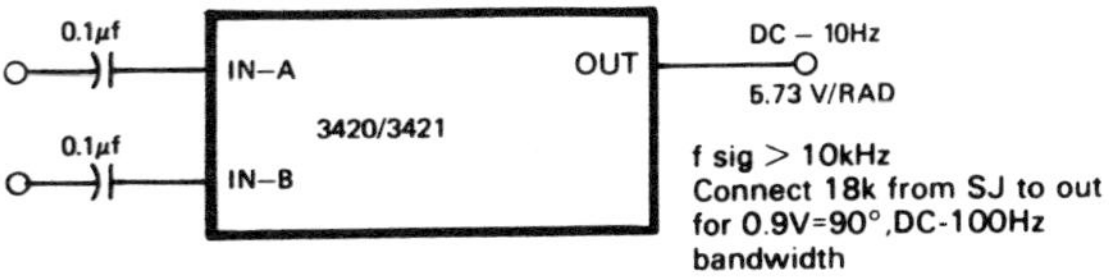

Phase demodulator (OE).

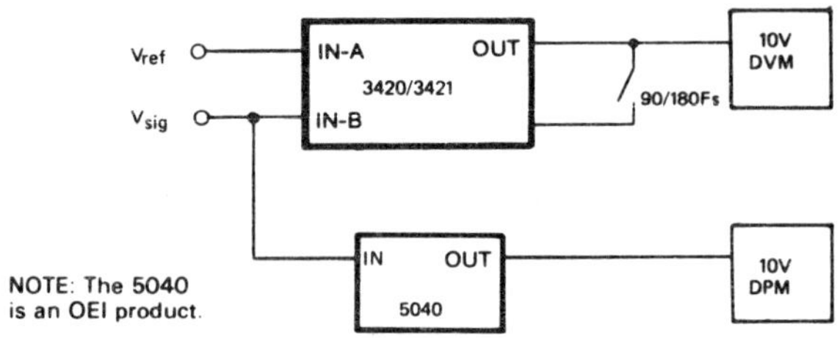

Rms—vector voltmeter (OE).

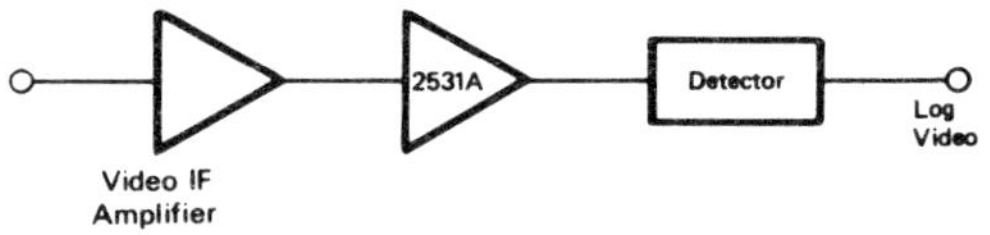

Video log response (OE).

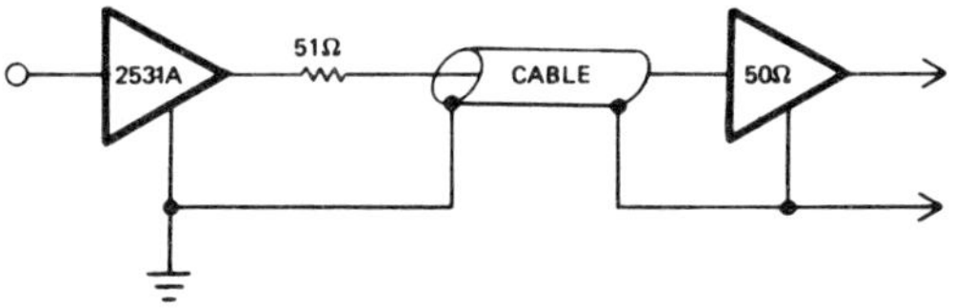

Log preamp and line driver (OE).

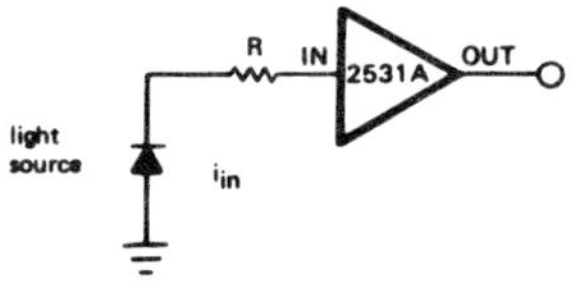

Light detection with photo diode (OE).

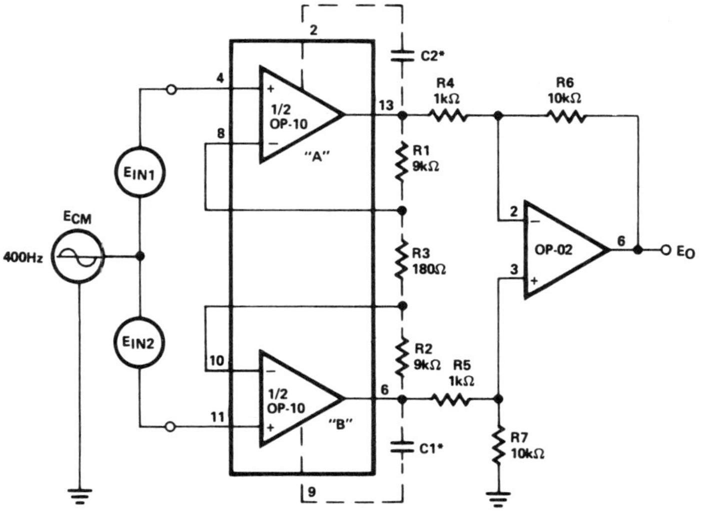

*SELECTED 5pF TO 100pF.

NOTES:

1. $E_O = (E_{IN1} - E_{IN2})\,A_{VCL}$
2. $A_{VCL} = \left(1 + \frac{R1 + R2}{R3}\right)\left(\frac{R6}{R4}\right) = 1000$
3. $CMRR = 20\ LOG\ \frac{E_{CM}\,(A_{VCL})}{E_O}$ @ DC
4. IF $\frac{R6}{R4} = \frac{R7}{R5}$, CMRR @ DC $\cong$ 120dB
5. AT 400Hz, CMRR IS A FUNCTION OF THE DIFFERENCE IN FREQUENCY RESPONSE OF SIDE "A" AND SIDE "B".
6. USE OF C1 OR C2, SELECTED USING THE PROCEDURE DESCRIBED, MATCHES THE FREQUENCY RESPONSE OF SIDE "A" AND SIDE "B", THEREBY MAXIMIZING CMRR AT 400Hz.

Triple op-amp instrumentation amplifier circuit (PM).

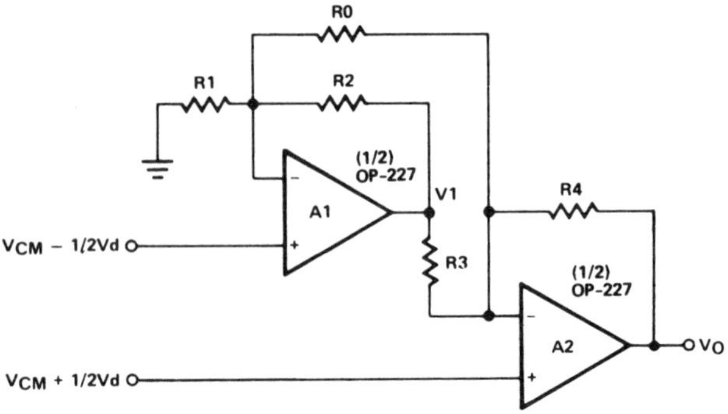

$$V_O = \frac{R4}{R3}\left[1 + \frac{1}{2}\left(\frac{R2}{R1} + \frac{R3}{R4}\right) + \frac{R2 + R3}{R0}\right] V_d + \frac{R4}{R3}\left(\frac{R3}{R4} - \frac{R2}{R1}\right) V_{CM}$$

Two-op-amp instrumentation amplifier configuration (PM).

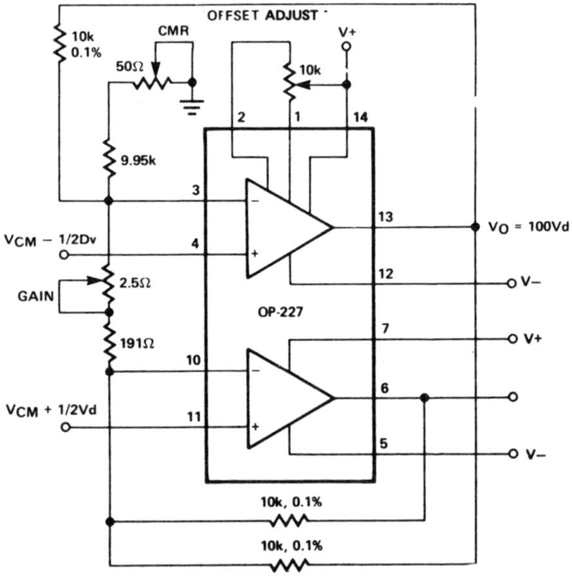

Two-op-amp instrumentation amplifier using OP-227 dual (PM).

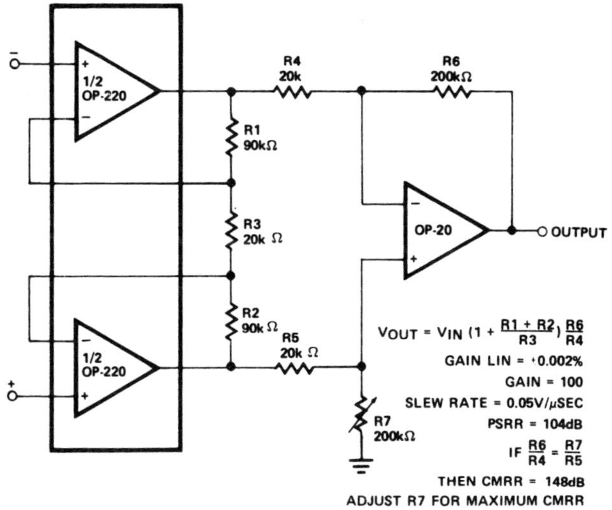

Instrumentation amplifier—3 op-amp design (PM).

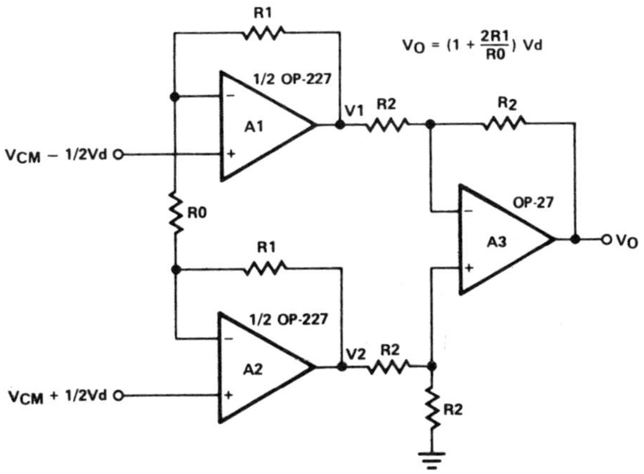

Three-op amp instrumentation amplifier using OP-227 and OP-27 (PM).

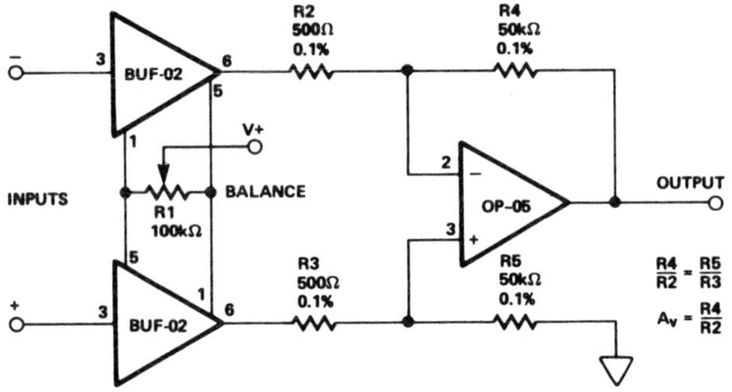

Differential input instrumentation amplifier (PM).

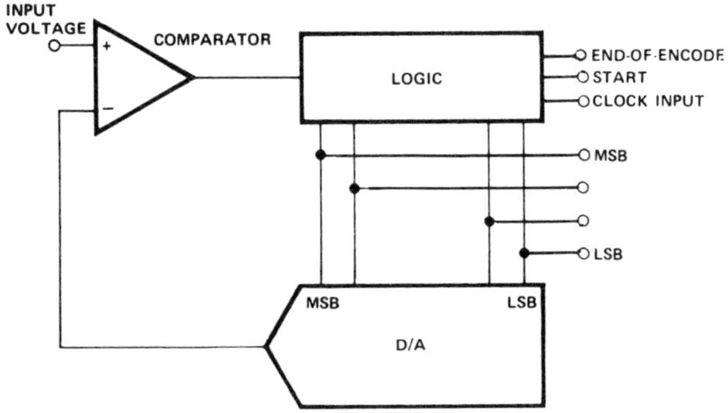

Basic feedback A/D converter (PM).

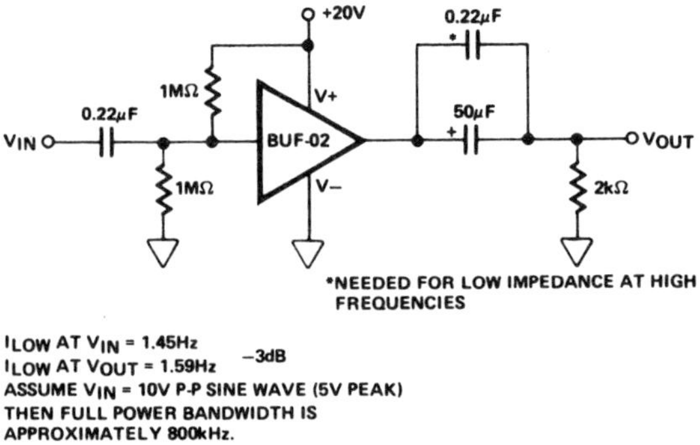

High-speed single-supply ac buffer (PM).

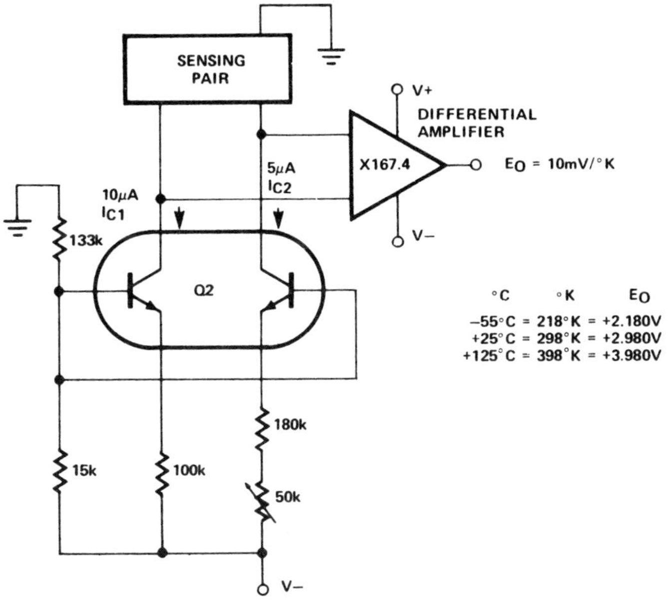

Basic temperature sensor (PM).

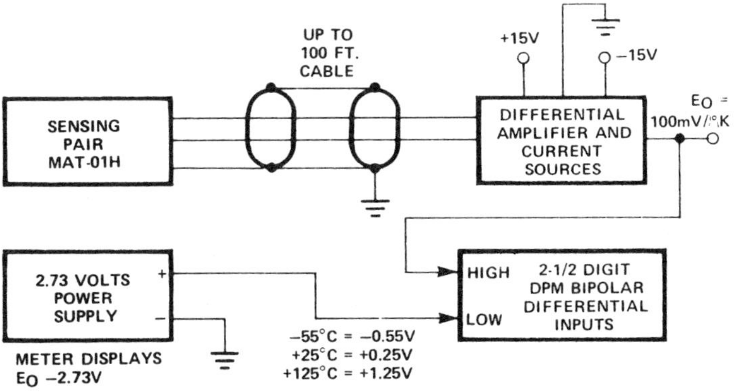

Digital thermometer with readout in °C (PM).

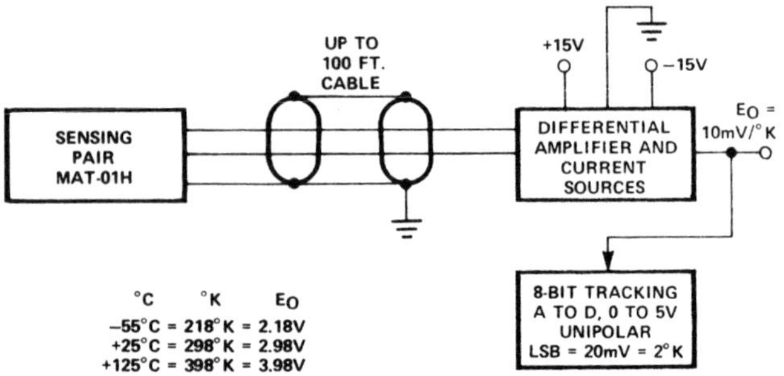

Binary-coded temperature readings with 2° resolution (PM).

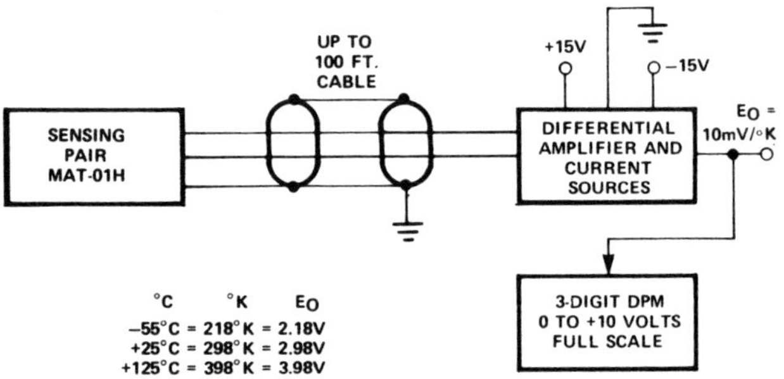

Basic digital thermometer with readout in degrees Kelvin (°k) (PM).

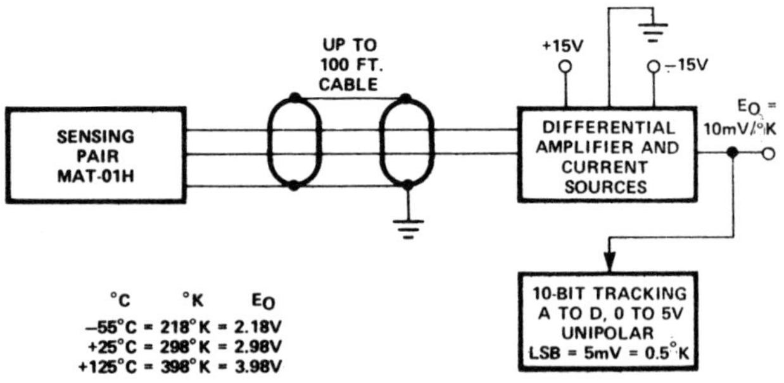

Binary-coded temperature readings with 5° resolution (PM).

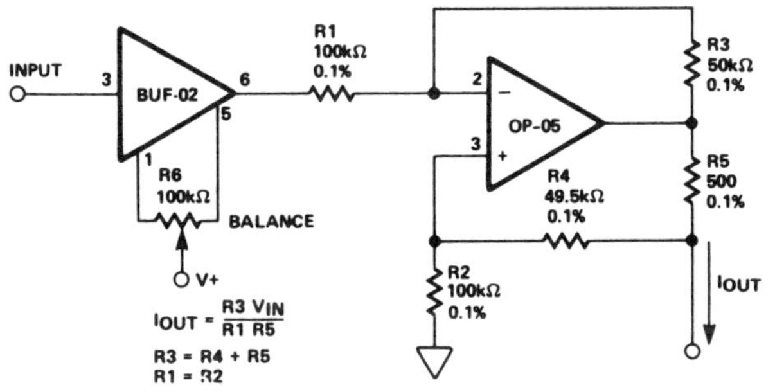

Bilateral current source (PM).

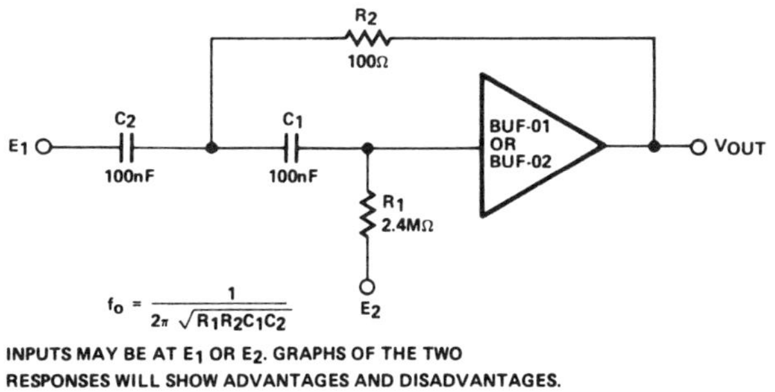

Tuned circuit (PM).

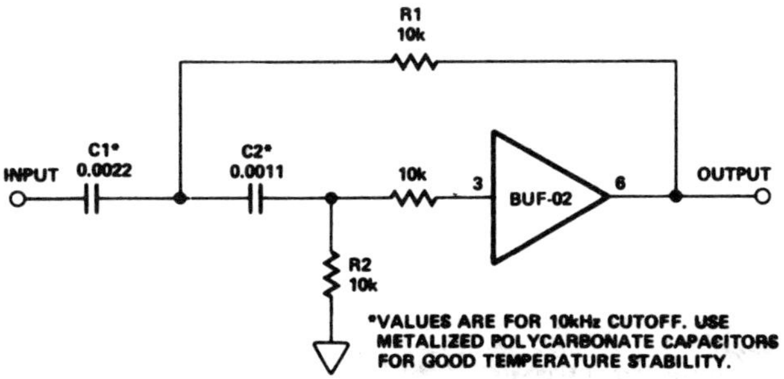

Second-order high-pass active filter (PM).

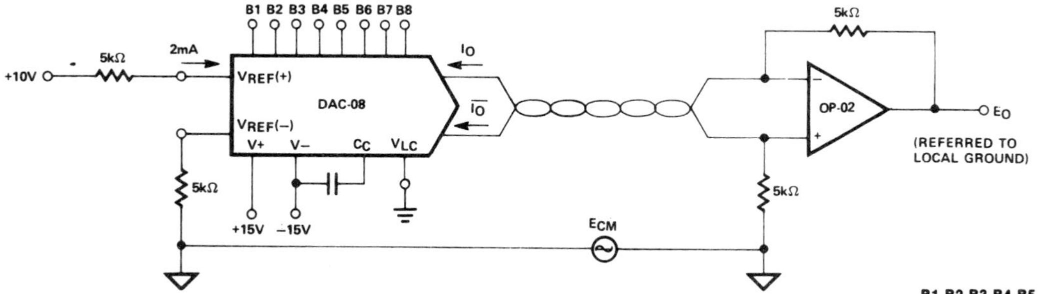

	B1	B2	B3	B4	B5	B6	B7	B8	E_O
POS FULL SCALE −LSB	1	1	1	1	1	1	1	1	+9.920
POS FULL SCALE 2LSB	1	1	1	1	1	1	1	0	+9.840
(+) ZERO SCALE	1	0	0	0	0	0	0	0	+0.040
(−) ZERO SCALE	0	1	1	1	1	1	1	1	−0.040
NEG FULL SCALE +2LSB	0	0	0	0	0	0	0	1	−9.840
NEG FULL SCALE +1LSB	0	0	0	0	0	0	0	0	−9.920

High noise immunity current to voltage conversion (PM).

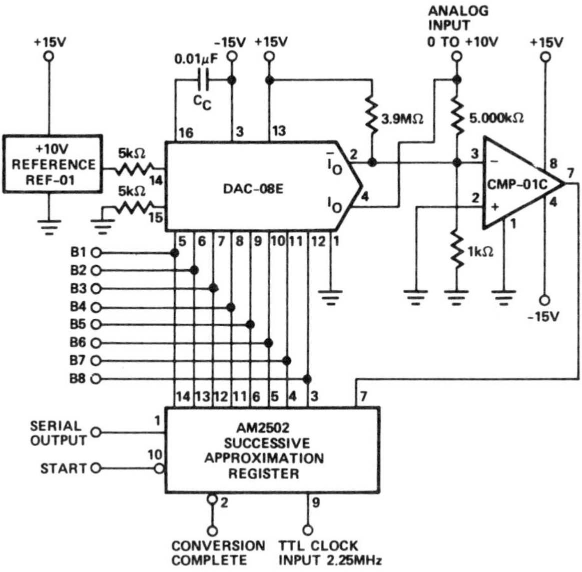

CONNECT "START" TO "CONVERSION COMPLETE" FOR CONTINUOUS CONVERSIONS.

3 IC Low cost A/D converter (PM).

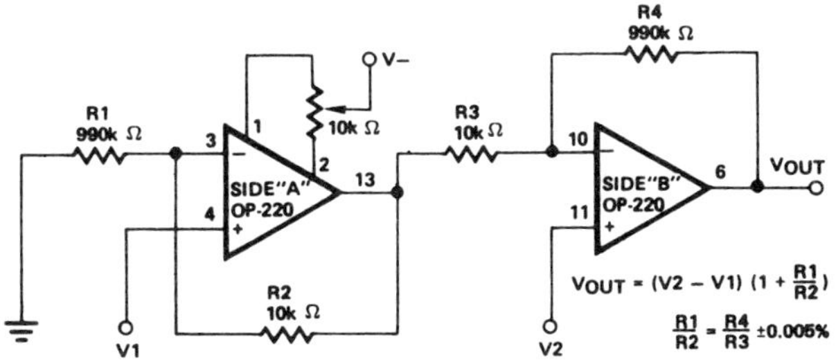

Y PACKAGE ONLY

Instrumentation amplifier 2 op-amp design (PM).

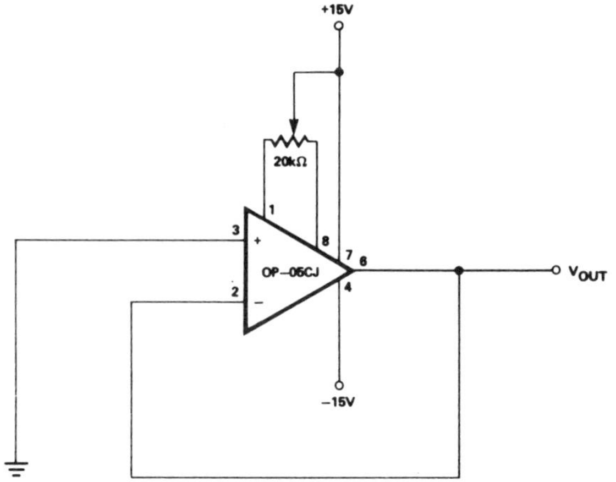

Zenerless precision millivolt source (PM).

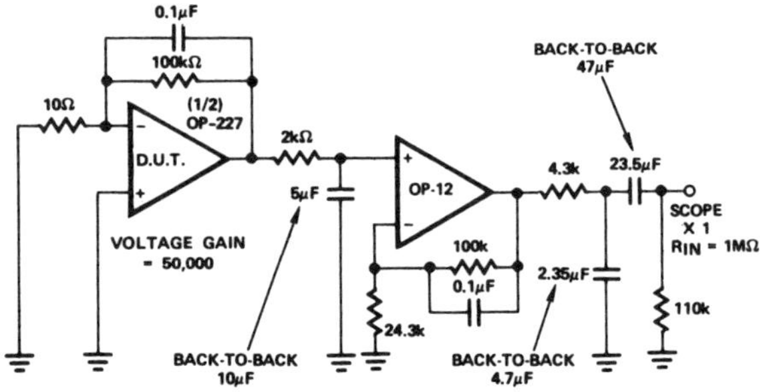

0.1 Hz to 10 Hz P-P noise circuit schematic (PM).

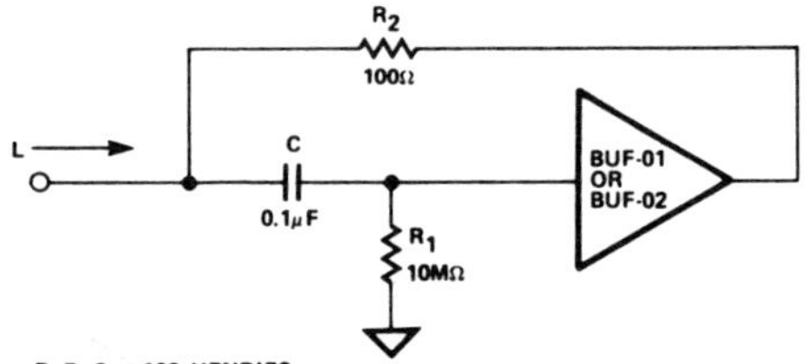

$L \cong R_1R_2C$ = 100 HENRIES
$R_S = R_2$ = 100Ω
$R_P = R_1$ = 10 MEGΩ
ASSUMING C_{STRAY} (ACROSS R_1) OF 5 pF THE UPPER FREQUENCY LIMIT IS APPROXIAMTELY 7kHz.
X_L = 100Ω AT f = 0.159Hz

Active inductor (PM).

Index

B

C

D

R

S